CONDITIONAL MEASURES AND APPLICATIONS

PURE AND APPLIED MATHEMATICS

A Program of Monographs, Textbooks, and Lecture Notes

MONOGRAPHS AND TEXTBOOKS IN
PURE AND APPLIED MATHEMATICS

Additional Volumes in Preparation

CONDITIONAL MEASURES AND APPLICATIONS

M. M. Rao
Department of Mathematics
University of California
Riverside, California

Marcel Dekker, Inc.   New York • Basel • Hong Kong

Library of Congress Cataloging-in-Publication Data

Rao, M. M. (Malempati Madhusudana)
　　Conditional measures and applications / M.M. Rao.
　　　p.　cm. -- (Monographs and textbooks in pure and applied
　　mathematics ; 177)
　　Includes bibliographical references and index.
　　ISBN 0-8247-8884-2 (alk. paper)
　　1. Probability measures.　I. Title.　II. Series.
　　QA273.6.R36　1993
　　519.2--dc20
　　　　　　　　　　　　　　　　　　　　　　　　93-24444
　　　　　　　　　　　　　　　　　　　　　　　　CIP

The publisher offers discounts on this book when ordered in bulk quantities.
For more information, write to Special Sales/Professional Marketing at the
address below.

This book is printed on acid-free paper.

MARCEL DEKKER, INC.
270 Madison Avenue, New York, New York　10016

Current printing (last digit):
10　9　8　7　6　5　4　3　2　1

PRINTED IN THE UNITED STATES OF AMERICA

To the memory of Professors

A.N. Kolmogorov and S. Bochner

whose fundamental contributions
form a basis for the following work

PREFACE

In presenting a successful solution of Hilbert's sixth problem dealing with the axiomatic foundations of the theory of probability, A.N. Kolmogorov took a further step in 1933 and gave a general basis of conditional expectations from which the concept of conditional probability was obtained. It includes the earlier case of conditioning relative to a nonnull event and hence also a countable partition of the underlying space. The generalization led to a significant growth of probability theory, especially Markov processes, martingales, sufficiency, and Bayesian inference in statistics as well as ergodic theory and quantum mechanics, among others.

Although the generalized concept of conditional expectation (and probability) is natural and contains the intuitive content of the elementary case, its true meaning will not be fully appreciated unless the abstract character of the new formulation is fully analyzed. In fact there is a deep distinction between the countable partition case and the general case. The problem is particularly visible when the conditioning is given by a nondiscrete random variable so that it takes an uncountable number of values. Then the determination of conditional expectation (and consequently probability) is often difficult, and *ad hoc* methods lead to different answers for the same question.

In spite of this situation, in many applications of interest there is no book (reference or text) devoted primarily to this subject in any language illuminating its structure, although many books contain a portion of a chapter on conditioning. It is the purpose of this volume to help bridge this gap by presenting a detailed account of the subject from a broad perspective together with some important applications from different areas.

Several examples and counterexamples are presented to illustrate the difficulties inherent in the general viewpoint. Also, a number of applications and some unresolved problems of the subject are discussed in detail. The following chapter-by-chapter outline will explain the content and the broad coverage attempted in this book.

The first chapter is essentially a motivation and starts with the elementary case treating conditioning relative to discrete random variables. Even here nontrivial questions can be posed and solved, as illustrated by the "secretary selection" problem. This leads to an abstract formulation and the Kolmogorov model is then introduced in Chapter 2. It is shown that the elementary case is subsumed and then some properties of the general concept as well as conditional independence are discussed.

In contrast to the elementary case, the general formulation presents nontrivial computational problems for evaluating both the conditional expectations and probabilities. These difficulties are discussed in detail in Chapter 3. As yet there are no algorithms to calculate conditional expectations. Some general solutions to this problem are proposed here. Attempts to bypass these questions, including a new axiomatic approach due to A. Rényi, are discussed in Chapter 4 and some consequences and characterizations of the new concept are also given. It is found that the earlier problems are not resolved in the new approach.

An extended treatment of regular conditional probabilities of the Kolmogorov model is presented in Chapter 5. It includes different types of conditions for the existence of regular conditional measures and some methods of exact evaluation of a class of conditional expectations (conditioned by general random variables) making it a pivotal chapter of the book. Thus these five chapters contain the fundamental theory of conditional measures that everybody interested in the subject should know.

Using the general definition, the Fisherian concept of sufficiency is treated in Chapter 6 for both the dominated and the undominated families of probability measures. Relations between these and certain vector-valued measures (of unbounded variation) that arise in this study are also included to clarify and better understand the nature of this subject. Chapter 7 is devoted to an abstraction of Kolmogorov's definition wherein several characterizations of conditional expectations and probabilities are given. It is then noted that this abstraction contains Rényi's approach when conditional measures are viewed as vector measures. Further, a general method for exact evaluation of these expectations, complementing the work of Chapter 5, is presented.

Chapter 8 is concerned with products of conditional probabilities, extending the fundamental works of Kolmogorov (1933) and Bochner (1947, 1955). This plays an important role in the existence theory of such key areas as (quite general) Markov processes and martingales, among others. The relation of these product conditional measures to the disintegration problem is also discussed. The rest of the book considers various (novel) applications.

The general conditioning is illustrated in Chapter 9 with an extended account of martingale theory and Markov processes. In the former case, the Kolmogorov formulation can be used without restrictions while the latter needs regularity of conditional measures almost from the start. Evaluation of these probabilities is important in real-world applications as well as in translating the properties of measures to the sample path analysis. Some unsettled problems exist here. Then, in Chapter 10, regular conditional measures as kernel functions and their use in potential theory are exemplified. The similarity of the latter with Reynolds operators and applications to bistochastic operators as well as the role of conditioning in the structural analysis of contractive projections are considered. In the final chapter, certain other results in contemporary analysis are discussed. These concern characterizations of averaging operators, mimicking conditioning on spaces of function algebras, and (noncommutative) conditional expectations in operator algebras on a Hilbert space that are useful in areas such as quantum mechanics. Also sufficiency and martingale convergence in this setup are treated. Throughout, several open problems and directions for further investigations of this subject are provided. More detailed summaries appear at the beginning of each of the chapters.

As the above outline shows, the work in the book can be divided into three interlocking parts: (i) the first six chapters treat the foundations and the basic material that everybody dealing with probability and statistical theory should know, (ii) the next two chapters consider an extended (abstract) account of the subject, and (iii) the final three chapters are devoted to a few applications of central importance in analysis. There is also some material appearing for the first time, especially in the latter parts of Chapters 6, 7, and 8. Further, I have included the general treatment for reference purposes as well as to entice researchers into considering the problems raised here.

The presentation is sufficiently detailed so that the book can be used for a semester (or a two-quarter) graduate course or seminar on the subject covering the first six chapters and Chapter 9 and sampling the rest as time and interests permit. For such a course, the prerequisite is a standard real analysis course—or at least a concurrent study of the latter subject is recommended for a proper appreciation. An acquaintance with some elementary probability and statistics will be useful although not strictly necessary. However, a great deal of detail, with adequate motivation, is included to facilitate a self-study by mature scientists to whom this subject may be new. But, even someone without the mathematical preparation can appreciate the issues involved by skipping the proofs and following the general discussion.

Many references to the work are given, but they are limited to those explicitly cited in the text. A bibliographical section at the end of each chapter includes discussions of historical developments of the subject. I have tried to assign due credit to individual contributions and hope that I have been successful in this task. Since many of the results covered here are scattered in the journal literature, I have tried to unify them. This gives a better perspective of the subject.

For cross-references, all items in the book are serially numbered. For instance, 5.3.7 is the seventh item of Section 3 of Chapter 5. In a particular section (chapter) the corresponding section (and chapter) number is omitted. Also, chapter and section titles appear at the top of the pages, to help locate a result.

The book has been in preparation for over five years. In fact the dissatisfaction with computational and some other problems of conditioning is already indicated in my book *Probability Theory with Applications* (1984). However, very few people seem to have spent much effort on them. Tjur (1974) presented a differential geometric approach to the computational problem, but it is not a solution to the Kolmogorov model and the same is true of the Rényi (1955) method. I indicated the difficulties explicitly under the name "paradoxes" in a paper (Rao, 1988). A class of solutions motivated by the ideas of Yeh (1974, 1975) were extended recently (Rao, 1993) and are included in this book. These are the only exact methods available for the computational problem, as far as I know.

The final preparation of this work is facilitated by a UCR sabbatical leave in 1991–92 spent at the Institute for Advanced Study in Princeton, and the spring of 1992 at MSRI in Berkeley. It is also partially supported by the ONR grant no. N00014-90-J-1705. An earlier draft of the first five chapters was typed by Mrs. Eva Stewart, and the revisions as well as the final version of the whole work were patiently and ably prepared by Ms. Jan Patterson. I am deeply appreciative of these institutions and individuals for this support and help. Finally, I will feel more than satisfied if this book helps the reader to better appreciate the intricacies of the conditioning concept as applied to both the theory and practice of our subject.

M.M. Rao

CONTENTS

Chapter 3: Computational Problems Associated with Conditioning   63

Chapter 4: An Axiomatic Approach to Conditional Probability   90

Chapter 5: Regularity of Conditional Measures   122

Chapter 10: Applications to Modern Analysis   334

Chapter 11: Conditioning in General Structures   368

References   398

Notation   411

Index   413

CHAPTER 1: THE CONCEPT OF CONDITIONING

In this chapter, a motivation for the concept of conditioning together with conditional probabilities and expectations in the case of countable partitions are presented. Elementary properties and the need for a generalization of the countable partition case are discussed. Most of the forthcoming work in the subject has its origins in these simple ideas and they serve as its natural basis.

1.1 Introduction

Let us start with a probability model describing a given experiment. This consists of a triple (Ω, Σ, P) where the points of Ω represent all possible outcomes of the experiment, Σ contains all events of conceivable interest to the experimenter, and P is a probability measure on Σ describing or measuring the uncertainties of the events. For mathematical convenience Σ is enlarged to become a σ-algebra, usually the smallest such that contains all the events of interest to the experimenter. Also P is taken to be σ-additive and $P(\Omega) = 1$. The σ-additivity is again stipulated for mathematical analysis although it is also a natural assumption in many cases. This model basically follows the Kolmogorov axiomatic setup. It has enabled a solid foundation, giving a phenomenal growth to probability theory ever since its introduction in 1933. Several earlier attempts, including R. von Mises' basic

contributions in this direction, were not satisfactory. Regarding such
an axiomatic basis, we note that D. Hilbert has proposed it as part of
the sixth problem in his famous compendium of mathematical problems
at the International Congress of Mathematics in 1900 in Paris.

For an appreciation of the situation, let us quote a relevant passage
here. Hilbert writes:

> Investigations of the foundations of geometry suggest the prob-
> lem: *To treat in the same manner, by means of axioms, those
> physical sciences in which mathematics plays an important part;
> first of all, the theory of probability and mechanics,* (italics orig-
> inal). As to the axioms of the theory of probability, it seems to
> me desirable that their logical investigation should be accompa-
> nied by a rigorous and satisfactory development of the method
> of mean values in mathematical physics and in particular in the
> kinetic theory of gases.

In including this problem in the compendium, it appears that Hilbert
was influenced by a popular lecture, due to a less-known scientist named
Bohlmann, given for some high school teachers on insurance problems
containing a brief account of the axioms of probability which apparently
was not satisfactory. Indeed several attempts, including those by von
Mises, have been unsuccessful since most of these tried to incorporate
the interpretation of the practical aspect of "randomness" as part of
the system. The breakthrough came with A. Kolmogorov in 1933 when
he was able to separate the latter from the desired axioms for a rigorous
development of the subject. In fact, Kolmogorov has gone a step further
and presented a very general definition of conditional expectation from
which the conditional probability function was immediately obtained.
In the elementary case, one defines the conditional probability first and
then a conditional expectation is obtained from it *while in the general
case the order is reversed.* This turns out to be an important and
necessary change in the view-point for such an extension.

To understand the problems clearly, we shall discuss here the elementary case in detail and show how this leads to the desired abstraction
in the next chapter. A quick and plausible argument in this transition
often leads to confusion and even to paradoxes. We illustrate some of
the latter in Chapter 3 and pinpoint the sources of trouble. Attempts

at new axiomatic methods of conditioning and their (limited) use will be discussed in Chapter 4. Then the fundamental Kolmogorov model and its ramifications as well as important applications will occupy the rest of this book. The treatment throughout is detailed. It is hoped that the subject is better understood and appreciated as a result.

Before proceeding further, it must be noted that the physical and philosophical interpretations of results with Kolmogorov's model (as with others) are still in need of resolution to the satisfaction of all practicing scientists. In particular the question of randomness in nature and its verification by mathematical modeling are still incompletely interpreted. However, this type of problem is not unique for probability theory and it relates to several philosophical questions in mathematical sciences in general. Consequently, they will be set aside in our treatment. Following the contemporary practices in other branches of mathematics, we shall concentrate on the results obtained from the logical investigations as outlined in Hilbert's formulation.

Thus the next section is devoted to the classical methods of conditioning and then its immediate consequences are included in the balance of this chapter. This will give a concrete basis for various extensions envisaged later on.

1.2 Conditional probability given a partition

If (Ω, Σ, P) is a probability triple (or space) modeling an experiment as described at the beginning of the preceding section, then the probability of an event A, (i.e., $A \in \Sigma$) denoted $P(A)$, is a measure of uncertainty associated with A. Thus $P(A) = 1$ if A is certain to occur and $P(A) = 0$ signifies its impossibility (then calling A a "null" event), while $0 < P(A) < 1$ gives a measure of the intermediate situation. Now let A, B be a pair of events of which B is known to have occurred. If these events relate to an experiment, then it is reasonable that the uncertainty of A should be influenced by the occurrence of B and hence the joint occurrence of A and B would be more uncertain than that of B alone; i.e., $P(A \cap B) \leq P(B)$ and by normalization therefore one takes the conditional probability of A given that B has occurred, as $P(A \cap B)/P(B)$. For this to be meaningful, we must insist that $P(B) > 0$. These considerations, which are found satisfactory in applications, lead to the following precise description of the concept.

Definition 1. If (Ω, Σ, P) is a probability space, $B \in \Sigma$ with $P(B) > 0$, then the conditional probability given B, denoted $P(\cdot|B)$, is taken as:

$$P_B(A) = P(A|B) = P(A \cap B)/P(B) \, , \quad A \in \Sigma, \ B \in \Sigma. \qquad (1)$$

Since $P(A|B)$ becomes an indeterminate form for $P(B) = 0$, such events are excluded in this definition. Although it appears reasonable not to include impossible events from given conditions, one finds on some reflection that there exist $B \neq \phi$ for which $P(B) = 0$, and they are an important part of the problems for investigation. Thus it is desirable to extend (1) to include these "exceptional" cases. We shall see that such an inclusion presents a nontrivial difficulty. Let us first analyze and study the simple case given by Definition 1 and some of its easy extensions to understand the subject.

The concept of conditioning formalized in (1) is reasonable in the sense that if A and B are independent events so that the occurrence of one has no influence over the presence or absence of the other, then $P(A|B) = P(A)$ should hold. This is true for (1) since the measures of uncertainty considered above translate to the independent events iff (= if and only if) $P(A \cap B) = P(A)P(B)$ so that $P(A|B) = P(A)$ follows from (1) under the supplementary condition that $P(B) > 0$. The concept of independence is fundamental for probability theory and in fact distinguishes it from the classical measure theory. In a sense we may also say that conditioning is basic and fundamental to probability although it is a somewhat more involved concept than independence.

For a finer analysis of the experiment under consideration one introduces *random variables*. These are mappings $f : \Omega \to \mathbb{R}$ such that (for technical reasons which however are automatic in most applications) $f^{-1}(I) \in \Sigma$ for each interval I of $\mathbb{R}$. Thus a random variable is a real valued function on Ω such that $\{\omega : f(\omega) \gtrless a\}$ are events for each $a \in \mathbb{R}$. Then the *expectation* of a random variable f on (Ω, Σ, P), denoted $E(f)$, is defined as:

$$E(f) = \int_\Omega f \ dP \qquad (2)$$

where the integral is taken, with a view to future analysis in the subject, as an absolute (or Lebesgue) integral, so that $E(f)$ *exists iff* $E(|f|) < \infty$. We can also introduce the conditional expectation of f given an event B with $P(B) > 0$, as follows.

Definition 2. Let f be a random variable on a probability triple (Ω, Σ, P). If an event B, with $P(B) > 0$, has taken place, then the *conditional expectation* of f given B, denoted $E(f|B)$ or $E_B(f)$, becomes (by (2) and (1))

$$E_B(f) = \int_\Omega f(\omega) \, dP(\omega|B) \quad \left(= \frac{1}{P(B)} \int_B f(\omega) dP(\omega)\right). \tag{3}$$

It is clear that $E_B(f)$ changes with B, in general, and $E_\Omega(f) = E(f)$. Also the conditional probability space $(\Omega, \Sigma, P(\cdot|B))$ is the same as $(B, \Sigma(B), P(\cdot|B))$ where $\Sigma(B) = \{A \cap B : A \in \Sigma\}$ is the *trace* of Σ on B and $P(B) > 0$. Further, the preceding concepts can be extended if the conditioning is not just for one event, but for a countable collection of events, say $\{A_i, i \geq 1\}$, forming a partition of Ω. To see this, first observe that if A and A^c satisfy $P(A) > 0, P(A^c) > 0$, then (1) and (3) give P_A, P_{A^c} and $E_A(f)$, $E_{A^c}(f)$ respectively. Thus the conditional expectation of f given a pair $\{A, A^c\}$ is generally a two valued function. Since $\{A, A^c\}$ forms a partition of Ω, one obtains an extension to the class $\mathcal{P} = \{A_n, n \geq 1\}$, $A_n \cap A_m = \phi$ for $n \neq m$, $P(A_n) > 0$, $n \geq 1$, $\cup_n A_n = \Omega$. Then $\{P_{A_n}, n \geq 1\}$ is a family of (conditional) probability functions on Σ, and the conditional expectation of f, with $E(|f|) < \infty$, given a partition $\mathcal{P}$ is (possibly) a countably valued function, denoted $E_{\mathcal{P}}(f)$, with values $\{E_{A_n}(f), n \geq 1\}$. Both these statements can be expressed in a compact form as follows. By definition of $E_{\mathcal{P}}(f)$

$$E_{\mathcal{P}}(f) = \sum_{n=1}^\infty \frac{1}{P(A_n)} \left(\int_{A_n} f dP\right) \chi_{A_n}, \tag{4}$$

where χ_A is the *indicator* function of A (i.e. $\chi_A(\omega) = 1$ or 0 according as $\omega \in A$ or $\omega \notin A$). From (1) and (3) we have $P_B(A) = E_B(\chi_A)$. Hence the corresponding expression for the conditional probability given a partition $\mathcal{P}$, denoted $P_{\mathcal{P}}(\cdot)$, is obtained from (4) as

$$P_{\mathcal{P}}(A) = E_{\mathcal{P}}(\chi_A) = \sum_{n=1}^\infty P_{A_n}(A)\chi_{A_n}. \tag{5}$$

Thus a conditional probability given a partition is (possibly) a countably valued function. It is noteworthy that this partition case of conditional probability is obtained from the conditional expectation in contrast to the single event case of (3) which is derived from (1). This change of roles will be the rule in the more general conditioning than a partition. To motivate that case, we represent (4) in an alternative form since, as will become clear later, a constructive definition such as (4) is not always possible for a conditioning that is more general than a partition, and the latter have to be considered in numerous applications. For instance, calculating a conditional expectation (or a probability distribution) of an integrable random variable given another random variable, involves a conditioning that need not be a partition. Thus the procedure becomes less intuitive. In anticipation of such cases, we restate (4) and (5) in a different form.

Let $\mathcal{B}$ be the smallest σ-algebra containing the given partition $\mathcal{P}$. Then the generators of $\mathcal{B}$ are of the form $A = \cup_{k \in J} A_k$ for subsets $J \subset N$, $A_k \in \mathcal{P}$. Since $E_{\mathcal{P}}(f)$ is clearly a random variable relative to $\mathcal{B}$, we may integrate it on each generator A of $\mathcal{B}$ to obtain (see (4))

$$
\begin{aligned}
\int_A E_{\mathcal{P}}(f)dP &= \sum_{n=1}^{\infty} \left(\int_{A_n} f dP / P(A_n) \right) \int_A \chi_{A_n} \, dP \\
&= \sum_{n \in J} \left(\int_{A_n} f dP \right) \frac{P(A_n \cap A)}{P(A_n)} , \quad \text{since } A_n \cap A_m = \phi \\
&\qquad \text{for } n \neq m , \\
&= \sum_{n \in J} \int_{A_n} f dP, \text{ since } A_n \subset A \quad \text{for } n \in J , \\
&= \int_A f dP.
\end{aligned}
\tag{6}
$$

This relation is important and the standard (Lebesgue) integration implies that (6) is then valid for all $A \in \mathcal{B}$. If $\nu_f : A \mapsto \int_A f dP$, $A \in \mathcal{B}$, then $\nu_f : \mathcal{B} \to \mathbb{R}$ is a σ-additive function (= signed measure) vanishing on P-null sets, so that ν_f is absolutely continuous relative to $P_{\mathcal{B}}$, the restriction of P to $\mathcal{B}$ (also written as $P|\mathcal{B}$). It may be noted that f need not be a random variable relative to $\mathcal{B}$ (although it is one for Σ) but $E_{\mathcal{P}}(f)$ is. The significance of this relation will become clear in the next chapter when we discuss the general case.

To gain facility and to motivate the desired formulation we consider here a few applications of the previously introduced elementary concept. In particular one may derive the so-called total probabilities and Bayes formulas of interest in applications. These and other relations are given in the following:

Proposition 3. *Let $\{A_k,\ 1 \leq k \leq n+1\}$ be events from (Ω, Σ, P). One has:*

(a) (product formula) *if $P(A_2 \cap \cdots \cap A_{n+1}) > 0$, then*

$$P(A_1 \cap \cdots \cap A_n | A_{n+1}) = \prod_{k=1}^{n} P(A_k | A_{k+1} \cap \cdots \cap A_{n+1}); \qquad (7)$$

(b) (total probability formula) *if $A_i \cap A_j = \phi$ for $i \neq j$, and $B \subset \bigcup_{i=1}^{n} A_i$, $P(A_i) > 0$, $i = 1, \ldots, n$, then*

$$P(B) = \sum_{i=1}^{n} P(A_i) P(B|A_i); \qquad (8)$$

(c) (Bayes formula) *if $A_i \cap A_j = \phi$ for $i \neq j$, $P(A_i) > 0$, $P(B) > 0$, and $B \subset \bigcup_{i=1}^{n} A_i$, then*

$$P(A_i|B) = \frac{P(A_i)P(B|A_i)}{\sum_{j=1}^{n} P(A_j)P(BA_j)}. \qquad (9)$$

Proof. All three parts are easy consequences of the definitions. Thus for (a), since $B = A_1 \cap \cdots \cap A_n$ is an event

$$P(B|A_{n+1}) = P(A_1 \cap \cdots \cap A_n | A_{n+1})$$
$$= \frac{P(A_1 \cap \cdots \cap A_{n+1})}{P(A_{n+1})}$$
$$= \prod_{i=1}^{n} \frac{P(A_i \cap \cdots \cap A_{n+1})}{P(A_{i+1} \cap \cdots \cap A_{n+1})}, \quad \text{since the denominators}$$
$$\text{are positive,}$$
$$= \prod_{i=1}^{n} P(A_i | A_{i+1} \cap \cdots \cap A_{n+1}).$$

For (b), the hypothesis implies

$$B = B \cap \bigcup_{i=1}^{n} A_i = \bigcup_{i=1}^{n} (B \cap A_i), \quad \text{(a disjoint union)}.$$

Hence

$$P(B) = \sum_{i=1}^{n} P(B \cap A_i) = \sum_{i=1}^{n} P(A_i) P(B|A_i).$$

Finally (c) follows from (1) and (b) since $P(B) > 0$ is also true here. Thus

$$P(A_i|A) = \frac{P(A_i \cap B)}{P(B)} = \frac{P(A_i)P(B|A_i)}{\sum\limits_{j=1}^{n} P(A_j)P(B|A_j)},$$

by (b). □

Discussion. Since $P(B|A_1 \cap A_2)$ is the conditional probability of B given A_1 and A_2, it can also be thought of as a conditioning with the (conditional) probability $P_{A_1}(\cdot)$ given A_2 so that

$$P(B|A_1 \cap A_2) = P_{A_1}(B|A_2)$$

which can be verified directly by (1). Consequently this alternative symbolism will be used when convenient. Using such a form, Bayes formula can be iterated for conditional probabilities and it then has an interesting interpretation in applications. We present this for a convenient reference:

Proposition 4. *If $A_1, \ldots, A_n$ are disjoint events, $P(A_i) > 0$, and B_1, B_2 are a pair of (possibly) different events such that $B_i \subset \cup_{j=1}^{n} A_j$, $i = 1, 2$, and $P(B_1 \cap B_2) > 0$, then we have for each $1 \leq i_o \leq n$,*

$$P(A_{i_o}|B_1 \cap B_2) = P_{B_1}(A_{i_o}|B_2)(= P_{B_2}(A_{i_o}|B_1))$$

$$= \frac{P_{B_1}(A_{i_o})P_{B_1}(B_2|A_{i_o})}{\sum\limits_{j=1}^{n} P_{B_1}(A_j)P_{B_1}(B_2|A_j)}, \tag{10}$$

where $P_{B_i}(\cdot) = P(B_i \cap \cdot)/P(B_i)$, $i = 1, 2$, are the usual conditional probability functions defined by (1).

The proof follows at once from an expansion of various symbols and need not be repeated. It is useful to give here a brief meaning of (10).

If A_i are events having $P(A_i)$ as their *prior* probability values (i.e., these are given from past experience and before the experiment at hand is performed) and if the outcome, after the experiment contains B so that $P(B|A_i)$, $i = 1, 2$, are determined for each A_i one can obtain the *posterior* probabilities of A_i after $B(= B_1 \cap B_2)$ is observed, so that $P(A_i|B)$ is obtainable, using (9). Instead, suppose two experiments are performed in succession and the events B_1 and B_2 are observed. Then one can calculate the posterior probability of A_i given B_1 and B_2 either directly by using (9) with $B = B_1 \cap B_2$ or using $P_{B_1}(A_i) = P(A_i|B_1)$ as the prior probability for the second experiment employing $P_{B_1}(\cdot)$ as the new function (or using $P_{B_2}(\cdot)$, if P_{B_1} is ignored for some reason) and working with (10). The above proposition shows that either procedure gives the same posterior probability, as one expects. The method can be iterated. Formulas (7) and (8) are often used in problems involving "Markov chain" dependence and similar structures. We shall illustrate the usefulness of the above method in two forms for a selection problem in Section 5 below to show how these simple alternative procedures lead to some nontrivial and interesting applications.

Part (a) of Proposition 3 leads to the concept of mutual independence of events. Suppose that $P(A_i|A_{i+1} \cap \cdots \cap A_{n+1}) = P(A_i)$, $i = 2, \ldots, n$. For such a case (7) becomes

$$P(A_1 \cap \cdots \cap A_m) = \prod_{i=1}^{m} P(A_i), \ 1 < m \le n. \tag{11}$$

When (11) holds then $A_1, \ldots, A_n$ will be termed *mutually independent* events without regard to any side conditions such as $P(A_i) > 0$. The number of equations to be satisfied here is $2^n - n - 1$. Now one may ask only that (11) be true for each pair of events. Then there are $\binom{n}{2}$ equations and if $n > 2$ these will imply fewer restrictions than demanded by the concept of (11). Thus the weaker concept is termed *pairwise independence*. Simple examples can be constructed to show that, except for $n = 2$, these two concepts are distinct. It should be emphasized that, although (9) is derived with the assumption of positive probability of the joint event $A_2 \cap \cdots \cap A_n$, no such condition is needed for (11), since this set of equations can and will be taken as the definition of mutual independence. The concept is a distinctive feature of probability theory as is the dependence relation of Definition 1. These two notions have no significance in Measure Theory.

We noted earlier that the formulas given by (7) - (10), although mere consequences of (1), are very useful in applications. That is why they are distinguished as propositions. In a similar manner we present some properties of conditional expectations, introduced in Definition 2, in the next section. They will also serve as a motivation for the desired generalizations, indicated before.

1.3 Conditional expectation: elementary case

The concept of elementary conditional expectation was introduced in Definition 2.2. Here we present a few properties that facilitate its immediate use.

Proposition 1. *Let $f : \Omega \to \mathbb{R}$ be a random variable on a probability space (Ω, Σ, P) and $\mathcal{P} \subset \Sigma$ be a partition of Ω (of positive probability for each member). If $\mathcal{P} = \{A_i, i \geq 1\}$, then the existence of $E_{\mathcal{P}}(f)$ implies that $E_A(f)$ exists for each subpartition $\mathcal{A} = \{A_{i_n}, n \geq 1\} \subset \mathcal{P}$, and one has*

$$E_A(f) = \frac{1}{P(A)} \int_A E_{\mathcal{P}}(f)dP = \sum_{n=1}^{\infty} E_{A_n}(f)P(A_{i_n}|A), \qquad (1)$$

with $A = \cup_n A_{i_n}$, and $E_A(\cdot)$, $E_{\mathcal{P}}(\cdot)$, are given by (3) and (4) of Section 1.2.

Proof. This is also an easy consequence of the definitions. Indeed by (2.4),

$$(E_{\mathcal{P}}(f))(\omega) = E_{A_n}(f)(\omega), \quad \omega \in A_n, \ n \geq 1,$$

since $\Omega = \cup_n A_n$ and each ω in Ω belongs to exactly one A_n. Therefore the existence of $E_{\mathcal{P}}(f)$ implies that of $E_{A_n}(f)$, $n \geq 1$, and hence of $E_A(f)$ or $E_{\mathcal{A}}(f)$. Further by (2.6), we have

$$\int_A E_{\mathcal{P}}(f)dP = \int_A f dP = P(A)E_A(f), \quad (\text{cf., (2.3)}). \qquad (2)$$

On the other hand,

$$\int_A f dP = \sum_{n=1}^{\infty} \int_{A_{i_n}} f dP, \quad (\sigma\text{-additivity of the integral})$$

$$= \sum_{n=1}^{\infty} P(A_{i_n})E_{A_{i_n}}(f), \quad \text{by (2.3)}$$

$$= \sum_{n=1}^{\infty} P(A_{i_n}|A)E_{A_{i_n}}(f)P(A), \qquad (3)$$

since $A_{i_n} \subset A$. It follows from (2) and (3), that (1) holds upon cancelling the positive factor $P(A)$. $\square$

The point of formula (1) is that it facilitates a calculation of $E_A(f)$ when the individual $E_{A_n}(f)$s are known. Such expressions are useful in several computations of relatively simple and interesting practical problems. The popularity of the definition $P_A(\cdot)$ of (2.1) stems from the fact that it behaves like the ordinary probability measure when $P(A) > 0$. This point is illuminated by the following simple result.

Proposition 2. *Let $A_1, A_2, \ldots$ be events from (Ω, Σ, P) such that $\lim_n A_n = A$ (so A is an event) and $P(A) > 0$. Then for any integrable random variable $f : \Omega \to \mathbb{R}$, we have*

$$\lim_{n \to \infty} E_{A_n}(f) = E_A(f), \tag{4}$$

pointwise, on Ω. In particular, if $\lim_n P(A_n) = 1$, then $\lim_n A_n = A$ exists, $P(A) = 1$, and we get $E_A(f) = E(f)$, with probability one.

Proof. Since $A_n \to A$ iff $\chi_{A_n} \to \chi_A$, as $n \to \infty$, pointwise, which follows from the definitions of respective convergences of these quantities, we have (by the bounded convergence)

$$0 < P(A) = \int_\Omega \chi_A \, dP = \lim_n \int_\Omega \chi_{A_n} \, dP = \lim_n P(A_n).$$

So there is an n_0 such that for $n \geq n_0$ we have $P(A_{n_0}) > 0$. Hence $E_{A_n}(f)$ and $E_A(f)$ are well-defined, $n \geq n_0$, as in the preceding result. On the other hand

$$E_{A_n}(f) = \frac{1}{P(A_n)} \int_{A_n} f \, dP = \frac{1}{P(A_n)} \int_\Omega f \chi_{A_n} \, dP$$
$$\to \frac{1}{P(A)} \int_A f \, dP, \text{ as } n \to \infty,$$

by the dominated convergence theorem. This gives (4).

For the last part, $\lim_n P(A_n) = 1$ implies $\chi_{A_n^c} \to 0$ with probability one. Hence $\chi_{A_n} \to 0$ and $A_n \to A$ with probability one and $P(A) = 1$. The rest is a simple consequence of this fact. $\square$

The last part of the above result has the following interesting supplement.

Proposition 3. *Let f and g be two random variables on (Ω, Σ, P) assuming at most countably many values with positive probability. If $E(f)$ exists, then $E(f|g = b_k)$ exists and defines a random variable taking the above (countable) values, and moreover*

$$E(E(f|g)) = E(f).$$

Here g takes the values $b_1, b_2, \ldots$ with positive probability.

Proof. Let $\{a_i, i \geq 1\}$ and $\{b_j, j \geq 1\}$ be the values taken by f and g respectively. If $B_k = [g = b_k]$, then $P(B_k) > 0$, and similarly for f. Then

$$
\begin{aligned}
E(f|B_k) &= \frac{1}{P(B_k)} \int_{B_k} f \, dP = \int_{\Omega} f(\omega) P_{B_k}(d\omega) \\
&= \sum_{i=1}^{\infty} a_i P(f = a_i, g = b_k)/P(g = b_k) \\
&= \varphi(b_k), \quad \text{(say)}.
\end{aligned}
\tag{6}
$$

Since $E(f)$ exists, this series converges absolutely because

$$
|\varphi(b_k)| \leq \frac{1}{P(B_k)} \sum_{i=1}^{\infty} |a_i| P(f = a_i, g = b_k),
$$

and so,

$$
\begin{aligned}
\sum_{k=1}^{\infty} |\varphi(b_k)| P(g = b_k) &\leq \sum_{i=1}^{\infty} \sum_{k=1}^{\infty} |a_i| \, P(f = a_i, g = b_k) \\
&= \sum_{i=1}^{\infty} |a_i| \sum_{k=1}^{\infty} P(f = a_i, g = b_k) \\
&= \sum_{i=1}^{\infty} |a_i| P(f = a_i) \\
&= E(|f|) < \infty.
\end{aligned}
\tag{7}
$$

Define $h : \Omega \to \mathbb{R}$ by the expression

$$
h = \sum_{k=1}^{\infty} \varphi(b_k) \, \chi_{B_k}.
\tag{8}
$$

Since g is a random variable, $B_k \in \Sigma$, and thus h is a random variable taking at most countably many (not necessarily distinct) values $\{\varphi(b_k), k \geq 1\}$. Further (8) implies that $h = \varphi(g)$. So this and (6) give $h(b_k) = E(f|g = b_k)$. Thus $h = E(f|g) = \varphi(g)$ is a random variable and from (7) one gets

$$
E(|h|) = \int_{\Omega} |\varphi(g)| dP \leq E(|f|) < \infty,
$$

and then

$$E(h) = \int_{\Omega} \varphi(g)dP = \sum_{k=1}^{\infty} \varphi(b_k)P(g = b_k),$$

$$= \sum_{k=1}^{\infty} \sum_{i=1}^{\infty} a_i P(f = a_i, g = b_k), \text{ (cf., (6))}$$

$$= \sum_{i=1}^{\infty} a_i \sum_{k=1}^{\infty} P(f = a_i, g = b_k), \text{ since the series is}$$

absolutely convergent ,

$$= \sum_{i=1}^{\infty} a_i P(f = a_i) = E(f).$$

This is precisely (5). □

It should be remarked that, for this argument, we need the condition that $P(g = b_k) > 0$ in addition to $E(|f|) < \infty$. However, in many applications, especially with continuous distributions of random variables, $P[b < g \leq b + \delta] \to 0$ as $\delta \downarrow 0$. In such cases the above method is no longer valid. We discuss later a few extensions together with the necessary machinery to take the resulting complications into account. A simple procedure can be used for some of these problems and it is illustrated in the next section.

1.4 Conditioning with densities

Let X, Y be a pair of random variables on (Ω, Σ, P) having an absolutely continuous (joint) distribution function $F_{X,Y}$ with density $f_{X,Y}$ so that

$$P(X < x, Y < y) = F_{X,Y}(x, y) = \int_{-\infty}^{x} \int_{-\infty}^{y} f_{X,Y}(u, v)dv\, du. \quad (1)$$

If F_X, F_Y are the marginal distributions of X, Y and f_X, f_Y are their respective densities then

$$F_X(x) = F_{X,Y}(x, +\infty) = \int_{-\infty}^{x} f_X(u)du = \int_{-\infty}^{x} \left(\int_{-\infty}^{\infty} f_{X,Y}(u, v)dv \right) du,$$

$$F_Y(y) = F_{X,Y}(+\infty, y) = \int_{-\infty}^{y} f_Y(v)dv = \int_{-\infty}^{y} \left(\int_{-\infty}^{\infty} f_{X,Y}(u, v)du \right) dv.$$

The conditioning concept of Definition 2.1, in this case, becomes

$$P([X \in A]|[Y \in B]) = \frac{P([X \in A] \cap [Y \in B])}{P(Y \in B])}$$

$$= \frac{\int_A \int_B f_{X,Y}(u, v)dudv}{\int_B f_Y(v)dv}. \quad (2)$$

To simplify (2), let $A = [a_1, a_2)$, $B = [b, b + h)$, $h > 0$. Then

$$P(a_1 \leq X < a_2 | b \leq Y < b + h) = \int_{a_1}^{a_2} \left(\frac{\int_b^{b+h} f_{X,Y}(u,v)dv}{\int_b^{b+h} f_Y(v)dv} \right) du. \quad (3)$$

Since both $P(b \leq Y < b + h)$ and $P(a_1 \leq X < a_2, b \leq Y < b + h)$ tend to zero as $h \downarrow 0$, the left side of (3) results in an indeterminate form. Note that these two probabilities become $P(Y = b)$ and $P(a_1 \leq X < a_2, Y = b)$ which are positive only if Y has a discrete part in its distribution at b. Since these are zero in the present case, one can apply a form of the L'Hôpital rule and evaluate (the right side of) the expression (3) and define it as the value of the left side. Thus one has

$$P(a_1 \leq X < a_2 | Y = b) = \lim_{h \downarrow 0} P([a_1 \leq X < a_2 | b \leq Y < b + h])$$

$$= \lim_{h \downarrow 0} \int_{a_1}^{a_2} \left(\frac{\frac{1}{h} \int_b^{b+h} f_{X,Y}(u,v)dv}{\frac{1}{h} \int_b^{b+h} f_Y(v)dv} \right) du$$

$$= \int_{a_1}^{a_2} \frac{f_{X,Y}(u,b)}{f_Y(b)} du. \quad (4)$$

Here we have taken the limits inside the integral and used a form of (Lebesgue's version of) the fundamental theorem of integral calculus. Now set

$$f_{X|Y}(u|b) = \begin{cases} f_{X,Y}(u,v)/f_Y(b) & , \text{ if } f_Y(b) > 0 \\ \alpha & , \text{ if } f_Y(b) = 0. \end{cases} \quad (5)$$

Here $\alpha \geq 0$ is an arbitrary number. Since $0 = f_Y(b) = \int_{-\infty}^{\infty} f_{X,Y}(u,b)du$, we must have $f_{X,Y}(u,b) = 0$ for almost all u, so that $f_{X|Y}(u|b)$ is indeterminate at b and we gave $\alpha \geq 0$ as an arbitrary value. This does not affect the definition of distributions. For simplicity, we take $\alpha = 0$ hereafter. It will be seen later that these indeterminate forms occur frequently in such situations and create difficult problems and the results depend on the method of taking limits. They will be analyzed. In the present context, using (5) we can express (4) as

$$P(a_1 \leq X < a_2 | Y = b) = \int_{a_1}^{a_2} f_{X|Y}(u|b)du,$$

and thus one *defines* the (horizontal) conditional density of X given $Y = b$ as $f_{X|Y}(\cdot|b)$. This method of conditioning will be given as:

Definition 1. Let X, Y be real random variables on (Ω, Σ, P) with an absolutely continuous distribution function whose joint density is $f_{X,Y}$. If f_Y is the marginal density of Y, then the conditional probability of X given $Y = b$ (a null event) is defined as

$$P([X \in A])|[Y = b]) = \int_A f_{X|Y}(u|b)du. \qquad (6)$$

where $f_{X|Y}(\cdot|b)$ is a (horizontal) conditional density of X given $Y = b$ and is the function defined by (5).

Since $f_{X|Y} \geq 0$ and $\int_{-\infty}^{\infty} f_{X|Y}(u|b)du = 1$ are immediate from (5), we are justified in calling $f_{X|Y}(\cdot|b)$ a density. However, it must be observed that the method of calculating the density in (4) is not universal, and the result depends on the manner in which the limit is evaluated. (Here it is along the x or horizontal axis.) This becomes in fact pronounced when conditioning is based on more than one random variable, the resulting (conditional) event has probability zero, and the limits are taken in other directions then the "horizontal". We postpone the analysis of such cases to Chapter 3 since conditioning with events of probability zero is a thorny problem. Also we discuss the compatibility of (6) with a generalization of Definition 2.1 there.

We now use the above "essentially natural" definition in evaluating corresponding conditional expectations. Thus if $E(X)$ exists, then one usually *defines* $E(X|Y = b)$ as

$$E(X|Y)(b) = E(X|Y = b) = \int_{-\infty}^{\infty} u f_{X|Y}(u|b)du. \qquad (7)$$

The existence of the integral follows as in (3.6). In fact, with the classical Fubini Theorem, the function $\varphi : b \mapsto E(X|Y)(b)$ of (7) is a random variable and depends on the values of Y so that $E(X|Y) =$

$\varphi(Y)$ is well-defined. Also

$$E(|\varphi(Y)|) = \int_{\Omega} |\varphi(Y)|dP, \quad \text{by definition},$$

$$= \int_{\mathbb{R}} |\varphi(b)|f_Y(b)db, \quad \text{by the image law (cf. Prop. 2 below)},$$

$$= \int_{\mathbb{R}} |\int_{\mathbb{R}} u\ f_{X|Y}(u|b)du\ |\ f_Y(b)db$$

$$\leq \int_{\mathbb{R}} \int_{\mathbb{R}} |u| \frac{f_{X|Y}(u|b)}{f_Y(b)}du\ f_Y(b)db, \quad \text{by (5)},$$

$$= \int_{\mathbb{R}} \int_{\mathbb{R}} |u|\ f_{X|Y}(u,b)du\ db$$

$$= \int_{\mathbb{R}} |u|\ f_X(u)du = \int_{\Omega} |X|dP = E(|X|) < \infty. \qquad (8)$$

This shows that (7) is meaningful, and $E(\varphi(Y))$ exists. Moreover, the same computation, without absolute values, shows that

$$E(\varphi(Y)) = E(E(X|Y)) = E(X). \qquad (9)$$

Thus the analog of Proposition 3.3 is valid for the continuous case. We shall see in the next chapter that this equation holds for all random variables X, Y for which $E(|X|) < \infty$, where it is shown that a more general definition of conditioning is possible.

First we need to establish the image law used above since it plays a basic role in the subject.

Proposition 2. (Fundamental Law of Probability) *Let X be a random variable (or n-vector), on (Ω, Σ, P) with distribution function F_X. Then for any continuous (or even a Borel) function $\varphi : \mathbb{R}^n \to \mathbb{R}, \varphi(X)$ is a random variable and*

$$E(\varphi(X)) = \int_{\Omega} \varphi(X)dP = \int_{\mathbb{R}^n} \varphi(x)dF_X(x) \qquad (10)$$

in the sense that if either integral exists then both exist and equality holds.

Proof. If $A \subset \mathbb{R}^n$ is an interval and $\varphi = \chi_A$, then $\varphi(X) = \chi_{X^{-1}(A)}$ where $X : \Omega \to \mathbb{R}^n$ is the given random vector. Thus if A is expressible as $\underset{i=1}{\overset{n}{\times}} [a_i, b_i)$ we then have on writing $X = (X_1, \ldots, X_n)$:

$$
\begin{aligned}
E(\varphi(X)) &= P([X \in A]) \\
&= P(a_1 \le X_1 < b_1, \ldots, a_n \le X_n < b_n) \\
&= \int_{a_1}^{b_1} \cdots \int_{a_n}^{b_n} dF_{X_1, \ldots, X_n}(x_1, \ldots, x_n) \\
&= \int \cdots \int_{\mathbb{R}^n} \chi_A \, dF_X \quad .
\end{aligned}
$$

Thus (10) holds in this case. By linearity of $E(\cdot)$ and of the integral on the last line, (10) also holds if $\varphi = \overset{n}{\underset{i=1}{\Sigma}} a_i \chi_{A_i}$, a simple function. Then by the monotone convergence theorem (10) is true for all Borel functions $\varphi \ge 0$, since every such φ is the pointwise limit of a monotone sequence of simple functions. But a general (Borel) φ is expressible as $\varphi^+ - \varphi^-$ where $\varphi^\pm \ge 0$, so that the result follows. □

The important point of (10) is that the (left) integral, on an abstract probability space, is equal to the concrete (Stieltjes) integral on $\mathbb{R}^n$. Also by Definition 2.2 we have

$$
E(\varphi(X)) = \int_{\mathbb{R}} u \, dF_{\varphi(X)}(u) \quad . \tag{11}
$$

But in this form, the right side of (11) demands that we first calculate the distribution of the random variable $Y = \varphi(X)$ which is more difficult than using F_X in (10). Now the preceding proposition shows that (11) can be evaluated, with (10), in terms of F_X itself. This is what we used in computing the bound in (8) for the special case with a conditional density. The general result implied by the identity (10) is enormously useful.

Another key observation here is that Proposition 3.3 or its extension (9) above, showed that $E(X|Y)$ always exists if $E(|X|) < \infty$. Is the converse of this statement true? Because of the ambiguity in our definition of $E(X|Y)$ for the general case, the question is not well-posed. However, when the densities exist and the procedure given here is meaningful can we infer from the existence of $E(|X||Y)(b)$, $b \in \mathbb{R}$, that $E(X)$ exists in the sense that $E(|X|) < \infty$? The following simple illustration shows that the answer even to this special question is in the negative.

Example 3. Let X, Y be a pair of random variables on (Ω, Σ, P) with an absolutely continuous joint distribution whose density $f_{X,Y}$ is given by

$$f_{X,Y}(x,y) = \begin{cases} \frac{1}{\pi} \exp\{-y(1+x^2)\}, & -\infty < x < \infty, y > 0, \\ 0, & \text{otherwise.} \end{cases} \tag{12}$$

Then $f_Y(u) = (\pi y)^{-\frac{1}{2}} e^{-y}, y > 0$, and $= 0$ for $y \leq 0$. Hence the conditional density of X given $Y = y$ is obtained from (5) as:

$$f_{X|Y}(u|y) = \begin{cases} (\pi/y)^{\frac{1}{2}} e^{-u^2 y}, & -\infty < u < \infty, y > 0, \\ 0, & \text{otherwise.} \end{cases} \tag{13}$$

Then for each $n \geq 0$ and $y > 0$, we have

$$E(X^n|Y = y) = \int_{\mathbb{R}} u^n f_{X|Y}(u|y) du$$
$$= \begin{cases} (\pi/y)^{1/2} 2 \int_0^\infty u^n e^{-u^2 y} \, du, & \text{if } n = 2m, m \geq 1, y > 0, \\ 0, & \text{if } n = 2m - 1, m \geq 1 \end{cases} \tag{14}$$

Thus $\varphi_n(y) = E(X^n|Y = y)$ exists for each $n \geq 1$. However, $E(X^n)$ does not exist for any $n \geq 1$, since $f_X(u) = [\pi(1+u^2)]^{-1}$ and

$$E(X^n) = \frac{1}{\pi} \int_{\mathbb{R}} \frac{u^n}{1+u^2} du,$$

by Proposition 2 above. Thus $E(X|Y)(y)$ directly defined by (14) cannot be a conditional expectation of X given Y since (9) does not hold.

This example shows that a conditional expectation cannot always be treated as an ordinary expectation and that (9) is *not* an identity. Moreover the case that the conditioning event having zero probability is a nontrivial (and new) element of the theory which asks for its own detailed analysis.

1.5 Conditional probability spaces: first steps

If (Ω, Σ, P) is a probability space, let $\Sigma^+ = \{A \in \Sigma : P(A) > 0\}$, so that a conditional measure $P_A(\cdot) = P(\cdot|A)$ exists for each $A \in \Sigma^+$. Hence $\{(\Omega, \Sigma, P, P_A) : A \in \Sigma^+\}$ is a family of conditional proability spaces. The point of such an extended view is to study the conditional family separately in order to extract or learn more about the "total" measure P. In this connection the following questions arise naturally; and our work should (and will) answer them.

(i) Can one obtain more knowledge of P by summing (or by some sort of integration procedure) the functions $P(B|\cdot)$ over Σ^+ for each $B \in \Sigma$?

(ii) If one considers $P(\cdot|\cdot) : \Sigma \times \Sigma^+ \to \mathbb{R}^+$ as a function having the properties of a conditional probability, given a priori, is there a unique measure function $P : \Sigma \to \mathbb{R}^+$ from which $P(\cdot|\cdot)$ can be derived?

(iii) Does the situation change materially if instead of Σ^+ one admits events having probability zero?

These are some of the key questions arising from the above work on the conditioning concept. An analysis and study to answer these problems will occupy a large part of the following pages. Let us first indicate a reason for considering these questions. If $B \in \Sigma$ and if there exist $A_k \in \Sigma^+$, disjoint, such that $B \subset \cup_{k=1}^n A_k$ for some n so that the event B is covered by a partition-like collection from Σ^+, then Proposition 2.3(b) implies that

$$P(B) = \sum_{k=1}^{n} P(B|A_k)P(A_k). \tag{1}$$

Thus any such $P(B)$ can be calculated with a knowledge of $P(\cdot|\cdot)$ on $\Sigma \times \Sigma^+$ if the "data" on Σ^+ is known. We illustrate this point with the following version of a classical (secretary) selection problem. It may be observed that an initial difficulty in such situations is to translate the word problem carefully into a mathematical one. Of course this is inherent in all applications, but it needs more work in our subject than perhaps in other parts of mathematics.

Example 1. Suppose there are n candidates for a (secretary) job, each having the desired minimal qualifications. Assume that they can be ordered or ranked according to their (verbal) ability in their interviews. At least i_0 of the candidates must be interviewed before a selection, based on rank, is made. The desirable procedure is that, as the candidates appear randomly, the best should not be left before the selection is completed. The problem therefore is to choose $1 \leq i_0 \leq n$ such that the probability p_{i_0} of selecting the best candidate is a maximum.

[Other optimality criteria are possible, and we illustrate with another after completing this case.] We solve the problem using a combinatorial argument.

Let B be the event that the best candidate is selected and that B_j is the event that the jth interviewed person is the best. Thus by Proposition 2.3,

$$p_{i_0} = P(B) = \sum_{j=1}^{n} P(B|B_j) \, P(B_j) , \tag{2}$$

and since by hypothesis at least i_0 candidates must appear before the experiment is terminated, we have $P(B|B_j) = 0$, $1 \leq j \leq i_0 - 1$; $P(B|B_{i_0}) = 1$ and because the best can appear at any place among the n, $P(B_j) = \frac{1}{n}$. On the other hand if $i_0 + 1 \leq j \leq n$, then the best candidate has not been selected in the first $j - 1$ interviews. This is the same as saying that the best among the first $(j - 1)$ cannot be among i_0 to $j - 1$ so that the relatively best candidate must be among the first $(i_o - 1)$ of them. (Since the interviews passed i_0, it follows that, in this case, i_0 is not the best.) Thus the probability of B, given that B_j has occurred (with $i_o \leq j - 1$) is the same as choosing the best among the first $i_o - 1$ places when $j - 1$ of them are available with equal chance of selection. Hence, for $i_0 + 1 \leq j \leq n$,

$$P(B|B_j) = (i_0 - 1)/(j - 1) \ .$$

Consequently (2) becomes

$$p_{i_0} = P(B_{i_0})P(B|B_{i_0}) + \sum_{j=1}^{i_0-1} P(B|B_j)P(B_j) + \sum_{j=i_0+1}^{n} P(B|B_j)P(B_j)$$

$$= \frac{1}{n} \cdot 1 + 0 + \frac{1}{n} \sum_{j=i_o+1}^{n} \frac{i_0 - 1}{j - 1}$$

$$= \frac{1}{n}[1 + (i_0 - 1) \sum_{j=i_0+1}^{n} \frac{1}{j - 1}] \ . \tag{3}$$

We now choose i_0 such that (3) is a maximum. There is a unique maximum, since $p_i - p_{i+1} = \frac{1}{n}(1 - \sum_{j=i+1}^{n} (j - 1)^{-1}) \geq 0$, if $\sum_{j=i+1}^{n} (j - 1)^{-1} \leq 1$ for $i \geq i_0$ and ≤ 0 for $i < i_0$. So we choose i_0 to satisfy this restriction. For instance, if $n = 15$, then $i_0 = 6$ and $p_{i_0} = 0.3894$. For the asymptotic result, the condition that $\sum_{j=i_0+1}^{n} (j - 1)^{-1} \leq 1$ and (3) together imply that as $n \to \infty, i_0 \to \infty$ at the same rate, and hence for large $n, p_{i_0} \sim \frac{i_0}{n}$. It is easily seen that $\sum\limits_{j=i_0}^{n} (j-1)^{-1} \sim \int_{i_0}^{n} \frac{dx}{x} = \log \frac{n}{i_0}$. Hence $\lim_{n \to \infty} p_{i_0} = e^{-1}$. The preceding argument can be converted into a more standard version (nontrivially) by describing the precise form of (Ω, Σ, P), and generalizing it. Instead of doing this we give another formulation with a more natural optimal rule (or criterion) which admits extensions. It is somewhat sophisticated.

Example 2. Let the selection procedure be the same as in the above problem where each candidate is given a number equal to the rank (i.e., 1 for the best and n for the least). If X denotes the actual ranking of the selected candidate, so that X takes values $1, 2, \ldots, n$, it is desired to find a method (or strategy) that makes the *expected* value of $X[E(X) =$ the expected rank] a minimum.

To solve this problem, we first describe the underlying triple (Ω, Σ, P). Let Ω denote all possible permutations of $(1, 2, \ldots, n)$ as points corresponding to the possible outcomes of the appearances of the candidates. Suppose the ranking is such that there are no ties. Thus Ω has $n!$ points, and Σ will be the power set of Ω. Since each permutation is equally likely we have $P(\{\omega\}) = \frac{1}{n!}, \omega = (i_1, i_2, \ldots, i_n) \in \Omega$. Let $X_j : \Omega \to \{1, 2, \ldots, n\}$ be defined by $X_j(\omega) = i_j$ where $\omega = (i_1, \ldots, i_n)$. Thus $X_j(\omega)$ is the jth candidate's rank in the observed appearance by ω. Let $Y_i(\omega)$ be the relative rank of the ith candidate for the "experiment" ω. Thus $Y_i(\omega)$ is the number of ranks $X_1(\omega), \ldots, X_{i-1}(\omega)$ that are strictly less than $X_i(\omega)$ plus one. Hence $P(Y_i = j) = i^{-1}$ for $1 \leq j \leq i$. The hypothesis now implies that $P(Y_{i_1} = j_1, \ldots Y_{i_n} = j_n) = P(Y_{i_1} = j) = i_1^{-1}$ if $i_1 \neq i_2$ or $\ldots$, or i_n, i.e., the $Y_1, \ldots, Y_n$ are independent. This implies therefore

$$P(X_i = k | Y_1 = j_1, \ldots Y_{i-1} = j_{i-1}, Y_i = \ell) = P(X_i = k | Y_i = \ell)$$
$$= \frac{\binom{k-1}{\ell-1}\binom{n-k}{i-\ell}}{\binom{n}{i}}. \tag{4}$$

Here the last expression is obtained by a reinterpretation of the conditional probability as choosing i objects from n such that $\ell - 1$ are from the upper or best ranked candidates $1, 2, \ldots, k - 1$, then the kth ranked one, and finally the remaining $(i - \ell)$ from the lower $(n - k)$

ranks. This probability is nonzero only if $1 \leq \ell \leq k \leq n - i + \ell$. Using this expression and Definition 2.2, we get

$$
\begin{aligned}
E(X_i | Y_i = \ell) &= \sum_{k=1}^{n} k P[X_i = k | Y_i = \ell] \\
&= \ell \sum_{k=1}^{n} \binom{k}{\ell} \binom{n-k}{i-\ell} / \binom{n}{i} \\
&= \ell \binom{n+1}{i+1} / \binom{n}{i} = \frac{n+1}{i+1} \ell,
\end{aligned}
\tag{5}
$$

where we used a binomial coefficient identity in the last line. This is not obvious. But it may be established by induction on i and ℓ, subject to the above conditions, using the following classical identity:

$$
\sum_{k=0}^{n-1} \binom{n-k}{i} = \binom{n+1}{i+1} \quad , \quad n \geq i \geq 0,
\tag{6}
$$

which itself is verified by induction on n.

The problem now is to find a method or "rule" that depends only on the observations already taken (i.e., the candidates already interviewed) but not on future values. Typically the rule is an integer valued random variable τ on $\{1, 2, \ldots, n\}$ such that the event $[\tau = j]$ depends on $Y_1, \ldots, Y_j$ but not on $Y_{j+1}, \ldots, Y_n$. Thus, for each such τ, we get from (5) and Proposition 2.3 that

$$
E(X_\tau | Y_\tau = \ell) = \sum_{i=1}^{n} E(X_i | Y_i = \ell) P(\tau = i).
\tag{7}
$$

Since X_τ is a bounded random variable (X_τ being a function of a function obtained from X_j and τ is clearly a random variable) we have by Proposition 3.3 and Equation (5) that

$$
E(X_\tau) = E(E(X_\tau | Y_\tau)) = E\left(\frac{n+1}{\tau+1} Y_\tau\right).
\tag{8}
$$

To minimize $E(X)$ is thus equivalent to minimizing $E(X_\tau)$ for various "stopping rules" τ. The procedure again is nontrivial, and one uses the so-called backward (or reversed) induction to solve the problem.

For a given n, we want to find a $\tau \geq i+1$ (since i candidates will be interviewed under any circumstances) such that $E(X_\tau)$ is a minimum.

Let $a_i(n)$ be this minimum value if i candidates are interviewed. First note that if $i = n - 1$ then $\tau = n$ with probability one, so that

$$a_{n-1}(n) = E\left(\frac{n+1}{n+1}Y_n\right) = E(Y_n) = \sum_{k=1}^{n} k \cdot \frac{1}{n} = \frac{n+1}{2}. \qquad (9)$$

Then to use the above stated induction for $i = n - 1, n - 2, \ldots, 1$, we have

$$a_{i-1}(n) = E[\min \left(\frac{n+1}{i+1} Y_i, a_i(n)\right)]$$

$$= \sum_{j=1}^{i} \min \left(\frac{n+1}{i+1} j, a_i(n)\right) \cdot \frac{1}{i}, \text{ by Proposition 4.2}. \qquad (10)$$

These are a set of difference equations with (9) as the boundary value. It is clear that $a_0(n) \le a_1(n) \le \cdots \le a_{n-1}(n) = \frac{n+1}{2}$ by (10) and (9), and thus the minimum value is $a_0(n)$. A closed expression for $a_0(n)$ is not simple. For instance, if $n = 4$ one can verify that $a_0(4) = a_1(4) = \frac{15}{8}$, $a_2(4) = \frac{25}{12}$, $a_3(4) = \frac{5}{2}$. It can be shown after a delicate analysis (given by Chow et al., 1964) that

$$\lim_{n \to \infty} a_0(n) = \prod_{i=1}^{\infty} (1 + 2i^{-1})^{1/i+1} = 3.8695 \quad . \qquad (11)$$

The work involves obtaining good upper and lower bounds to $a_0(n)$ which are shown to converge to the given limit here.

This problem indicates how the simple case of conditional probabilities and expectations given a partition, already lead to easily stated, useful and nontrivial applications. Indeed to analyze the class of optimal rules in more general contexts, it is necessary to treat conditioning for random variables which are not restricted to the discrete (or finite) values or even only finite sets of such variables. Thus one has to study conditioning that is not based on (countable) partitions. This will be the subject matter of our work beginning with the next chapter.

1.6 Bibliographical notes

Although the concepts of independence and dependence were present in some form since the early times of our civilization when probability was regarded as a relative frequency, it is the notion of independent events that was better understood and successfully employed in the subject. This was intuitive and hence used since even before the time of G. Cardano in the 16th century and employed effectively by J. Bernoulli in his establishment of the (weak) law of large numbers. Since dependence was understood negatively as that relation which is not independence, there was difficulty in making the concept precise. It is only natural that classes of dependent events can be considered. Thus Definition 2.1 is an attempt to make this precise, and the manner in which the concept contributed to extend the classical probability theory is illustrated in this chapter. The first systematic treatment of chain dependence is due to A.A. Markov who introduced it in 1906. His studies gave a great incentive to extend the independence theory to larger classes. This and the general successful axiomatization of probability by A.N. Kolmogorov in 1933 has also led to a generalization of the elementary conditioning and enriched the subject. This extension will primarily be analyzed and applied to several problems in this book. For a recent exposition of the overview of the classical evolution of the subject, see the Encyclopedia article by this author (1987).

The example in Section 4 is essentially a reformulation of the one due to P. Ennis (1973). The secretary problem was treated rigorously by Y. S. Chow et al. (1964), and various other applications and related questions on the subject were given in the book by Chow, Robbins and Sigmund (1971), and the simpler case of the first example was discussed in DeGroot (1986). Some other problems and applications have been treated by DeGroot (1970), and Shiryayev (1978). For related applications and a bibliographical discussion, one may refer to Gilbert and Mosteller (1966). These works exemplify applications to interesting practical problems that involve conditioning in the very formulations and then exhibit adroit combinatorial procedures together with the analytical difficulties. We shall treat the subject from a broad perspective to understand the general structure of the concept and illustrate its potential in some key applications. This is the main purpose of the following pages of this book.

CHAPTER 2:   THE KOLMOGOROV FORMULATION AND
ITS PROPERTIES

This chapter is devoted to the general concept of conditioning, which includes the earlier one using countable partitions, and its basic properties. We present a detailed account when one extends the standard methods of integration for such a generalized conditional probability function. This treatment further clarifies the relation between the two definitions given in the preceding chapter for the case with densities and for the general case.

2.1 Introduction of the general concept

As seen in the last chapter, an extension of conditioning from a single event to a countable partition is a relatively easy procedure (cf., (1.2.5)). However, such a "constructive" definition is impossible if the conditioning event has arbitrarily small (or zero) probability, which happens in many applications. But a clue for a generalization is present in the (alternative) form given in (1.2.6). The new idea is nontrivial and in fact the procedure could not have been possible, if certain related developments in real analysis (or measure theory) have not taken place. The reference here is to the 1930 Nikodým version of differentiation of set functions, now known as the (abstract) Radon-Nikodým theorem. With this result at hand, Kolmogorov was able to recognize immediately its potential and present an appropriate and fundamental generalization of the concept of conditioning within the next two years. The suitable procedure is such that, in contrast to the single event case, the conditional expectation is obtained first and then the corresponding probability concept is derived from it. As a result, the simple and intuitive appeal of conditioning, so clear in the elementary case, is lost. Recognizing this fact is a first step in understanding the nature of the general concept of conditioning. Indeed we have to make a serious effort to study its structure.

Equation (1.2.6) motivates the following direct method. Suppose $f : \Omega \to \mathbb{R}$ is an *integrable* random variable on (Ω, Σ, P) and $\mathcal{B} \subset \Sigma$ is any σ-algebra. Then the mapping

$$\nu_f : A \mapsto \int_A f dP, \ A \in \mathcal{B} \ , \tag{1}$$

is a well-defined σ-additive real function on $\mathcal{B}$ (a signed measure) which vanishes on P null sets of $\mathcal{B}$, so that ν_f is absolutely continuous relative to $P_{\mathcal{B}} (= P|\mathcal{B})$, the restriction of P to $\mathcal{B}$. This is expressed as $\nu_f << P_{\mathcal{B}}$. Hence by the previously noted abstract Radon-Nikodým theorem, there is a $P_{\mathcal{B}}$-unique integrable random variable $\tilde{f}$ on $(\Omega, \mathcal{B}, P_{\mathcal{B}})$ such that

$$\nu_f(A) = \int_A \tilde{f} \ dP_{\mathcal{B}} \ , \ \ A \in \mathcal{B} \ . \tag{2}$$

Note that f itself need not be a random variable on $(\Omega, \mathcal{B}, P_{\mathcal{B}})$. Also if $\mathcal{B}$ is the σ-algebra generated by $\mathcal{P}$, denoted by $\sigma(\mathcal{P})$ where $\mathcal{P}$ is a countable partition, then the essential uniqueness in (2) and a comparison with (1.2.6) together imply that $\tilde{f} = E_{\mathcal{P}}(f)$. Thus the special concept is included in this generalization and so we are justified in calling $\tilde{f}$ the conditional expectation of f relative to $\mathcal{B}$ and denote it $E^{\mathcal{B}}(f)$ or $E(f|\mathcal{B})$. Let us state it precisely:

Definition 1. Let (Ω, Σ, P) be a probability space and $f : \Omega \to \mathbb{R}$ an integrable random variable on (Ω, Σ, P). If $\mathcal{B} \subset \Sigma$ is any σ-algebra, then the *conditional expectation* of f relative to $\mathcal{B}$, denoted $E^{\mathcal{B}}(f)$, is a random variable on $(\Omega, \mathcal{B}, P_{\mathcal{B}})$ satisfying the system of equations (2) with $\tilde{f} = E^{\mathcal{B}}(f)$ there. Then the *conditional probability* on Σ relative to $\mathcal{B}$, is defined as

$$P^{\mathcal{B}}(A) = E^{\mathcal{B}}(\chi_A), \ A \in \Sigma, \tag{3}$$

so that $P^B(\cdot)$ also satisfies the system of equations

$$\int_B P^B(A)dP_B = \int_B \chi_A \, dP = P(A \cap B), \quad A \in \Sigma, \ B \in \mathcal{B}. \qquad (4)$$

Any member of the P_B-equivalence class, again denoted by $E^B(f)$ or $P^B(A)$, is termed a *version* of the conditional expectation or probability respectively.

It may be observed that the existence of such a general function $E^B(f)$, $f \in L^1(\Omega, \Sigma, P)$, essentially depends on the Radon-Nikodým theorem. For this it actually suffices if ν_f of (1) is just σ-additive. In view of the counterexample 1.4.3, one needs the condition that either $E(f^+) < \infty$ or $E(f^-) < \infty$, for the σ-additionity of ν_f. We use this fact in later computations. Since ν_f has finite variation iff f is P-integrable, we deduce that $E^B(f)$ is integrable iff f is, i.e. iff $|E(f)| < \infty$.

In order to use the Radon-Nikodým theorem, and hence the general concept of conditioning, it is important that $\mathcal{B}$ be a σ-algebra (or at least a σ-ring when $|E(f)| < \infty$). Thus the partition of the elementary case is strengthened (or restricted) to this algebraic structure. This abstract generalization has the following additional advantage. If (Ω, Σ, P) is a nonfinite measure space such that $P_B (= P|\mathcal{B})$ is σ-finite, then the Randon-Nikodým theorem is still true so that the (generalized) conditional expectation $E^B(f)$ of any P-integrable function f exists. *Consequently all properties, discussed below, are valid if P, instead of being a probability measure, is a measure such that P_B is σ-finite (or if there are many ordered σ-algebras then $\mathcal{B}$ is the smallest on which P_B has this property).* This point will be of interest in connection with the analysis of Chapter 4 on a new axiomatic treatment of the subject proposed by A. Rényi.

It should be observed that the abstract generalization of conditioning as formulated by Kolmogorov in 1933, conforms with the expressed wishes of Hilbert. However, there is no easy or simple recipe to actually calculating the conditional expectation or probability, given a σ-algebra if the latter is not generated by a partition. We consider this aspect in detail and explore the computational difficulties in Chapter 3.

As is evident from (1)–(3), a conditional expectation and probability (or their versions) are functions and not constants. Indeed, the mappings $E^{\mathcal{B}} : f \mapsto E^{\mathcal{B}}(f)$ and $P^{\mathcal{B}} : A \mapsto P^{\mathcal{B}}(A)$ from $L^1(\Omega, \Sigma, P)$ and Σ respectively, take their values in the Lebesgue space $L^1(\Omega, \mathcal{B}, P_{\mathcal{B}})$. Moreover, $f \geq 0$ implies $E^{\mathcal{B}}(f) \geq 0$, a.e., and $0 \leq P^{\mathcal{B}}(A) \leq 1$, a.e., $A \in \Sigma$.

Some simple consequences of Definition 1 are as follows:

Proposition 2. *The operator $E^{\mathcal{B}}$ is linear, positivity preserving, contractive in the sense that $\|E^{\mathcal{B}}(f)\|_1 \leq \|f\|_1$, and faithful, i.e., $E^{\mathcal{B}}(f) \geq 0$ a.e. for $f \geq 0$, and $= 0$ iff $f = 0$ a.e. Moreover $E^{\mathcal{B}}(f) = f$ a.e. iff f is a random variable on $(\Omega, \mathcal{B}, P_{\mathcal{B}})$ itself, and in particular $E^{\mathcal{B}}$ is the identity if $\mathcal{B} = \Sigma$. So $E^{\mathcal{B}}(f) = E(f)$ a.e., if $\mathcal{B} = \{\phi, \Omega\}$.*

Proof. By (2) it follows that $f \geq 0$ implies $E^{\mathcal{B}}(f) = \tilde{f} \geq 0$ a.e., and replacing f there by $a_1 f_1 + a_2 f_2$, $f_i \in L^1(\Omega, \Sigma, P)$, $a_i \in \mathbb{R}$, $i = 1, 2$ we get by the linearity of the integral

$$\int_B E^{\mathcal{B}}(a_1 f_1 + a_2 f_2) dP_{\mathcal{B}} = \int_B (a_1 f_1 + a_2 f_2) dP$$

$$= a_1 \int_B f_1 \, dP + a_2 \int_B f_2 \, dP$$

$$= a_1 \int_B E^{\mathcal{B}}(f_1) dP_{\mathcal{B}} + a_2 \int_B E^{\mathcal{B}}(f_2) dP_{\mathcal{B}}$$

$$= \int_B [a_1 E^{\mathcal{B}}(f_1) + a_2 E^{\mathcal{B}}(f_2) dP_{\mathcal{B}}, \quad B \in \mathcal{B}.$$

The extreme integrands are random variables relative to $\mathcal{B}$, and B is arbitrary in $\mathcal{B}$. Hence they must agree a.e. This shows that $E^{\mathcal{B}}$ is a linear operator on $L^1(\Omega, \Sigma, P)$ with range $L^1(\Omega, \mathcal{B}, P_{\mathcal{B}})$. Also from the inequality $-|f| \leq f \leq |f|$ one gets

$$-E^{\mathcal{B}}(|f|) \leq E^{\mathcal{B}}(f) \leq E^{\mathcal{B}}(|f|), \quad a.e. \tag{5}$$

Hence

$$|E^{\mathcal{B}}(f)| \leq E^{\mathcal{B}}(|f|), \quad a.e., \tag{6}$$

and integration yields

$$\|E^{\mathcal{B}}(f)\|_1 = \int_\Omega |E^{\mathcal{B}}(f)| dP_{\mathcal{B}} \leq \int_\Omega E^{\mathcal{B}}(|f|) dP_{\mathcal{B}}$$

$$= \int_\Omega |f| dP = \|f\|_1.$$

The last assertion of the proposition is a direct consequence of the definition of $E^{\mathcal{B}}$ in (2). $\square$

For a better understanding of the abstract concept of conditioning we now present an alternative definition and show, by specialization, how Definition 1.2.2 is naturally included in the present version. Thus let (Ω, Σ, P) be a probability space and $(S, \mathcal{S})$ be a measurable space, i.e., S is a set and $\mathcal{S}$ is a σ-algebra on it. Let $h : \Omega \to S$ be a $(\Sigma, \mathcal{S})$ random variable in the sense that $h^{-1}(A) \in \Sigma$ for each $A \in \mathcal{S}$. Consider the image probability function $\mu = P \circ h^{-1} : \mathcal{S} \to \mathbb{R}^+$ so that $\mu(A) = P(h^{-1}(A))$, $A \in \mathcal{S}$. Recall that, by definition, $h^{-1}(C) = \phi$ for all $C \subset S - h(\Omega)$, i.e., the (complete) inverse image of any set in S is unambiguous.

Define $\mathcal{B} = h^{-1}(\mathcal{S})$ so that $\mathcal{B}$ is a σ-algebra contained in Σ (and is determined by h). Let $P_{\mathcal{B}} = P|\mathcal{B}$. If $f : \Omega \to \mathbb{R}$ is any P-integrable random variable, consider as before

$$\nu_f : B \mapsto \int_B f\, dP, \ B \in \mathcal{B}.$$

Then ν_f is a signed measure, and if $\tilde{\nu}_f = \nu_f \circ h^{-1} : \mathcal{S} \to \mathbb{R}$, then $\tilde{\nu}_f$ is also σ-additive and $\tilde{\nu}_f << \mu$ (and $\nu_f << P_{\mathcal{B}}$). Since μ is a probability there is, by the Radon-Nikodým theorem, a μ-integrable function $g_f : S \to \mathbb{R}$ (μ-unique) such that

$$\tilde{\nu}_f(A) = \int_A \tilde{g}_f\, d\mu \ , \ A \in \mathcal{S},$$
$$= \nu_f(h^{-1}(A)) = \int_{h^{-1}(A)} f\, dP. \tag{7}$$

In this form $\tilde{g}_f$ is often called the *conditional expectation of f given h* and denoted by

$$\tilde{g}_f(s) = E^{\mathcal{B}}(f)(s) = E(f|h = s), \ s \in S. \tag{8}$$

To deduce Definition 1 from this form, let $S = \Omega$, $\mathcal{S} = \mathcal{B} \subset \Sigma$ and $h =$ identity. Then $E^{\mathcal{B}}(f)$, defined by (8), is identical with that given by (2). Also if $S = \mathbb{N}$, the natural numbers, $\mathcal{S} =$ the power set of $\mathbb{N}$, and $\mathcal{P} = \{A_n, n \geq 1\} \subset \Sigma$, is a partition of Ω, let $h : \Omega \to \mathbb{N}$

be defined by $h(A_n) = n$. Then $h^{-1}(\mathcal{S})$ is the σ-algebra generated by $\mathcal{P}$. If $\mu(\{n\}) = P(h^{-1}(\{n\})) = P(A_n)$, then $(\mathbb{N}, \mathcal{S}, \mu)$ is a probability space; so (7) and (8) become

$$g_f(n) \cdot \mu(\{n\}) = \int_{h^{-1}(\{n\})} f\, dP = \int_{A_n} f\, dP. \tag{9}$$

With the notation of (8), we have $\tilde{g}_f(n) = E(f | h = n)$ and since $[h = n] = A_n$, one can use $E(f | A_n) = E_{A_n}(f)$ which is the notation of Definition 1.2.2. Thus (9) becomes

$$E_{A_n}(f) = \frac{1}{P(A_n)} \int_{A_n} f\, dP, \tag{10}$$

which is precisely (1.2.3). This shows that Definition 1 is the desired extension of the elementary case. The following observation of the general concept based on (8) and (10) is recorded for a clarification.

Suppose $\mathcal{B} \subset \Sigma$ is a σ-algebra and $A \in \mathcal{B}$ is an atom, i.e., $A_1 \subset A$, $A_1 \in \mathcal{B}$ implies $P_{\mathcal{B}}(A_1) = 0$, or $= P_{\mathcal{B}}(A)$. Then (8) and (10) give

$$E_A(f) = \frac{1}{P(A)} \int_A f\, dP = \frac{1}{P(A)} \int_A E^{\mathcal{B}}(f)\, dP_{\mathcal{B}}. \tag{11}$$

Thus, $E_A(f)$ being a $\mathcal{B}$-measurable function, we have

$$\int_A E^{\mathcal{B}}(f)\, dP_{\mathcal{B}} = \int_A E_A(f)\, dP_{\mathcal{B}},$$

and by their essential uniqueness on $\mathcal{B}(A) = \{A \cap B : B \in \mathcal{B}\}$, we can identify $E^{\mathcal{B}}(f)$ and $E_A(f)$ so that $E^{\mathcal{B}}(f)$ is the average of f on A relative to the measure P for each $\mathcal{B}$-atom A $(P(A) > 0)$. This fact is sometimes expressed by saying that f is "smoothed" on A by replacing it with its average. Thus $E^{\mathcal{B}}(f)$ generally takes fewer values than f. Motivated by the above alternative view of conditioning by a function g, one often denotes, if $\mathcal{B}_g$ is the σ-algebra generated by g, $E^{\mathcal{B}_g}(f)$ as $E(f | g)$ and if $g = (f_1, \ldots, g_n)$ is a random vector, then as $E(f | g_1, \ldots, g_n)$. *Note that $E(f | g = s)$ is simply $E(f | g)(s)$.* This notation can also be justified further on using the next assertion. It shows that the random variable $E(f | g)$ is a function of g. This is made precise by the following result, termed a *Doob-Dynkin lemma.*

Proposition 3. *Let (Ω_i, Σ_i), $i = 1, 2$, be measurable spaces and $f :$ $\Omega_1 \to \Omega_2$ be (Σ_1, Σ_2) measurable, i.e., $\mathcal{A} = f^{-1}(\Sigma_2) \subset \Sigma_1$. Then any function $g : \Omega_1 \to \mathbb{R}$ is $\mathcal{A}$-measurable (i.e., random variable relative to $\mathcal{A}$) iff $g = h \circ f$ for some measurable $h : \Omega_2 \to \mathbb{R}$.*

Proof. Since any function of the form $h \circ f$ is $\mathcal{A}$-measurable because $(h \circ f)^{-1}(\mathcal{B}) = f^{-1}(h^{-1}(\mathcal{B})) \subset f^{-1}(\Sigma_2) = \mathcal{A}$, it is only necessary to consider the converse. Here $\mathcal{B}$ is the Borel σ-algebra of $\mathbb{R}$.

First suppose g is a simple random variable relative to $\mathcal{A}$. Thus $g = \sum_{i=1}^{n} a_i \chi_{A_i}$, $A_i \in \mathcal{A}$. Then there exist $B_i \in \Sigma_2$ such that $A_i = f^{-1}(B_i)$. Define $h = \sum_{i=1}^{n} a_i \chi_{B_i}$, so that $h : \Omega_2 \to \mathbb{R}$ is a random variable for Σ_2 and in this representation one may assume both the $\{A_i, 1 \leq i \leq n\}$ and $\{B_i, 1 \leq i \leq n\}$ are disjoint. Now one has

$$
\begin{aligned}
g(\omega) &= \sum_{i=1}^{n} a_i \chi_{A_i}(\omega) \\
&= \sum_{i=1}^{n} a_i \chi_{f^{-1}(B_i)}(\omega) = \sum_{i=1}^{n} a_i \chi_{B_i}(f(\omega)) \\
&= \left(\sum_{i=1}^{n} a_i \chi_{B_i} \right)(f(\omega)) = (h \circ f)(\omega), \quad \omega \in \Omega .
\end{aligned}
\tag{12}
$$

This shows that the result is true for simple g. If g is any random variable for $\mathcal{A}$, then (by the structure theorem) there exist simple $\mathcal{A}$-measurable functions g_n such that $g_n \to g$ pointwise. Hence by the special case, there exist simple random variables for Σ_2, say h_n, such that $g_n = h_n \circ f$, $n \geq 1$. Let $\Omega_0 = \{\omega \in \Omega_2 : \liminf_n h_n(\omega) = \limsup_n h_n(\omega)\}$. Since Σ_2 is a σ-algebra, and the h_n are Σ_2-measurable, one deduces that $\Omega_0 \in \Sigma_2$ and $f(\Omega_1) \subset \Omega_0$, using the fact that $g_n \to g$ on Ω_1. If we define $h : \Omega_2 \to \mathbb{R}$ by the equation:

$$
h(\omega) = \begin{cases} \liminf_n h_n(\omega) & , \omega \in \Omega_0 \\ 0 & , \omega \notin \Omega_0 \end{cases} ,
\tag{13}
$$

then h is a Σ_2-random variable and $g = h \circ f$ as desired. $\square$

Taking $\Omega_2 = \mathbb{R}$, $\Sigma_2 = \mathcal{B}$ (the σ-algebra of Borel sets of $\mathbb{R}$) we get $E(f|g) = h \circ g$ by the above proposition with $h : \mathbb{R} \to \mathbb{R}$ now being a Borel function. This entitles us to say that the conditional expectation of an integrable function f relative to a random variable g, is a Borel function of g. Moreover, in the notation of (8)

$$E^{\mathcal{B}_h}(f)(s) = E(f|h = s) = \varphi(h(s)), \ s \in S, \tag{14}$$

for some Borel function $\varphi : \mathbb{R} \to \mathbb{R}$. Note that the middle expression in (14) is, by definition, the left most expression. The point of (14) is that the equation is always well-defined, a.e., *without regard to the measure of the event* $\{\omega : h(\omega) = s\}$, and this is meaningful in the general theory. In contrast, the same statement *cannot* be made in the special (or "constructive") case of Definition 1.2.2. Thus we need to analyze this aspect in more detail. This is done in the next chapter.

2.2 Basic properties of conditional expectations

We have already seen that a conditional expectation is a positive contractive linear operation on $L^1(\Omega, \Sigma, P)$, the real Lebesgue space of equivalence classes of P-integrable functions. However, several other properties, including some less familiar ones, can also be given quickly. We start with the following:

Proposition 1. *Let* (Ω, Σ, P) *be a probability space,* $\mathcal{B} \subset \Sigma$ *a* σ-algebra *and* $\{X, Y, XY\} \subset L^1(\Omega, \Sigma, P)$. *Then we have*

(i) *(averaging identity)* $E^{\mathcal{B}}(XE^{\mathcal{B}}(Y)) = E^{\mathcal{B}}(X)E^{\mathcal{B}}(Y)$, a.e., *and in particular if* Y *is a* $\mathcal{B}$-measurable function then

$$E^{\mathcal{B}}(XY) = YE^{\mathcal{B}}(X), \ \text{a.e.,}$$

(ii) *(commutativity) if* $\mathcal{B}_1 \subset \mathcal{B}_2 \subset \Sigma$ *are* σ-algebras then

$$E^{\mathcal{B}_1}(E^{\mathcal{B}_2}(X)) = E^{\mathcal{B}_2}(E^{\mathcal{B}_1}(X)) = E^{\mathcal{B}_1}(X), \quad \text{a.e.,}$$

and, in particular,

$$E(E^{\mathcal{B}}(X)) = E(X) \tag{1}$$

for any integrable random variable and σ-algebra $\mathcal{B} \subset \Sigma$.

Proof. (i) It suffices to establish the second part since the first is then an immediate consequence of this. The argument is somewhat unmotivated. By (1.2) we need to show that the desired result is equivalent to verifying the truth of the following system of equations:

$$\int_A Y E^{\mathcal{B}}(X) dP_{\mathcal{B}} = \int_A E^{\mathcal{B}}(XY) dP_{\mathcal{B}}$$
$$= \int_A XY \, dP \, , \, A \in \mathcal{B}. \qquad (2)$$

But if $Y = \chi_B$, $B \in \mathcal{B}$, then (2) becomes

$$\int_A \chi_B E^{\mathcal{B}}(X) dP_{\mathcal{B}} = \int_{A \cap B} E^{\mathcal{B}}(X) dP_{\mathcal{B}} = \int_{A \cap B} X dP = \int_A \chi_B X dP.$$
$$(3)$$

Since $A \cap B \in \mathcal{B}$, the equation (3) is true for any P-integrable X. By linearity of the integral, (3) holds if χ_B is replaced by any simple $\mathcal{B}$-random variable Y. Then by the monotone convergence theorem the same result holds if $X \geq 0$ and $Y \geq 0$ since there exist $\mathcal{B}$-simple $0 \leq Y_n \uparrow Y$. Finally expressing $X = X^+ - X^-$ and $Y = Y^+ - Y^-$ and applying the result obtained in the preceding sentence, we can conclude that (2) holds as stated; and at the same time infer that $Y E^{\mathcal{B}}(X)$ is also P-integrable when XY is.

(ii) Since $\mathcal{B}_1 \subset \mathcal{B}_2$ and $E^{\mathcal{B}_1}(X)$ is a $\mathcal{B}_1$-(and hence $\mathcal{B}_2$-) random variable, we get by (i)

$$E^{\mathcal{B}_2}(E^{\mathcal{B}_1}(X)) = E^{\mathcal{B}_1}(X) E^{\mathcal{B}_2}(1) = E^{\mathcal{B}_1}(X) \, , \, \text{a.e.,}$$

because by (1.2) and the essential uniqueness of the Radon-Nikodým derivative we get $E^{\mathcal{B}}(1) = 1$ a.e., for any σ-algebra $\mathcal{B} \subset \Sigma$. Thus the last half of (ii) holds. Regarding the first part, we again go to the defining property and proceed as in the proof of (2). For any $A \in \mathcal{B}_1 \subset \mathcal{B}_2$, we have

$$\int_A E^{\mathcal{B}_1}(E^{\mathcal{B}_2}(X)) dP_{\mathcal{B}_1} = \int_A E^{\mathcal{B}_2}(X) dP_{\mathcal{B}_2}$$
$$= \int_A X dP = \int_A E^{\mathcal{B}_1}(X) dP_{\mathcal{B}_1} .$$

Since the extreme integrands are $\mathcal{B}_1$-measurable and $A \in \mathcal{B}_1$ is arbitrary, they can be identified (i.e., equal a.e.). This shows that the given identity is valid.

Finally writing $\mathcal{B}_1 = \{\phi, \Omega\}$, $\mathcal{B}_2 = \mathcal{B} \subset \Sigma$, a σ-algebra, one gets (1) on recalling that $E^{\mathcal{B}_1}(X) = E(X)$, a.e., $\quad \square$

Remark. If P is replaced by a nonfinite measure μ having the property that μ_B is σ-finite, then we can apply the Radon-Nikodým theorem for all random variables $X \geq 0$ even when X is not μ-integrable. It again follows from (1.2) that $E^B(X)$ is well-defined and $E^B(1) = 1$ a.e. Thus although $E(1) = +\infty$, we *always have* $E^B(1) = 1$ if μ_B is σ-finite. This is a significant by-product of the general theory. Thus (1) holds for all positive random variables but *not* for all random variables as the counterexample of Section 1.4 shows.

It may be noted that (15) is true if either X^+ or X^- is integrable and this is the best possible condition for conditional expectations. Also we should mention that part (ii) of the above proposition fails if the inclusion relation $\mathcal{B}_1 \subseteq \mathcal{B}_2$ is not satisfied. This can be verified by examples.

The operator E^B inherits several properties, familiar in the elementary case, but fails for some. The next result contains the positive statements.

Proposition 2. *Let* $\{X_n, n \geq 1\}$ *be a sequence of random variables on* (Ω, Σ, P) *and* $\mathcal{B} \subset \Sigma$ *be a* σ-*algebra. Then the following assertions hold:*

(i) (monotone convergence) $X_n \geq X_0$, $E(|X_0|) < \infty$ *and* $X_n \leq X_{n+1} \Rightarrow$

$$E^B(\lim_n X_n) = \lim_n E^B(X_n), \text{ a.e.,}$$

(ii) (Fatou's lemma) $X_n \geq X_0$, $E(|X_0|) < \infty [X_n \leq \tilde{X}_0, E(|\tilde{X}_0|) < \infty] \Rightarrow$

$$E^B(\liminf_n X_n) \leq \liminf_n E^B(X_n), \text{ a.e.,}$$

$$[E^B(\limsup_n X_n) \geq \limsup_n E^B(X_n), \text{ a.e.,}]$$

(iii) (dominated convergence) $X_n \to X$ a.e., $|X_n| \leq Y$ a.e., $E(Y) < \infty \Rightarrow \lim_n E^B(X_n) = E^B(X)$ a.e., *and in* $L^1(P)$-*mean.*

Proof. (i) Replacing X_n by $X - X_0 \geq 0$, we may assume that the sequence is nonnegative and increasing. To use the same technique as in the last proposition, observe that $E^{\mathcal{B}}(X_n - X_0)$ exists and is ≥ 0 a.e., since $E^{\mathcal{B}}(X_0)$ exists. Hence

$$E^{\mathcal{B}}(X_n - X_0) + E^{\mathcal{B}}(X_0) = E^{\mathcal{B}}(X_n), \quad \text{a.e.,}$$

exists (cf., also the preceding remark). Similarly if $X = \lim_n X_n$, which clearly exists a.e., and $X \geq X_0$ a.e., then $E^{\mathcal{B}}(X)$ exists a.e. Consequently, for all $A \in \mathcal{B}$ one has

$$\int_A E^{\mathcal{B}}(X)dP_{\mathcal{B}} = \int_A XdP = \int_A \lim_n X_n dP$$
$$= \lim_n \int_A X_n dP$$
$$= \lim_n \int_A E^{\mathcal{B}}(X_n)dP_{\mathcal{B}}$$
$$= \int_A \lim_n E^{\mathcal{B}}(X_n)dP_{\mathcal{B}} \ ,$$

by the classical monotone convergence theorem and the fact that $E^{\mathcal{B}}(X_n)$ is increasing a.e. Since the extreme integrands on either side of the equality are random variables for $\mathcal{B}$, and A in $\mathcal{B}$ is arbitrary, the integrands can be identified. It is precisely the desired assertion.

(ii) This part follows from (i) as in the classical case. Thus letting $Y = \liminf_n X_n$ so that $Y \geq X_0$ a.e., if $Y_n = \inf\{X_m : m \geq n\}$ then $X_0 \leq Y_n \uparrow Y$, a.e. Hence by (i)

$$E^{\mathcal{B}}(Y) = \lim_n E^{\mathcal{B}}(Y_n), \quad \text{a.e.,}$$
$$= \liminf_n E^{\mathcal{B}}(Y_n)$$
$$\leq \liminf_n E^{\mathcal{B}}(X_n), \quad \text{a.e.}$$

The paranthetical statement is similarly proved, or it is deduced from the above by considering $-X_n$ in place of X_n.

(iii) The hypothesis implies $-Y \leq X_n \leq Y$ and $E(Y) < \infty$, so that the sequence satisfies the conditions of both parts of (ii). Since

$X_n \to X$ a.e., we have $X = \liminf_{n} X_n = \limsup_{n} X_n$, a.e., and by (ii)

$$
\begin{aligned}
E^{\mathcal{B}}(X) &= E^{\mathcal{B}}(\liminf_{n} X_n) \\
&\leq \liminf_{n} E^{\mathcal{B}}(X_n) \\
&\leq \limsup_{n} E^{\mathcal{B}}(X_n) \\
&\leq E^{\mathcal{B}}(\limsup_{n} X_n) = E^{\mathcal{B}}(X), \text{ a.e.}
\end{aligned}
$$

This shows that $\lim_{n} E^{\mathcal{B}}(X_n^n)$ exists a.e. and equals $E^{\mathcal{B}}(X)$, a.e.

Finally for the mean convergence, note that

$$
\begin{aligned}
E(|E^{\mathcal{B}}(X_n) - E^{\mathcal{B}}(X)|) &= E(|E^{\mathcal{B}}(X_n - X)|) \\
&\leq E(E^{\mathcal{B}}(|X_n - X|)), \text{ by (1.6)}, \\
&= E(|X_n - X|), \text{ by (1)}, \\
&\to 0, \quad \text{as } n \to \infty,
\end{aligned}
$$

by the classical dominated convergence theorem itself. □

The mutual independence concept for events defined in (1.2.11) extends to functions as follows. Random variables X, Y are *independent* on (Ω, Σ, P) if for any real bounded Borel functions φ, ψ, we have

$$
E(\varphi(X)\psi(Y)) = E(\varphi(X))E(\psi(Y)).
$$

A similar equation should hold for a finite collection $X_1, \ldots, X_n$ to be mutually independent. An arbitrary collection of random variables will be termed an independent set if each finite subcollection is independent. Similarly a random variable X and an algebra $\mathcal{B}$ are termed independent if X, χ_B are independent for each $B \in \mathcal{B}$.

The following assertion for conditional expectations is of interest.

Proposition 3. *If X is an integrable random variable on (Ω, Σ, P) which is independent of a σ-algebra $\mathcal{B} \subset \Sigma$, then $E^{\mathcal{B}}(X) = E(X)$, a.e.*

Proof. For any $A \in \mathcal{B}$, we have

$$\int_A E^{\mathcal{B}}(X)dP_{\mathcal{B}} = \int_A XdP, \text{ by definition,}$$

$$= \int_{\Omega} \chi_A XdP$$

$$= E(\chi_A)E(X), \text{ by independence of } X \text{ and } \chi_A,$$

$$= E(X) \, P_{\mathcal{B}}(A)$$

$$= \int_A E(X)dP_{\mathcal{B}}.$$

Since the extreme integrands are random variables relative to $\mathcal{B}$, they can be identified to get $E^{\mathcal{B}}(X) = E(X)$ a.e. as desired. $\square$

We next establish some classical integral inequalities for conditional expectations since they are important for applications.

Theorem 4. *Let,* $X, Y \in L^p(\Omega, \Sigma, P)$, *and* $Z \in L^q(\Omega, \Sigma, P)$, *where* $p^{-1} + q^{-1} = 1$, $p \geq 1$. *Then we have for any σ-algebra $\mathcal{B} \subset \Sigma$*

(i) *(conditional Hölder's inequality)*

$$E^{\mathcal{B}}(|XZ|) \leq [E^{\mathcal{B}}(|X|^p)]^{1/q} [E^{\mathcal{B}}(|Z|^q)]^{1/q} \, , \text{ a.e.,}$$

(ii) *(conditional Minkowski's inequality) for $p \geq 1$,*

$$[E^{\mathcal{B}}(|X + Y|^p)]^{1/p} \leq [E^{\mathcal{B}}(|X|^p)]^{1/p} + [E^{\mathcal{B}}(|Y|^p)]^{1/p} \, , \text{a.e.,}$$

(iii) *(conditional Jensen's inequality) if $\varphi : \mathbb{R} \to \mathbb{R}$ is a continuous convex function such that either $\varphi \geq 0$ or $\varphi(X)$ is integrable, then*

$$\varphi(E^{\mathcal{B}}(X)) \leq E^{\mathcal{B}}(\varphi(X)), \text{ a.e.} \tag{4}$$

Consequently, $E^{\mathcal{B}}$ is a positive linear contraction on all $L^p(\Omega, \Sigma, P)$, $1 \leq p \leq \infty$.

Proof. (i) The result is immediate if either $E^{\mathcal{B}}(|X|)^p)$ or $E^{\mathcal{B}}(|Z|^q)$ is zero a.e. (note that both are finite a.e.). Thus let $N_x^p = E^{\mathcal{B}}(|X|^p)$ and similarly N_z^q be defined and $0 < N_x^p, N_z^q < \infty$ a.e. Each of these quantities is a $\mathcal{B}$-measurable function. To establish the desired inequalities one needs to use a trick as in the classical case. Here we recall the classical result that a twice differentiable convex (concave) function $\Psi : \mathbb{R} \to \mathbb{R}$

is one for which the second derivative $\Psi''(x) \geq 0 (\Psi''(x) \leq 0)$. Taking $\Psi(x) = -\log x$, $x > 0$, we note that $\Psi(\cdot)$ is convex and hence for $0 < \alpha < 1$, $\beta = 1 - \alpha$,

$$\log (\alpha x + \beta y) \geq \alpha \log x + \beta \log y = \log x^\alpha y^\beta, \ x > 0, \ y > 0.$$

It follows by the strictly increasing property of the logarithm function that

$$x^\alpha y^\beta \leq \alpha x + \beta y \ . \tag{5}$$

Turning to our proof, if $p = 1$, then $q = +\infty$, so that $||Z||_\infty = k < \infty$, and since (cf., (1.6)) $E^B(|Z|) \leq k$ a.e., we have

$$|E^B(XZ)| \leq E^B(|XZ|)$$
$$\leq k \ E^B(|X|) = k \ N_x^1, \ \text{a.e.} \tag{6}$$

But $||Z||_\infty = k$ implies that $A_\epsilon = [|Z| > k - \epsilon]$ is an event for each $\epsilon > 0$, and $P(A_\epsilon) > 0$. Then choose a B in $\mathcal{B}$ such that $B \supset A_\epsilon$ a.e. This is clearly possible. [In fact, the class of all sets of $\mathcal{B}$ that contain A_ϵ has an infimum in $\mathcal{B}$, and this can be shown to be valid even if $P_\mathcal{B}$ is σ-finite.] Then

$$k \geq N_z^\infty \geq E^B(|Z|) \geq E^B(|Z|\chi_B)$$
$$\geq E^B(|Z|\chi_{A_\epsilon}) \geq k - \epsilon, \ \text{a.e.} \tag{7}$$

Since $\epsilon > 0$ is arbitrary we get $N_\epsilon^\infty = k$, and so (6) implies (i) in this case. Next let $1 < p < \infty$, and put $\alpha = 1/p$, $\beta = 1/q$, $x^\alpha = |X(\omega)|/N_x$, $y^\beta = |Z(\omega)|/N_z$ in (5). The result then becomes

$$\frac{|XZ|}{N_x N_z} (\omega) \leq \frac{1}{p} \frac{|X|^p}{N_x^p} (\omega) + \frac{1}{q} \frac{|Z|^q}{N_z^q} (\omega) , \ a.a.(\omega) \ . \tag{8}$$

Applying E^B to both sides of this functional inequality, and using Proposition 1(i), (8) becomes

$$E^B(|XZ|)/N_x N_z \leq \frac{1}{p} + \frac{1}{q} = 1 \ , \text{a.e.,} \tag{9}$$

and this gives (i) in all cases.

(ii) With the same argument as in the first part of (i), it follows at once that the result holds if $p = 1$ or $p = +\infty$, because of the numerical triangle inequality. The result is also trivial if $E^{\mathcal{B}}(|X+Y|^p) = 0$, a.e. The inequality

$$|X+Y|^p \leq [2^p \max (|X|,|Y|)^p] \leq 2^p[|X|^p + |Y|^p]$$

implies that we need to consider only the case that $0 < E^{\mathcal{B}}(|X+Y|^p) < \infty$, a.e., with $1 < p < \infty$. With this assumption, let $q = p/(p-1)$, and then

$$\begin{aligned}
E^{\mathcal{B}}(|X+Y|^p) &\leq E^{\mathcal{B}}(|X+Y|^{p-1}(|X|+|Y|)), \text{a.e.} \\
&= E^{\mathcal{B}}(|X+Y|^{p-1}|X|) + E^{\mathcal{B}}(|X+Y|^{p-1}|Y|) \\
&\leq [E^{\mathcal{B}}(|X+Y|^{(p-1)q})]^{1/q}[(E^{\mathcal{B}}(|X|^p))^{1/p} + \\
&\quad + (E^{\mathcal{B}}(|Y|^p))^{1/p}],
\end{aligned}$$

by (i). Cancelling a non zero factor on both sides, we get (ii) in this case also. Thus the argument is practically the classical one.

(iii) Let $\varphi : \mathbb{R} \to \mathbb{R}$ be a continuous convex function as given. Then by the support line property, $\varphi(x) \geq ax + b$, $x \in \mathbb{R}$; for some real numbers a, b. Replacing x by $X(\omega)$ here, $\omega \in \Omega$, and then integrating,

$$aE(X) + b \leq E(\varphi(X)) < \infty$$

so that $E(X^+)$ or $E(X^-)$ is finite. Thus $E^{\mathcal{B}}(X)$ exists. Next we use a classical characterization of such a convex φ. Namely, φ is the upper envelope of a countable collection of such lines: $a_n x + b_n$, $n \geq 1$. Since

$$E^{\mathcal{B}}(\varphi(X)) \geq a_n E^{\mathcal{B}}(X) + b_n, \text{ a.e.}, n \geq 1, \tag{10}$$

we get, on taking the pointwise supremum on the countable collection of the right side of (10), $E^{\mathcal{B}}(\varphi(X)) \geq \varphi(E^{\mathcal{B}}(X))$, a.e. This gives (4).

Finally, if $\varphi(x) = |x|^p, p \geq 1$, we get, on integrating (4),

$$\begin{aligned}
\|E^{\mathcal{B}}(X)\|_p^p &= E(|E^{\mathcal{B}}(X)|^p) \\
&\leq E(E^{\mathcal{B}}(|X|^p)) \\
&= E(|X|^p) = \|X\|_p^p, \text{ by (1).}
\end{aligned}$$

Taking the pth root, this implies the contractivity of $E^{\mathcal{B}}$ on all $L^p(\Omega, \Sigma, P)$ spaces. $\square$

A simple application of Proposition 1 (ii) [and Prop. 4 (iii)], gives the following useful extension of a classical inequality due to C. R. Rao and D. Blackwell who obtained it in the middle 1940s independently for $\varphi(x) = x^2$. Thus let $\varphi : \mathbb{R} \to \mathbb{R}^+$ be a continuous convex function. If Y is a random variable such that $E(\varphi(2Y)) < \infty$, then for each σ-algebra $\mathcal{B} \subset \Sigma$, $Z = E^{\mathcal{B}}(Y)$ is well defined, and

$$E(\varphi(Y - a)) \geq E(\varphi(Z - a)) \tag{11}$$

where $a = E(Y)$. In fact, by the convexity of φ

$$E(\varphi(Y - a)) = E\left(\varphi\left(\frac{2Y - 2a}{2}\right)\right)$$
$$\leq \frac{1}{2} \left[E(\varphi(2Y)) + \varphi(2a) \right] < \infty$$

and by (2) $E(E^{\mathcal{B}}(Z - a)) = E(Y - a) = 0$. Note that using once again the support line property of φ as in the last part of the preceding proof, we conclude that $E^{\mathcal{B}}(Y) = Z$ is well-defined. Next with (4),

$$E^{\mathcal{B}}(\varphi(Y - a)) \geq \varphi(E^{\mathcal{B}}(Y) - a) = \varphi(Z - a), \text{ a.e.} \tag{12}$$

Taking expectations on both sides of (12), we get (11). If $\varphi(x) = x^2$, then $E(\varphi(Y - a)) = $ Var Y, and $E(\varphi(Z - a)) = $ Var Z. Thus (12) implies

$$\text{Var } Y \geq \text{Var}(E^{\mathcal{B}}(Y)) . \tag{13}$$

Now, if $E^{\mathcal{B}}(Y - E^{\mathcal{B}}(Y))^2$ is termed the conditional variance of Y given $\mathcal{B}$, denoted $\text{Var}^{\mathcal{B}}(Y)$, then (13) can be expressed as follows:

$$\text{Var}^{\mathcal{B}} Y = E^{\mathcal{B}}(Y - E^{\mathcal{B}}(Y))^2 = E^{\mathcal{B}}(Y^2) - (E^{\mathcal{B}}(Y))^2 .$$

Hence

$$\begin{aligned}
\text{Var } Y &= E(Y - a)^2 \\
&= E(E^{\mathcal{B}}[Y - E^{\mathcal{B}}(Y) + E^{\mathcal{B}}(Y) - a]^2) \\
&= E(\text{Var}^{\mathcal{B}} Y) + \text{Var } E^{\mathcal{B}}(Y) \geq \text{Var } E^{\mathcal{B}}(Y) , \tag{14}
\end{aligned}$$

since $E^{\mathcal{B}}(Y - E^{\mathcal{B}}(Y)) = 0$, a.e. Thus we have a direct proof of (13) in this special case. The inequality (13) or (14) says that the conditioned variable has possibly a smaller variance than the original one. The significance of this result in applications is that, if Y is an "unbiased" estimator of a parameter 'a' (i.e., $E(Y) = a$), then it can be improved (i.e., will have smaller variance) if we replace it by its conditional mean relative to any σ-algebra $\mathcal{B}$ (i.e., take $E^{\mathcal{B}}(Y)$ as an unbiased estimator of a). Here $\mathcal{B}$ is "free" and we may choose it according to convenience. We see later on (in Chap. 6) how the best $\mathcal{B}$ can be chosen in many problems when a property called "sufficiency" of the estimator is available.

The preceding results indicate that the operator $E^{\mathcal{B}}$ has nearly the same behavior as the ordinary expectation or integral. But this is deceptive, and not a true analogy. The first significant difference is obtained when it is recognized that the classical Vitali convergence theorem for $E(\cdot)$ (i.e. for the Lebesgue integrals) is not valid for a conditional expectation $E^{\mathcal{B}}$. More precisely we have the following:

Theorem 5. *There exists a probability space* (Ω, Σ, P), *a sequence of uniformly integrable random variables* $X_n : \Omega \to \mathbb{R}$, $n \geq 1$, *such that* $X_n \to X$, *a.e., and a* σ-*algebra* $\mathcal{B} \subset \Sigma$, *for which*

$$P[\lim_{n\to\infty} E^{\mathcal{B}}(X_n) = E^{\mathcal{B}}(X)] = 0 \ .$$

The classical Vitali theorem states that under the hypothesis of Theorem 5 (i.e. for any uniformly integrable X_n, $n \geq 1$ and $X_n \to X$ a.e.), we always have $\lim_{n\to\infty} E(X_n) = E(X)$. To get a corresponding result in the conditional case, we need to strengthen the hypothesis on X_n, $n \geq 1$ (or restrict the admitted probability spaces (Ω, Σ, P)) considerably. Here we omit the proof of Theorem 5 since it is not essential for our further work, but present a restricted conditional version of the Vitali theorem.

It is convenient to introduce a new concept:

Definition 6. If $\{X_n, \ n \geq 1\}$ is an integrable sequence of random variables on (Ω, Σ, P) and $\mathcal{B} \subset \Sigma$ is a σ-algebra, then the sequence is *conditionally uniformly integrable* relative to $\mathcal{B}$, provided

$$\lim_{k\to\infty} E^{\mathcal{B}}(|X_n| \chi_{[|X_n| \geq k]}) = 0, \text{ a.e. }, \tag{15}$$

uniformly in n.

If $\mathcal{B} = (\phi, \Omega)$, then the above concept reduces to the ordinary uniform integrability. We can now establish the following:

Theorem 7. (Conditional Vitali). *Let $\{X_n, \geq 1\}$ be a sequence of random variables on (Ω, Σ, P) such that $X_n \rightarrow X$ a.e. Then (a) the uniform integrability of $X_n s$ implies that $E^{\mathcal{B}}(X_n) \rightarrow E^{\mathcal{B}}(X)$ in $L^1(\Omega, \mathcal{B}, P_{\mathcal{B}})$-mean for any σ-algebra $\mathcal{B} \subset \Sigma$; and*

(b) the conditional uniform integrability of the sequence relative to a σ-algebra $\mathcal{B} \subset \Sigma$, implies that $E^{\mathcal{B}}(X_n) \rightarrow E^{\mathcal{B}}(X)$, a.e. and also in $L^1(\Omega, \mathcal{B}, P_{\mathcal{B}})$-mean.

Proof. (a) This part is an easy consequence of the classical Vitali theorem. Thus for a σ-algebra $\mathcal{B} \subset \Sigma$,

$$E(|E^{\mathcal{B}}(X_n) - E^{\mathcal{B}}(X)|) \leq E(E^{\mathcal{B}}(|X_n - X|)), \quad \text{by Theorem 4(iii)},$$
$$= E(|X_n - X|) \rightarrow 0,$$

as $n \rightarrow \infty$ by the stated classical theorem. Note that by Fatou's lemma X is integrable so that $E^{\mathcal{B}}(X)$ is well-defined and the initial step is meaningful. The a.e. convergence of $\{E^{\mathcal{B}}(X_n), n \geq 1\}$ does not necessarily hold as implied by Theorem 5 and the positive statement is the next part.

(b) By hypothesis the given sequence is conditionally uniformly integrable relative to $\mathcal{B}$. So the same property holds for $\{X_n^{\pm}, n \geq 1\}$. Consequently we have

$$U_k = \sup_{n \geq 1} E^{\mathcal{B}}(X_n^- \chi_{[X_n^- > k]}) \rightarrow 0 \text{ a.e., as } k \rightarrow \infty .$$

Consider

$$E^{\mathcal{B}}(X_n) = E^{\mathcal{B}}(X_n^+) - E^{\mathcal{B}}(X_n^- \chi_{[X_n^- \leq k]}) - E^{\mathcal{B}}(X_n^- \chi_{[X_n^- > k]})$$
$$\geq E^{\mathcal{B}}(X_n^+) - E^{\mathcal{B}}(X_n^- \chi_{[X_n^- \leq k]}) - U_k$$
$$\geq E^{\mathcal{B}}(X_n \chi_{[X_n^- \leq k]}) - U_k .$$

Hence applying Proposition 2(ii) we get

$$\liminf_n E^{\mathcal{B}}(X_n) \geq \liminf_n E^{\mathcal{B}}(X_n \chi_{[X_n^- \leq k]}) - U_k$$
$$\geq E^{\mathcal{B}}(\liminf_n X_n \chi_{[X_n^- \leq k]}) - U_k , \qquad (16)$$

since the X_n are bounded below by k. But now we have $U_k \to 0$, a.e., as $k \to \infty$. So (16) implies

$$\liminf_n E^{\mathcal{B}}(X_n) \geq E^{\mathcal{B}}(\liminf_n X_n) \, , \quad \text{a.e.} \tag{17}$$

Replacing X_n by $-X_n$ and noting that $V_k = \sup_{n \geq 1} E^{\mathcal{B}}(X_n^+ \chi_{[X_n^+ > k]})$ goes to zero a.e., as $k \to \infty$, by (15) we get the inequality (17) for this sequence also. Hence

$$\limsup_n E^{\mathcal{B}}(X_n) = - \liminf_n E^{\mathcal{B}}(-X_n)$$
$$\leq E^{\mathcal{B}}(- \liminf_n (-X_n))$$
$$= E^{\mathcal{B}}(\limsup_n X_n), \quad \text{a.e.} \tag{18}$$

Since $X_n \to X$ a.e., we get $\limsup_n X_n = \liminf_n X_n$ a.e. So (17) and (18) imply

$$E^{\mathcal{B}}(X) \leq \liminf_n E^{\mathcal{B}}(X_n) \leq \limsup_n E^{\mathcal{B}}(X_n) \leq E^{\mathcal{B}}(X), \quad \text{a.e.} \tag{19}$$

Thus $E^{\mathcal{B}}(X_n) \to E^{\mathcal{B}}(X)$ a.e. in addition to the mean convergence. The latter follows from (a) since $\mathcal{B}_0 = (\phi, \Omega) \subset \mathcal{B}$ and $E^{\mathcal{B}_1}(X_n \chi_{[|X_n| \geq k]}) \to 0$ uniformly in n as $k \to \infty$ for any σ-algebra $\mathcal{B}_0 \subset \mathcal{B}_1 \subset \mathcal{B}$ if it holds for $\mathcal{B}$ as a consequence of (15). This proves (b) in all parts. $\square$

2.3 Conditional probabilities in the general case

As observed at the beginning of this chapter, a conditional probability function must be obtained from the corresponding expectation. Thus according to Definition 1.1, the conditional probability on (Ω, Σ, P) relative to a σ-algebra $\mathcal{B} \subset \Sigma$, denoted $P^{\mathcal{B}}(\cdot)$, is given by the set of equations,

$$P^{\mathcal{B}}(A) = E^{\mathcal{B}}(\chi_A) \, , \quad A \in \Sigma \ .$$

This is defined even if $P_{\mathcal{B}}$ is only σ-finite, and $0 \leq P^{\mathcal{B}}(A) \leq 1$, a.e., is always true. We can deduce a number of properties of $P^{\mathcal{B}}(\cdot)$ from those of $E^{\mathcal{B}}(\cdot)$. They are listed as follows.

Proposition 1. *Let (Ω, Σ, P) be a measure space and $\mathcal{B} \subset \Sigma$ be a σ-algebra such that $P_\mathcal{B}$ is σ-finite. Then there is a $P_\mathcal{B}$-unique function $P^\mathcal{B} : \Sigma \to \mathbb{R}^+$ satisfying the set of equations:*

$$\int_B P^\mathcal{B}(A) dP_\mathcal{B} = P(A \cap B) \,, \ A \in \Sigma, \ B \in \mathcal{B}, \ P(B) < \infty \ . \qquad (1)$$

Further $P^\mathcal{B}$ has the following properties a.e. (P).

(i) $0 \le P^\mathcal{B}(A)$, $A \in \Sigma$, (ii) $P^\mathcal{B}(\Omega) = 1$, and $P^\mathcal{B}(A) = 0$ if $P(A) = 0$, (iii) $P^\mathcal{B}(B) = \chi_B$, $B \in \mathcal{B}$, (iv) if $\{A_n, \ n \ge 1\} \subset \Sigma$ is a monotone sequence with limit A, then $P^\mathcal{B}(A_n) \to P^\mathcal{B}(A)$, (v) if $\{A_n, n \ge 1\} \subset \Sigma$ is a disjoint sequence, then $P^\mathcal{B}(\cup_n A_n) = \Sigma_n P^\mathcal{B}(A_n)$, and if $P(\Omega) < \infty$, then this series also converges in L^p-mean for $1 \le p < \infty$ (but not for $p = +\infty$).

Proof. The fact that $P^\mathcal{B}(\cdot)$ exists, $P_\mathcal{B}$-unique and satisfies (1) is a consequence of the Radon-Nikodým theorem as shown by equations (1.2) and (1.3) which hold even if $P_\mathcal{B}$ is σ-finite. Since $E^\mathcal{B}$ is order-preserving, faithful, $E^\mathcal{B}(1) = 1$ a.e., and $E^\mathcal{B}(\chi_B) = \chi_B$ a.e. for all $B \in \mathcal{B}$, we get the properties (i)-(iii). Regarding (iv), if $A_n \subset A_{n+1}$, $n \ge 1$, then the conditional monotone convergence (cf., Proposition 2.2) yields

$$\lim_n P^\mathcal{B}(A_n) = \lim_n E^\mathcal{B}(\chi_{A_n})$$
$$= E^\mathcal{B}(\chi_A) = P^\mathcal{B}(A) \,, \text{ a.e.} \qquad (2)$$

If $A_n \supset A_{n+1}$, $n \ge 1$, then by (1) for each $B \in \mathcal{B}$, $P(B) < \infty$,

$$\lim_n \int_B P^\mathcal{B}(A_n) dP_\mathcal{B} = \lim_n P(A_n \cap B) = P(A \cap B) \,. \qquad (3)$$

Since $P(B) < \infty$ and $P^\mathcal{B}(A_n) \le 1$ a.e., the limit and integral on the left side of (3) can be interchanged, by the classical bounded convergence theorem. Hence

$$\int_B \lim_n P^\mathcal{B}(A_n) dP_\mathcal{B} = P(A \cap B) = \int_B P^\mathcal{B}(A) dP_\mathcal{B} \,. \qquad (4)$$

But this system of equations holds for all $B \in \mathcal{B}$ of finite measure. By the σ-finiteness of $P_\mathcal{B}$ and the fact that the integrands in (4) are $\mathcal{B}$-measurable imply that they can be identified. This proves the validity of (iv) in general.

Finally consider (v). The argument is similar to the above. In fact, by the classical monotone convergence theorem, if $B \in \mathcal{B}$, of finite measure,

$$\int_B \sum_{n=1}^{\infty} P^B(A_n)dP_B = \sum_{n=1}^{\infty} \int_B P^B(A_n)dP_B$$

$$= \sum_{n=1}^{\infty} P(A_n \cap B), \text{ by } (1),$$

$$= P(\bigcup_{n=1}^{\infty} (A_n \cap B)), \text{ since } P \text{ is } \sigma\text{-additive},$$

$$= \int_B P^B(\bigcup_{n=1}^{\infty} A_n)dP_B, \text{ by } (1) . \qquad (5)$$

Since P_B is σ-finite, $B(\in \mathcal{B})$ is arbitrary of finite measure, and the extreme integrands are $\mathcal{B}$-measurable, they can be identified. With $P(\Omega) < \infty$, the last statement follows from the classical bounded convergence theorem. Indeed,

$$||P^B(\bigcup_{k=1}^{\infty} A_k) - \sum_{k=1}^{n} P^B(A_k)||_p^p = \int_\Omega [P^B(\bigcup_{k>n} A_k)]^p \, dP_B , \text{ by } (5),$$

$$\leq \int_\Omega E^B(\chi_{\bigcup_{k>n} A_k})^p \, dP_B,$$

$$\text{by Theorem } 2.4(\text{iii}) ,$$

$$= P(\bigcup_{k>n} A_k) \to 0, \text{ as } n \to \infty.$$

For $p = \infty$, this fails since $||\chi_{\bigcup_{k>n} A_k}||_\infty = 1$ for all n. Thus we have all the assertions. $\square$

Remark. It is important to note that all the statements of the preceding proposition hold only a.e., and not everywhere. The stronger statements can be made, in general, only if $\mathcal{B}$ is generated by a countable partition. In this case a fixed null set can be selected and on its complement the assertions hold everywhere so that we will be effectively in the elementary case. Thus in the general case one must note this difficulty dealing with conditional measures. This point and the difficulties will be illustrated further later on.

We now present an extension of the classical Fubini theorem with the ideas of conditioning as it proves to be of interest in a number of applications.

Let (Ω, Σ, μ) be a measure space, and consider a mapping $P(\cdot, \cdot) : \Sigma \times \Omega \to \mathbb{R}^+$ satisfying the following conditions: (i) $P(\cdot, \omega)$ is a probability measure for each $\omega \in \Omega - N, \mu(N) = 0$, (ii) $P(A, \cdot)$ is a $\mathcal{B}$-random variable for a σ-algebra $\mathcal{B} \subset \Sigma$, $A \in \Sigma$, and (iii) $\int_B P(A, \omega) \, d\mu_{\mathcal{B}}(\omega) = \mu(A \cap B)$, $B \in \mathcal{B}$, $A \in \Sigma$. If $\mu_{\mathcal{B}}$ is σ-finite and $P(A, \cdot)$ is written for the function $P^{\mathcal{B}}(A)$ in the above propositon, then it has all the properties that qualify it to be a conditional probability relative to $\mathcal{B}$. However, not every conditional probability satisfies (i)-(iii), and especially (i) can be false. To distinguish the general one from that satisfying conditions (i)-(iii), the latter is called a *regular conditional probability*. It follows that each elementary conditioning with $\mathcal{B}$ generated by a countable partition (the case considered in Chapter 1) is regular. The general nonregular case can be illustrated by using Theorem 2.5 as follows.

Fix $\omega_0 \in \Omega - N$ and set $P_{\omega_0}^{\mathcal{B}}(A) = P(A, \omega_0)$, $A \in \Sigma$. If each $P^{\mathcal{B}}$ is a regular conditional measure, then for each μ-integrable random variable X one has

$$E^{\mathcal{B}}(X)(\omega_0) = \int_\Omega X(\omega) dP_{\omega_0}^{\mathcal{B}}(\omega) \qquad (6)$$

to be well-defined and $E^{\mathcal{B}}(\cdot)$ is a conditional expectation of X relative to $\mathcal{B}$. Indeed this holds if $X = \chi_A$, $A \in \Sigma$, and by linearity it holds for all simple and then for all μ-integrable random variables by the classical Lebesgue limit theorems since we are assuming that $P_{\omega_0}^{\mathcal{B}}(\cdot)$ is a probability measure. Thus setting $E^{\mathcal{B}}(X) = 0$ on N, we get a standard function:

$$E^{\mathcal{B}}(X) = \int_\Omega X dP^{\mathcal{B}} \, , \quad \text{a.e.} \qquad (7)$$

However, Theorem 2.5 implies that there are probability spaces (Ω, Σ, μ) and σ-algebras $\mathcal{B}_0 \subset \Sigma$ and a sequence of uniformly integrable random variables $X_n \to X$ a.e., such that

$$\lim_{n \to \infty} E^{\mathcal{B}_0}(X_n) = \lim_{n \to \infty} \int_\Omega X_n dP^{\mathcal{B}_0}$$

$$\neq \int_\Omega X dP^{\mathcal{B}_0} = E^{\mathcal{B}_0}(X) \qquad (8)$$

the inequality holding a.e. But if $P^{\mathcal{B}_0}$ were regular, the classical Vitali theorem should ensure the equality in (8). This shows that our supposition that $P^{\mathcal{B}}$ is always regular for each $\mathcal{B}$ is not valid and hence a conditional measure cannot always be treated as an ordinary measure. Therefore we need to make a special study of regular conditional measures and find suitable hypothesis on $P^{\mathcal{B}}$ in order that regularity holds. This will be done in Chapter 5.

Since the elementary case shows that the class of regular conditional measures is nonempty (and by Chap. 5 there are lots of nonelementary ones), the following result which extends the classical Fubini theorem is of interest in applications.

Proposition 2. *Let (Ω, Σ) be a measurable space and $(S, \mathcal{S}, \mu)$ be a probability space. Suppose that there is a mapping $Q(\cdot, \cdot) : \Sigma \times S \to \mathbb{R}^+$ such that (i) $Q(\cdot, s)$ is a probability for each $s \in S - S_0$, $\mu(S_0) = 0$, and (ii) $Q(A, \cdot)$ is a measurable function for $\mathcal{S}$, $A \in \Sigma$. Then there is a probability measure P on the product measurable space $(\Omega \times S, \Sigma \otimes \mathcal{S})$ such that for each random variable $X \geq 0$, we have*

$$\int_{\Omega \times S} X dP = \int_S [\int_\Omega X(\omega, s) Q(d\omega, s)] \, \mu(ds) , \qquad (9)$$

where both sides are infinite or both are finite and equal.

Proof. Consider the class $\Sigma \times \mathcal{S} = \{A \times T : A \in \Sigma, T \in \mathcal{S}\}$. It is a semi-ring containing $\Omega \times S$ so that it is termed a semi-algebra, using the standard terminology of real analysis. Define P on it by means of the equation:

$$P(A \times T) = \int_T Q(A, s) \, \mu(ds), \quad A \in \Sigma, \ T \in \mathcal{S} . \qquad (10)$$

Since μ and Q are bounded, P is well-defined and takes values in $[0, 1]$ with $P(\Omega \times S) = 1$. We assert that P is σ-additive on $\Sigma \times \mathcal{S}$. For, let $\{A_n \times T_n, \ n \geq 1\}$ be a disjoint sequence from $\Sigma \times \mathcal{S}$ with union $A \times T$ also in it. Then one has

$$\chi_{A \times T}(\omega, s) = \chi_A(\omega) \chi_T(s)$$
$$= \sum_{n=1}^{\infty} \chi_{A_n \times T_n}(\omega, s) = \sum_{n=1}^{\infty} \chi_{A_n}(\omega) \chi_{T_n}(s) . \qquad (11)$$

Note that for each $(\omega, s) \in A \times T$, exactly one term on the right of (11) is nonzero. Now fixing an $s \in S$, and considering the series above as a function in $\omega \in \Omega$, we get on integrating it relative to $Q(\cdot, s)$

$$Q(A, s) \chi_T(s) = \sum_{n=1}^{\infty} Q(A_n, s) \, \chi_{T_n}(s) , \qquad (12)$$

by the monotone convergence theorem (applicable to $Q(\cdot, s)$). Next treating (12) as a function of s, and integrating it relative to μ, one has by the same (monotone) theorem

$$P(A \times T) \;=\; \int_T Q(A, s)\mu(ds) \;=\; \overset{\infty}{\underset{n=1}{\Sigma}} \, P(A_n \times T_n) \ .$$

This shows that $P : \Sigma \times \mathcal{S} \to [0, 1]$ is σ-additive. Hence by the classical Hahn extension theorem, P has a unique σ-additive extension to the σ-algebra $\Sigma \otimes \mathcal{S}$ generated by $\Sigma \times \mathcal{S}$. We denote this extended function by the same symbol. Thus P on $\Sigma \otimes \mathcal{S}$ is a probability measure.

To complete the demonstration, it is only necessary to show that (9) holds for this P. In fact, this is true by construction above if $f = \chi_{A \times T}$. By linearity of the integrals on both sides of (9) it holds if $f = \Sigma_{i=1}^n a_i \chi_{A_i \times T_i}$ so that the assertion is true for all $\Sigma \otimes \mathcal{S}$ simple functions. But $\Sigma \times \mathcal{S}$ is dense in $\Sigma \otimes \mathcal{S}$ for the semi-metric ρ defined on $\Sigma \otimes \mathcal{S}$ by $\rho(E, F) = P(E \Delta F)$ by the standard results in analysis. Thus we can approximate every $\Sigma \otimes \mathcal{S}$ simple function with one of $\Sigma \times \mathcal{S}$. From this it follows immediately that (9) holds for $\Sigma \otimes \mathcal{S}$-simple functions. Finally, every random variable $f : \Omega \times S \to \mathbb{R}^+$, for $\Sigma \otimes \mathcal{S}$, is a pointwise limit of an increasing sequence of such simple $f \geq 0$, so that by the monotone convergence criterion (9) holds for all positive random variables. $\square$

Taking $S = \Omega, \mathcal{S} \subset \Sigma$, and μ a probability on Σ, $Q(\cdot, \cdot)$ becomes a regular conditional probability. Thus the above proposition gives an extension of the classical Fubini theorem. In fact if $Q(\cdot, s)$ is independent of s, then the result reduces to the Fubini theorem on finite measure spaces. But then the σ-finite case is a simple extension.

It is desirable to have a result of the above type for any conditional measure Q on Σ without being regular. Indeed it is possible to have such an extension, but because of the discussion between (7) and (8) we cannot expect one if the classical theory of (Lebesgue) integration is demanded. The desired method of integration for such function space valued measures and the corresponding result will be given, with a suitable preparation, in a later chapter. For now the stated version suffices.

We present an applicaiton of the above proposition to certain measures called *mixtures* in the literature. Thus let $(S, \mathcal{S}, \mu)$ be a probability space in which μ concentrates on a countable set of points $\{s_n,\ n \geq 1\}$ with the corresponding masses $\{p_n,\ n \geq 1\}$. Then the mapping $Q(\cdot, \cdot) : \Sigma \times S \to \mathbb{R}^+$ of the above proposition gives, on writing $Q(\cdot, s_n)$ as $Q_n(\cdot)$, a measure P as:

$$P(A \times T) = \sum_{s_n \in T} p_n Q_n(A) , \quad \sum_{n \geq 1} p_n = 1 . \tag{13}$$

Such a P is a *discrete mixture* of Q_n and $\{p_n, n \geq 1\}$. Other specializations are possible. For instance, let $\Omega = \mathbb{R}$, $\Sigma =$ Borel σ-algebra of $\mathbb{R}$, $S = \mathbb{R}$, $\mathcal{S} \subset \Sigma$. Suppose that $Q(A_x, y) = F(x, y)$ for each $A_x = (-\infty, x)$, $x \in \mathbb{R}$, where F is a distribution function, i.e., a monotone nondecreasing and left continuous function with $\lim_{\substack{x \to \infty \\ y \to \infty}} F(x, y) = 1$ and $\lim_{x \to -\infty} F(x, y) = 0$, $\lim_{y \to -\infty} F(x, y) = 0$. In this case μ is any probability function on Σ, and we have

$$P(A_x \times T) = \int_T \int_{A_x} d_u F(u, y) \mu(dy)$$

$$= \int_T F(x, y) \mu(dy) , \tag{14}$$

so that P is a *continuous mixture*. We now give an example to show that the above illustrations appear in natural applications.

Example 3. Suppose that $X_1, X_2, \ldots$ are a sequence of random variables representing the assets (or liabilities) of individuals chosen for inspection from a "homogeneous population." If the recorder stops observation at the Nth individual when a (telephone) message arrives, then the total recorded value will be $S_N = \sum_{i=1}^{N} X_i$ (in monetary units). Assuming the sequence comes from a random selection, so that the X_i are independent and identically distributed, and also assuming that the X_i have two moments finite, it is desired to find the variance of S_N (to know the "spread" of the distribution). Here the message arrivals are random and have a given (usually Poisson) distribution.

We translate the problem to one of mixtures, (using (13)) and thus the probability measure P governing the S_N can be found. Our hypothesis implies that S_N has two moments finite. (The conditions should be contrasted with the counterexample of Section 1.4.)

Let $(\mathbb{N}, \mathcal{P}, \mu)$ be a probability space with $\mathbb{N} = \{1, 2, \ldots\}$, $\mathcal{P} = $ power set, and μ is a probability on $\mathbb{N}$. Suppose that $N : \mathbb{N} \to \mathbb{N}$ is a random variable such that $\mu(N = n) = p_n$. Let (Ω, Σ) be a measurable space (describing the desired incomes of individuals in the homogeneous populations under consideration) and define $Q(\cdot, \cdot) : \Sigma \times \mathbb{N} \to \mathbb{R}^+$ as the conditional measure such that for each $n \in \mathbb{N}$, $Q(\cdot, n)$ gives the probability distribution of X_n. Then the probability $P : \Sigma \times \mathcal{P} \to \mathbb{R}^+$ governing S_n is given by (13) so that

$$P(A \times B) = \sum_{n \in B} Q(A, n) p_n, \quad A \in \Sigma \ .$$

Consequently we have

$$E(S_N) = \int_{\Omega \times \mathbb{N}} S_n(\omega) dP = \sum_{n=1}^{\infty} p_n \int_{\Sigma} S_n(\omega) \, Q(d\omega, n) \ , \qquad (15)$$

and similarly

$$E(S_N^2) = \sum_{n=1}^{\infty} p_n \int_{\Omega} S_n^2(\omega) Q(d\omega, n) \ . \qquad (16)$$

Since by assumption the X_n are identically distributed with mean α and variance σ^2, and since X_n, X_m are independent, we have

$$\int_{\Omega} S_n(\omega) Q(d\omega, n) = \sum_{k=1}^{n} \int_{\Omega} X_k(\omega) Q(d\omega, k)$$

$$= \sum_{k=1}^{n} \alpha = n \, \alpha \ ,$$

and

$$\int_{\Omega} S_n^2(\omega) Q(d\omega, n) = \sum_{k=1}^{n} [\int_{\Omega} X_k^2(\omega) Q(d\omega, k) + n(n-1) \, \alpha^2],$$

by independence of X_k's $E(X_k X_{k'}) = E(X_k) E(X_{k'}) = \alpha^2$ for $k \neq k'$,

$$= n \int_{\Omega} X_1^2(\omega) Q(d\omega, 1) + n(n-1) \, \alpha^2 \ .$$

But $\sigma^2 = E(X_1 - \alpha)^2 = E(X_1^2) - \alpha^2$. Hence

$$\int_{\Omega} S_n^2(\omega) Q(d\omega, n) = n \, \sigma^2 + n \, \alpha^2 + n(n-1) \, \alpha^2$$

$$= n \, \sigma^2 + n^2 \, \alpha^2 \ .$$

Thus (15) and (16) give

$$E(S_N) = \sum_{n=1}^{\infty} n \, \alpha \, p_n = \alpha \, E(N) = E(X_1)E(N) ,$$

and

$$E(X_N^2) = \sum_{n=1}^{\infty} p_n(n\sigma^2 + n^2\alpha^2)$$
$$= \sigma^2 E(N) + \alpha^2 E(N^2) .$$

Hence

$$\text{Var } S_N = E(S_N - E(S_N))^2 = E(S_N^2) - (E(S_N))^2$$
$$= \sigma^2 E(N) + \alpha^2 E(N^2) - (\alpha E(N))^2$$
$$= \sigma^2 E(N) + \alpha^2 \text{ Var } N$$
$$= E(N)\text{Var } X_1 + (E(X_1))^2 \text{ Var } N .$$

Higher moments can be calculated with similar work.

2.4 Remarks on the inclusion of previous concepts

It was already seen in relations (1.7) and (1.8) that the elementary case of Definition 1.2.2 (cf. also the corresponding relations (1.2.4) and (1.2.5)) is included in the general concept due to Kolmogorov. However, the procedure with densities given by Definition 1.4.1 contained a limit process and it is not clear that the latter concept is also included in the general study. After all, the same name (or conditioning) should not be used in different places unless the definitions are compatible, i.e., they agree with each other. Thus it is necessary to consider the density case. We now show this to be consistent with the general definition.

Let $f : \mathbb{R}^2 \to \mathbb{R}^+$ be a Borel function such that $\int \int_{\mathbb{R}^2} f(x,y)dx \, dy = 1$. Thus f is a density with marginals f_1, f_2 where $f_1(x) = \int_{\mathbb{R}} f(x,y)dy$ and $f_2(y) = \int_{\mathbb{R}} f(x,y)dx$. Let P be the probability measure induced by f:

$$P(A \times B) = \int_A \int_B f(x,y)dx \, dy , \tag{1}$$

$A \subset \mathbb{R}$, $B \subset \mathbb{R}$ are Borel sets. Then P has a unique extension to be a probability on $\mathcal{B}^2$, the Borel σ-algebra of $\mathbb{R}^2$. Writing $\mathbb{R}^2 = \mathbb{R}_1 \times$

$\mathbb{R}_2 (\mathbb{R}_i = \mathbb{R})$, let $\mathcal{B}_i$ be the (component) Borel σ-algebras of $\mathbb{R}_i$ so that $\mathcal{B}^2 = \mathcal{B}_1 \otimes \mathcal{B}_2$. We now produce a pair of random variables (X_1, X_2) on $\mathbb{R}^2$ whose (joint) distribution is induced by P. This is actually a simple consequence of a general existence theorem due to Kolmogorov but this special case can be obtained directly, although (unfortunately) with little motivation. Thus let (X_1, X_2) be coordinate functions on $\mathbb{R}^2$ so that if $\omega = (x_1, x_2) \in \mathbb{R}^2$, then $X_1(\omega) = x_1, X_2(\omega) = x_2$. It is simple to verify that X_1, X_2 are random variables for $\mathcal{B}^2$, and then we have

$$
\begin{aligned}
P(\{\omega : X_i(\omega) < u\}) &= \int\limits_{\{\omega : X_i(\omega) < u\}} \int f(x,y) dx \, dy \\
&= \begin{cases} \int_{-\infty}^{u} \int_{-\infty}^{\infty} f(x,y) dy dx, & i = 1 \\ \int_{-\infty}^{\infty} \int_{-\infty}^{u} f(x,y) \, dy dx, & i = 2, \end{cases} \\
&= \begin{cases} \int_{-\infty}^{u} f_1(x) dx, & i = 1 \\ \int_{-\infty}^{u} f_2(y) dy, & i = 2 \; . \end{cases}
\end{aligned}
\tag{2}
$$

Thus X_1, X_2 have the desired structure with (marginal) densities f_1 and f_2 respectively. Next let $\mathcal{C}_i = X_i^{-1}(\mathcal{B}_i) \subset \mathcal{B}^2, i = 1, 2$. The $\mathcal{C}_i$ are the cylinder σ-algebras of $\mathcal{B}^2$ and the latter is easily seen to be generated by $\mathcal{C}_1 \cup \mathcal{C}_2$. To use the elementary definition of conditioning with densities, we consider

$$
f_{X_1 | X_2}(x_1 | x_2) = \begin{cases} \frac{f(x_1, x_2)}{f_2(x_2)}, & \text{if} \quad f_2(x_2) \neq 0 \\ 0, & \text{if} \quad f_x(x_2) = 0 \; . \end{cases}
\tag{3}
$$

We have already noted that $f_{X_1 | X_2}(\cdot | x_2)$ is a density for each $x_2 \in \mathbb{R}$. Now define for each $A \in \mathcal{B}^2$ and $\omega \in \mathbb{R}^2$ the mapping

$$
P(A | \omega) = \int_{A(\pi_2(\omega))} f_{X_1 | X_2}(x_1 | x_2) dx_1, \quad \omega = (x_1, x_2),
$$

where $A(\pi_2(\omega))$ is the $\pi_2(\omega)$ section of A with $\pi_i : \mathbb{R}^2 \to \mathbb{R}_i$ being the ith coordinate projection. Thus we may express the above as:

$$
P(A | \omega) = \int_{A(\pi_2(\omega))} f_{X_1 | X_2}(x_1 | \pi_2(\omega)) dx_1 \; .
\tag{4}
$$

It is clear that $P(\cdot | \cdot) : \mathcal{B}^2 \times \mathbb{R}^2 \to \mathbb{R}^+$ is well-defined, and to see that $P(\cdot | \omega)$ is a measure only its σ-additivity need be verified. So let

$\{A_n, \, n \geq 1\} \subset \mathcal{B}^2$ be a disjoint sequence with union A. Then for each fixed $\omega \in \mathbb{R}^2$ we have

$$\pi_1(A(\pi_2(\omega))) = \pi_1(\bigcup_{n=1}^{\infty} A_n(\pi_2(\omega)))$$

$$= \bigcup_{n=1}^{\infty} \pi_1(A_n(\pi_2(\omega))).$$

Moreover the right side is a disjoint union. This is clear if the A_n are horizontally laid out, but if they are vertically stacked, only one of them will be nonempty, as seen by drawing a picture. The other possibilities are verified similarly. Since the integral used in (4) is σ-additive, we get

$$P(\bigcup_n A_n | \omega) = \int_{\pi_1(\bigcup_n A_n(\pi_2(\omega)))} f_{X_1|X_2}(x_1|\pi_2(\omega))dx_1$$

$$= \sum_{n=1}^{\infty} \int_{\pi_1(A_n(\pi_2(\omega)))} f_{X_1|X_2}(x_1|\pi_2(\omega))dx_1$$

$$= \sum_{n=1}^{\infty} P(A_n|\omega) . \tag{5}$$

It remains to show that $P(A|\omega)$ is a version of $P^{\mathcal{C}_2}(\cdot)$. By the essential uniqueness of the Radon-Nikodým integrand, we only have to verify that $P(\cdot|\cdot)$ of (4) satisfies the functional equation (1.4). Let $B \in \mathcal{C}_2$. Being a cylinder, it can be represented as $B = \mathbb{R} \times B_2$ for some $B_2 \in \mathcal{B}_2$. Hence for each $A \in \mathcal{B}^2$, we have

$$\int_B P(A|\omega)P_{\mathcal{C}_2}(d\omega) = \int_{\mathbb{R} \times B_2} \left[\int_{A(\pi_2(\omega))} f_{X_1|X_2}(x_1|\pi_2(\omega))dx_1 \right] P_{\mathcal{C}_2}(d\omega)$$

$$= \int_{\mathbb{R}} \int_{B_2} \left[\int_{A(\pi_2(\omega))} f_{X_1|X_2}(x_1|x_2)dx_1 \right] f(x_1,x_2)dx_1 dx_2 ,$$

by definition of P and Proposition 1.4.2,

$$= \int_{B_2} \int_{A(\pi_2(\omega))} f_{X_1|X_2}(x_1|x_2)dx_1 \left(\int_{\mathbb{R}} f(u,x_2)du \right) dx_2$$

$$= \int_{B_2} \int_{A(x_2)} f_{X_1|X_2}(x_1|x_2)f_2(x_2)dx_1 dx_2 ,$$

by the Tonelli theorem,

$$= \int_{B \cap A} f(x_1,x_2)dx_1 dx_2, \text{ by (3)},$$

$$= P(A \cap B) . \tag{6}$$

This is (1.4) and hence $P(\cdot|\cdot)$ defined by (4) is a version of $P^{\mathcal{C}_2}(\cdot)$. Thus the definition of conditional density in (3) is included in the general (Kolmogorov) concept.

The preceding computations extend without change if $\mathbb{R}^2$ is replaced by $\mathbb{R}^{m+n}$. Hence the conditional densities, given with the method of (4), will be candidates for defining regular (versions) of conditional probabilities. Are there other methods of defining such functions? This is not easy to answer. In fact, we shall see in the next chapter that an actual calculation of such conditional densities is not simple; especially when the conditioning is made by a random vector. Note that in Section 1.4 we used a form of the L'Hôpital rule in evaluating certain quantities in obtaining a conditional density. There are problems in removing the indeterminacy in such situations, and we shall analyze the matter in some detail, in the next chapter, since the methods are not entirely intuitive.

2.5 Conditional independence and related concepts

One can pursue the study of conditional measures by extending some of the absolute concepts to the general case. Thus one of the natural notions is independence. This can be extended to the present (conditional) context, without imposing any regularity restrictions as follows.

Definition 1. Let (Ω, Σ, μ) be a measure space, and $\{\mathcal{B}, \mathcal{B}_\alpha : \alpha \in I\}$ be σ-algebras contained in Σ such that μ_B is σ-finite. If the cardinality of I is at least two, then $\{\mathcal{B}_\alpha, \alpha \in I\}$ is called a *conditionally independent* family relative to (or given) $\mathcal{B}$ if for each distinct set of indices $\alpha_1, \ldots, \alpha_n$ of I and each $A_{\alpha_i} \in \mathcal{B}_{\alpha_i}$, one has

$$P^{\mathcal{B}}\left(\bigcap_{i=1}^{n} A_{\alpha_i}\right) = \prod_{i=1}^{n} P^{\mathcal{B}}(A_{\alpha_i}) \quad \text{a.e.,} \quad n \geq 2 . \tag{1}$$

If $\{X_\alpha, \ \alpha \in I\}$ is a family of random variables on (Ω, Σ, μ) in the above, then it is conditionally independent relative to $\mathcal{B}$, whenever the σ-algebras $\{\mathcal{B}_\alpha = X_\alpha^{-1}(\mathcal{R}), \ \alpha \in I\}$ are conditionally independent given $\mathcal{B}$, i.e., (1) holds for them. Here $\mathcal{R}$ is the Borel σ-algebra of $\mathbb{R}$.

Note that if $\mathcal{B} = \{\phi, \Omega\}$ and that $\mu_B(\Omega) = 1$ we have the usual (unconditional) concept defined before. It will be useful to present different

sets of statements equivalent to the conditional independence and give
some applications, so that one will have a better understanding of the
extension. Observe that the smaller the σ-algebra $\mathcal{B}$ is, the more strin-
gent the conditions (1) become. In particular, if $\mathcal{B}_\alpha = \mathcal{B}$, $\alpha \in I$, then
the system of equations (1) reduce to a tautology since $P^{\mathcal{B}}(\chi_A) = \chi_A$,
a.e., for all $A \in \mathcal{B}$. Thus there is interest only when $\mathcal{B}$ is distinct from
all the $\mathcal{B}_\alpha, \alpha \in I$. The following result illuminates this observation.

Proposition 1. *Let $\mathcal{B}, \mathcal{B}_1, \mathcal{B}_2$ be σ-subalgebras of (Ω, Σ, μ) with $\mu_\mathcal{B}$ σ-*
finite (or μ itself is a probability). Then the following are equivalent
statements.

(i) $\mathcal{B}_1, \mathcal{B}_2$ are conditionally independent given $\mathcal{B}$.
(ii) For each $B_1 \in \mathcal{B}_1$, $P^{\alpha_2}(B_1) = P^{\mathcal{B}}(B_1)$ a.e., with $\alpha_2 = \sigma(\mathcal{B} \cup \mathcal{B}_2)$.
(iii) For each $B_2 \in \mathcal{B}_2$, $P^{\alpha_1}(B_2) = P^{\mathcal{B}}(B_2)$ a.e., with $\alpha_1 = \sigma(\mathcal{B} \cup \mathcal{B}_1)$.
(iv) For each $X : \Omega \to \mathbb{R}^+$, $\mathcal{B}_1$-measurable, $E^{\alpha_2}(X) = E^{\mathcal{B}}(X)$, a.e.
(v) For each $X : \Omega \to \mathbb{R}^+$, $\mathcal{B}_2$-measurable, $E^{\alpha_1}(X) = E^{\mathcal{B}}(X)$, a.e.

[In (iv) and (v) the σ-algebras α_1, α_2 are as defined in (iii) and (ii)
above.]

Proof. (i) $\Rightarrow$ (ii) Assume that $\mathcal{B}_1, \mathcal{B}_2$ are conditionally independent
given $\mathcal{B}$. We need to verify that $P^{\alpha_2}(B_1)$ and $P^{\mathcal{B}}(B_1)$ satisfy the functional equation on α_2 for each $B_1 \in \mathcal{B}_1$. Since the σ-algebra α_2 is
generated by the sets of the form $\{B \cap B_2 : B \in \mathcal{B}, B_2 \in \mathcal{B}_2\}$, it suffices
to verify the desired equation for the generators. Thus

$$
\int_{B \cap B_2} P^{\alpha_2}(B_1) dP_{\alpha_2} = \int_{B \cap B_2} E^{\alpha_2}(\chi_{B_1}) dP_{\alpha_2}
$$

$$
= \int_B \chi_{B_1 \cap B_2} dP
$$

$$
= \int_B E^{\mathcal{B}}(\chi_{B_1 \cap B_2}) dP_\mathcal{B}
$$

$$
= \int_B P^{\mathcal{B}}(B_1) P^{\mathcal{B}}(B_2) dP_\mathcal{B}, \quad \text{by hypothesis,}
$$

$$
= \int_B E^{\mathcal{B}}(\chi_{B_2} P^{\mathcal{B}}(B_1)) dP_\mathcal{B}, \quad \text{by Proposition 2.1(i),}
$$

$$
= \int_{B \cap B_2} P^{\mathcal{B}}(B_1) dP_{\alpha_2}, \quad \text{since } \mathcal{B} \subset \alpha_2 .
$$

The extreme integrands are α_2-measurable and $B \cap B_2$ is an arbitrary
generator of α_2. Hence we can identify them, giving (ii).

(ii) $\Rightarrow$ (i) To prove the conditional independence of $\mathcal{B}_1, \mathcal{B}_2$ relative to $\mathcal{B}$, consider for any $B_i \in \mathcal{B}_i$, $i = 1, 2$, with $\alpha_2 = \sigma(\mathcal{B} \cup \mathcal{B}_2)$,

$$
\begin{aligned}
P^{\mathcal{B}}(B_1 \cap B_2) &= E^{\mathcal{B}}(\chi_{B_1 \cap B_2}) \\
&= E^{\mathcal{B}}(E^{\alpha_2}(\chi_{B_1 \cap B_2})) \text{ , by Proposition 2.1(ii),} \\
&\qquad \text{since } \mathcal{B} \subset \alpha_2 \text{ ,} \\
&= E^{\mathcal{B}}(\chi_{B_2} E^{\alpha_2}(\chi_{B_1})) \text{ , since } B_2 \in \alpha_2 \text{ ,} \\
&= E^{\mathcal{B}}(\chi_{B_2} P^{\alpha_2}(B_1)) \\
&= E^{\mathcal{B}}(\chi_{B_2} P^{\mathcal{B}}(B_1)) \text{ , a.e., by hypothesis ,} \\
&= P^{\mathcal{B}}(B_1) P^{\mathcal{B}}(B_2) \text{ , a.e.}
\end{aligned}
$$

Hence (i) is true so that (i) $\Leftrightarrow$ (ii).

(i) $\Leftrightarrow$ (iii). By interchanging the subscripts 1 and 2 in the preceding demonstration, we get this part with an identical argument.

(ii) $\Leftrightarrow$ (iv) Expressing (ii) in terms of conditional expectations we get

$$
E^{\alpha_2}(\chi_{B_1}) = E^{\mathcal{B}}(\chi_{B_1}), \text{ a.e., } B_1 \in \mathcal{B}_1 \text{ .} \tag{2}
$$

Hence by linearity of the conditional expectation operator, (2) holds for all $\mathcal{B}_1$-simple random variables. Then by Proposition 2.2(i) and the fact that any $\mathcal{B}_1$-measurable $X \geq 0$ is a pointwise limit of an increasing sequence X_n of $\mathcal{B}_1$ measurable simple random variables, we get (iv). That (iv) $\Rightarrow$ (ii) is trivial.

(iii) $\Leftrightarrow$ (v). Replacing the subscripts 2 for 1 and 1 for 2 in the preceding part, we get this equivalence. $\square$

To see the usefulness of this conditional independence concept, and for some motivation for further work, we present a simple illustration.

Example 2. Let $\Omega = \mathbb{N}$, $\Sigma = \mathcal{P}(\mathbb{N})$ (the power set of the natural numbers), $S = \{B \subset \mathbb{N}, 0 < |B| < \infty\}$ where $|B|$ denotes the cardinality of the set B. Thus (Ω, Σ) is our measurable space and $(S, \mathcal{P}(S), |\cdot|)$ is also a measure space where $\mathcal{P}(S)$ is the power set of S. Let $Q(\cdot, \cdot) : \Sigma \times S \to \mathbb{R}^+$ be a conditional measure of the type described in Proposition 3.2, i.e., $Q(A, B) = |A \cap B|/|B|$, $|B| > 0$. It is easily verified that $Q(\cdot, \cdot)$ has the desired properties of the proposition. Let $p \geq 1$ be an integer and $A_p = \{kp : k \in \mathbb{N}\}$, $B_n = \{1, 2, \ldots, n\}$. Then for each n that is divisible by p, we get

$$Q(A_p, B_n) = \frac{n/p}{n} = \frac{1}{p} \ .$$

Thus if $p_1, \ldots, p_k$ are relatively prime and n is divisible by the product $p_1 p_2 \cdots p_k$, then $Q(A_{p_i}, B_n) = 1/p_i$, $i = 1, \ldots, k$. Further

$$A_{p_1} \cap \cdots \cap A_{p_k} = A_{p_1 p_2 \cdots p_k} \ ,$$

$$Q(A_{p_1} \cap \cdots \cap A_{p_k}, B_n) = \frac{1}{p_1 p_2 \cdots p_k}$$

$$= \prod_{i=1}^{k} Q(A_{p_i}, B_n) \ . \tag{3}$$

Thus such sets of numbers $A_{p_1}, \ldots, A_{p_k}$ are conditionally independent given B_n (in the elementary sense) where n is divisible by $p_1 p_2 \ldots, p_k$. Although this is a very simple example, such a concept has interest in analytic number theory. Here it serves as a motivation for our further work.

The monotone class argument, which we omit, gives immediately the following extension of Definition 1 above.

Corollary 3. *Let $X_1, \ldots, X_n$ be random variables on a probability space (Ω, Σ, P) and $\mathcal{B} \subset \Sigma$ be a σ-algebra. Then $X_1, \ldots, X_n$ are conditionally independent given $\mathcal{B}$ iff for each event $A_{x_i} = \{\omega : x_i(\omega) < x_i\}$, we have*

$$P^{\mathcal{B}} \left(\bigcap_{i=1}^{n} A_{x_i} \right) = \prod_{i=1}^{n} P^{\mathcal{B}}(A_{x_i}), \text{ a.e., } x_i \in \mathbb{R} \ . \tag{4}$$

Expanding this theme further, we can give a generalization of the Kolmogorov zero-or-one law as follows.

Proposition 4. *Let $\{X_n, \ n \geq 1\}$ be a sequence of random variables on a probability space (Ω, Σ, P) which are conditionally independent given a σ-algebra $\mathcal{B} \subset \Sigma$. If $\sigma(X_k, \ k \geq n)$ denotes the smallest σ-algebra containing the events $\{X_k^{-1}(I), \ k \geq n, \ I \subset \mathbb{R} \text{ interval }\}$, and $\mathcal{T} = \cap_{n=1}^{\infty} \sigma(X_k, \ k \geq n)$, called the tail σ-algebra of the X_n sequence, then $\mathcal{T}$ is essentially contained in $\mathcal{B}$ in the sense that for each $T \in \mathcal{T}$, there is a $B \in \mathcal{B}$ such that $P(B \triangle T) = 0$. In particular if $\mathcal{B} = (\phi, \Omega)$,* *so that the X_ns are mutually independent, then each tail event (i.e., each member of $\mathcal{T}$) has probability zero-or-one. (This last part is the classical Kolmogorov zero-or-one law.)*

Proof. It follows from definition (cf., (1)) that $\sigma(X_1, \ldots, X_n)$ and $\sigma(X_k, k \geq n+1)$ are conditionally independent given $\mathcal{B}$ for each $n \geq 1$, by the preceding corollary. Hence $\sigma(\cup_n \sigma(X_1, \ldots, X_n)) = \sigma(X_n, n \geq 1)$ and $\mathcal{T}$ are conditionally independent given $\mathcal{T}$. But $\mathcal{T} \subset \sigma(X_n, n \geq 1)$. So $\mathcal{T}$ is conditionally independent of itself given $\mathcal{B}$ and so for $A \in \mathcal{T}$, $P^{\mathcal{B}}(A \cap A) = (P^{\mathcal{B}}(A))^2$ a.e. Thus $P^{\mathcal{B}}(A) = \chi_B$ a.e. for some $B \in \mathcal{B}$. But by the essential uniqueness of the Radon-Nikodým derivative, for each $C \in \mathcal{B}$, one has

$$P(A \cap C) = \int_C P^{\mathcal{B}}(A) dP_{\mathcal{B}} = \int_C \chi_B dP_{\mathcal{B}}$$
$$= P(B \cap C) . \tag{5}$$

Taking first $C = \Omega - B$ and then $C = \Omega - A$, (5) yields $P(A - B) = 0 = P(B - A)$, so that $P(A \Delta B) = 0$. Hence $A = B$, a.e.

The last statement is immediate since $\mathcal{B} = (\phi, \Omega)$ implies $P^{\mathcal{B}}$ is a probability and every element of $\mathcal{T}$ is equivalent to either ϕ or Ω so that it has probability either zero-or-one. $\square$

We shall illustrate several important applications of the preceding concepts from time to time in the following chapters.

Before concluding this chapter we present another related concept of dependence, and the Hewitt-Savage zero-or-one law which complements the Kolmogorov version given above.

Definition 5. (a) A sequence $\{X_n, n \geq 1\}$ of random variables on a probability space (Ω, Σ, P) is called *symmetrically dependent* (or *exchangeable*) if $\{X_1, \ldots, X_n\}$ and $\{X_{i_1}, \ldots, X_{i_n}\}$ are identically distributed for each $n \geq 1$, where $(i_1, \ldots, i_n)$ is a permutation of $(1, \ldots, n)$.

(b) Let Π be the set of all finite permutations of the natural numbers $\mathbb{N} = \{1, 2, \ldots\}$ so that $\pi \in \Pi$ iff $\pi(\mathbb{N})$ permutes only a finite subset of $\mathbb{N}$ leaving the rest unchanged. If $X = \{X_1, X_2, \ldots\}$, let $\pi(X)$ denote the permuted sequence $\{X_{i_1}, \ldots, X_{i_n}, X_{n+1}, X_{n+2}, \ldots\}$ whenever $\pi = (i_1, \ldots, i_n, n+1, n+2, \ldots)$ changing the first n-indices (and similarly other elements of Π). The class $\Sigma_0 = \{X^{-1}(A) : X^{-1}(A) = (\pi(X))^{-1}(A),$ a.e., $\pi \in \Pi, A \in \mathcal{B}^\infty\}$ is a σ-algebra, called the class of *permutable* or *symmetric* events of X. Here $\mathcal{B}^\infty = \otimes_{i=1}^\infty \mathcal{B}_i,$ $\mathcal{B}_i = \mathcal{B}$ is the Borel σ-algebra of $\mathbb{R}$.

A sequence $\{X_n,\ n \geq 1\}$ is said to have a common conditional distribution on (Ω, Σ, P) relative to a σ-algebra $\mathcal{B} \subset \Sigma$, if $P^{\mathcal{B}}(X_n^{-1}(A))$ is independent of n for any Borel set $A \subset \mathbb{R}$. Since $E(P^{\mathcal{B}}(X_n^{-1}(A))) = P(X_n^{-1}(A))$, this implies that the X_n are identically distributed, but the former concept is more restrictive than the latter. Also note that a collection of conditionally independent random variables with a common distribution, both relative to a σ-algebra $\mathcal{B} \subset \Sigma$, is symmetrically dependent. To see this, let $A_i \subset \mathbb{R}$ be a Borel set. Then

$$P^{\mathcal{B}}\left(\bigcap_{k=1}^n X_k^{-1}(A_k)\right) = \prod_{k=1}^n P^{\mathcal{B}}(X_k^{-1}(A_k)), \text{ a.e., by (4),}$$

$$= \prod_{k=1}^n P^{\mathcal{B}}(X^{-1}(A_k)), \text{ a.e., by the hypothesis of}$$

$$\text{common distribution}.$$

Hence

$$P^{\mathcal{B}}\left(\bigcap_{k=1}^n X_{i_k}^{-1}(A_{i_k})\right) = \prod_{k=1}^n P^{\mathcal{B}}(X_1^{-1}(A_k))$$

$$= P^{\mathcal{B}}\left(\bigcap_{k=1}^n X_k^{-1}(A_k)\right), \text{ a.e.}$$

This implies the assertion. What is interesting here is that B. de Finetti has proved the (nontrivial) converse of this statement by determining the precise σ-algebra $\mathcal{B}$ relative to which the result is true. We state it as follows.

Theorem 6. (de Finetti) *Let* $\{X_n,\ n \geq 1\}$ *be a symmetrically dependent sequence of random variables on a probability space* (Ω, Σ, P). *Then they are conditionally independent with a common distribution, relative to the σ-algebra Σ_0 of permutable events of* $\{X_n,\ n \geq 1\}$. *The finite dimensional distributions are therefore representable as*

$$P\left[\bigcap_{k=1}^n \{\omega : X_k(\omega) < x_k\}\right] = E\left(\prod_{k=1}^n P^{\Sigma_0}\{\omega : X_k(\omega) < x_k\}\right). \quad (6)$$

We shall not present a proof of this result, since it is not essential for our work. However we illustrate the representation (6) by the following interesting consequence. The last part is called the Hewitt-Savage zero-or-one law for which we include an independent proof below.

Proposition 7. *Let $\{X_n, \; n \geq 1\}$ be a symmetrically dependent sequence of random variables on (Ω, Σ, P). If $\mathcal{T} = \cap_{n=1}^{\infty} \sigma(X_k, \; k \geq n)$ is the tail σ-algebra, then $\mathcal{T}$ is essentially contained in Σ_0, i.e., for any A in $\mathcal{T}$ there is a $B \in \Sigma_0$ such that $P(A \triangle B) = 0$. In particular, if $\{X_n, \; n \geq 1\}$ is also an independent sequence, then $\mathcal{T}$ is degenerate, i.e., $A \in \mathcal{T}$ implies $P(A) = 0$ or 1. (In the last part $\mathcal{T}$ may be replaced by Σ_0 as the second proof below shows.)*

Proof. By Theorem 6, $\{X_n, \; n \geq 1\}$ is a conditionally independent sequence relative to Σ_0, and then by Proposition 4 $\mathcal{T}$ is contained essentially in Σ_0.

For the last part, X_ns are also independent. Hence they are conditionally independent relative to the trivial algebra $\{\phi, \Omega\}$. But by Proposition 4, $\mathcal{T}$ is contained in (hence essentially equal to) this $\{\phi, \Omega\}$, as asserted. $\quad\square$

We now present an alternative demonstration of the last part, by showing that Σ_0 itself is degenerate (so that $\mathcal{T}$ is also), following an independent elementary argument essentially due to Feller.

Alternative proof (of the last part). Recall that $\rho : \Sigma \times \Sigma \to \mathbb{R}^+$ defined by $\rho(A, B) = P(A \triangle B)$ is a semimetric on Σ and (Σ, ρ) is a complete (semimetric) space in which the operations $\cap, \cup$ and $\triangle$ are (uniformly) continuous. These properties are standard facts in measure theory. Also $\sigma(X_1, \ldots, X_n) \subset \Sigma_0$ so that $\cup_{n=1}^{\infty} \sigma(X_1, \ldots, X_n) \subset \Sigma_0$ and is dense in the latter for the ρ-topology. Then there exist $A_n \in \sigma(X_1, \ldots, X_n)$ such that $\rho(A, A_n) \to 0$ as $n \to \infty$. By definition of $\sigma(X_1, \ldots, X_n), A_n$ may be taken as $A_n = \{\omega : (X_1, \ldots, X_n)(\omega) \in B_n\}$ for some Borel set $B_n \subset \mathbb{R}^n$. Consider a permutation such that $\tilde{A}_n = \{\omega : (X_{2n}, X_{2n-1}, \ldots, X_{n+1})(\omega) \in B_n\}$. Then A_n and $\tilde{A}_n$ are independent events because of the hypothesis of independence of the X_n's. Let $\tau : \Omega \to \Omega$ be a mapping induced by the above permutation which reverses just $(n+1, \ldots, 2n)$ leaving the other indices unchanged. Then $\tau A_n = \tilde{A}_n$, and τ is one-to-one, measurable, and measure preserving. Also

$$X = (X_1, X_2, \dots) \text{ and } \tilde{X} = (X_1, \dots, X_n, X_{2n}, \dots, X_{n+1}, X_{2n+1}, \dots)$$

are identically distributed so that $\tau A = A$ a.e. Hence

$$P(A, \tilde{A}_n) = \rho(\tau A, \tau A_n) = \rho(A, A_n) \to 0, \text{ as } n \to \infty \ . \qquad (7)$$

Thus $A_n \cap \tilde{A}_n \to A \cap A = A$ in ρ-semimetric, and the (uniform) continuity of the operation $\cap$. However $P(A_n \cap \tilde{A}_n) = P(A_n)P(\tilde{A}_n)$, by the independence of X_ns. Since $A_n \to A$ (in ρ) implies $P(A_n) \to P(A)$, we get

$$\begin{aligned} P(A) = P(A \cap A) &= \lim_{n\to\infty} P(A_n \cap \tilde{A}_n) \\ &= \lim_{n\to\infty} P(A_n)P(\tilde{A}_n) = (P(A))^2 \ . \qquad (8) \end{aligned}$$

It follows that $P(A) = 0$ or 1. $\square$

The preceding results indicate several developments in different directions of the subject when a general conditioning concept is at our disposal. But the applications to specific problems involve some explicit calculations of the conditional probabilities and expectations. This is nontrivial. We devote the next chapter in considering some of these and discuss some new problems awaiting satisfactory solutions.

2.6 Bibliographical notes

The general conditioning concept, involving a new and basic idea (more complex than the elementary case but extending the latter) was introduced by Kolmogorov (1933) in his *Foundations*. It was further elaborated, and regular conditional probabilities were discussed by Doob (1953). A somewhat more detailed account of the subject was thereafter included in the book by Loève (1955). All these studies show that conditioning in the general case is not simple, and the occassional counterexamples served only to deepen the mystery of the subject. Among the published works thus far, the uneasiness resulting from the nonuniqueness inherent in conditioning in the general case and some paradoxes were explained in some detail only in the present author's book (1984). The situation still needs a further clarification, and this will be discussed in the next chapter and then some important developments follow in the ensuing work.

The Doob-Dynkin lemma is a result first given in Doob (1953) for the finite dimensional case, and the abstract reformulation of it is included in Dynkin (1961). The nonexistence of regular conditional probability functions were known from the late 1940s, and those results depended on the choice axiom (or at least on the ultrafilter axiom). But Theorem 2.5 follows from a result of Blackwell and Dubins (1963), as discussed in the author's book (1984), and this does not explicitly use the above axiom.

The concept of conditional independence as well as its effective use was apparently first given by Markov (1906) in his seminal study of chain dependence. The symmetric dependence (and exchangeability) is due to de Finetti (1939) and a proof of Theorem 5.7 is to be found in this paper. A detailed argument has also been included in Chow and Teicher (1978). The zero-or-one law for symmetrically dependent random variables is due to Hewitt and Savage (1955) and our alternative argument follows Feller (1966).

Several applications, characterizations of conditional measures as well as expectations, and integration relative to conditional proability functions, will occupy a major portion of this book. Indeed, the last third of the work is devoted precisely to such results and extensions of the subject. These include martingales, Markov processes as well as noncommutative conditioning. The latter plays an important role in quantum mechanics and large parts of abstract analysis.

CHAPTER 3:   COMPUTATIONAL PROBLEMS ASSOCIATED
WITH CONDITIONING

Although the general concept of conditioning of the preceding chapter is abstractly shown to be well-defined and has all the desirable properties, actual computation of conditional probabilities satisfying the Kolmogorov definition is a nontrivial task. In this chapter it is shown that, in cases when the conditioning event has probability zero, there are real problems of well-posedness to use the L'Hôpital rule of evaluation, and the latter method leads to surprising paradoxes. These come from natural applications and are exemplified with works of Borel, and Kac and Slepian. Methods of resolution of these difficulties and the reasons for such ambiguities in the first place as well as the appropriate procedures for correct solutions through the differentiation theory are presented here in considerable detail. For a class of regular conditional probabilities, unambiguous computational methods can be given and these will be presented in Chapter 5 where a thorough discussion of regularity is undertaken.

3.1 Introduction

We have seen in the preceding chapters that the conditional probability $P(\cdot|B)$ of (Ω, Σ, P) is well-defined by the elementary formula (1.2.1) if $P(B) > 0$; and the same is true of $P^B(\cdot)(\omega)$, $\omega \in B$, in the general case by (2.1.2). The latter is a solution of the set of functional equations (2.1.3), but no recipe is given for a calculation of $P^B(\cdot)$. Its existence and uniqueness are derived from the Radon-Nikodým theorem. The standard properties of conditioning, noted in the elementary case have been verified in the general case also. As seen in Section 2.4, a rigorous proof of conditioning directly given for densities, is not a trivial matter. Since the elementary case is not immediately applicable when the conditioned event has probability zero, it is natural to try an argument based on L'Hôpital's rule of (multivariable) calculus to evaluate such a conditional probability. If this is not satisfactory, one should try other, often more sophisticated, methods to evaluate this quantity which in reality is a Radon-Nikodým density of the (abstract) differentiation theory. Since the latter is more involved, one may like to consider the first method. As shown below, this leads to nonunique solutions.

Thus we start with the simple case. It will be seen that there is a fundamental difference between conditioning with a single event (of positive probability), and conditioning with a random variable or equivalently conditioning by a class of events generating a σ-algebra. If the single event has probability zero, it will become clear that there is an inadequacy of formulation to use the first method; but that the general case is always valid. However, an explicit evaluation involves less elementary computations. The latter leads to a separate treatment of the subject whose detailed study is not included in most books. This will be considered here.

To understand and treat the problem better we begin with some interesting examples amplifying the difficulties and then analyze the underlying reasons.

3.2 Some examples with multiple solutions: paradoxes

Although we discussed an aspect of conditional probability in Section 1.4, including a difficulty with the associated expectation, only a simple method was indicated to evaluate probabilities when the conditioned event has probability zero. This prescription is not sufficient to throw light on some deeper troubles. First we illustrate these matters with two classes of examples having multiple solutions.

Let (X, Y) be a pair of random variables on a probability space (Ω, Σ, P) with an absolutely continuous distribution $F_{X,Y}$ whose density is denoted by $f_{X,Y}$. Suppose that it is desired to find the conditional probability of the event $A = [X < x]$, given that $B = [Y = y]$ has occurred. Since $P(B) = 0$, the elementary definition (1.1.1) takes an indeterminate form. If $f_{X|Y}(\cdot|y)$ is the corresponding conditional density of X given $Y = y$, then, as noted before, one may use an abstract form of the L'Hôpital rule and calculate $P(A|B)$ from (1.4.4). In fact, such a calculation was employed to solve a problem raised by E. Borel. An explanation was indicated by A. Kolmogorov. It will now be elaborated. The problem is shown to have nonunique solutions, resulting in "paradoxes".

(a) *The Borel-Kolmogorov type example.* Let X, Y be independent random variables with a common distribution given by

$$P[X < x] = P[Y < x] = \begin{cases} 1 - e^{-x}, & x > 0 \\ 0, & x \le 0, \end{cases} \tag{1}$$

Let $a > 0$ and $Z = (X - a)/Y$. Thus Z is a random variable, and if $\alpha \in \mathbb{R}$, $B = [Z = \alpha]$, then $P(B) = 0$. If $A = [Y < y]$, it is desired to find $P(A|B)$. To use (1.4.4) we need to find the joint density of $f_{Y,Z}$ and its marginal f_Z so that $f_{Y|Z}(\cdot|\alpha)$ can be obtained and then the desired probability calculated. With a familar change of variables technique one can easily get $f_{X,Z}$ from $f_{X,Y}$ by setting $y = y$ and $x = a + zy$ so that the Jacobian is y, and hence

$$f_{Y,Z}(y, z) = \begin{cases} y \exp[-(yz + a) - y], & y > 0 \text{ and } yz > -a \\ 0, & \text{otherwise.} \end{cases} \tag{2}$$

Hence $f_Z(z) = \int_{\mathbb{R}} f_{Y,Z}(y, z)dy$, $f_{Y|Z}(y|z) = f_{Y,Z}(y, z)/f_Z(z)$, so that for any $\alpha \ge 0$, after a simple computation one has for all $a > 0, y > 0$ and $\alpha y > -a$ [and the latter is automatic since $y > 0, \alpha > 0$]:

$$f_{Y|Z}(y|\alpha) = \begin{cases} y(1 + \alpha)^{-2} \exp[-y(1 + \alpha)], & y > 0, \\ 0, & \text{otherwise.} \end{cases} \tag{3}$$

And this gives for $x > 0$,

$$\begin{aligned} P(A|B) &= \int_0^x f_{Y|Z}(y|\alpha)dy \\ &= (1 + \alpha)^4 [1 - e^{-x(1+\alpha)} \{1 + x(1 + \alpha)\}]. \end{aligned} \tag{4}$$

But the desired conditional probability may also be obtained as follows. Since the event $B = [Z = a] = [X - Y\alpha = a]$, we can calculate $P(A|B)$, by letting $U = X - Y\alpha$, and considering the conditional density of X given $U = a$. Thus a similar procedure as in the preceding case yields

$$f_{Y,U}(y, u) = \begin{cases} \exp[-y(1+\alpha) - u], & y > 0, \ \alpha y + u > 0 , \\ 0, & \text{otherwise.} \end{cases} \tag{5}$$

Hence $f_{Y|U}(y|a)$ is obtained for $a > 0$, $\alpha \geq 0$, with a simple computation, [for $a > 0$, $\alpha \geq 0$, in (5), $\alpha y > -a$ which is automatic for $y > 0$]:

$$f_{Y|U}(y|a) = \begin{cases} (1+\alpha) \exp[-y(1+\alpha)], & y > 0 , \\ 0, & \text{otherwise.} \end{cases}$$

Then the desired conditional probability is again calculated as

$$P(A|B) = \int_0^x f_{Y|U}(y|a)dy, x > 0 ,$$
$$= (1 - e^{-x(1+\alpha)}), x > 0 . \tag{6}$$

However, the probabilities given by (4) and (6) are clearly not equal. It follows that $P(A|B)$ has multiple values, resulting in a paradox! While discussing an entirely similar situation Kolmogorov (1933, page 51) briefly states: "the concept of a conditional probability with regard to an isolated given hypothesis [namely B here] whose probability equals zero is inadmissible." Similar examples with an analogous remark can be found at many places in the literature. See for instance DeGroot (1986, page 173). Since such problems appear frequently enough in probability and statistical practice, they cannot be dismissed.

Some readers may feel that the above examples (and similar ones) are somewhat artificial. So we now consider a different but a natural class having uncountably many solutions when we again use the L'Hôpital approximation method that is so prevalent in the literature. The following examples were first noted in a somewhat different context by Kac and Slepian (1959), but they serve the present purpose well.

(b) *The Kac-Slepian examples.* As in the preceding case, the problem is to calculate $P(A|B)$ when $P(B) = 0$, by some approximation using a sequence of events $B_i \downarrow B$ with $P(B_i) > 0$ for each i. This means

$$P(A|B) = \lim_i P(A|B_i) = \lim_i \frac{P(A \cap B_i)}{P(B_i)}, \tag{7}$$

provided the limit exists. It should however be verified that the so defined quantity on the left side is a conditional probability in the sense of Definition 2.1.1. Since this is not obvious, we include the details.

First note that $P(\cdot|B_i)$ is a probability on the σ-algebra Σ for each i. By assumption $\lim_i P(A|B_i)$ exists for each $A \in \Sigma$ and the fixed sequence $\{B_i, i \geq 1\}$. Hence as a consequence of a classical result (the Vitali-Hahn-Saks theorem, the special version we are using was originally due to Nikodým), one deduces that $P(\cdot|B) : \Sigma \to (0,1]$ is a probability measure. So for each bounded random variable X, the mapping $B \mapsto \int_\Omega X(\omega)P(d\omega|B)$ is well-defined. If this is denoted $\tilde{E}_B(X)$, we need to show that it satisfies Kolmogorov's definition in the sense that it is a version of *some* conditional expectation, "for B". Let $\mathcal{B}$ be the smallest σ-algebra containing the class $\{B_i, i \geq 1, B\}$. Then $\tilde{E}_{B_i}(X)$ is defined, and is a constant on each B_i so that it and $\tilde{E}_B(X)$ are $\mathcal{B}$-measurable. Further, on each B_{i_0} (a generator) we get

$$\int_{B_{i_0}} X dP = \int_{B_{i_0}} \tilde{E}_{B_i}(X) dP = \int_{B_{i_0}} \left(\int_\Omega X dP_{B_i} \right) dP$$

$$\to \int_{B_{i_0}} \left(\int_\Omega X dP_B \right) dP, \text{ as } i \to \infty,$$

by the first part of this paragraph,

$$= \int_{B_{i_0}} \tilde{E}_B(X) \, dP . \tag{8}$$

Since B_{i_0} is a generator of $\mathcal{B}$, (8) implies $\tilde{E}_B(X)$ is a version of $E^{\mathcal{B}}(X)$, and hence our assertion is established. Thus we may denote $\tilde{E}_B(X)$ as $E(X|B)$, in the notation of the work in Chapter 1.

Let us now show with a detailed computation, using the approximations of the first method, that $P(\cdot|B)$ with $P(B) = 0$ does not have a unique value, which thus leads to paradoxes. Later we isolate the underlying reasons for this phenomenon.

Suppose that $\{X_t, t \in \mathbb{R}\}$ is a real stationary Gaussian process with mean zero and having a derivative almost everywhere (or even in mean) at each t. This implies, $E(X_t) = 0, r(s,t) = E(X_s X_t) = \rho(s-t)$ and as $h \to 0$ (X_t being a real random variable)

$$\frac{X_{t+h} - X_t}{h} \to Y_t, \text{ a.e. (or in mean of order 2).}$$

Thus $\{Y_t, t \in \mathbb{R}\}$ is the derived process. Here the nonuniqueness of Y_t on null sets causes no difficulty for our illustration below. Assume further that $\rho(t) \to 0$ as $|t| \to \infty$, and ρ is twice continuously differentiable. Such a process can be shown to exist using a form of the Kolmogorov fundamental existence theorem. [See e.g., Sect. 8.3 later, or any book on probability theory. In fact $\rho(t) = e^{-t^2}$ satisfies the above conditions, and it will be adequate for the present discussion.] A consequence of these assumptions is that the above X_t-process is "ergodic". This means that there is a one-to-one measure preserving $\tau : \Omega \to \Omega$ such that $X_t = X_0 \circ \tau^t$, and each member of the class $\{A \in \Sigma : P(\tau(A)) = P(\tau^{-1}(A))\}$ has probability either zero or one.

From our assumptions it follows immediately that $\{Y_t, X_t;\ t \in \mathbb{R}\}$ is a vector normal (or Gaussian) process and $E(X_t Y_t) = 0$ so that the X_t and Y_t are also independent, $t \in \mathbb{R}$. [The latter fact can be established for any continuous Gaussian ergodic process X_t with an a.e. (or in mean) derived process Y_t, but for simplicity we keep the above assumptions in force.] Our problem now is this: find the conditional probability density of the "slope" Y_0 given that $X_0 = a$. In other words, if the process is known to pass through the point 'a' at time 0, then find the conditional density of its slope Y_0 at that instant. Since $P[X_0 = a] = 0$ we are in a situation similar to that of subsection (a). It will now be shown that there are infinitely many solutions to this problem by using the method of approximations.

Since Y_0 is also a nondegenerate normal random variable with mean zero, let its variance be $\alpha^2 > 0$. We now find a class of approximations for the event $B = [X_0 = a]$. Since X_0 is normal with mean zero and variance $\sigma^2 > 0$, it is clear that $P(B) = 0$. Let $\delta > 0$ and m be real numbers. Consider

$$A_\delta^m = \{\omega : X_t(\omega) \text{ passes through the line } y = a + mt \text{ of length}$$
$$\delta \text{ for some } t \}$$
$$= \{\omega : X_t(\omega) = a + mt \text{ for some } 0 \le t \le \delta(1 + m^2)^{-\frac{1}{2}} \} .$$

Clearly $A_\delta^m \downarrow B$ as $\delta \downarrow 0$ for each m. We now calculate, with (7), the right side of the following:

$$P[Y_0 < y|B] = \lim_{\delta \downarrow 0} P[T_0 < y|A_\delta^m], \ |m| < \infty . \tag{9}$$

The ensuing work shows that the limit exists.

Let $p_{Y_0}(\cdot)$ and $f_{X_0}(\cdot)$ be the probability density functions of Y_0 and X_0 so that $p_{Y_0}(y) = (2\pi\alpha^2)^{-\frac{1}{2}} \exp(-y^2/2\alpha^2)$, and $f_{X_0}(x) = (2\pi\sigma^2)^{-\frac{1}{2}} \exp(-x^2/2\sigma^2)$. By independence of X_0 and Y_0, $f_{X_0,Y_0}(u,v) = f_{X_0}(u) \cdot f_{Y_0}(v)$. However, A_δ^m and Y_0 are not necessarily independent for $\delta > 0$ and $|m| < \infty$. Now to evaluate the right side of (9), consider

$$\lim_{\delta \downarrow 0} P[Y_0 < y|A_\delta^m]$$

$$= \lim_{\delta \downarrow 0} \frac{P[Y_0 < y, A_\delta^m, Y_0 > m] + P[Y_0 < y, A_\delta^m, Y_0 \le m]}{P[A_\delta^m, Y_0 > m] + P[A_\delta^m, Y_0 \le m]}$$

$$= \lim_{\delta \downarrow 0} \frac{\int_{-\infty}^y \left(\int_{A_\delta^m \cap [Y_0 > m]} P[X_t \in du, Y_0 \in dv] \right)}{\int_{A_\delta^m \cap [Y_0 > m]} P[X_t \in du, \ Y_0 \in dv] + \int_{A_\delta^m \cap [Y_0 \le m]} P(X_t \in du, \ Y_0 \in dv)}$$

$$+ \lim_{\delta \downarrow 0} \frac{\int_{-\infty}^y \int_{A_\delta^m \cap [Y_0 \le m]} P[X_t \in du, Y_0 \in dv]}{\int_{A_\delta^M \cap [Y_0 > m]} P[X_t \in du, \ Y_0 \in dv] + \int_{A_\delta^m \cap [Y_0 \le m]} P[X_t \in du, Y_0 \in dv]}$$

$$= I_1 + I_2 \text{ (say)}.$$
$$\tag{10}$$

We simplify the two limits in (10) separately. Since $|\frac{X_t - (a - mt)}{t} - Y_0| \to 0$, a.e., as $\delta \downarrow 0$, we have

$$I_1 = \lim_{\delta \downarrow 0}$$

$$\frac{\int_{-\infty}^y \frac{1}{\delta} \int_{a-(v-m)\delta}^a f_{X_0,Y_0}(u,v)dudv}{\frac{1}{\delta} \int_m^\infty \int_{a-(v-m)\delta}^a f_{X_0,Y_0}(u,v)dudv + \frac{1}{\delta} \int_{-\infty}^m \int_a^{a-(v-m)\delta} f_{X_0,Y_0}(u,v)dudv}.$$
$$\tag{11}$$

Note that we used the fact when $Y_0 > m$, the relevant approximation holds only when X_0 satisfies the inequality $a - (v - m)\delta \le X_0 \le a$. Next we see that the limits exist for the numerator and denominator of (11) separately, and then they can be evaluated individually. Thus using the independence of X_0 and Y_0, the numerator of (11) becomes

$$\lim_{\delta\downarrow 0}\int_{-\infty}^{y}f_{Y_0}(v)(\frac{1}{\delta}\int_{a-(v-m)\delta}^{a}f_{X_0}(u)du)dv$$

$$=\int_{-\infty}^{y}f_{Y_0}(v)(v-m)(\lim_{\delta\downarrow 0}\frac{1}{(v-m)\delta}\int_{a-(v-m)\delta}^{a}f_{X_0}(u)du)dv$$

$$=\int_{-\infty}^{y}f_{y_0}(v)(v-m)(-f_{X_0}(a))dv,\ \text{by the classical Lebesgue theorem}$$

on differentiation of integrals,

$$=-f_{X_0}(a)\int_{-\infty}^{y}(v-m)f_{Y_0}(v)dv. \tag{12}$$

Similarly, with two such evaluations, the denominator of (11) gives

$$\lim_{\delta\downarrow 0}(\text{denominator of (11)})$$

$$=\int_{m}^{\infty}(v-m)f_{Y_0}(v)(-f_{X_0}(a))dv+\int_{-\infty}^{m}(v-m)f_{Y_0}(v)f(a)dv$$

$$=-f_{X_0}(a)[\int_{m}^{\infty}(v-m)f_{Y_0}(v)dv-\int_{-m}^{m}(v-m)f_{Y_0}(v)dv+$$

$$\int_{m}^{\infty}(v+m)f_{Y_0}(v)dv]\ ,\ \text{since}\ f_{y_0}(\cdot)\ \text{is symmetric about 0}\ ,$$

$$=-f(a)[2\int_{m}^{\infty}vf_{Y_0}(v)dv+m\int_{-m}^{m}f_{Y_0}(v)dv]$$

$$=-f(a)[\sqrt{\frac{2\alpha^2}{\pi}}e^{-m^2/2\alpha^2}+m\int_{-m}^{m}f_{Y_0}(v)dv]\ . \tag{13}$$

Hence using (12) and (13) in (11), we get

$$I_1=\frac{\int_{-\infty}^{y}(v-m)e^{-v^2/2\alpha^2}\ dv}{2\alpha^2 e^{-m^2/2\alpha^2}+m\int_{-m}^{m}e^{-v^2/2\alpha^2}\ dv}\ . \tag{14}$$

With a similar calculation one finds

$$I_2=\frac{\int_{-\infty}^{y}(m-v)e^{-v^2/2\alpha^2}\ dv}{2\alpha^2 e^{-m^2/2\alpha^2}+m\int_{-m}^{m}e^{-v^2/2\alpha^2}\ dv}\ . \tag{15}$$

Substitution of (14) and (15) in (10) yields the desired limit as

$$\lim_{\delta\downarrow 0}P[Y_0<y|A_{\delta}^m]=\int_{-\infty}^{y}\frac{|v-m|e^{-v^2/2\alpha^2}\ dv}{2\alpha^2 e^{-m^2/2\alpha^2}+\int_{-m}^{m}e^{-v^2/2\alpha^2}\ dv}\ . \tag{16}$$

From (10) and (16) we finally get the evaluation of (9) as:

$$F_{Y_0|X_0}(y|B) = P[Y_0 < y|B]$$

$$= \int_{-\infty}^{y} f_{Y_0|X_0}^{m}(v|a)dv \; , \; \text{(say)} \tag{9'}$$

where $f_{Y_0}^{m}(\cdot|\cdot)$ is the integrand of the right side of the main integral in (16), which thus depends on m. Since there are uncountably many values of m in $|m| < \infty$, we get that many distinct solutions to the left side conditional probability. Consequently there is no single answer to our problem, and it results in a *paradox in a very striking manner!*

Some special values of m may be recognized as commonly used solutions without a justification. They are as follows:

(i) Letting $m \to \infty$ in (9') so that we take the limits on the *vertical line*, called a "vertical window or v.w." we get

$$P[Y_0, y|B]_{v.w.} = \int_{-\infty}^{y} e^{-v^2/2\alpha^2} (2\pi\alpha^2)^{-\frac{1}{2}} dv \; . \tag{17}$$

This corresponds to the fact that Y_0 and X_0 are independent. But this *explanation ignores part of the information that Y_0 is obtained as a limit* (a.e. or in mean) of the quotients $[\frac{1}{t}(X_t - X_0)]$ as $t \downarrow 0$. This is traditionally used as though it is the only solution of the problem when in fact it is not, as seen here!

(ii) Letting $m \to 0$ in (9') so that the limit is taken *horizontally*, called a "horizontal window or h.w.", we get

$$P[Y_0 < y|B]_{h.w.} = \int_{-\infty}^{h} |v|e^{-v^2/2\alpha^2} (2\alpha^2)^{-1} dv \; . \tag{18}$$

This solution, which is clearly different from the preceding one, seems to have some special relation with the "mean recurrent time" problem studied in statistical mechanics.

It may be noted that these uncountably many distinct values for $P[Y < y|B]$, obtained above by the "L'Hôpital type approximations," do not exhaust the possible solutions of our problems. For instance, the following is another result which is different from all the others above.

Suppose that B_δ is the set of sample paths of the process that reach the point $(0, a)$ in the plane through a circle of radius $\delta > 0$. Thus

$$B_\delta = \{\omega : (X_t(\omega) - a)^2 + t^2 < \delta^2, \text{ for some } 0 < t < \delta\} \; . \tag{19}$$

Again one has $B \subset B_\delta$ and $\lim_{\delta \downarrow 0} B_\delta = B$. Since $\frac{X_t - a}{t} \to Y_0$ a.e. as $t \to 0$, one has for this "circular window or c.w." method

$$P[Y_0 < y|B]_{c.w.} = \lim_{\delta \downarrow 0} \frac{P[Y_0 < y, B_\delta]}{P(B_\delta)}$$

$$= \lim_{\delta \downarrow 0} \frac{\int_{-\infty}^{y} \int_{a-\delta(1+v^2)^{\frac{1}{2}}}^{a+\delta(1+v^2)^{\frac{1}{2}}} P(X_t \in du, \, Y_0 \in dv)}{\int_{-\infty}^{\infty} \int_{(a-\delta(1+v^2)^{\frac{1}{2}})}^{a+\delta(1+v^2)^{\frac{1}{2}}} P(X_t \in du, Y_0 \in dv)}$$

$$= \lim_{\delta \downarrow 0} \frac{\int_{-\infty}^{y} \frac{1}{\delta} \int_{a-\delta(1+v^2)^{\frac{1}{2}}}^{a+\delta(1+v^2)^{\frac{1}{2}}} f_{Y_0}(v)dv \cdot f_{X_0}(u)du}{\int_{-\infty}^{\infty} \frac{1}{\delta} \int_{a-\delta(1+v^2)^{\frac{1}{2}}}^{a+\delta(1+v^2)^{\frac{1}{2}}} f_{Y_0}(v)dv \cdot f_{X_0}(u)du}$$

$$= \frac{\int_{-\infty}^{y} f_{Y_0}(v)(1+v^2)^{\frac{1}{2}} 2 f_{X_0}(a)}{\int_{-\infty}^{\infty} f_{Y_0}(v) 2 f_{X_0}(a)dv}$$

$$= \frac{\int_{-\infty}^{y} (1+v^2)^{\frac{1}{2}} e^{-v^2/2\alpha^2} \, dv}{\int_{-\infty}^{\infty} (1+v^2)^{\frac{1}{2}} e^{-v^2/2\alpha^2} \, dv} .$$

This is different from all the values given by (16). It is thus clear that using other geometrical figures one can obtain still distinct values for the probability sought for, and yet there is no particular technical claim here for one to be more representative than others. [See Sec. 4 below, and, regarding the correct approach to such problems, see Secs. 5.6, 5.7 and 7.5 later.]

To understand the distinction further we now include an outline of the corresponding solutions to some useful questions on the same type of Gaussian processes that are considered above. These relate to "large deviations" and we again follow Kac and Slepian. The problems under discussion have practical importance and yet one can only obtain different (or nonunique) answers by using different methods of calculations.

As before let $\{X_t, t \in \mathbb{R}\}$ be an ergodic, pathwise differentiable real stationary Gaussian process with mean zero and covariance function ρ. It will be assumed, for convenience, that $\rho(0) = 1$, so that ρ becomes a correlation function of the process. Further suppose that ρ admits the following Taylor expansion:

$$\rho(t) = \int_{-\infty}^{\infty} e^{it\lambda} \, dF(\lambda) = \int_{-\infty}^{\infty} \cos t\lambda \, dF(\lambda)$$

$$= 1 - \frac{\alpha_0 t^2}{2} + o(t), \, a_0 > 0 . \tag{20}$$

Here $F(\cdot)$ is called the spectral distribution of the process and α_0 is its second moment. Also $\{X_t, t \in \mathbb{R}\}$ is said to have a *level crossing* of height 'a' at t_0 if in each neighborhood of t_0 there are points t_1, t_2 such that $(X_{t_1} - a)(X_{t_2} - a) < 0$ a.e., and an *upcrossing* if $X_t \leq a$ in $(t_0 - \varepsilon, t_0)$ and $X_t \geq a$ in $(t_0, t_0 + \varepsilon)$ a.e. for some $\varepsilon > 0$. By reversing the inequalities in the latter statement, one gets the *down-crossing* definition. If c_a, u_a (and d_a) are the numbers of level and up (down) crossings, respectively, then $c_a = u_a + d_a$ if $c_a < \infty$ a.e., and when $\alpha_0 < \infty$ in (20),

$$E(u_a) = E(d_a) = \frac{1}{2}E(c_a) ,$$

and under the hypothesis (20),

$$E(u_a) = \frac{1}{2\pi}\sqrt{\alpha_0}e^{-a^2/2} . \tag{21}$$

A proof of this result is not simple, and a somewhat simplified argument together with references to earlier contributors may be found in Cramér and Leadbetter ((1967), Sec. 10.3). But one also has

$$P[X_t \geq a] = 1 - P[X_t < a]$$
$$= \frac{1}{2}[1 - \sqrt{\frac{2}{\pi}}\int_0^a e^{-u^2/2}\,du] . \tag{22}$$

Then the expected number of times, per unit interval, X_t crosses the level $X_t = a$, is the ratio of (22) to (21), by definition. If this ratio is $\theta(= \theta(a))$, then

$$\theta = \frac{\pi e^{a^2/2}}{\sqrt{\alpha_0}}[1 - \sqrt{\frac{2}{\pi}}\int_0^a e^{-u^2/2}\,du]$$
$$= \sqrt{\frac{2}{\alpha_0}}e^{a^2/2}\int_a^\infty e^{-u^2/2}\,du \sim \sqrt{\frac{2\pi}{\alpha_0}} \cdot \frac{1}{a} , \tag{23}$$

for large enough a. Using this result, which is given for an over all picture of the problem, we present distinct solutions for the probabilities of large positive excursions. This will show the difficulties propagated by the conditioning calculations of (16).

Thus the problem is to obtain the behavior of the probabilities of the process:

$$\Delta(t, \theta) = (X_{\theta t} - a)/\theta, \quad \text{as} \quad a \to \infty$$

given that $\Delta'(0,\theta) = \frac{\partial \Delta(t,\theta)}{\partial t}(0) \geq 0$ and $\Delta(0,\theta) = 0$ a.e. Since the conditioning event has probability zero, and since we need to find the conditional distribution of the above quantity, the difficulties discussed above reappear here. We give two solutions of the problem, again following Kac and Slepian, using two of the many "windows" (namely the v.w. and h.w.) included in (16).

Suppose we use the h.w. method. Then for each finite set $t_1, \ldots, t_n$ of time points, we have

$$\lim_{a \to \infty} P[\Delta(t_1, \theta) < y_i, i = 1, \ldots, n | \Delta'(0, \theta) \geq 0, \Delta(0, \theta) = 0]_{\text{h.w.}}$$

$$= P(\sqrt{\alpha_0} t_i Y - \sqrt{\frac{\alpha_0 \pi}{2}} t_i^2 < y_i, i = 1, \ldots, n), \quad (24)$$

where $Y \geq 0$ is a random variable with the distribution

$$P[Y < y] = \sqrt{\frac{2}{\pi}} \int_0^y e^{-u^2/2} \, du, \ y \geq 0 . \quad (25)$$

Thus the limit in (24) is a singular n-dimensional *half-normal distribution*.

If we use the v.w. method instead, the general limit in (24) exists, has a similar form but now $Y \geq 0$ is a random variable with the so-called *Rayleigh distribution*:

$$P[Y < y] = \int_0^y u \, e^{-u^2/2} \, du, \ y \geq 0 . \quad (26)$$

Since the numerical conditional probabilities in these two methods (with (25) and (26)) are different, the answers to our problem are different being distinct *even* asymptotically. It is clear that several other answers will result if we use other values of m in (16), or the c.w. method.

For the same process, if we calculate the asymptotic distribution of the first return time, given that $\Delta(0, \theta) = 0$ and $\Delta'(0, \theta) \geq 0$ a.e. we get the following values. We again use the h.w. and v.w. methods for this illustration. Thus, assuming that the correlation function is four times continuously differentiable, let

$$P_a(T) = P[\Delta(t, \theta) \geq 0 \text{ for } 0 \leq t \leq T | \Delta'(0, \theta) \geq 0, \ \Delta(0, \theta) = 0] .$$

Then (i), with the h.w. method, one finds

$$\lim_{a \to \infty} P_a(T) = \begin{cases} e^{-\pi T^2/4}, & T \geq 0 \\ 0, & T < 0, \end{cases} \tag{27}$$

and (ii), with the v.w. method, one gets

$$\lim_{a \to \infty} P_a(T) = \begin{cases} 1 - \int_0^T e^{-(\pi x^2)/4} \ dx, & T \geq 0 \\ 0, & T < 0. \end{cases} \tag{28}$$

The details of these calculations are not essential for our purposes. They can be found in Kac and Slepian (1959), and will not be detailed here. The point of these results is that, in general, both the finite as well as the asympotic values of the conditional probabilities are critically dependent on the method of calculations, and thus their significance and utility are diminished. Several related problems are discussed in Cramér and Leadbetter (1967), where some interesting "natural reasons" for using h.w. method in preference to others are advanced. But the *essential and key point to remember here is the nonuniqueness of the solutions in these cases* and no one answer is inherently superior to the others. Thus the method of calculation, which usually is not reported in such results, should also be explicitly given for completeness and accuracy of the conclusions drawn from them.

In the next section we discuss the reasons for these paradoxes and show how they may be resolved in the general Kolmogorov model for conditioning. In a later chapter we present some characterizations of conditional expectations and of probabilities which will illuminate their structure and exhibit their functional analytic character.

3.3 Explanation of paradoxes

The preceding discussion does not show how one of the many solutions may be selected over the others, since there are no technical reasons for such a procedure. To see the problem more clearly we need to turn to the abstract theory of Chapter 2 which asserts the essential uniqueness of the conditional expectation and of probability in each situation, without additional hypotheses.

In order to analyze the difficulty, consider the paradoxes of the subsection (b) of the preceding section. For each m, the approximating sets $\{A_\delta^m, \delta > 0\}$ are events with the property that $\cap_{\delta>0} A_\delta^m = B$. Let B^m denote the σ-algebra generated by such collections of events. Then $B \in \cap_{m\geq0} B^m$ where $\delta > 0$ and $m \geq 0$, may be restricted to go through a dense set of reals in these computations. As shown in the next section, $B^m = \sigma(A_\delta^m, \delta > 0)$ has a "differentiation basis" (to be defined precisely there) relative to which the probability measure (here a Gaussian one) can be differentiated so that $(Y_0 = X'(0))$, $a \in \mathbb{R}$,

$$P^{B^m}(Y_0 < y)(a) = F_{Y_0}(y|B)$$

$$= \int_{-\infty}^y f_{Y_0|X_0}^m(v|a)dv \ . \tag{1}$$

Since for each m, the generated σ-algebra B^m is different, the corresponding conditional probability measure $P^{B^m}(\cdot)$ which is automatically regular (see Chap. 5 later), is naturally different as assured by the general theory. Thus the only requirement that $[X_0 = a] = B \in \cap_{m\geq0} B^m$, can not determine a unique conditional distribution. Indeed using the c.w. method with $\{B_\delta, \delta > 0\}$ there and other geometrical figures, one can obtain many other different (from the preceding ones) conditional probability measures since the generated σ-algebra of these new families are different from B^m, although each contains B. Thus we must conclude that the calculation of the conditional distribution of Y_0, through L'Hôpital's type approximation, for $B = [X_0 = a]$, is not *well-posed*. This statement applies also to the calculation of conditional probabilities for the problem in subsection (a) of the last section. The unicity of solution is thus obtainable on an auxiliary specification of the limit procedure. For instance, the "horizontal window" condition has apparently a natural place in statistical physics, according to Smoluchowski (as reported by Kac and Slepian). Later some conditioning procedures will be discussed axiomatically in the next chapter, but further light will be shed in the following section.

Let us now consider how well-posedness of the problem and the resulting unicity will result in a given situation. This implies that first the conditional probability be defined uniquely and then its value corresponding to an event such as $B = [X_0 = a]$ be determined. In the general theory, $P[X_0' < y|B]$ is by definition:

$$P[X_0' < y|X_0 = a] = P[X_0' < y|X_0](a)$$
$$= P^{\mathcal{B}(X_0)} (X_0' < y)(a)$$
$$= \varphi(X_0 = a) , \qquad (2)$$

for a Borel function $\varphi : \mathbb{R} \to \mathbb{R}$, by the Doob-Dynkin lemma. It follows that the correct value of the left side conditional probability is that given by the last line of (2). The general Kolmogorov theory assures that this value is unique outside a P-null set. The paradoxes of the last section arise when the left side is given an extended meaning based on the elementary case and a subsequent use of the L'Hôpital approximation. Such an evaluation introduces *ad hoc* limit procedures and the end result has little, if any, relation with the (correct) right side value of (2). Note that the σ-algebras $\mathcal{B}^m$ (and others like that) are all *different* from (and generally sub algebras of) $\mathcal{B}(X_0)$, the σ-algebra generated by X_0.

The different methods and examples of the preceding section show that if the conditioning random variable is two or higher dimensional and the given probability distribution is not discrete, then the conditional probabilities obtained through the above (L'Hôpital type) approximations always lead to nonunique (hence ill-posed) situations. Note, however, that Proposition 2.2.1 (ii) implies that for *any* σ-algebra $\mathcal{B}$ and *integrable* random variable X, we must have $E(E^{\mathcal{B}}(X)) = E(X)$. Thus while the unconditional values, using any of the *ad hoc* methods described above, are unaltered, one can recognize that the conditional values may become nonunique and therefore unreliable when computed with methods which are simple extensions of the elementary case.

The underlying problem discussed here is generally ignored or slurred over in most books on the subject. It was only explicitly noted (as far as I can determine) by Neuts (1973, page 220) as "technically a very deep one," who then gave a heuristic approach to the problem at hand, based again on a use of L'Hôpital's method described above, to evaluate the left side of (2). Thus a rigorous analysis included here with the "anatomy of the paradoxes" and their resolution appears to be necessary for a proper understanding of the subject and of the structure of conditioning itself.

3.4 Some methods of computation

The preceding section shows that a correct function giving the conditional expectation or probability is the one provided by the general Definition 2.1.1. Consequently, for any conditioning σ-algebra, it is necessary to find methods of computing $P^{\mathcal{B}}(A)$ or $E^{\mathcal{B}}(X)$ which are equivalent to evaluating the Radon-Nikodým derivative. Thus in most cases this is technical and additional work is needed. We present some "good" conditions to accomplish this as it takes the center stage now. But this is known to be a key problem in the theory of differentiation of set functions. There does not seem to be any escape from it, if multiple solutions or paradoxes or ill-posedness be avoided. We present this in a graded manner with enough commentary to ease some dreary technical detail. Special methods will be given in Chapters 5 and 7 for some useful computational problems. The rest of this section may be skipped on a first reading.

The approximations used for the examples of Section 2 clearly show that one is considering converging sequences or nets of events in evaluating conditional probabilities. For a general procedure, these should not be *ad hoc*, but should form part of a differentiation basis in the following precise sense.

Definition 1. (a) Let (Ω, Σ, μ) be a measure space and $A \subset \Omega$ be a set. For each $\omega \in A$, let $\{B_i^{\omega}, i \in I\} \subset \Sigma$ be a (not necessarily countable) family with a directed index set I, such that $B_i^{\omega} \to \omega$ in the Moore-Smith sense (i.e., the family is a *converging net*) and that for each cofinal sequence J of I, $B_j^{\omega} \to \omega$ also. [Here cofinal means, for each $i \in I$, there is a $j \in J$ such that $j > i$.] Let $\mathcal{D}$ be the collection of all $\{B_i^{\omega} : 0 < \mu(B_i^{\omega}) < \infty, i \in I, \omega \in A\}$. Then $\mathcal{D}$ is called a *differentiation* or a *derivation basis* on A.

(b) A converging net $\{B_i^{\omega}, i \in I\}$ is called *contracting* to ω, if there is an $i_0 \in I$ such that $\omega \in B_i^{\omega}$ for all $i > i_0$, where '$>$' is the direction of I.

The pointwise differentiation of a set function, such as $A \mapsto E(f\chi_A)$, relative to a measure (such as the Lebesgue measure in $\mathbb{R}^n$) is defined in the usual manner. Thus if $\nu : \Sigma \to \mathbb{R}$ is a signed measure and

$\{B_i^\omega, i \in I\}$ an ω-converging sequence for $\omega \in A \subset \Omega$, then consider

$$(D^*\nu)(\omega) = \sup\{\limsup_i \frac{\nu(B_i^\omega)}{\mu(B_i^\omega)} : \text{ all nets } B_i^\omega \to \omega\}, \tag{1}$$

and similarly

$$(D_*\nu)(\omega) = \inf\{\liminf_i \frac{\nu(B_i^\omega)}{\mu(B_i^\omega)} : \text{ all nets } B_i^\omega \to \omega\}. \tag{2}$$

The values $(D^*\nu)(\omega)$ and $(D_*\nu)(\omega)$ are often called the *upper* and *lower* *derivates* of ν, relative to μ, at $\omega \in A$. If $(D^*\nu)(\omega) = (D_*\nu)(\omega)$, denoting the common value by $(D\nu)(\omega)$, we say that ν is differentiable relative to μ at ω of A. In case all the B_i^ω-sequences are part of a universal sequence (e.g., if there is a monotone sequence) then the "sup" and "inf" in (1) and (2) are redundant and can be omitted. In any case, each such sequence B_i^ω of (1) and (2) is also termed a *deriving sequence*.

The preceding concepts will be elaborated so that they may be easier to use in applications.

Definition 2. (a) If $\mathcal{D}$ is a differentiation basis of a set $A \subset \Omega$, then a family $\mathcal{F} \subset \mathcal{D}$ is a *fine covering* of A, if for each ω in A, there is some $\{B_i^\omega, i \in J\} \subset \mathcal{F}$ converging to ω.

(b) If $\mathcal{D}$ is a differentiation basis in (Ω, Σ, μ) for $A \subset \Omega$ of finite outer measure, then $\mathcal{D}$ is said to have the *Vitali property* relative to μ, provided for any fine covering $\mathcal{F} \subset \mathcal{D}$ of A and $\varepsilon > 0$ and a measurable cover $\tilde{A}$ of A, there is at most a countable collection $\mathcal{C} \subset \mathcal{F}$ such that the following two conditions hold:

(i) $\mu(\tilde{A} \Delta V) = 0$ where $V = \cup\{B : B \in \mathcal{C}\}$, Δ=symmetric difference,

(ii) the μ-measure of the overlap of members of $\mathcal{C}$ is $< \varepsilon$. More explicitly, if $\varphi_{\mathcal{C}}(\omega)$ is the number of sets of $\mathcal{C}$ to which ω belongs, then for V of (i),

$$\int_V (\varphi_{\mathcal{C}}(\omega) - 1) \, d\mu\omega < \varepsilon. \tag{3}$$

If the members of $\mathcal{C}$ are disjoint (or their pairwise intersection is μ-null so that (3) is true for any $\varepsilon > 0$) the corresponding $\mathcal{D}$ is said to have the *strong Vitali property* (modulo μ-null sets).

To have an alternative view of the above property we state the following result due to R. de Possel.

Theorem 3. *Let (Ω, Σ, μ) be a measure space and $\mathcal{D} \subset \Sigma$ be a deriva-tion basis. Then $\mathcal{D}$ has a Vitali property iff for any A in Σ and $\omega \in A$, there is an ω-converging net $\{B_i^\omega, i \in I\} \subset \mathcal{D}$ such that*

$$\lim_i [\mu(B_i^\omega \cap A)/\mu(B_i^\omega)] = \chi_A(\omega) , \quad \text{a.a. } (\omega) . \tag{4}$$

Thus an indefinite integral of each measurable set A can be differ-entiated and the derivative thus obtained is χ_A in the presence of a Vitali property. The result may seem intuitively clear but its proof is not very simple. We can find the significance of the above assertion from the following fact.

Theorem 4. *Let (Ω, Σ) be a measurable space, so that Σ is a σ-algebra of Ω. Let μ, ν be two finite measures on Σ. Let ν be μ-continuous, (denoted $\nu \ll \mu$). If $\mathcal{D} \subset \Sigma$ is a derivation basis on Ω, having the Vitali property relative to both ν and μ, then we have*

$$(D\nu)(\omega) = \frac{d\nu}{d\mu}(\omega) , \quad \text{a.e. } (\mu) . \tag{5}$$

The importance of this result is that we may apply for such derivation bases $\mathcal{D}$ of ν and μ, the ratio $\frac{\nu(A)}{\mu(A)}$ and recover the Radon-Nikodým density $f = \frac{d\nu}{d\mu}$, where $\nu(A) = \int_A f d\mu$, without any ambiguity, in the limit. The importance of this assertion becomes clear when it is realized that an arbitrary differentiation procedure applied to an indefinite integral such as $\nu : A \mapsto \int_A f d\mu$, one can get $(D^*\nu)(\omega) = +\infty$ a.e. and *not* $f(\omega)$. If $\Omega = \mathbb{R}$, and $\Sigma =$ Borel σ-algebra, then some of the problems of the above construction (of the Vitali system) are more familiar, as shown below, although the result even in this case (the classical Lebesgue differentiation of integrals) is nontrivial. The difficulties quickly multiply in higher dimensions.

To illuminate our discussion, we present a few results in this di-rection, refering the reader to the survey of Hayes and Pauc (1970) for details. Further work on the latter subject will digress us too far, and also is not crucial for our later theory except to understand its nontrivial structure.

Suppose that ν is a signed measure on (Ω, Σ) and $\mu : \Sigma \to \mathbb{R}^+$ is a finite measure (a σ-finite extension is not difficult). If $\nu \ll \mu$, then from the classical Radon-Nikodým theory we have

$$\nu(A) = \int_A f d\mu, \quad \text{or} \quad f = \frac{d\nu}{d\mu} , \quad \text{a.e. } (\mu) .$$

Let $f_n = f_{\chi_{[|f| \leq n]}} + n\, \chi_{[f>n]} - n\, \chi_{[f<-n]}$, and $\nu_n : A \mapsto \int_A f_n \, d\mu$. Then $|\nu_n(A)| \leq |\nu|(A)$ for all n, and $\frac{d\nu_n}{d\mu} = f_n$ a.e. (μ), where $|\nu|(\cdot)$ is the variation measure of ν. We then have:

Theorem 5. *Let (Ω, Σ, μ) be a finite measure space, ν a signed measure on Σ, and $\mathcal{D} \subset \Sigma$ a derivation basis for μ. If $\mathcal{D}$ has the Vitali property relative to $|\nu|(\cdot)$ on Ω in addition to μ, then $D\nu_n$ exists for each n for $\mathcal{D}$ and $D\nu_n = f_n$ a.e. In fact, $D\nu = f$ a.e. if $\mathcal{D}$ has the Vitali property relative to $|\nu|$ on Ω in addition to the same property for μ on Ω.*

Many other results and refinements with additional structure for (Ω, Σ) or a topology for Ω are known. We now present an extension of Lebesgue's theorem in $\mathbb{R}^n$ for a comparison in which Vitali property is not explicitly invoked. This result is due to B. Jessen, J. Marcinkiewicz and A. Zygmund (1935).

Let $\mathcal{I}(\mathcal{J})$ be the collection of all closed (open) nondegenerate intervals in $\mathbb{R}^n$ so that $I \in \mathcal{I}(J \in \mathcal{J})$ iff $I = \underset{i=1}{\overset{n}{\times}} [a_i, b_i](J = \underset{i=1}{\overset{n}{\times}} (a_i, b_i))$ with $a_i < b_i, i = 1, \ldots, n$. If $\delta(A) = \sup\{||x - y|| : x, y \text{ in } A\}$ is the diameter of $A \subset \mathbb{R}^n$, $||x||$ being the Euclidean length of the vector $x \in \mathbb{R}^n$, and if there is a sequence $B_i^x \in \mathcal{I}(\mathcal{J})$ such that $x \in B_i^x$ and $\underset{i}{\lim} \delta(B_i^x) = 0$ for each $x \in \mathbb{R}^n$, then $(\mathcal{I}, \delta)$ as well as $(\mathcal{J}, \delta)$ form differentiation bases in $\mathbb{R}^n$, called *D-bases* (or Denjoy bases). We then have the desired result by the above authors as follows. (For an extension, see Sec. 7.5.)

Theorem 6. *Let $\mathcal{B}$ be the Borel σ-algebra of $\mathbb{R}^n$, $\mu : \mathcal{B} \to \bar{\mathbb{R}}^+$ be the Lebesgue measure and $(\mathcal{I}, \delta)$ a D-basis. Let $f : \mathbb{R}^n \to \mathbb{R}$ be a measurable function such that f and $|f|(\log^+ |f|)^{n-1}$ are Lebesgue integrable. If $\nu_f : A \mapsto \int_A f d\mu$, $A \in \mathcal{B}$, then $D\nu_f = f$ a.e. where the differentiation is carried out relative to the basis $(\mathcal{I}, \delta)$.*

The importance of the integrability condition in $\mathbb{R}^n$, $n \geq 2$, stems from the fact that there are μ-integrable functions for which the corresponding ν_f are not derivable relative to either of the D-bases $(\mathcal{I}, \delta)$ or $(\mathcal{J}, \delta)$. Moreover, as shown by an example due to S. Saks, the integrability condition in the above result is the best possible. In fact, Saks actually showed that on $L^1(I^2, \mathcal{B}, \mu)$, where I^2 is the closed unit square in $\mathbb{R}^2$, the following set

$$\{f \in L^1(I^2, \mathcal{B}, \mu) : (D^* \nu_f)(x) = +\infty \text{ at each } x \in I^2, \ \nu_f(A)$$
$$= \int_A f d\mu, \ A \in \mathcal{B}\}$$

is full, i.e., is of the second category in the Banach space $L^1(I^2, \mathcal{B}, \mu)$.

Since in most of the above results the Vitali property of a derivation procedure appears, it is natural to ask for a characterization of measure spaces admitting such Vitali systems. To present the best condition we recall that a measure space (Ω, Σ, μ) is called *Carathéodory regular* if it is complete and has no further extensions by the Carathéodory process, i.e., if we define for each $A \subset \Omega$

$$\mu^*(A) = \inf\{\Sigma_{i=1}^{\infty} \mu(B_i) : \cup_{i=1}^{\infty} B_i \supset A, \ B_i \in \Sigma\} \tag{6}$$

then the class of all μ^*-measurable sets is Σ itself. For instance Lebesgue measure is Carathéodory regular, and any measure space can be extended by using (6) for a Carathéodory regular space with atmost one such extension. The measure μ of such a space (i.e. (Ω, Σ, μ) is Carathéodory regular) is *decomposable* (also called *strictly localizable*) iff there exists a collection $\{A_i, i \in I\} \subset \Sigma$ such that (a) $0 < \mu(A_i) < \infty$, $i \in I$, with $N = \Omega - \cup_{i \in I} A_i$ satisfying $\mu(N) = 0$, and (b) $B \in \Sigma, 0 < \mu(B) < \infty$ implies $\{i \in I : \mu\{A_i \cap B) > 0\}$ is atmost countable. Thus every σ-finite space is decomposable but there are non σ-finite decomposable spaces. (A Haar measure space on a locally compact group is an example of the latter. On these matters one may consult, e.g., the author's text (1987) for details.) With this preparation, we have:

Theorem 7. *Let (Ω, Σ, μ) be a Carathéodory regular measure space. Then there exists a differentiation basis with the Vitali property on Ω iff μ is decomposable. Consequently every complete σ-finite measure space, which can be taken to be Carathéodory regular, has always a differentiation basis with the stated property. [Hence on such measure spaces there is a differentiation basis which may be used to obtain $(D\nu_f)(\omega) = f(\omega)$, a.e. for each μ-integrable f where $\nu_f(A) = \int_A f d\mu, \ A \in \Sigma$.]*

The strength of this result is the existence of a basis with the Vitali property, and the weakness is that there is no recipe to construct such a system routinely. For instance, in Theorem 6 the D-basis $(\mathcal{I}, \delta)$, also called an *interval basis*, does not necessarily have the Vitali property and so the integrability condition has to be strengthened. The point of this discussion is that a *Vitali system* (i.e., a basis with the Vitali property) is the best if one can directly employ, and if other bases are used then we must first verify the corresponding additional conditions in order for the differentiation procedure to give the Radon-Nikodým density. It is thus clear that a calculation of the latter is a nontrivial problem which additionally enters as a crucial ingredient of the theory of conditional probabilities and their densities when the latter exist.

Let us illustrate the last point for a general conditioning, thereby presenting the means through which the paradoxes of Section 2 above can be clearly understood and eliminated. Thus if (Ω, Σ, P) is a probability space, $\mathcal{B} \subset \Sigma$, a σ-subalgebra, and $P_{\mathcal{B}} = P|\mathcal{B}$ is the restriction then for any integrable random variable $X : \Omega \to \mathbb{R}$, we have from Chapter 2

$$Q^X(B) = \int_B X dP = \int_B E^{\mathcal{B}}(X) dP_{\mathcal{B}}, \ B \in \mathcal{B} . \tag{7}$$

So $Q^X(\cdot)$ is an indefinite integral on $(\Omega, \mathcal{B}, P_{\mathcal{B}})$ of $E^{\mathcal{B}}(X)$, a random variable for $\mathcal{B}$. Taking $X = \chi_A$, $A \in \Sigma$, we get

$$Q^X(B) = Q^A(B) = \int_B P^{\mathcal{B}}(A) \, dP_{\mathcal{B}}$$

and if $\{B_i, i \in I\}$ is a Vitali cover of Ω from $\mathcal{B}$, forming a differentiation basis, relative to both Q^A and $P_{\mathcal{B}}$, then Theorem 5 above implies that for each $\omega \in \Omega$, for a cofinal subnet $\{B_j^\omega, \ j \in J\}$ we get

$$(DQ^A)(\omega) = \lim_j \frac{Q^A(B_j^\omega)}{P_{\mathcal{B}}(B_i^\omega)} = P^{\mathcal{B}}(A)(\omega), \ \text{a.e.} \tag{8}$$

Thus in this procedure if the basis is a Vitali system, then the procedure gives a version of the conditional probability $P^{\mathcal{B}}(A)(\omega)$ for a.a.(ω), and all $A \in \Sigma$. So there can be no ambiguity or multiple answers. It should be noted that the σ-algebra $\mathcal{B}$ has no restrictions in this procedure, but the approximating net is not an arbitrary one that decends to $\{\omega\}$. Indeed it must obey the conditions of the differentiation procedure. Moreover, when these hypotheses are met the relation (8) produces

with $\mathcal{B} = \sigma(X_0)$. Consequently the results obtained using the latter (including the "h.w." or "v.w." procedures) do not give $P^{\mathcal{B}}(X'_0 < y)(a)$ in Kolmogorov's sense. Such approximations and the corresponding results form values of a different class of conditional distributions somewhat in the sense proposed by A. Rényi (1955). We study the latter in the next chapter and compare it with the Kolmogorov model that has been the basis for most of the work in the literature. A further consequence of traditional computations in applications will now be examined.

Since the computational problem is an essential ingredient in such areas as Bayesian inference (both estimation and decision procedures) it is useful to discuss the (adverse) impact of the preceding analysis on this subject. To introduce the problem briefly, let $X_1, \ldots, X_n$ be a set of random variables (for observation) which has a joint distribution $F_n(x_1, \ldots, x_n | \theta)$ depending on a parameter θ. For simplicity, let it be absolutely continuous, or discrete, with density $f_n(x_1, \ldots, x_n | \theta)$, so that

$$F_n(x_1, \ldots, x_n | \theta) = \int_A \cdots \int f_n(t_1, \ldots, t_n | \theta) d\mu_n(t_1, \ldots, t_n) , \qquad (4)$$

where $A = \underset{i=1}{\overset{n}{\times}}(-\infty, x_i)$ and μ_n is either the Lebesgue measure on $\mathbb{R}^n$ or the counting measure. In the Bayesian analysis one assumes that θ is a value of a random variable Θ, having a (prior) probability distribution assumed absolutely continuous with density $\zeta(\theta)$ where Θ takes values in $T \subset \mathbb{R}^k$. Thus

$$h_n(t_1, \ldots, t_n, \theta) = f_n(t_1, \ldots, t_n | \theta) \zeta(\theta) \qquad (5)$$

denotes the joint density of the random vector $(X_1, \ldots, X_n, \Theta)$ in $\mathbb{R}^n \times T$. Let $\tilde{P}$ be the induced measure by h_n of (5). Then the conditional density of Θ given $X_1 = x_1, \ldots, X_n = x_n$, called the *posterior density*, is defined as:

$$\zeta_n(\theta | x_1, \ldots, x_n) = h_n(x_1, \ldots, x_n, \theta) / \int_{\mathbb{R}^n} \cdots \int h_n(x_1, \ldots, x_n, \theta) d\mu_n .$$

Hence we have

$$\int_A \zeta_n(\theta | x_1, \ldots, x_n) d\theta = P(\Theta \in A | X_1 = x_1, \ldots, X_n = x_n) , \qquad (5)$$

which is regarded as the posterior probability of Θ with values in A, given that $X_1 = x_1, \ldots, X_n = x_n$ are the observed values of the X_is. Now let $\mathcal{B} = \sigma(X_1, \ldots, X_n)$, and consider the set $\tilde{A} = \mathbb{R}^n \times A$. Then we would compare the right side of (5) with the conditional probability $\tilde{P}^{\mathcal{B}}(\tilde{A})(x_1, \ldots, x_n)$ where $\tilde{P}^{\mathcal{B}}(\cdot)$ is the function defined in the Kolmogorov model (Definition 2.1.1). If the X_is have a continuous distribution, then the preceding work implies, when $n \geq 2$, the identification of $\tilde{P}^{\mathcal{B}}(\tilde{A})(x_1, \ldots, x_n)$ with the right side of (5) need not hold. On the other hand, the work of Section 2.4 shows that $\zeta_n(\cdot|x_1, \ldots, x_n)$ defines a conditional probability measure on some space with $X_1, \ldots, X_n$ and Θ as certain coordinate variables of $\mathbb{R}^{n+k}$. In general, however, from the joint probability $\tilde{P}$ of $(X_1, \ldots, X_n, \Theta)$ to calculate the conditional probability function $\tilde{P}^{\mathcal{B}}(\Theta \in A)(x_1, \ldots, x_n)$ one has to use the *appropriate* differentiation basis (with Vitali conditions) and the L'Hôpital's type approximations [common in practice] lead to (different) values dependent on the methods employed. Since for a correct evaluation of the Kolmogorov model one needs to have an "admissible" basis, we now present a simple case which qualifies to be the Vitali system. It is a specialization of a general condition called a "halo" in differentiation theory.

Consider the event $A_0 = [X_1 = x_1, \ldots, X_n = x_n]$ with $\tilde{P}(A_0) = 0$. Let $\{A_k, k \geq 1\}$ be a sequence of Borel sets in $\mathbb{R}^n$, such that $A_k \subset B(x, \alpha_k)$, where $\alpha_k > 0$ and $B(x, \alpha_k)$ is an open ball of radius α_k with center at $x = (x_1, \ldots, x_n)$. Here x need not be in A_k. This sequence is said to satisfy a *special halo property* if there exists a $\beta > 0$, independent of the A_k, such that

$$\mu(A_k) \geq \beta \mu(B(x, \alpha_k)), \quad k \geq 1, \tag{6}$$

where μ is a Borel measure on $\mathbb{R}^n$. [We take μ as the Lebesgue measure of $\mathbb{R}^n$ in the above examples.] The classical differentiation theory then guarantees that the special halo property implies the validity of the density theorem for a family of such bounded open sets $\{A_k, k \geq 1\}$. We state this for reference as follows:

Proposition 1. *Let μ be a Borel measure in $\mathbb{R}^n$ (so it is σ-finite), and let $\mathcal{O}$ be the class of open sets of finite positive μ-measure. Suppose also that for each $G \in \mathcal{O}$, the class of all homothetical sets to G are in $\mathcal{O}$. Then the special halo property is sufficient (and is also necessary) for* *the validity of the density theorem for $\mathcal{O}$ so that*

$$\lim_{k \to \infty} \frac{\mu(A_k^x \cap A)}{\mu(A_k^x)} = \chi_A \text{ , } a.e. \text{ } (\mu) \text{ ,} \tag{7}$$

for each Borel set A, and $\{A_k^x, \; k \geq 1\} \subset \mathcal{O}$ is a sequence such that $A_k^x \to x$, as $k \to \infty$.

An example of such a sequence $\{A_k, \; k \geq 1\}$ is a nested family of open cubes of $\mathbb{R}^n$ shrinking to x, or of its open rectangles the ratio of whose minimum and maximum sides is bounded away from zero and infinity as $k \to \infty$. Under these conditions we get in (5)

$$\tilde{P}^{\mathcal{B}}(A) = P(\Theta \in A | X_1 = x_1, \dots, X_n = x_n) \text{ , } a.e., \tag{8}$$

where $\mathcal{B} = \sigma(X_1, \dots, X_n)$. Thus one cannot use arbitrary decreasing sequences to evaluate the above type of probabilities. This aspect is especially significant when one is discussing such topics as the "Bayesian confidence intervals" in inference questions. In these cases one has to use appropriate Vitali systems to get unique solutions to the problems under consideration.

We next turn to alternative procedures that do not immediately confront the differentiation questions. Such a method was proposed by Rényi (1955) axiomatically and we shall analyze its strengths and shortcomings in the following chapter.

3.6 Bibliographical notes

In the literature *ad hoc* methods were proposed for the calculation of conditional probabilities and this resulted in paradoxes. The examples given for the Borel-Kolmogorov analysis appear briefly in Kolmogorov's (1933) book, and the ones in the text are motivated by an elementary modification of them given in DeGroot (1986). The second set of examples which exposed the difficulties vividly are from Kac and Slepian (1959), and we included their detailed analysis from a different perspective in the text. The difficulties inherent in such calculations were presented by the author in a lecture in 1983 and were given in (1988), and more details are included here for completeness. The statements related to Carathéodory regularity and other measurability results can be found in any standard book. (Royden, 1968, and for an account emphasizing these points, see the text by Rao, 1987.) The problem of precise and correct (unambiguous) formulations of conditioning were not thoroughly treated in the literature for too long a period. There is no real escape from using differentiation theory if one wants to obtain correct conditional probabilities using the Kolmogorov model. T. Tjur (1974) has used differential geometric methods but with L'Hôpital's approximations. The only other place where an attempt was made through differentiation method is Pfanzagl (1979), but he has given only a weaker version. Even the stated result again proceeds by reference to the L'Hôpital type approximation of Kac and Slepian. It is mentioned that the end result depends on a "differentiation system."

Because of these difficulties and the fact that Kolmogorov's model plays a central role in the current theory of probability and its applications, we have given a detailed discussion in Sections 4 and 5. In order not to digress further, several results from differentiation theory are given without proofs. However, most of the latter can be found in a comprehensive survey by Hayes and Pauc (1970). Although the notation and terminology are somewhat hard in this book to follow, one has to study them for a real understanding. A more readable, but shorter, treatment is available in Bruckner's memoir (1971).

It may be noted that for a qualitative study with conditioning, where explicit computations are not essential from the beginning, one can avoid differentiation theory. This can be seen in such areas as martingales and some aspects of Markov processes. However, the computational problems cannot be neglected in applications like statistical inference where the conclusions depend on explicit calculations of conditional probabilities. Lack of attention results in nonuniqueness (or ill-posedness) of solutions. It adds to the already existing confusion and controversy in these studies. One hopes that the present extended treatment of the problem will help reduce, and even eliminate, these conceptual (and philosophical) difficulties. For this purpose, we devote the following two chapters to other aspects of these problems.

CHAPTER 4: AN AXIOMATIC APPROACH TO CONDITIONAL PROBABILITY

An alternative to Kolmogorov's definition of conditioning is an axiomatic method with a view to simplifying the problems seen in the preceding chapters. This approach, due to A. Rényi, is analyzed in this chapter. Also the structure of such a conditional probability, its representation as the ratio of two functions determined by a sigma-finite measure, and some of its consequences are considered. Finally the difficulties noted in the last chapter surrounding the paradoxes and the position of this new approach in the general theory are discussed.

4.1 Introduction

After seeing the problems in implementing the general concept of conditioning introduced by Kolmogorov, it is natural to investigate the possibilities of a direct and perhaps a simpler method which is analogous to the original (absolute) probability model. Such a plan was advanced by A. Rényi in 1955. This continues the ideas indicated in Section 1.5. Recall that the elementary concept of conditioning as well as the "Borel paradox" and associated difficulties were already in the literature before 1930. Although this is a good enough reason to search for a simple model, there are some interesting practical considerations for taking a conditional probability as a basic building block of the probability structure. Let us indicate a motivation for it.

The ordinary (or absolute) probability model consists of the triple (Ω, Σ, P) describing an experiment. Thus the points of Ω represent all possible outcomes of an experiment and Σ is the smallest σ-algebra containing the class of events that are of conceivable interest to the experimenter. Also P is originally given (or decided by the experiment) as a probability on the basic class of events. Then P is extended to Σ uniquely, by various technical devices, completing the description of a model. However, an experimenter in reality can describe the values of P on the basic class generally from previous experience. This means, if A is an event (of the basic class) then the probability of A is describable only on the basis of an earlier (known) event B (or a class of events $\mathcal{B}$). Thus the number $P(A|B)$, depending both on A and B, becomes the probability that can be determined. It is this $P(\cdot|B)$ which is an additive function and which often changes as B varies (due to different types of experiences or experimenters), that is realistically prescribable. Consequently, if $\mathcal{B}$ is the collection of known events prior to the experiment (but related to it), and $\mathcal{A}$ is the class of interesting events to the experimenter, then $P(\cdot|\cdot) : \mathcal{A} \times \mathcal{B} \to \mathbb{R}^+$ will be the basic (probability) function so that it becomes the starting "measure instrument." Thus one takes Σ as the smallest σ-algebra containing both $\mathcal{A}$ and $\mathcal{B}$, and the basic model becomes a family of spaces $\{(\Omega, \Sigma, P(\cdot|B)), B \in \mathcal{B}\}$ which satisfies a natural "consistency" or connecting condition that relates its various members. This feeling is sometimes expressed by the statement that *every probability is, in reality, conditional.*

Such a modification in point of view motivates an analog of Kolmogorov's successful axiomatization of the abstract probability model. Naturally, if $\mathcal{B}$ is a singleton it should agree with the abstract case noted earlier, and hopefully should be simpler than the conditioning concept of the preceding chapters, i.e., of Kolmogorov's. An analog of the abstract method to fulfill these desirable goals has been considered by A. Rényi towards the middle 1950s and we shall present his approach and then analyse it in the next three sections. Although it was not possible to incorporate all the desired features, the method has some interesting properties which we now describe. In Section 5, we show how a further extension is needed to include those examples considered in the preceding chapter. A "natural" generalization will bring in the possibilities of multiple solutions again!

4.2 Axiomatization of conditioning based on partitions

Recalling the notion of conditional probability as a ratio (Definition 1.2.1), and remembering a number of properties that followed (Proposition 1.2.3), one should consider a "new" function in which some of the key features are reflected. We observe that in $P(A|B) = P(A \cap B)/P(B)$, it was assumed that $P(B) > 0$. This is reasonable since we do not generally base the occurrance of an event A on an "impossible" event B. Thus the possible class of conditions $\mathcal{B}$ should not contain ϕ, the impossible event. With this background, we introduce a function $P(\cdot|\cdot)$ axiomatically as follows.

Let Ω be a nonempty point set, Σ a σ-algebra of subsets of Ω and $\mathcal{B} \subset \Sigma$ a class of sets. Suppose $P(\cdot|\cdot) : \Sigma \times \mathcal{B} \to \mathbb{R}^+$ is a mapping which obeys the following three axioms:

Axiom I. $A \in \Sigma, B \in \mathcal{B} \Rightarrow 0 \leq P(A|B) \leq 1$, and $P(B|B) = 1$.

Axiom II. $P(\cdot|B) : \Sigma \to \mathbb{R}^+$ is σ-additive for each $B \in \mathcal{B}$.

Axiom III. Given $A \in \Sigma$, $B \in \mathcal{B}$ and $C \in \mathcal{B}$, one has

(i) $P(A|B) = P(A \cap B|B)$, and

(ii) $A \subset B \subset C \Rightarrow P(A|B)P(B|C) = P(A|C)$.

Such a function $P(\cdot|\cdot)$, obeying the three axioms, is called a *conditional probability*, in the sense of Rényi. In this chapter we omit the last qualifying phrase unless the Kolmogorov definition, denoted $P^{\mathcal{B}}(\cdot)$ when $\mathcal{B}$ is a σ-algebra, is also involved in a comparison. It is clear that the elementary case $P(A|B) = P(A \cap B)/P(B)$ derived from a given probability space, satisfies Axioms I - III. Here Axiom III is the distinguishing feature of this new concept. It is not implied by the other axioms, and we shall give two equivalent forms for later applications. Also Axioms I and II imply that $P(B|B) = 1$ so that $\phi \notin \mathcal{B}$ since otherwise $P(\phi|\phi) = 1$ and $P(\phi|\phi) = 0$ must hold simultaneously. Thus the possibility of adjoining ϕ to $\mathcal{B}$ is prevented by these two axioms.

The desired equivalent forms of Axiom III are given by:

Proposition 1. *When Axioms I, II hold, then the following are equivalent:*

(i) *Axiom* III *is true.*

(ii) *Axiom* III$_1$: $A \in \Sigma, \{B,C\} \subset \mathcal{B}$ *with* $B \cap C \in \mathcal{B} \Rightarrow$

$$P(A|B \cap C)P(B|C) = P(A \cap B|C). \tag{1}$$

(iii) *Axiom* III$_2$: $A \in \Sigma, \{B,C\} \subset \mathcal{B}$ *with* $B \subset C \Rightarrow$

$$P(A|B)P(B|C) = P(A \cap B|C). \tag{2}$$

Proof. Assume Axiom III. If $A \in \Sigma, \{B,C\} \subset \mathcal{B}$ with $B \cap C \in \mathcal{B}$, then by III, one has

$$
\begin{aligned}
P(A|B \cap C)P(B|C) &= P(A \cap B \cap C|B \cap C)\, P(B \cap C|C) \\
&= P(A \cap B \cap C|C), \text{ using III (ii)} \\
&= P(A \cap B|C)\,.
\end{aligned}
$$

So Axiom III$_1$ holds.

Suppose Axiom III$_1$ holds. Letting $B = C$ in (1), one gets III (i) since $P(B|B) = 1$. If $A \subset B \subset C$ and $\{B,C\} \subset \mathcal{B}$, then (1) gives III (ii). So Axiom III holds. Thus (III) $\leftrightarrow$ (III$_1$).

That Axiom (III$_1$) $\Rightarrow$ (III$_2$) is clear.

Conversely under III$_2$, for any $A \in \Sigma$ and $\{B,C\} \subset \mathcal{B}$ with $B \cap C \in \mathcal{B}$,

$$
\begin{aligned}
P(A \cap B|C) &= P(A \cap B \cap C|C), \text{ since Axiom I applied to (2)} \\
&\quad \text{with } B = C \text{ there, gives Axiom III(i)}, \\
&= P(A|B \cap C)P(B \cap C|C), \text{ by (2)}, \\
&= P(A|B \cap C)P(B|C), \text{ by Axiom I}\,.
\end{aligned}
$$

This is (III$_1$) so that Axioms III$_1$ and III$_2$ are also equivalent. $\square$

A simplification implied in these axioms is that conditioning is considered in terms of a single event at a time, and the elementary case is in view. Since $P(\cdot|\cdot)$ is a set function of two variables, one of which varies on a class that is not a ring, we include certain other properties to gain some feeling for its structure. The next result deals with the monotonicity property of this function.

Proposition 2. (i) *If* $\{A_n, n \geq 1\} \subset \Sigma$ *is a monotone sequence and* $B \in \mathcal{B}$, *then* $P(A_n|B)$ *is monotone in the same sense and* $\lim_n P(A_n|B) = P(\lim_n A_n|B)$.

(ii) *If* $\{B_n, n \geq 1\} \subset \mathcal{B}$ *is monotone,* $A \in \Sigma$ *and* $B_n \supset A$, *then* $B_n \subset B_{n+1}$ *implies* $P(A|B_n) \leq P(A|B_{n+1})$. *Moreover, if* $\lim_n B_n = B \in \mathcal{B}$, *and either* (a) $P(B|B_0) > 0$ *for some* $B_0 \in \mathcal{B}$, *where* $B_n \supset B_{n+1}$ *and* $B_{n_0} \subset B_0$ *for an* n_0, *or* (b) $B_n \subset B_{n+1}$, *then*

$$\lim_n P(A|B_n) = P(A|B), \ A \in \Sigma .$$

Proof. (i) Since $P(\cdot|B)$ is an ordinary probability measure on Σ, by Axiom II, the conclusion is a standard result for finite measures.

(ii) Let $B = \lim_n B_n = \cap_n B_n$, and $A \subset B \subset B_n$. Then

$$
\begin{aligned}
P(A|B_{n+1}) &= P(A \cap B_{n+1}|B_{n+1}), \text{ by Axiom III(i)}, \\
&= P(A \cap B_n|B_{n+1}), \text{ since } B_{n+1} \supset B_n \supset A , \\
&= P(A|B_n \cap B_{n+1})P(B_n|B_{n+1}), \text{ by (1)}, \\
&\leq P(A|B_n)P(B_{n+1}|B_{n+1}), \text{ by Axiom II and } B_n \subset B_{n+1} , \\
&= P(A|B_n), \text{ Axiom I}.
\end{aligned}
$$

Thus $P(A|B_n)$ is increasing, if $B_n \supset B_{n+1} \supset A$.

(a) Suppose the additional hypothesis holds, so that $B_{n+1} \subset B_n \subset B_0 \in \mathcal{B}$, $n \geq n_0$. Then for $n \geq n_0$,

$$
\begin{aligned}
P(A|B_n) &= P(A|B_n \cap B_0), \text{ since } B_n \subset B_0 , \\
&= P(A|B_n \cap B_0)P(B_n|B_0)/P(B_n|B_0), \text{ since } P(B_n|B_0) > 0, \\
&= \frac{P(A \cap B_n|B_0)}{P(B_n|B_0)} , \text{ by (1)} , \\
&\to \frac{P(A \cap B|B_0)}{P(B|B_0)} , \text{ Axiom II, and } P(B|B_0) > 0, \\
&= P(A|B \cap B_0), \text{ by (1)}, \\
&= P(A|B), \text{ since } B \subset B_0 .
\end{aligned}
$$

(b) If $B_n \subset B_{n+1}$, then $\lim_n P(B_n|B) = P(B|B) = 1$, so that $P(B_n|B) > 0$ from some n_0 onwards. Also for any $A \in \Sigma$

$$P(A|B_n)P(B_n|B) = P(A \cap B_n|B), \text{ by (2)} .$$

Hence using (i) and Axiom III(i), we get

$$P(A|B_n) = \frac{P(A \cap B_n|B)}{P(B_n|B)}$$
$$\rightarrow \frac{P(A \cap B|B)}{P(B|B)} = P(A|B) .$$

This establishes (ii) also. □

The next property is an analog of the total probabilities formula, given in Proposition 1.2.3(b) in the case that $P(\cdot|\cdot)$ is representable as a ratio.

Proposition 3. *Let* $\{B_n, n \geq 1\} \subset \mathcal{B}$, *be a disjoint sequence and* $B \subset \overset{\infty}{\underset{n=1}{\cup}} B_n$ *for some* $B \in \mathcal{B}$. *If* $B \cap B_n \in \mathcal{B}$, $n \geq 1$, *then*

$$P(A|B) = \overset{\infty}{\underset{n=1}{\Sigma}} P(A|B \cap B_n)P(B_n|B), A \in \Sigma . \qquad (3)$$

Proof. Consider a general term on the right side:

$$P(A|B \cap B_n)P(B_n|B) = P(A|B_n \cap B)P(B_n \cap B|B), \text{ by Axiom III(i)},$$
$$= P(A \cap B_n \cap B|B), \text{ by (2) } .$$

Consequently

$$\overset{\infty}{\underset{n=1}{\Sigma}} P(A|B \cap B_n)P(B_n|B) = \overset{\infty}{\underset{n=1}{\Sigma}} P(A \cap B_n \cap B|B)$$
$$= P(A \cap B \cap \overset{\infty}{\underset{n=1}{\cup}} B_n|B), \text{ since } B_n\text{'s}$$
$$\text{are disjoint and Axiom II applies },$$
$$= P(A \cap B|B), \text{ since } B \subset \underset{n}{\cup} B_n ,$$
$$= P(A|B), \text{ by Axiom III(i) } .$$

This gives (3). □

The formula presented here, although simple, will have an interesting analog in the abstract study to be discussed later in Section 7.4. The following consequence is recorded for reference. It will also be invoked in Section 5 below.

Corollary 4. *Suppose $\mathcal{B}$ is closed under countable unions and let $\{B_n, n \geq 1\} \subset \mathcal{B}$ be a disjoint sequence. If $P(A|B_k) \leq \lambda P(A'|B_k)$ $k \geq 1$, and $\lambda \geq 0$, where A, A' are a pair of events, then*

$$P(A|\underset{k}{\cup} B_k) \leq \lambda P(A'|\underset{k}{\cup} B_k) \ .$$

This relation is immediate from (3) by a substitution of the given inequality with $B = \underset{k}{\cup} B_k \in \mathcal{B}$ there.

We intend to consider some consequences of the above properties of these conditional probability functions and then discuss the difficulties with multiple solutions of problems of the preceding chapter. However, these are better appreciated if we analyze the structure of $P(\cdot|\cdot)$ in terms of functions of a single variable. So we devote the next section to this aspect.

4.3 Structure of the new conditional probability functions

If (Ω, Σ) is a measurable space, so that Ω is a point set and Σ is a σ-algebra of subsets of Ω, let $\mathcal{B} \subset \Sigma$ be a class and $P(\cdot|\cdot) : \Sigma \times \mathcal{B} \to \mathbb{R}^+$ be a conditional probability space as defined in the preceding section. Hereafter, $(\Omega, \Sigma, \mathcal{B}, P(\cdot|\cdot))$ will be termed a *conditional probability space*, so that $\phi \notin \mathcal{B}$. If there is given a probability measure $Q : \Sigma \to \mathbb{R}^+$ such that $P(A|B) = Q(A \cap B)/Q(B)$, $B \in \mathcal{B}$, $A \in \Sigma$, then it is an example; and $\mathcal{B}$ will also be called a *set of conditions*. In the opposite direction, it is desired to find necessary restrictions, if any, under which $P(\cdot|\cdot)$ is representable as such a ratio. We follow Rényi (1970), Chapter 2.

The next result presents a general solution of this problem:

Theorem 1. *Let $(\Omega, \Sigma, \mathcal{B}, P(\cdot|\cdot))$ be a conditional probability space. Suppose $\mathcal{B}$ is closed under finite unions and contains sufficiently many elements to cover each member of Σ in the sense that each $A \in \Sigma$ is included in a finite or countable union of B_is from $\mathcal{B}$. Suppose also that $P(B|C) > 0$ for $B \subset C$ and $\{B, C\} \subset \mathcal{B}$. Then there exists a σ-finite measure $\mu : \Sigma \to \bar{\mathbb{R}}^+$ such that $\mu(\mathcal{B}) \subset \mathbb{R}^+ - \{0\}$, and*

$$P(A|B) = \frac{\mu(A \cap B)}{\mu(B)} \ , \quad A \in \Sigma \ , \ B \in \mathcal{B} \ . \tag{1}$$

Moreover, μ is unique except for a positive multiplicative factor, and then $\mathcal{B}$ can be replaced by a (possibly) larger set of conditions $\tilde{\mathcal{B}} = \{A \in \Sigma : 0 < \mu(A) < \infty\}$, in (1).

Proof. Define $\mu : \mathcal{B} \to \mathbb{R}^+$ by the (unfortunately unmotivated) equation

$$\mu(B) = \frac{P(B|B \cup B_0)}{P(B_0|B \cup B_0)} \, , \quad B \in \mathcal{B} \, , \tag{2}$$

for an arbitrarily fixed $B_0 \in \mathcal{B}$. This is meaningful since $\mathcal{B}$ is closed under finite unions, and the argument stops here if the latter hypothesis is not included. We next extend this definition to the *trace* or *restriction* σ-algebra $\Sigma(B) = \{A \cap B : A \in \Sigma\}$, by setting

$$\mu(A) = P(A|B)\mu(B), \quad A \in \Sigma(B) \, . \tag{3}$$

Definitions (2) and (3) are the key parts of the proof and the rest is a technical detail of verification that μ is the desired function for (1). Now $\mu(\cdot)$ is clearly a measure on $\Sigma(B)$ since $P(\cdot|B)$ is. We assert that $\mu(\cdot)$, which depends on B_0, is uniquely defined on $\Sigma(B_1) \cap \Sigma(B_2)$ for any $B_i \in \mathcal{B}, i = 1, 2$, so that it can be extended to Σ by a kind of piecing together which is standard in such problems. Since by (2) it is clear that $\mu(\mathcal{B}) \subset \mathbb{R}^+ - \{0\}$, the extended μ will be σ-finite on Σ and then we will deduce from (3) that μ is the desired function. Uniqueness will then follow without much difficulty. Let us fill in the details.

Let $B_i \in \mathcal{B}, i = 1, 2$, and $A \in \Sigma(B_1) \cap \Sigma(B_2)$, so that $A \subset B_1 \cap B_2$ in particular. We first verify that $\mu(A)$ defined by (3) is unambiguous, i.e., for any B_1 and B_2 and A as above, we assert that

$$P(A|B_1)\mu(B_1) = P(A|B_2)\mu(B_2) \, . \tag{4}$$

To use Axiom III$_2$, and (2), let $B_3 = B_0 \cup B_1 \cup B_2 (\in \mathcal{B})$, so that

$$\mu(B_i) = P(B_i|B_3)/P(B_0|B_3) \, , \quad i = 1, 2, \tag{5}$$

since $B_i \subset B_3$. Hence

$$\begin{aligned}
P(A|B_1)\mu(B_1) &= \frac{P(A|B_1)P(B_1|B_3)}{P(B_0|B_3)} \\
&= \frac{P(A \cap B_1|B_3)}{P(B_0|B_3)} \, , \text{ by Axiom III}_2 \text{ again,} \\
&= \frac{P(A|B_3)}{P(B_0|B_3)} \, , \text{ since } A \subset B_1 \cap B_2, \\
&= \frac{P(A \cap B_2|B_3)}{P(B_0|B_3)} \\
&= P(A|B_2)\frac{P(B_2|B_3)}{P(B_0|B_3)} \, , \text{ by Axiom III}_2 \, , \\
&= P(A|B_2)\mu(B_2), \text{ by (5) } \, .
\end{aligned}$$

This shows that (4) holds.

To see that μ is defined on Σ, we now use a familiar argument from real analysis. Thus let $\tilde{\Sigma} = \{A \in \Sigma : A \subset B \text{ for some } B \in \mathcal{B}\}$. The additional hypothesis on $\mathcal{B}$ implies that $\tilde{\Sigma}$ is closed under finite unions, differences, and $\phi \in \tilde{\Sigma}$. It follows that $\tilde{\Sigma}$ is a ring and contains $\mathcal{B}$. If $A_n \in \tilde{\Sigma}$, disjoint, and $\underset{n}{\cup} A_n \in \tilde{\Sigma}$, then $(\underset{n}{\cup} A_n) \in \Sigma(B)$ for some $B \in \mathcal{B}$, by definition of $\tilde{\Sigma}$. Since by (3) μ is a measure on $\Sigma(B)$ (because of Axiom II for $P(\cdot|B)$), this implies that $\mu : \tilde{\Sigma} \to \mathbb{R}^+$ is a measure. But each element of Σ is covered by a (perhaps) countable collection of B_ns from $\mathcal{B}$, so that Σ is generated by $\tilde{\Sigma}$, and $\Omega = \underset{n}{\cup} A_n$, $A_n \in \tilde{\Sigma}$, disjoint. It follows that μ has a unique extension to Σ on which it is σ-finite. Denote the extension by the same symbol. Then (3) becomes

$$P(A|B) = \frac{\mu(A \cap B)}{\mu(B)}, \quad A \in \Sigma \ , \ B \in \mathcal{B} \ , \tag{6}$$

since $A \cap B \in \Sigma(B)$ and hence is finite. This gives (1).

Regarding the essential uniqueness of μ, let ν be another (σ-finite) measure so that $\nu(\mathcal{B}) \subset \mathbb{R}^+ - \{0\}$, and

$$P(A|B) = \frac{\nu(A \cap B)}{\nu(B)} = \frac{\mu(A \cap B)}{\mu(B)} \quad B \in \mathcal{B}, \ A \in \Sigma \ . \tag{7}$$

It is clear that $\nu = \alpha\mu$, $\alpha > 0$ satisfies (7). On the other hand, (7) implies

$$\nu(A) = \alpha(B) \, \mu(A), \ A \in \Sigma(B) \ , \tag{8}$$

with $\alpha(B) = \nu(B)/\mu(B)$. Since $\mathcal{B}$ is closed under finite unions and each member of Σ has a countable cover from $\mathcal{B}$ by hypothesis, there exist $B_n \in \mathcal{B}$, $B_n \uparrow \Omega$ and (8) holds for each B_n. Let $\alpha_n = \alpha(B_n)$. Since $\Sigma(B_n) \subset \Sigma(B_{n+1})$, we have

$$\nu(A) = \alpha_n \mu(A), \ \nu(\tilde{A}) = \alpha_{n+1} \, \mu(\tilde{A}) \ , \tag{9}$$

for all $\tilde{A} \in \Sigma(B_{n+1})$, $A \in \Sigma(B_n)$. It then follows from (9) that

$$\nu(A) = \alpha_n \mu(A) = \alpha_{n+1} \, \mu(A), \ A \in \Sigma(B_n) \ . \tag{10}$$

In particular, taking $A = B_n$, and noting that $0 < \mu(B_n) < \infty$, $n \geq 1$, we deduce from (10) that $\alpha_n = \alpha_{n+1} (= \alpha(\text{say}))$. Hence $\nu = \alpha\mu$, on

$\tilde{\Sigma}$ and by the unique extension of a σ-finite measure to $\sigma(\tilde{\Sigma}) = \Sigma$, the same relation holds on Σ. Thus μ of (1) is essentially unique.

Finally let $\tilde{B}$ be as in the statement. Then $\tilde{B}$ is the same collection for any μ for which (1) holds by the uniqueness established above. It follows that $B \subset \tilde{B}$, and since μ is a measure on Σ, $P(\cdot|\cdot)$ can be extended to $\Sigma \times \tilde{B}$ by the definition

$$\tilde{P}(A|B) = \frac{\mu(A \cap B)}{\mu(B)}, \ B \in \tilde{B} - B, \ A \in \Sigma .\tag{11}$$

Since $\tilde{P}|\Sigma \times B = P$, and $\tilde{B}$ satisfies the same hypothesis as B does, the procedure of the last paragraph yields that $\tilde{P}(\cdot|\cdot)$ is a conditional probability function satisfying Axioms I - III. It extends $P(\cdot|\cdot)$ and admits an essentially unique ratio representation as (1). $\square$

Remarks. 1. The above argument uses the hypothesis that B is closed under finite unions and has sufficiently many elements to cover each member of Σ. However, if the hypothesis on finite unions is dropped, then the essential uniqueness of μ in (1) fails and if "$P(B|C) > 0$ for $B \subset C$" in B is omitted, then the representation (1) with a single σ-finite μ need not hold. In this case, either one has to admit an (possibly uncountable) infinity of such measures or else admit such measures that take values in a non-Archimedean field. These points emerge from the works of Rényi (1956) and Krauss (1968). The former result will be given in Proposition 4 below. Representations of conditional measures under various weakened hypotheses were investigated by Császár (1955). It will be clear from our later work that these infinite collections correspond to the theme of "disintegration of measures." The results of Propositions 3 and 4 below will motivate that subject.

2. The last part of the above theorem shows that the conditional probability space $(\Omega, \Sigma, B, P(\cdot|\cdot))$ under the given hypothesis can be assumed to have been derived from a *complete* σ-finite space (Ω, Σ^*, μ), Σ^* being the μ-completion of Σ. Then replacing B by $\tilde{B}$ there, one has $(\Omega, \Sigma^*, \tilde{B}, P(\cdot|\cdot), \mu)$ as an extended or "full" conditonal measure space. Thus one may assume, under the hypothesis of this theorem, that the space to be full. Note tht $\tilde{B}$ does *not* have μ-equivalence class of μ-null sets.

The preceding result can be recast in a somewhat different terminology that has some independent interest.

Proposition 2. *Let $\{(\Omega, \Sigma, P_A), A \in \mathcal{B}\}$ be a family of probability spaces indexed by the collection $\mathcal{B} \subset \Sigma$ where $\mathcal{B}$ is closed under finite unions and contains countable covers of elements of Σ. If $P_B(B) = 1, B \in \mathcal{B}$, and $P_A(B) > 0$ for $B \subset A, B \in \mathcal{B}$, suppose that $P_C(A \cap B)/P_C(B)$, $A \in \Sigma$, $B \in \mathcal{B}$, is independent of $C \in \mathcal{B}$. Then*

$$P_B(A) = \mu(A \cap B)/\mu(B) , \quad A \in \Sigma , \tag{12}$$

for an essentially unique σ-finite measure $\mu : \Sigma \to \bar{\mathbb{R}}^+$ and $\mu(\mathcal{B}) \subset \mathbb{R}^+ - \{0\}$.

Proof. It is sufficient to observe that $P(\cdot|\cdot) : \Sigma \times \mathcal{B} \to \mathbb{R}^+$ defined by

$$P(A|B) = \frac{P_C(A \cap B)}{P_C(B)} , \tag{13}$$

satisfies the hypothesis of Theorem 1, and (12) follows from (13) on setting $B = C$ in the latter. $\square$

We now consider a weakening of the hypothesis of Theorem 1 and show that the representation of $P(\cdot|\cdot)$ can be given by an infinite collection of "ratios" using a family of measures. As a motivation for this case, let us first establish the easy direct part, namely such a family under suitable conditions determines a conditional probability function. After this, the converse part will be given.

Proposition 3. *Let (Ω, Σ) be a measurable space and $\{\mathcal{A}_\tau, \tau \in T\}$ be a filtering (upwards) family of σ-subalgebras where T is a directed index set, so that $\tau_1 < \tau_2$ implies $\mathcal{A}_{\tau_1} \subseteq \mathcal{A}_{\tau_2}$. Let $\mathcal{B}_\tau \subset \mathcal{A}_\tau, \tau \in T$, be a family such that $\mathcal{B}_{\tau_1} \cap \mathcal{B}_{\tau_2} = \phi$ if $\tau_1 \neq \tau_2$. Let $\mu_\tau : \Sigma \to \bar{\mathbb{R}}^+$, $\tau \in T$, be a set of σ-finite measures with the property that $\mu_\tau(\mathcal{B}_\tau) \subset \mathbb{R}^+ - \{0\}$, $\tau \in T$, and for $\tau_1 < \tau_2, \mu_{\tau_2}(\mathcal{B}_{\tau_1}) \subset \{0, \infty\}$. Then the mapping $P(\cdot|\cdot)$ defined on the algebra $\mathcal{A} = \bigcup_{\tau \in T} \mathcal{A}_\tau$, and the family $\mathcal{B} = \bigcup_{\tau \in T} \mathcal{B}_\tau$, by the equation*

$$P(A|B) = \frac{\mu_\tau(A \cap B)}{\mu_\tau(B)} , \quad A \in \mathcal{A}, \ B \in \mathcal{B}_\tau , \tag{14}$$

(for some τ) denoted $P(\cdot|\cdot) : \mathcal{A} \times \mathcal{B} \to \mathbb{R}^+$, is a conditional probability function, in the sense of Section 2, generated by the family $\{\mu_\tau, \ \tau \in T\}$.

Proof. The result follows on verifying that the $P(\cdot|\cdot)$ of (14) satisfies Axioms I, II and III$_2$, since III $\Leftrightarrow$ III$_2$ by Proposition 2.1. But Axioms

I and II are immediate. As for III_2, one uses the fact that $\{B,C\} \subset B \Rightarrow B \in B_{\tau_1}$, $C \in B_{\tau_2}$ and $B \subset C$ is possible iff $\tau_1 = \tau_2$ by the definition of B_τ. The result thus holds in this case also. Since $P(\cdot|B)$ is a probability on $\mathcal{A}$, it has a unique extension to $\sigma(\mathcal{A})$. If the latter is a proper subset of Σ, we may set $\tilde{T} = T \cup \{\tau_0\}$ with $\tau \in \tilde{T}$, $\tau \neq \tau_0$ then $\tau < \tau_0$, so that taking $B_{\tau_0} = \Sigma - \sigma(\mathcal{A})$, we may adjoin B_{τ_0} to the class $\{B_\tau, \tau \in T\}$, $\tilde{B} = B \cup B_{\tau_0}$, and assume $\sigma(\mathcal{A}) = \Sigma$ for convenience. Then the result again holds on $\Sigma \times \tilde{B}$. $\square$

The opposite direction needs more work. Here is the desired representation.

Proposition 4. *Let $(\Omega, \Sigma, B, P(\cdot|\cdot))$ be a conditional probability space such that B is closed under finite unions. Then there is a linearly ordered index set T and a family $\{\mu_\tau, \tau \in T\}$ of σ-finite measures on Σ representing $P(\cdot|\cdot)$ in the following sense:*

  (i) *$\mu_\tau(B) \subset \mathbb{R}^+ - \{0\}$, $\tau \in T$,*
  (ii) *for each $B \in B$, there is a $\tau \in T$ such that $0 < \mu_\tau(B) < \infty$ and if $\tau' \in T$, then $\mu_{\tau'}(B) = 0$ or $+\infty$ accordingly as $\tau' > \tau$ or $\tau' < \tau$,*
  (iii) *for each $B \in B$ there is a unique $\tau \in T$ such that*

$$P(A|B) = \frac{\mu_\tau(A \cap B)}{\mu_\tau(B)} , \quad A \in \Sigma . \tag{15}$$

Remark. T reduces to a point iff for each $\{B,C\} \subset B$, $B \subset C$ implies $P(B|C)$ is strictly positive.

Proof. The idea of proof is to express B as a union of a nonempty collection $\{B_\tau, \tau \in T\}$ with members which are pair-wise disjoint and each B_τ satisfies conditions similar to those of Theorem 1 so that the same method may be employed to deduce (15). Thus to construct the B_τ, we introduce a suitable equivalence relation on B so as to divide it into the desired mutually exclusive classes. This is the key item in the proof.

In order to introduce a relation, if B_1, B_2 are in B, (so $B_1 \cup B_2 \in B$) let us define the *indicatrix* $i(B_1, B_2)$ by the equation

$$i(B_1, B_2) = \frac{P(B_1|B_1 \cup B_2)}{P(B_2|B_1 \cup B_2)} \left(= \frac{1}{i(B_2, B_1)} \right) . \tag{16}$$

Evidently $0 \le i(B_1, B_2) \le \infty$. Note that $i(\cdot, \cdot)$ is never indeterminate, since by Axioms I and II,

$$1 = P(B_1 \cup B_2 | B_1 \cup B_2) \le P(B_1 | B_1 \cup B_2) + P(B_2 | B_1 \cup B_2) ,$$

and since $P(\cdot | B)$ is subadditive, being a measure. Next define $B_1 \sim B_2$ (i.e., of the same parity) iff $0 < i(B_1, B_2) < \infty$. We assert that ' $\sim$ ' is the desired equivalence relation. Indeed, $B_1 \sim B_1$ and $B_1 \sim B_2 \Rightarrow B_2 \sim B_1$ following at once from (16) (so '$\sim$' is reflexive and symmetric), we need only verify its transitivity. Thus let $B_i \in \mathcal{B}$, $i = 1, 2, 3$ such that $B_1 \sim B_2$ and $B_2 \sim B_3$. For any $\{B_1, B_2, C, D\} \subset \mathcal{B}$ with $C \supset D$, Axioms I and III imply

$$\frac{P(B_1 | D)}{P(B_2 | D)} = \frac{P(B_1 | C)}{P(B_2 | C)} . \tag{17}$$

Next let $C = B_1 \cup B_2 \cup B_3$ and $D = B_1 \cup B_3$ or $B_1 \cup B_2$ in (17). We then have the following string of equalities:

$$
\begin{aligned}
i(B_1, B_3) &= \frac{P(B_1 | B_1 \cup B_3)}{P(B_3 | B_1 \cup B_3)} \\
&= \frac{P(B_1 | C)}{P(B_3 | C)} , \text{ since } C \supset D \text{ in } \mathcal{B} , \\
&= \frac{P(B_1 | C)}{P(B_2 | C)} \frac{P(B_2 | C)}{P(B_3 | C)} \\
&= \frac{P(B_1 | B_1 \cup B_2)}{P(B_2 | B_1 \cup B_2)} \frac{P(B_2 | B_2 \cup B_3)}{P(B_3 | B_2 \cup B_3)} , \text{ since } B_i \cup B_j \subset C \\
&\quad \text{ for } 1 \le i, \ j \le 3 , \\
&= i(B_1, B_2) I(B_2, B_3) . \tag{18}
\end{aligned}
$$

Hence $0 < i(B_1, B_3) < \infty$ also holds so that $B_1 \sim B_3$ and transitivity obtains. Thus '$\sim$' is an equivalence relation in $\mathcal{B}$ under $i(\cdot, \cdot)$.

Now divide $\mathcal{B}$ into $\{\mathcal{B}_\tau, \ \tau \in T\}$, the equivalence classes under '$\sim$'. If $B_j \in \mathcal{B}_{\tau_j}$, $i = 1, 2$, we say $\tau_1 < \tau_2$ iff $i(B_1, B_2) = 0$, (and $\tau_1 > \tau_2$ iff $i(B_1, B_2) = [i(B_2, B_1)]^{-1} = +\infty$). Note that '$<$' depends only on the equivalence class but not on individual members B_j. This follows from (18). To see that '$<$' is a linear ordering on $\mathcal{B}$, observe that for any B_1, B_2 in $\mathcal{B}$ with $B_1 \in \mathcal{B}_{\tau_1}$, $B_2 \in \mathcal{B}_{\tau_2}$, $\tau_1 \ne \tau_2$, we get $i(B_1, B_2) = 0$ or $+\infty$ according as $\tau_1 > \tau_2$ or $\tau_1 < \tau_2$. Thus to use the argument of Theorem 1, it should be verified that each $\mathcal{B}_\tau$ is closed under finite unions. Indeed, if $\{B_1, B_2\} \subset \mathcal{B}_\tau$, so that $B_1 \sim B_2$, we have, since $B_1 \cup B_2 \in \mathcal{B}$, by using (16),

$$i(B_1 \cup B_2, B_1) = \frac{P(B_1 \cup B_2 | B_1)}{P(B_1 | B_1 \cup B_2)} = \frac{1}{P(B_1 | B_1 \cup B_2)}, \quad (19)$$

because $P(A|B) = P(A \cap B|B)$. But $0 < i(B_1, B_2) < \infty$ so that (19) and (16) imply that $P(B_j | B_1 \cup B_2) > 0, j = 1, 2$ and $B_1 \cup B_2 \sim B_1 \sim B_2$. Hence $B_1 \cup B_2 \in \mathcal{B}_\tau$. Then by Theorem 1, $P(\cdot|\cdot) : \Sigma \times \mathcal{B}_\tau \to \mathbb{R}^+$ admits a representation of the type given in (1). So there exists $\mu_\tau : \Sigma \to \bar{\mathbb{R}}^+$, σ-finite, such that $\mu_\tau(\mathcal{B}_\tau) \subset \mathbb{R}^+ - \{0\}$, and

$$P(A|B) = \frac{\mu_\tau(A \cap B)}{\mu_\tau(B)} , A \in \Sigma, B \in \mathcal{B}_\tau . \quad (20)$$

The measures $\{\mu_\tau, \tau \in T\}$ are essentially unique, as noted for (1). The last parenthetical statement follows from Theorem 1 in one direction, and the reverse direction is obtained by reviewing once again the above argument. $\square$

Remark. If $\mathcal{B}$ is not closed under finite unions, then $P(\cdot|\cdot)$ can be determined in many ways as seen from the following:

Example 5. Let $(\mathbb{R}, \Sigma, m)$ be the Lebesgue real line, $\mathcal{B}_+, \mathcal{B}_-$ be given by

$$\mathcal{B}_+ = \{A \in \Sigma : A \subset \mathbb{R}^+, \ 0 < m(A) < \infty\}$$
$$\mathcal{B}_- = \{A \in \Sigma : A \subset (-\infty, 0), 0 < m(A) < \infty\} ,$$

let $m_+(A) = m(A)$ if $A \in \Sigma(\mathbb{R}^+)$; $m_-(B) = m(B)$ if $m_+(B) = 0$, and $m_-(B) = +\infty$ if $m_+(B) > 0$, $B \in \Sigma$. Set $P(A|B) = m(A \cap B)/m(B)$, for all $B \in \mathcal{B} = \{B \in \Sigma : 0 < m(B) < \infty\}$. Then $P(\cdot|\cdot)$ is a conditional probability function; but we also have

$$P(A|B) = \begin{cases} \frac{m_+(A \cap B)}{m_+(B)} , & \text{if } B \in \mathcal{B}_+, \\ \frac{m_-(A \cap B)}{m_-(B)} , & \text{if } B \in \mathcal{B}_- . \end{cases}$$

So $P(\cdot|\cdot)$ is generated by (m_+, m_-) as well as by m, and the properties of the preceding representation of Theorem 1 or Proposition 4 do not hold.

4.4 Some applications

Since the very concept of conditional probability incorporates the "regularity" as Axiom II, one can develop analogs of the classical probability theorems for the family $\{(\Omega, \Sigma, P(\cdot|B)), B \in \mathcal{B}\}$, by introducing directly the conditional distributions, densities, characteristic functions and the like. We indicate a few of these here, and compare them with the Kolmogorov model studied in Chapter 2.

Let $X : \Omega \rightarrow \mathbb{R}$ be a random variable on $(\Omega, \Sigma, \mathcal{B}, P(\cdot|\cdot))$. Recall that by definition, $\{\omega : X(\omega) \in I\} = X^{-1}(I) \in \Sigma$ for each interval I, so that the concept involves only (Ω, Σ). Thus for each $B \in \mathcal{B}$, the conditional distribution of X given B, denoted $F_X(\cdot|B)$, is defined as:

$$F_X(x|B) = P[X < x|B] , \quad x \in \mathbb{R} . \tag{1}$$

Then the *image conditional law* can be obtained as follows. If $\mathcal{R}$ is the Borel σ-algebra of $\mathbb{R}$, let $\tilde{\mathcal{B}} = \{D \in \mathcal{R} : X^{-1}(D) \in \mathcal{B}\}$. In case $\tilde{\mathcal{B}} \neq \phi$, we set

$$\tilde{P}_X(C|D) = P(X^{-1}(C)|X^{-1}(D)), \quad C \in \mathcal{R}, \ D \in \tilde{\mathcal{B}} . \tag{2}$$

Then $(\mathbb{R}, \mathcal{R}, \tilde{\mathcal{B}}, \tilde{P}_X(\cdot|\cdot)$ is the induced (or image) conditional probability space of X. For instance, if (Ω, Σ, μ) is a σ-finite space $\mathcal{B} = \{B \in \Sigma : 0 < \mu(B) < \infty\}$, and $X : \Omega \rightarrow \mathbb{R}$ is a random variable let $F_X([a,b)) = F_X(b) - F_X(a) = \mu(\omega : a \leq X(\omega) < b)$. If F_X is assumed finite on bounded intervals (i.e., is *locally bounded*), than

$$\tilde{P}_X(a \leq X < b|c \leq X < d) = \frac{F_X(d) - F_X(c)}{F_X(b) - F_X(a)} , \tag{3}$$

for all $-\infty < a \leq c < d \leq b < \infty$. Other examples, if X is a vector random variable, can be similarly constructed.

To complete the discussion on these conditional measures, we now show that their products are also of the same kind so that they can be used in applications as though one is dealing essentially with ordinary probability measures. Thus if $(\Omega_i, \Sigma_i, \mathcal{B}_i, P_i(\cdot|\cdot))$, $i = 1, 2$ are a pair of conditional measure spaces, then their product is defined as follows. Set $\Omega = \Omega_1 \times \Omega_2$, $\mathcal{B} = \mathcal{B}_1 \times \mathcal{B}_2$ (cartesian products), and $\Sigma = \Sigma_1 \otimes \Sigma_2$ (the smallest σ-algebra containing the class $\Sigma_1 \times \Sigma_2$). We define $P(\cdot|\cdot)$ as:

$$P(A|B) = P_1(A_1|B_1)P_2(A_2|B_2), \quad A = A_1 \times A_2 \in \Sigma_1 \times \Sigma_2,$$

$$B = B_1 \times B_2 \in \mathcal{B} . \tag{4}$$

For fixed B, the right side of (4) is an ordinary product measure and $P(\Omega|B) = 1$ by Axioms I and II. Hence the standard product measure theory implies $P(\cdot|B)$ is a bounded σ-additive function on $\Sigma_1 \times \Sigma_2$ so that it has a unique σ-additive extension to Σ which satisfies (4), for each $B \in \mathcal{B}$. Since $P(B|B) = 1$, we need only verify Axiom III to conclude that this $P(\cdot|\cdot)$ is a conditional measure. Instead we show that the equivalent Axiom III$_2$ holds.

Let $A = A_1 \times A_2 \in \Sigma_1 \times \Sigma_2$, $B = B_1 \times B_2 \in \mathcal{B}$ and $C = C_1 \times C_2 \in \mathcal{B}$, such that $B \subset C$. The latter condition is true iff $B_1 \subset C_1$ and $B_2 \subset C_2$ as seen easily by drawing a picture. (Here the fact that neither B nor C is empty is also used.) Hence

$$
\begin{aligned}
P(A|B)P(B|C) &= P_1(A_1|B_1)P_1(B_1|C_1)P_2(A_2|B_2)P_2(B_2|C_2) \\
&= P_1(A_1 \cap B_1|C_1)P_2(A_2 \cap B_2|C_2), \text{ by Axiom III}_2 \\
&\quad \text{applied to } P_i(\cdot|\cdot), i = 1, 2, \\
&= P((A_1 \cap B_1) \times (A_2 \cap B_2)|C_1 \times C_2), \text{ by (4),} \\
&= P((A_1 \times A_2) \cap (B_1 \times B_2)|C_1 \times C_2), \text{ since} \\
&\quad A_i \cap B_i \in \Sigma_i, i = 1, 2, \\
&= P(A \cap B|C), \ A \in \Sigma_1 \times \Sigma_2, \ B \subset C \text{ in } \mathcal{B} . \quad (5)
\end{aligned}
$$

Thus Axiom III$_2$ holds in this case. Since for fixed B, C, both sides of (5) are bounded measures on $\Sigma_1 \times \Sigma_2$ the same relation must be true on Σ, by the classical theory. Thus $P(\cdot|\cdot)$ is a conditional product probability function.

The above result extends to a finite number of factors by induction. But an extension for infinite number of factors is also not difficult. We state this and add the detail for completeness.

Let $\{(\Omega_i, \Sigma_i, \mathcal{B}_i, P_i(\cdot|\cdot)), i \geq 1\}$ be a family of conditional probability functions and set $\mathcal{B} = \underset{i=1}{\overset{\infty}{\times}} \mathcal{B}_i, (\Omega, \Sigma) = (\underset{i=1}{\overset{\infty}{\times}} \Omega_i, \underset{i=1}{\overset{\infty}{\otimes}} \Sigma_i)$, the latter being the classical product measurable space. Thus $\underset{i}{\otimes} \Sigma_i$ is the σ-algebra generated by the cylinder sets A of $\Omega = \underset{i}{\times} \Omega_i$, where $A = A_1 \times \cdots \times A_n \times \underset{i>n}{\times} \Omega_i$ for some $n \geq 1$ and $A_i \in \Sigma_i$. Then $\mathcal{B} \subset \Sigma$ and since $P_i(\Omega_i|B_i) = 1$, we may define

$$
P(A|B) = \prod_{i=1}^{n} P_i(A_i|B_i) , \quad (6)
$$

for an n-cylinder $A = A_1 \times \cdots \times A_n \times \underset{i>n}{\times} \Omega_i$, $B_i \in \mathcal{B}$. The classical measure theory implies that $P(\cdot|B)$ has a unique extension to be a probability measure on Σ for each $B \in \mathcal{B}$, satisfying (6) for all cylinders in Σ. It is seen immediately that Axioms I and II hold for $P(\cdot|\cdot)$, and we again verify that III holds which needs some computation. Equivalently, consider Axiom III$_1$:

$$P(A|B \cap C)P(B|C) = P(A \cap B|C), \tag{7}$$

for $A \in \Sigma, \{B, C\} \subset \mathcal{B}$ with $B \cap C \in \mathcal{B}$. Again we may restrict A to range over cylinder sets which generate Σ because $P(\cdot|B \cap C)$ as well as $P(\cdot \cap B|C)$ are bounded measures.

Thus let A be an n-cylinder and consider the right side of (7). Since $P_i(\Omega_i|(B \cap C)_i) = 1$, one has

$$P(A|B \cap C) = \prod_{i=1}^{n} P_i(A_i|(B \cap C)_i) \tag{8}$$

where $(B \cap C)_i$ is the ith component of $B \cap C = (\overset{\infty}{\underset{i=1}{\times}} B_i) \cap (\overset{\infty}{\underset{i=1}{\times}} C_i)$. With the argument as in the case of products, it is easy to see that $B \cap C = \overset{\infty}{\underset{i=1}{\times}} (B_i \cap C_i)$ so that $(B \cap C)_i = B_i \cap C_i$. Hence (8) becomes

$$P(A|B \cap C) = \prod_{i=1}^{n} P_i(A_i|B_i \cap C_i), \tag{9}$$

for each n-cylinder A in Σ. If we set $p_n = \prod_{i=1}^{n} P_i(B_i|C_i)$, then $p_n \downarrow p$ where by definition $p = P(B|C)$. If $p = 0$ then the left side of (7) vanishes since $A \cap B \subset B$ so that

$$0 \leq P(A \cap B|C) \leq P(B|C) = 0,$$

$P(\cdot|C)$ being a measure. Thus (7) holds in this case. Let $p > 0$. Then the right sides of (7) and (9) give

$$P(A \cap B|C) = (\prod_{i=1}^{n} P_i(A_i \cap B_i|C_i)) \prod_{i>n} P_i(B_i|C_i)$$

$$= [\prod_{=1}^{n} P_i(A_i|B_i \cap C_i)P_i(B_i|C_i)] \prod_{i>n} P_i(B_i|C_i),$$

by Axiom III$_2$ applied to each $P_i(\cdot|\cdot)$,

$$= [\prod_{i=1}^{n} P_i(A_i|B_i \cap C_i)] \prod_{i} P_i(B_i|C_i),$$

$$= P(A|C \cap C)P(B|C).$$

This shows that (7) is true in general, and $P(\cdot|\cdot)$ extends uniquely to be an infinite product conditional probability function.

We summarize the above result in the following:

Proposition 1. *Let* $\{(\Omega_i, \Sigma_i, \mathcal{B}_i, P_i(\cdot|\cdot), i \geq 1\}$ *be a sequence of conditional probability spaces. If* $(\Omega, \Sigma) = \overset{\infty}{\underset{i=1}{\times}} (\Omega_i, \Sigma_i)$ *is the product measurable space,* $\mathcal{B} = \overset{\infty}{\underset{i=1}{\times}} \mathcal{B}_i$ *is the cartesian product, then there exists a unique conditional probability function* $P(\cdot|\cdot) : \Sigma \times \mathcal{B} \to \mathbb{R}^+$ *such that for each n-cylinder A in $\Sigma, B \in \mathcal{B}$, we have*

$$P(A|B) = \prod_{i=1}^{n} P_i(A_i|B_i), \tag{10}$$

and

$$P(B|C) = \prod_{i=1}^{\infty} P_i(B_i|C_i). \tag{11}$$

Although it is possible to extend this for uncountable products, just as in the classical theory with easy modifications, we shall not pursue the subject further as we have no need for such a result. The corresponding product theory for the Kolmogorov definition will be discussed in detail later in Chapter 8.

We now proceed to some other results on functions of random variables involving (conditional) expectations among others. So let us state the concept precisely for $(\Omega, \Sigma, \mathcal{B}, P(\cdot|\cdot))$. Thus if $X : \Omega \to \mathbb{R}$ is a random variable, then its *conditional expectation, given $B \in \mathcal{B}$*, is defined as:

$$E(X|B) = \int_{\Omega} X P(d\omega|B), \tag{12}$$

provided $\{\int_\Omega |X| P(d\omega|B) : B \in \mathcal{B}\} \subset \mathbb{R}^+$. Since we did not assume enough to warrant that P is representable as a ratio, there is no absolute expectation here. Comparing this with Example 1.4.3, one can say that this concept is slightly weaker than the earlier one even in the ratio case. There is no need to spell out the details here.

It is clear that $E(\cdot|B)$ is a positive linear operation on the vector space of real bounded measurable functions on (Ω, Σ). The standard limit theorems can be extended, using conditional characteristic functions. The latter is defined by:

$$\varphi_B(t) = \varphi(t|B) = E(e^{itX}|B), \ B \in \mathcal{B}, \tag{13}$$

for each random variable X on $(\Omega, \Sigma, \mathcal{B}, P(\cdot|\cdot))$. As a sample result we give a (conditional) strong law of large numbers. This, however, is obtained from a classical strong limit theorem which we now state.

Proposition 2. (Kolmogorov) *Let $X_1, X_2, \ldots$ be a sequence of mutually independent random variables with means zero, and variances $\sigma_1^2, \sigma_2^2, \cdots$, all defined on a probability space. Then the series, $\sum\limits_{k=1}^{\infty} X_k$ converges a.e. if $\sum\limits_{i=1}^{\infty} \sigma_k^2 < \infty$.*

A proof of this result is available in most books on probability theory (see, for instance, Rao 1984, p. 51). We need the following consequence of the above result.

Corollary 3. *If $X_1, X_2, \cdots$ is a sequence of mutually independent random variables on (Ω, Σ, P) with nonnegative means $\alpha_1, \alpha_2, \ldots$, and variances $\sigma_1^2, \sigma_2^2, \ldots$, such that*

$$(i) \sum\limits_{i=1}^{\infty} \alpha_k = +\infty, \ and \ (ii) \sum\limits_{k=1}^{\infty} (\sigma_k^2/\alpha_k^2) < \infty, \tag{14}$$

then we have

$$\lim\limits_{n} \frac{1}{a_n} \sum\limits_{i=1}^{n} X_k = 1, \ a.e., \tag{15}$$

where $a_n = \sum\limits_{i=1}^{n} \alpha_i$.

Proof. Let $Y_n = (X_n - \alpha_n)/a_n$. Then $E(Y_n) = 0$ and variance of Y_n is σ_n^2/a_n^2. By (14) and Proposition 2, $\sum\limits_{n=1}^{\infty} Y_n$ converges a.e., i.e.,

$$\lim\limits_{n} \sum\limits_{k=1}^{n} (X_k - \alpha_k)/a_k \text{ converges a.e.}$$

By a form of the Kronecker's lemma (verified below) this implies

$$\lim_n \frac{1}{a_n} \sum_{k=1}^{n} (X_k - \alpha_k) = 0, \quad a.e., \tag{16}$$

so that (16) gives (15). $\square$

The form of *Kronecker's lemma* used above is this: If $c_n \in \mathbb{R}, 0 < \beta_n \uparrow \infty$ and $\sum_{n=1}^{\infty} (c_n/\beta_n)$ converges, then $\lim_n \frac{1}{\beta_n} \sum_{k=1}^{n} c_k = 0$.

Proof of lemma. Let $s_n = \sum_{k=n+1}^{\infty} (c_k/\beta_k)$, and $R_n = \sum_{k=1}^{n} c_k$. Then

$$s_{n-1} - s_n = c_n/\beta_n.$$

Hence

$$R_n = \sum_{k=1}^{n} \beta_k (s_{k-1} - s_k) = s_0 \beta_1 - s_n \beta_n + \sum_{k=1}^{n-1} s_k (\beta_{k+1} - \beta_k),$$

on rearranging the order of summation.

Consequently,

$$\frac{R_n}{\beta_n} = s_0 \frac{\beta_1}{\beta_n} - s_n + \frac{1}{\beta_n} \sum_{k=1}^{n-1} s_k (\beta_{k+1} - \beta_k). \tag{17}$$

Since $s_n \to 0$ and $0 < \beta_n \uparrow \infty$, so that $\frac{\beta_k}{\beta_{n+1}} \to 0$, as $n \to \infty$, the first two terms are negligible, and it suffices to show that the last term of (17) satisfies

$$\lim_n \frac{1}{\beta_n} \sum_{k=m}^{n} s_k (\beta_{k+1} - \beta_n) = 0, \tag{18}$$

for any fixed but arbitrary m. But $s_n \to 0$, so for given $\varepsilon > 0$, choose $m = n_0$ such that $|s_n| < \varepsilon$ for $n \geq n_0$. Then the expression (18) becomes:

$$-\varepsilon(1 - \frac{\beta_{n_0}}{\beta_n}) \leq \frac{1}{\beta_n} \sum_{k=n_0}^{n} s_k (\beta_{k+1} - \beta_n) \leq \varepsilon(1 - \frac{\beta_{n_0}}{\beta_n}).$$

Letting $n \to \infty$, this gives (18) since $\varepsilon > 0$ is arbitrary. $\square$

As a consequence of the Corollary we can give the conditional strong law of large numbers. First let us recall the independence concept in the present context which is similar to that of Definition 2.5.1.

A sequence $\{X_n, n \geq 1\}$ of random variables on $(\Omega, \Sigma, \mathcal{B}, P(\cdot|\cdot))$ is *mutually (conditionally) independent relative to* $P(\cdot|B)$ for each $B \in \mathcal{B}$, if for any intervals $I_i \subset \mathbb{R}, i = 1, 2, \cdots, n$, and $X_{j_1}, \ldots, X_{j_n}, n \geq 1$ we have

$$P(X_{j_i} \in I_i, i = 1, \ldots, n | B) = \prod_{i=1}^{n} P(X_{j_i} \in I_i | B), B \in \mathcal{B},$$

where $(j_1, \ldots, j_n)$ is any finite subset of the numbers $\{1, 2, \ldots\}$.

The following is a form of the strong law of large numbers:

Proposition 4. *Let $X_1, X_2, \ldots$ be a sequence of mutually (condition-ally) independent random variables on $(\Omega, \Sigma, \mathcal{B}, P(\cdot | \cdot))$. If $I = (a, b)$ is a nondegenerate interval such that $B_n = X_n^{-1}(I) \in \mathcal{B}, n \geq 1$, and $B_n \subset C$ for some $C \in \mathcal{B}$, suppose the following conditions hold:*

(i) $E(X_n | B_n) = \alpha_n > 0, 0 < Var(X_n | B_n) = \sigma_n^2 < \infty, n \geq 1$,

(ii) if $p_n = P(B_n | C)$, then $\sum_{n=1}^{\infty} p_n = +\infty$, and $\sum_{n=1}^{\infty} p_n \alpha_n = +\infty$,

(iii) $\lim_{n} (\sum_{i=1}^{n} p_i \alpha_i / \sum_{i=1}^{n} p_i) = a_o$ exists,

(iv) $\sum_{i=1}^{\infty} [p_i(\sigma_i^2 + (1 - p_i)\alpha_i^2) / \sum_{j=1}^{i} p_j \alpha_j] < \infty$.

Then

$$P[\lim_{n} \frac{\sum_{k=1}^{n} X_k^I}{N_n^I} = a_o | C] = 1 , \tag{19}$$

where N_n^I is the number of the X_is among the first n that visit the interval I and X_k^I is the truncation of X_k to I, so that $X_k^I = X_k \chi_{B_k}$.

Proof. This result will be deduced from the preceding one. The action is all relative to C or $P(\cdot | C)$. We now use Axiom III$_2$ and make the following calculation for X_k^I:

$$E(X_k^I | C) = \int_{B_k} X_k(\omega) P(d\omega | C)$$
$$= \int_{\Omega} X_k(\omega) P(d\omega | B_k) P(B_k | C), \text{ since}$$
$$P(A \cap B | C) = P(A | B) P(B | C) \text{ for } B \subset C,$$
$$= p_k E(X_k | B_k) = p_k \alpha_k .$$

Similarly

$$E((X_k^I)^2|C) = \int_{B_k} X_k^2(\omega)P(d\omega|C)$$

$$= p_k \int_{\Omega} X_k^2 P(d\omega|B_k)$$

$$= p_k(\sigma_k^2 + \alpha_k^2) .$$

Hence

$$\mathrm{Var}(X_k^I|C) = p_k(\sigma_k^2 + \alpha_k^2) - p_k^2\alpha_k^2$$

$$= p_k(\sigma_k^2 + (1 - p_k)\alpha_k^2).$$

Thus by Corollary 3, on using (i) and (iv) of the hypothesis for the mutually (conditionally) independent sequence $\{X_k^I, k \geq 1\}$ one has

$$\lim_n \frac{\sum\limits_{k=1}^{n} X_k^I}{\sum\limits_{k=1}^{n} p_k\alpha_k} = 1, a.e. \tag{20}$$

But $\{\chi_{B_k}, k \geq 1\}$ is likewise an independent sequence with $E(\chi_{B_n}|C) = p_n$, and $\mathrm{Var}(\chi_{B_k}|C) = p_k(1 - p_k)$. Thus (ii) and Corollary 3 imply

$$\lim_n \frac{\sum\limits_{k=1}^{n} \chi_{B_k}}{\sum\limits_{k=1}^{n} p_k} = 1, \ a.e. \tag{21}$$

Since $N_n^I = \sum\limits_{k=1}^{n} \chi_{B_k}$, (20), (21) and (iii) of the hypothesis give (on dividing (20) by (21) and taking limits) (19) immediately. □

Remark. The result is a partial extension of the classical Glivenko-Cantelli theorem on empirical distributions. Indeed, if the X_k are (in addition) indentically distributed so that $P(X_k \in I|C) = p, k \geq 1$, then (19) says that the conditional relative frequency of the given sequence of random variables tends to the true value p with the conditional probability 1 given C.

Regarding the events of $\mathcal{B}$ as parameters one may similarly obtain several classical limit theorems. Here we present an extension of the Lévy continuity theorem from this point of view. In fact we shall give a uniform limit theorem, relative to $\mathcal{B}$ following Parzen (1954) who obtained several other extensions in the same manner. To make the discussion intelligible let us first state the classical version.

Theorem 5. (Lévy) *Let $F_1, F_2, \ldots$, be a sequence of distribution functions and $\varphi_1, \varphi_2, \ldots$ be the corresponding characteristic functions, [i.e., take $B = \Omega$ in (1) and (13)]. Then there is a distribution function F such that $F_n(x) \to F(x)$ at every continuity point x of F iff $\varphi_n(t) \to \varphi(t), t \in \mathbb{R}$, provided the limit $\varphi(\cdot)$ is also continuous at $t = 0$. When the latter holds, then φ is the characteristic function of F.*

To present the corresponding result in the conditional case, with a stronger uniformity condition, we introduce a relevant concept here.

Definition 6. Let $F_n(\cdot|B)$ and $F(\cdot|B)$ be conditional distribution functions, given $B \in \mathcal{B}$. Then $F_n(\cdot|B) \to F(\cdot|B)$ *uniformly* if (i) for each pair of points $x_1, x_2(x_1 < x_2)$ and $\varepsilon > 0$, there is an $n_0(= n_0(x_1, x_2, \varepsilon))$ such that for each $B \in \mathcal{B}$ we can find $x_1 < x < x_2$ satisfying

$$|F_n(x|B) - F(x|B)| < \varepsilon, n > n_0,$$

and (ii) $\lim\limits_{x \to \pm\infty} \lim\limits_{n \to \infty} [\lim\limits_{y \to \pm\infty} F_n(y|B) - F_n(x|B)] = 0$, uniformly in B.

Recall that a family $f_\alpha : \mathbb{R} \to \mathbb{R}, \alpha \in I$, of functions is *equicontinuous* at $x \in \mathbb{R}$, if for each $\varepsilon > 0$, a neighborhood U_x of x exists such that

$$\sup\{\{\sup |f_\alpha(x) - f_\alpha(y)| : y \in U_x\} : \alpha \in I\} < \varepsilon \ .$$

In other words, for each generalized sequence $x_\tau \to x, f_\alpha(x_\tau) \to f_\alpha(x)$, uniformly in $\alpha \in I$. If this condition holds for all $x \in \mathbb{R}$, the family $\{f_\alpha, \alpha \in I\}$ is equicontinuous on $\mathbb{R}$.

With these concepts the desired extension is given by:

Theorem 7. (Parzen) *Let $\{F_n(\cdot|B), F(\cdot|B)\}_{B \in \mathcal{B}}$ be conditional distribution functions with the corresponding characteristic functions $\{\varphi_n(\cdot|B), \varphi(\cdot|B)\}_{B \in \mathcal{B}}$. Then $F_n(\cdot|B) \to F(\cdot|B)$ uniformly in $B \in \mathcal{B}$, in the sense of Defintion 6, iff (i) $\varphi_n(t|B) \to \varphi(t|B), t \in \mathbb{R}$, as $n \to \infty$, uniformly in $B \in \mathcal{B}$, and (ii) $\{\varphi(\cdot|B), B \in \mathcal{B}\}$ is equicontinuous at $t = 0$. If these conditions hold, then $\varphi_n(t|B) \to \varphi(t|B)$ uniformly both in $t \in I$ and $B \in \mathcal{B}$, where $I \subset \mathbb{R}$ is any bounded open interval.*

It is interesting to note that Parzen has obtained this and related extensions in 1952 before Rényi formulated his new axiomatic treatment of the subject. Except for the works (and talks) by Rényi, Császár and a related account by Krauss referred to in this chapter, there appears to be very little advance in the utilization of this new method. The known applications are based on the classical results, with suitable uniformity conditions added.

We shall omit the proof of the above theorem, which is somewhat lengthy, since several classical formulas and ideas have to be recalled. The reader is referred to the original accessible paper in which all the details are spelled out.

Now we proceed to the problems and paradoxes arising in the application of conditional probabilities examined in the last chapter in light of the new axiomatic set up. There one sees that those difficulties are not eliminated even with the new method. In fact, further restrictions on this model are necessary and there is a close analogy between the latter and the Vitali conditions discussed previously.

4.5 Difficulties with earlier examples

Several of the difficulties associated with calculating conditional probabilities using the Kolmogorov model were illustrated in Section 3.2 for some important, realistic, and practical problems. The earliest of these is the so-called "Borel paradox." Can we find a better or easier explanation of the difficulty with Rényi's new model? The problem in the classical case arose from the fact that the conditioning event has probability zero and hence an approximation argument was tried in evaluating the numerical value. But the "conditions", $\mathcal{B}$, in Rényi's model explicitly exclude such an event and so a direct application is not possible. To illustrate some of these questions, especially for the Borel problem, Rényi (1955) suggested a refinement of his model which we now discuss.

The following set of conditions under the name "Cavalieri space" has been introduced by Rényi. However, because of its close resemblence with the classical strong Vitali system in differentiation theory (see Hayes and Pauc, 1970, p. 27) we propose to call it simply the Vitali type condition to emphasize the analogy between these concepts and also for a comparison with the work of Section 3.4. Thus a conditional probability space $(\Omega, \Sigma, \mathcal{B}, P(\cdot|\cdot))$ is said to satisfy a *Vitali type condition* if one has:

(V.1) there is a family $\{B_t, a \leq t < b\} \subset \mathcal{B}$

(V.2) $a \leq \alpha < \beta \leq b \Rightarrow \bigcup_{\alpha \leq t < \beta} B_t = B_\alpha^\beta \in \mathcal{B}$

(V.3) $s \neq t$ in [a,b) $\Rightarrow B_s \cap B_t = \emptyset$,

(V.4) for $A, \tilde{A}$ in $\Sigma, \lambda \geq 0, P(A|B_t) \leq \lambda P(\tilde{A}|B_t), a \leq \alpha \leq t < \beta \leq b \Rightarrow P(A|B_\alpha^\beta) \leq P(\tilde{A}|B_\alpha^\beta)$.

If (V.4) is replaced by the following stronger condition then the space will be said to satisfy a *strict Vitali type condition*:

(V.5) for each $A \in \Sigma$, the function $t \mapsto P(A|B_t)$ is Borel measurable on $[a, b)$ and there is a real valued increasing and continuous function F on $[a, b)$ such that for $a \leq \alpha < \beta \leq b$,

$$P(A|B_\alpha^\beta) = \int_\alpha^\beta P(A|B_t) \, dF(t)/[F(\beta) - F(\alpha)] \, . \tag{1}$$

In particular if the derivative F' exists everywhere on [a,b) and (1) holds, then we say that the model satisfies a *stricter Vitali type condition*.

It can be shown that there are conditional probability spaces that do not satisfy a strict Vitali type condition (as can be seen from Proposition 3.4) and also there are spaces which satisfy a strict but not stricter Vitali type condition. Note that (V.4) follows easily from (V.5), justifying the sharpening used for the latter. We now record a few simple consequences of the above conditions which incidentally were introduced to "resolve" the Borel paradox in the first place.

Consequences. 1. If in $\{B_t, t \in I\} \subset \mathcal{B}, I$ is countable (e.g. $I = \mathbb{N}$) and (V.1) – (V.3) hold then (V.4) is automatic.

Indeed, from the identity (2.3), if $B = \bigcup_n B_n \in \mathcal{B}$, then for $A, \tilde{A}$ in Σ,

$$P(A|B) = \sum_n P(A|B_n)P(B_n|B), \text{ since } B_n \subset B$$

$$\leq \lambda \sum_n P(\tilde{A}|B_n)P(B_n|B), \text{ by hypothesis,}$$

$$= \lambda P(\tilde{A}|B), \text{ by (2.3) } .$$

Thus (V.4) comes into force only for an uncountable index I.

2. If $P(A|B_t) = \lambda P(\tilde{A}|B_t)$, or $P(A|B_t) = \lambda$, all t, then the same holds when B_t is replaced by B_α^β, for any given $A \in \Sigma$.

For, by (V.4) we have $P(A|B_\alpha^\beta) \leq \lambda P(\tilde{A}|B_\alpha^\beta)$ and if $\lambda = 0$ there is equality. If $\lambda > 0$, then the hypothesis becomes $P(\tilde{A}|B_t) \leq \frac{1}{\lambda}P(A|B_t)$ for all t. So one has $P(\tilde{A}|B_\alpha^\beta) \leq \frac{1}{\lambda}P(A|B_\alpha^\beta)$. From these two inequalities we deduce that

$$\lambda P(\tilde{A}|B_\alpha^\beta) \leq P(A|B_\alpha^\beta) \leq \lambda P(\tilde{A}|B_\alpha^\beta) \, . \tag{2}$$

Thus there is equality. If $P(A|B_t) = \lambda$ for all t, then from (2) with $\tilde{A} = B_\alpha^\beta$, and Axiom I, we get $P(\tilde{A}|B_\alpha^\beta) = 1$ so that the second statement follows.

3. If (V.5) also holds, then for almost all t (F-measure) in $[\alpha, \beta) \subset [a, b)$, we have (since $B_t^t = B_t$)

$$\lim_{h \downarrow 0} P(A|B_t^{t+h}) = P(A|B_t), \quad A \in \Sigma . \tag{3}$$

In fact, this follows from (1) when we write it as

$$\lim_{h \downarrow 0} P(A|B_t^{t+h}) = \lim_{h \downarrow 0} \frac{\int_t^{t+h} P(A|B_\tau) dF(\tau)}{F(t+h) - F(t)}$$
$$= P(A|B_t), \quad \text{a.e.} [F] ,$$

where we used the Lebesgue differentiation theorem in the last step.

4. Under (V.5) we always have $P(B_t|B_\alpha^\beta) = 0$, for almost all $\alpha \leq t < \beta$, in F-measure.

For, by Axiom III, $P(B_{t_0}|B_t) = P(B_{t_0} \cap B_t|B_t) = 0$ or 1 accordingly as $t \neq t_0$ or $t = t_0$. Since F is continuous, $P(B_t|B_\alpha^\beta) = 0$.

5. Let $X : \Omega \to \mathbb{R}$ be a random variable, on (Ω, Σ), and $B_t = X^{-1}(\{t\}), t \in \mathbb{R}$. Suppose that $\{B_t, t \in \mathbb{R}\} \subset \mathcal{B}$. (This assumption clearly restricts the random variables admitted.) Then for any $\alpha \leq a < b \leq \beta$ and for the conditional probability space $(\Omega, \Sigma, \mathcal{B}, P(\cdot|\cdot))$ we have with $I = [\alpha, \beta) \supset J = [a, b), A = X^{-1}(J), B = X^{-1}(I)$,

$$P(X \in J | X \in I) = P(A|B) = P(A \cap B|B)$$
$$= \int_J P(A|B_t) \, dF(t) / \int_I dF(t)$$
$$= \int_b^a P(A \cap B_t|B_t) dF(t)/(F(\beta) - F(\alpha))$$
$$= \frac{F(b) - F(a)}{F(\beta) - F(\alpha)} . \tag{4}$$

Here F is only locally bounded on $\mathbb{R}$ and, for instance $F(x) = x$ is possible. (This may be compared with Sect. 5.3 in Kolmogorov, 1933.) It has some utility in applications in which unbounded Fs are allowed.

We now consider the Borel paradox in this set up and present a solution. The Borel problem in its original form is as follows. Let Ω be the surface of the unit sphere in $\mathbb{R}^3$, and Σ be the σ-algebra of the Borel sets on this surface. Choose a fixed meridian circle passing through the north and south poles, and take the equator as the initial line so that the longitude is traced by $\theta(0 \leq \theta < \pi)$ and the latitude by $\varphi(0 \leq \varphi < 2\pi)$. The probability measure on Σ is the uniform measure. Then the problem is to find the conditional probability function of φ for a given θ. It is "reasonable" to expect that this is again a uniform distribution. Kolmogorov has shown that this conditional probability (using his definition, cf. Chapter 2) is given by

$$P(\alpha \leq \varphi < \beta | \theta) = \frac{1}{4} \int_\alpha^\beta |\cos \varphi| d\varphi , \qquad (5)$$

so that it is not uniform. Since $P(\theta = a) = 0$ for each 'a', the ratio definition can not be used directly to obtain (5). It is in this connection that Kolmogorov states: "a conditional probability relative to an isolated hypothesis whose probability equals zero in inadmissible", (for such a calculation). Since these sets are excluded from the collection $\mathcal{B}$ of Rényi's method, we now present his alternative argument using the Vitali type condition, and it will be seen that the result so obtained is different from that of (5). However, this will illuminate the difficulties discussed in the work of Section 3.4 as well as justify the reasons for the introduction of a Vitali type condition in the present study.

Since the position of a point on the sphere is given by (x, y, z), we convert the problem by considering its spherical polar coordinates. Thus let $x = \sin \theta \cos \varphi, y = \sin \theta \sin \varphi$, and $z = \cos \theta$ where $0 \leq \varphi < 2\pi, 0 \leq \theta < \pi$, so that (θ, φ) uniquely determine the position. Here Ω is the rectangle $[0, \pi) \times [0, 2\pi)$, and Σ is its Borel σ-algebra. Let $\mathcal{B} = \{B_{a,c}^{b,d} : B_{a,c}^{b,d} = [a, b) \times [c, d) \subset \Omega\}$. If P denotes the uniform probability of the problem on Ω, we take $P(\cdot | \cdot)$ as $P(A|B) = P(A \cap B)/P(B)$ where $P(B) = \frac{\text{area of (B)}}{2\pi^2}$. Thus

$$P(A|B_{a,c}^{b,d}) = \frac{\int \int_{A \cap B_{a,c}^{b,d}} \sin \theta \, d\theta \, d\varphi}{\int \int_{B_{a,c}^{b,d}} \sin \theta \, d\theta \, d\varphi}. \qquad (6)$$

We now construct a family $\{B_t, 0 \leq t < 2\pi\} \subset \mathcal{B}$, relative to which a strict Vitali type condition holds for $P(\cdot | \cdot)$, with a suitable monotone function F.

Indeed, let $B_t = B_{t,0}^{t,\pi}$ and choose a measure μ_t on Σ such that $\mu_t(B_t) = 1$, and (the existence of μ_t will be clear from the following work):

$$P(A|B_{t,c}^{t,d}) = \frac{\mu_t(A \cap B_{t,c}^{t,d})}{\mu_t(B_{t,c}^{t,d})}, \ 0 \le c < d < \pi \ . \tag{7}$$

Set $P(B_{\theta,0}^{2\pi,x}|B_t) = G(x,t)$, and assume that $G(\cdot,\cdot)$ is jointly continuous in (x,t) and $G(\cdot,t)$ is strictly increasing. We assert that $(\Omega, \Sigma, \mathcal{B}, P(\cdot|\cdot))$ is a conditional probability space satisfying a strict Vitali type condition iff

$$\mu_t(A) = \frac{1}{2} \int_{A \cap B_t} \sin \theta d\theta, \ A \in \Sigma, \ 0 \le t < 2\pi \ . \tag{8}$$

Observe that for each t, B_t represents a meridian on the sphere. By construction $P(\cdot|\cdot)$ satisfies conditions (V.1) – (V.3). We now show that (V.5) holds iff (8) is true. Thus let $0 \le \lambda < 1$ be fixed. Define the number x_t^λ such that $A_t^\lambda = [0, x_t^\lambda] \subset [0, \pi)$ and $P(A_t^\lambda|B_t) = \lambda$. By the continuity of G, the function $t \mapsto x_t^\lambda$ is continuous for each λ, and if we set

$$A^\lambda = \cup\{A_t^\lambda : 0 \le t < 2\pi\} \ .$$

then $A^\lambda \in \mathcal{B} \subset \Sigma$. As noted in (2), $P(A^\lambda|B_t) = \lambda$ for all t. Consequently, we get $P(A^\lambda|B_a^b) = \lambda$ by the same result, where $B_a^b = B_{a,0}^{b,\pi}$ for $0 \le a < b \le 2\pi$. Hence (7) becomes with (6) and $A = A^\lambda$ there,

$$\lambda = P(A^\lambda|B_a^b)$$
$$= \frac{\int_a^b (\int_0^{x_t^\lambda} \sin \theta d\theta) dt}{\int_a^b \int_0^\pi \sin \theta d\theta dt}, \ \text{since } A^\lambda \cap B_a^b = B_a^b. \tag{9}$$

But the denominator is $= 2(b - a)$, and (9) gives

$$\int_a^b \int_0^{x_t^\lambda} \sin \theta \ d\theta dt = 2\lambda(b - a), \ 0 \le a < b \le 2\pi \ .$$

By the arbitrariness of b, this implies (with the Lebesgue differentiation theorem again)

$$\int_0^{x_t^\lambda} \sin \theta \ d\theta = 2\lambda, \ 0 \le t < 2\pi \ . \tag{10}$$

With (7) and the fact that $\lambda = P(A^\lambda|B_t)$, we get

$$\lambda = P(A^\lambda|B_t) = \frac{\mu_t(A^\lambda \cap B_t)}{\mu_t(B_t)} = \frac{1}{2} \int_0^{x_t^\lambda} \sin \theta d\theta \ .$$

Since $\mu_t(B_t) = 1$, we get $\mu_t(A) = \frac{1}{2}\int_A \sin\theta \, d\theta, A \in \Sigma(B_t)$. But if μ_t is defined by (8), then (9) and (10) are true, and this shows that there is a choice of μ_t (only one under the conditions) in order that the conditional probability space satisfying a strict Vitali type condition, solves the Borel problem. The latter hypothesis (and the solution) obtains iff the distribution on each meridian is given by (8). Note that by (8) and (9) one has

$$\lambda = P(A^\lambda|B_a^b) = \frac{1}{b-a}\int_a^b \int_{A^\lambda \cap B_t} \sin\theta \, d\theta \, dt$$

$$= \frac{\int_a^b P(A^\lambda|B_t)dt}{\int_a^b dt}, \tag{11}$$

and this same expression holds if A^λ is replaced by $A \in \Sigma$. This also implies that the stricter Vitali condition holds with $F(x) = x$ in (1). Thus it again depends on an auxiliary measure.

Finally we present some comments on this solution.

1. To have a (unique) solution, the Rényi model should also satisfy the Vitali type condition on $\mathcal{B}$. It is tailored to the problem at hand and does not suggest easily other modifications in applying it to different situations. For instance how should this be modified to yield a (unique) solution in the case of the Kac-Slepian paradox? Although the strict Vitali type condition is on $\mathcal{B}$ which is not a ring, its employment (with possibly small changes) will give the ratio representation (Proposition 3.4 or Theorem 3.1). We may also restate and complete the class into a Vitali system as discussed at length in Section 3.4. Then the desired evaluation of such a conditional probability becomes a part of differentiation theory, with $\mathcal{B} \subset \Sigma$ being a σ-ring and the measure space (Ω, Σ, μ) playing a basic role in the construction. Here μ can be σ-finite (or more generally "decomposable").

2. One of the points stressed by Rényi about his representation theorem is that μ is non-finite, but only σ-finite in contrast to the classical Kolmogorov (ordinary) probability model. However, this is not a significant difference. Indeed, in Kolmogorov's definition of conditioning, only that $\mu_\mathcal{B} = \mu|\mathcal{B}$ be σ-finite (or decomposable) is required, since it depends on an employment of the Radon-Nikodým theorem which in fact is the basic ingredient. Thus

$$\mu(B \cap A) = \int_B P^B(A)(\omega)\mu_\mathcal{B}(d\omega), \quad B \in \mathcal{B}, \ A \in \Sigma \,,$$

and one always has $0 \leq P^B(A) \leq 1$ a.e. $[\mu]$ and $P^B(\cdot)$ is σ-additive as well as monotone.

Further comparison and abstraction of the Kolmogorov (conditional) model will be discussed in Chapter 7. It can be inferred from the work there that an abstract version of the classical formulation has the most desirable properties. It appears that the computational problems noted in Chapter 3 are the basic ones awaiting for a satisfactory solution.

4.6 Bibliographical notes

As noted at the beginning of the chapter, the work described here is primarily based on Rényi's seminal paper (1955), on his later extension (1956), and on streamlining in his book (1970). For some modifications of the axioms and consequences one should consult Császár (1955) and also Krauss (1968). The motivation for this new axiomatic approach appears to be a desire to include some problems involving σ-finite spaces in applications, giving rise to standard "regular" conditional probability measures. The Rényi extension, as stated by the author himself, is to include the basic absolute (unconditional) model (Ω, Σ, P) of Kolmogorov's. Since the conditional probability derived from such a measure space is really obtained from an application of the Radon-Nikodým theorem, which is valid for σ-finite (and more general) spaces, this generalization by itself is not the key part. However, the new axiomatic setup *includes* regularity of conditional measures in its formulation, and this is not assumed in the Kolmogorov model. For many problems of the latter model, this property is not available, although for a considerable number of results of practical interest one can find it. This aspect of Kolmogorov's definition will be treated fully in the next chapter.

An interesting point to be noted in the Rényi formulation is that it is necessary to place restrictions by stipulating the Vitali type conditions to include a discussion for the kinds of paradoxes discussed in Chapter 3. But the latter are not resolved in the new axiomatic method either, since here the null sets are simply disallowed in B. But this excludes many important applications. So an extension of Rényi's formulation is needed. In discussing the Bayes formula for densities, he was forced to allow such "null" events into B, and the author presents an *ad hoc* procedure (1955, p. 319) and talks of using "generalized" densities and distributions. This, however, is not that illuminating, and it needs a further development.

The Rényi theory is more of an extension of the elementary condtional model of Chapter 1, and the available treatment in the literature indicates that the new model is not adequate for some important applications. Indeed, many of the functional relations and other structural properties (e.g., iterated conditional expectations, conditioning with a partition etc.) are not easily included in this new method. On the other hand, the classical limit theorems are extendable, using the conditioning class $\mathcal{B}$ as a parameter set and related developments are possible. This however seems to use the basic (unconditional) theory as the main step and the new addition appears to be not so fundamental, although interesting. It is somewhat surprizing that in his later major work on the subject (1970), Rényi, after describing this in Chapter 2, did not employ it again in the theory or applications in the rest of that book.

The new limit theory with some conditioned classes as parameters has been investigated by Parzen (1954) in a more general context, and also prior to Rényi's formulation. We have included just one of his results here, but it clearly shows the importance of conditional limit theorems in asymptotic analysis. It seems unnoticed that the restrictions of the Vitali type used by Rényi under the name "Cavalieri spaces" are close to abstract differentiation theory and its Vitali conditions. We shall see in Chapter 7, that an abstract formulation of Kolmogorov's definition (identifying it as an intrinsic part of vector measure theory) will unify both the new and classical notions. In the special treatments, we see that Kolmogorov's conditioning and Rényi's formulation may yield *different numerical answers* for problems in which the conditioning event has measure zero (where Rényi's ratio representaiton also holds). Thus any further development of this new axiomatic method should take these problems into account, and show how it relates to the Kolmogorov conditioning method. We therefore consider the latter from now on, and compare both of them later in Chapter 7.

Finally we note that there has been some recent effort on an extension of Rényi's conditional probability spaces enlarging both Σ and the set of conditons $\mathcal{B}$, using transfinite methods. For these one may refer to the papers by Kamiński (1985) and Aniszezyk, Burzyk and Kamiński

(1987). As yet these results could not resolve the paradoxes of the type raised by the Kac-Slepian examples discussed in the preceding chapter. In fact one should compare the hypothesis and the resulting procedures of the "Cavalieri spaces" imposed by Rényi and the L'Hôpital type methods employed in many practical problems including those illustrated in the previous chapter. Neither of these ideas belongs to the basic setup of the models and both are *ad hoc* in that the solutions depend on the methods used. In the following chapter we present for a class of problems, for which the regularity of conditional probabilities is available from the original formulation, a computational procedure that is part of the Kolmogorov conditioning method itself and the solutions are unique. These later procedures and applications will be given in Sections 5.6, 5.7 and 7.5 which essentially contain all the currently available results on this problem.

CHAPTER 5: REGULARITY OF CONDITIONAL MEASURES

In this chapter various restrictions on the underlying probability space are given for the (Kolmogorov) conditional probability functions to behave like scalar or ordinary measures. Existence results for such measures, some easily recognizable sufficient conditions for the purpose, and then the related problem of disintegrating a measure as a "convex combination" of regular conditional measures are treated in detail. The latter results have a close relationship with the topology of the underlying space and this is discussed. These existence results are fundamental for theoretical developments. However, no algorithm is available for their constructions in general. But, for a broad class of regular conditional probabilities (and expectations) we include, in the last two sections, constructive computational methods with illustrations to (Brownian motion) processes and functionals on them.

5.1 Introduction

Different problems involving Borel type paradoxes presented in Sections 3.2 and 4.5 indicate that the Rényi axiomatic formulation of conditioning did not indicate any easier methods for answering the associated questions when the conditioning event has probability zero. On the other hand the Kolmogorov set up admits a unique solution, in principle, for these problems through the Doob-Dynkin lemma, although an algorithm (or a method) of explicit calculation is not available. The solution is given as $P^{\mathcal{B}}(A)(\omega) = P(A|X)(\omega) = g(A)(\omega)$ for each event A and $\mathcal{B} = \sigma(X)$, the σ-algebra generated by the conditioning random variable X. But it was noted that $P^{\mathcal{B}}(\cdot)(\omega)$ is not always a measure, although $P^{\mathcal{B}} : \Sigma \to L^p(\mathcal{B}), p \geq 1$, is σ-additive making it a vector measure. This last property will be examined closely in Chapter 7. Since we have seen instances where $P^{\mathcal{B}}(\cdot)(\omega)$ is a genuine probability (e.g., when $\mathcal{B}$ is generated by a partition) and since the latter is the most desirable property for the usual applications so that the classical (Lebesgue) integration methods and results can be freely utilized, we first examine this aspect of conditioning in detail in this chapter. Indeed we present several results for $P^{\mathcal{B}}(\cdot)(\omega)$ to behave as a measure for each ω in $\Omega - N, P(N) = 0$. This is the regularity property of the conditional probability, and it has been studied by several authors.

The subject will be treated in three stages. After finding some general conditions for the existence of such regular measures, we consider the basic probability triple (Ω, Σ, P) with Ω abstract as well as with topology. In both cases the aim is to obtain $P^{\mathcal{B}}(\cdot)(\omega)$ to be a measure. The first case includes the triple to be perfect, compact, pure and such classes. The second stage is to consider topological aspects which show that a probability can be "disintegrated" into (proper) regular conditional measures. Various classes of these results are included here. Both these studies take up the next four sections. The last two sections are then devoted to computations of a class of conditional expectations for certain stochastic processes. Specific statistical applications will be given in the next chapter. Unless the contrary is explicitly stated *the Kolmogorov model is the basic set up in all that follows.*

5.2 Existence of regular conditional probabilities: generalities

If (Ω, Σ, P) is a probability space, $\mathcal{B} \subset \Sigma$ is a σ-algebra and $P^{\mathcal{B}}(\cdot)$ is the conditional probability function on Σ relative to $\mathcal{B}$, then it satisfies the following system of equations:

$$\int_B P^{\mathcal{B}}(A)(\omega)dP_{\mathcal{B}}(\omega) = P(A \cap B), \; A \in \Sigma, \; B \in \mathcal{B} , \qquad (1)$$

where $P_{\mathcal{B}} = P|\mathcal{B}$, by Definition 2.1.1. Recalling the preliminary work of Section 2.2, $P^B(\cdot)$ is σ-additive a.e. [P], but generally $P^B(\cdot)(\omega)$: $\Sigma \to \mathbb{R}^+$ need not be a measure. If this is a measure for all $\omega \in \Omega - N$, $P(N) = 0$, then $P^B(\cdot)$ is termed a *regular conditional probability*, so that $P^B(A)(\cdot)$ is B-measurable and satisfies (1). Since such a measure need not always exist, it is desirable to find conditions on B or (and) on the basic triple (Ω, Σ, P) in order that a regular conditional measure exists. The partition case treated in Chapter 1 is seen to admit a solution. Thus we need to find more general cases for which the desired solution still exists. In particular we intend to show that the image of a conditional probability function by a random variable, called the *conditional probability distribution*, often admits such a regular version.

The first question to be treated is a kind of converse to the elementary case of Section 1.3 on conditioning. More precisely, if B is a countably generated σ-algebra from Σ of (Ω, Σ, P), can we decompose the measure space so that the partition based conditional expectation and probability are obtained? In this case the regularity of P^B is automatic. Under a mild restriction, a positive solution of our special (converse) problem, essentially due to Blackwell (1942), will be presented.

Recall that a set $A \in B$ is an *atom* if for each $B \in B$, either $A \cap B = A$ or $= \phi$. Since $P^B(E)$ is B-measurable for each $E \in \Sigma$, it is clear that the σ-algebra B' generated by $(\{\omega : P^B(E) < a\}, E \in \Sigma, a \in \mathbb{R}^+)$ satisfies $B' \subset B$, and each B-atom is also a B'-atom. Such atoms are termed P^B-*indecomposable*. With these concepts, we have the following result on the structure of some regular conditional probability functions.

Theorem 1. (Blackwell) *Let (Ω, Σ, P) be a probability space with $B \subset \Sigma$ as a σ-algebra containing a countably generated σ-algebra whose atoms are P^B indecomposable. If P^B is regular then Ω and the conditional probability function P^B admit the following representations:*

$$\Omega = N \cup \bigcup_{\alpha \in I} A_\alpha, \ P(N) = 0 \tag{2}$$

$$P^B(E) = \sum_{\alpha \in I} P_{A_\alpha}(E) \chi_{A_\alpha}, \ a.e., \ E \in \Sigma, \ P_{A_\alpha}(A_\alpha) = 1, \tag{3}$$

where $P_{A_\alpha}(\cdot)$ is the elementary conditional probability function and the A_α are disjoint. In particular, if Σ, B or B' is countably generated, then Ω and a regular P^B admit representations (2) and (3).

Remark. Although in (2) and (3) there can be continuum many terms, it follows that for each $E \in \Sigma$, the set $\{\alpha \in I : P(E \cap A_\alpha) > 0\}$ is at most countable. When P^B is regular with (2) and (3) holding, we have for each P-integrable $X : \Omega \to \mathbb{R}$, $E^B(X)$ can be obtained from the elementary formula for $E_{A_\alpha}(X)$, as (cf. Eqs. (3) and (4) in Sect. 2.1)

$$E^B(X) = \sum_{\alpha \in I} E_{A_\alpha}(X)\chi_{A_\alpha}, \quad a.e.,$$

in which at most countably many terms on the right are nonzero.

Proof. Let $\tilde{\mathcal{B}}(\subset \mathcal{B})$ be the σ-algebra generated by a countable class $\{F_n, n \geq 1\}$ whose atoms are P^B-indecomposable, as in the hypothesis. By adding finite unions and complements, if necessary, this collection of F_ns may and will be assumed to form an algebra. In order to derive the decomposition (2), we can obtain all the atoms of $\tilde{\mathcal{B}}$ from the generating set $\{F_n, n \geq 1\}$ as follows. Denote by I the set $\{+1, -1\}^N$ so that I consists of sequences of $+1, -1$. If $\alpha \in I$, let $\varepsilon_n(\alpha)$ denote the nth coordinate of α. Thus $\varepsilon_n(\alpha) = +1$ or -1 and we denote $F_n^{\varepsilon_n(\alpha)}$ to be F_n if $\varepsilon_n(\alpha) = +1$, and $= F_n^c$ if $\varepsilon_n(\alpha) = -1$. Then $B_\alpha = \cap_n F_n^{\varepsilon_n(\alpha)} \in \tilde{\mathcal{B}}$, and is an atom by construction. Also each F_n is a finite or a countable union of B_α's. So the σ-algebra generated by the collection $\{B_\alpha, \alpha \in I\}$ is $\tilde{\mathcal{B}}$ itself. Also $\Omega = \cup_{\alpha \in I} B_\alpha$, and $\mathrm{card}(I) = 2^{\aleph_0}$. We now obtain (2) and (3), using the hypothesis that each B_α is P^B-indecomposable.

By definition of P^B, for each $A \in \Sigma$, $P^B(A)$ is a.e. a constant on B_α since B_α is P^B-indecomposable. Hence

$$P(A \cap B_\alpha) = \int_{B_\alpha} P^B(A)dP_B$$

$$= P^B(A) \int_{B_\alpha} dP_B = P^B(A)P_B(B_\alpha), \quad a.e. \qquad (4)$$

Consequently $P^B(A) = P_{B_\alpha}(A)$, a.e. Also $P^B(F_n) = \chi_{F_n}$ a.e. Let $N_n = \{\omega : P^B(F_n) \neq \chi_{F_n}(\omega)\}$, and $N = \cup_n N_n$. Then N_n and hence N are P_B-null sets, and let $A_\alpha = B_\alpha - N$. It follows that $\Omega = N \cup \cup_\alpha A_\alpha$, and $P_{A_\alpha}(A_\alpha) = P^B(A_\alpha) = 1$ a.e. by (4). Since $P(\Omega) = 1$, we deduce that at most countably many A_α satisfy $P(A_\alpha) > 0$. Hence $P^B(A_\alpha) = \chi_{A_\alpha} = P_{A_\alpha}(A_\alpha) = 1$ by (4). Thus (2) and (3) hold as stated.

Finally if Σ is countably generated then so in $\mathcal{B}$ and the last statement follows from the fact that $\tilde{\mathcal{B}}(\subset \mathcal{B})$ is determined by the collection

$\{P^{\mathcal{B}}(A), A \in \Sigma\}$, and is countably generated. Since $P^{\mathcal{B}}$ is regular, the atoms of $\tilde{\mathcal{B}}$ are $P^{\mathcal{B}}$- indecomposable and $P^{\mathcal{B}}$ is a constant on them. $\square$

The earlier discussion in Section 2.3 implies that regular conditional probability functions need not exist if the basic triple (Ω, Σ, P) is such that the conditioning algebra $\mathcal{B} \subset \Sigma$ is not countably generated. But the last part of the above result shows that if Σ is countably generated then such functions exist. However this can be extended to various subalgebras of Σ, generated by classes (countable families) of random variables. The following is such a result, and is due to Doob (1953). This itself will be used in a number of applications. We analyze its hypothesis and generalize it, in the ensuing sections, to include many more applications.

Theorem 2. (Doob) *Let (Ω, Σ, P) be a probability space and $X : \Omega \to \mathbb{R}^n$ be a random variable such that $X(\Omega)$ is a Borel subset of $\mathbb{R}^n$. If $\mathcal{B}(\subset \Sigma)$ is a σ-algebra and $P^{\mathcal{B}}$ is the conditional probability function relative to $\mathcal{B}$, then $P^{\mathcal{B}}$ restricted to $\mathcal{S} \subset \Sigma$ has a version which is a regular conditional probability function, where $\mathcal{S} = \sigma(X)$ is the σ-algebra generated by X.*

Note that $\mathcal{S}$ is countably generated although $\mathcal{B}$ and Σ need not be, and thus $P^{\mathcal{B}}$ on Σ itself may not be regular.

Proof. The idea is to consider the image measure $P^{\mathcal{B}} \circ X^{-1}$ on the Borel σ-algebra $\mathcal{R}$ of $\mathbb{R}^n$ and, using the topology and separability of $\mathbb{R}^n$, establish the existence of a regular version on $\mathcal{R}$. Then one translates the result to $\mathcal{S} = X^{-1}(\mathcal{R})$ on using the hypothesis that $X(\Omega) \in \mathcal{R}$.

By definition of the inverse image, if $B_0 = X(\Omega)$, then for all $D \subset \mathbb{R}^n - B_0$, $X^{-1}(D) = \phi$ so that $X^{-1}(B_0) = X^{-1}(\mathbb{R}^n) = \Omega$. Define

$$\tilde{Q}_X(B, \omega) = P^{\mathcal{B}}(X^{-1}(B))(\omega), \ B \in \mathcal{R} .$$

Since $X^{-1}(\cdot)$ preserves all set operations, it is clear that $\tilde{Q}_X(\cdot, \omega)$ is σ-additive, and $\tilde{Q}_X(\mathbb{R}^n, \omega) = 1$ a.e. Suppose now that there is a version $Q_X(\cdot, \cdot) : \mathcal{R} \times \Omega \to \mathbb{R}^+$, of $\tilde{Q}_X$ which is a regular conditional measure. (This is verified below.) Then we can deduce the existence of a regular version of $P^{\mathcal{B}}$ on $\mathcal{S}$.

Indeed consider the trace $\mathcal{R}(B_0) = \{A \cap B_0 : A \in \mathcal{R}\}$ where $B_0 = X(\Omega) \in \mathcal{R}$. Then $X^{-1} : \mathcal{R}(B_0) \to \mathcal{S}$ is one-to-one and onto since

$X : \Omega \to B_0$ is onto. If $A \in \mathcal{S}$, then there is a unique $B_1 = X^{-1}(A)$, and with Q_X define $\nu : \mathcal{S} \times \Omega \to \mathbb{R}^+$ by $\nu(A, \omega) = Q_X(B_1, \omega)$, for all $\omega \in \Omega - N, P(N) = 0$, the null set coming with the version Q_X. Then ν is well-defined on $\mathcal{S} \times (\Omega - N)$, such that $\nu(\cdot, \omega) : \mathcal{S} \to \mathbb{R}^+$ is a probability for all $\omega \in \Omega - N$, and $\nu(A, \cdot)$ is $\mathcal{B}$-measurable for $A \in \mathcal{S}$. Moreover,

$$\nu(A, \omega) = Q_X(B_1, \omega) = P^{\mathcal{B}}(X^{-1}(B_1))(\omega) = P^{\mathcal{B}}(A)(\omega), \text{ a.a. } (\omega) .$$

Since $Q_X(B_0^c, \omega) = 0$, a.e., one has $P^{\mathcal{B}}(X^{-1}(D))(\omega) = 0$, a.a. (ω), for all $D \in \mathcal{R}(B_0^c)$. It follows that

$$\nu(A, \omega) = P^{\mathcal{B}}(A)(\omega), \text{ a.a. } (\omega), \ A \in \mathcal{S} . \tag{5}$$

Thus ν is a version of $P^{\mathcal{B}}$, and is a regular conditional measure on $\mathcal{S}$. It remains to show that there is a version Q_X as required above.

Let $r_1, r_2, \dots$, be a countable dense set of reals (e.g., rationals). Define the image measure, the distribution, of $X = (X_1, \dots, X_n)$ as:

$$F_n(r_{i_1}, \dots, r_{i_n}, \omega) = P^{\mathcal{B}}(\{\omega' : X_{i_j}(\omega') < r_{i_j}, 1 \le j \le n\})(\omega) ,$$

$\omega \in \Omega$. Thus by Proposition 2.3.1, there is a null set $N_n = N(r_1, \dots, r_n)$, such that if $N = \bigcup_n N_n$, $P(N) = 0$, and if $\omega \notin N$, $F_n(\dots; \omega)$ is a left continuous, nondecreasing, nonnegative function. Its definition can be extended to all $(x_1, \dots, x_n) \in \mathbb{R}^n$ by setting

$$F_n(x_1, \dots, x_n; \omega) = \begin{cases} \lim_{\substack{r_{i_k} \nearrow x_k \\ 1 \le k \le n}} F_n(r_{i_1}, \dots, r_{i_n}; \omega), & \text{if } \omega \notin N \\ P([X_1 < x_1, \dots, X_n < x_n]), & \text{if } \omega \in N. \end{cases} \tag{6}$$

Then $F_n(., \dots, .; \omega)$ is a distribution function on $\mathbb{R}^n$ for each $\omega \in \Omega$, and $F_n(x_1, \dots, x_n; .)$ is $\mathcal{B}$-measurable. Let

$$Q_X(B, \omega) = \int \cdots_{B} \int dF_n(x_1, \dots, x_n; \omega), \ B \in \mathcal{R} . \tag{7}$$

Then $Q_X(\cdot, \omega) : \mathcal{R} \to \mathbb{R}^+$ is a probability and $Q_X(B, \cdot)$ is $\mathcal{B}$-measurable. From definition of F_n, it follows immediately that

$$Q_X(B, \omega) = P^{\mathcal{B}}(X^{-1}(B))(\omega), \ B \in \mathcal{R}, \text{ a.a. } (\omega) . \tag{8}$$

Thus Q_X is a regular version. In fact, if

$$\mathcal{C} = \{B \in \mathcal{R} : Q_X(B,\omega) = P^{\mathcal{B}}(X^{-1}(B))(\omega), \text{ a.a. } \omega \in \Omega\}$$

then $\mathcal{C}$ contains the semi-ring of half-open intervals of $\mathbb{R}^n$ and is easily shown to be closed under monotone limits. Hence $\mathcal{C} = \mathcal{R}$, by the monotone class theorem. Thus Q_X is a regular version of $P^{\mathcal{B}} \circ X^{-1}$, establishing the earlier supposition. $\square$

Remark. Using the Kolmogorov existence theorem on determining a probability measure based on consistent families of finite dimensional distributions (see Chap. 8, Sect. 3, for a generalization of this result even for conditional measures), the above work may be extended without difficulty to countable families of random variables and then to "separable stochastic processes."

The last part of the above proof indicates that on topological measurable spaces having some distinguished classes of σ-algebras (such as $\mathcal{S}$), one may obtain regular versions of $P^{\mathcal{B}}$. We consider this aspect of the problem in the next section, and then reexamine a combination of (1) and (3) to obtain the "disintegration" of probability measures in the following sections.

5.3 Special spaces admitting regular conditional probabilities

Since a large part of classical stochastic analysis and its interpretation depends on using the conditional probability function as one of an ordinary set of measures, and since there exist objects which do not have this property, it is natural to devote considerable effort in finding such (regular) measures. It will become clear, by the work of Chapter 7, that much of the conditional measure theory extends to an abstract setting. However, in that context, the familiar (frequency) interpretation of the results or the use of Lebesgue's integral, is not as simple or possible as the regular case. This is why one often insists on the regularity aspects, and hence we proceed with this specialized study here in considerable detail.

For a motivation we reexamine the last two results. The point in both cases is to consider regularity on σ-algebras that are relatively small in the sense of their cardinality. This is clarified by the following well-known representation:

1. Let Σ be a countably generated σ-algebra of a nonempty set Ω. Then there exists a separable metric space S such that its Borel σ-algebra $\mathcal{B}$ is σ-isomorphic to Σ, i.e., there is a set mapping $\varphi : \mathcal{B} \to \Sigma$ such that φ preserves countable operations, is one-to-one, and is onto. On the other hand, the result of Theorem 2.2 goes in the opposite direction. More precisely:

2. If (Ω, Σ, P) is any probability space, $\mathcal{A} \subset \Sigma$ is a σ-algebra and $X : \Omega \to \mathbb{R}^n$ is a random vector such that $X(\Omega) = B_0 \in \mathcal{R}$, then $S = X^{-1}(\mathcal{R}(B_0)) \subset \Sigma$, and there is a version of $P^{\mathcal{A}}$ which is regular on $S, \mathcal{R}$ being the Borel σ-algebra (and $\mathcal{R}(B_0)$ its trace on B_0) of $\mathbb{R}^n$.

Here the additional condition on the range of X is used in finding a (generally) smaller S for the purpose. However, for 1., such a random variable can easily be constructed. For instance, if $A_1, A_2, \ldots$, generate Σ, then the (Marczewski) function

$$X(\omega) = \sum_{n=1}^{\infty} 2\chi_{A_n}(\omega)/3^n, \omega \in \Omega, \tag{1}$$

satisfies $X^{-1}(\mathcal{R}) = \Sigma$, although $X(\Omega)$ need not be, in general, a Borel subset of $\mathbb{R}$. But the requirement on $X(\Omega)$ is "nearly" satisfied if (Ω, Σ) is one of the familiar topological measurable spaces. For instance, it is a classical result of Luzin, stating that if (Ω, Σ) is a Borelian space, with Ω as a separable metric space, then for each random variable X, $X(\Omega) \subset \mathbb{R}$ is an analytic set (i.e., the latter is the range of a continuous function on $\mathbb{N}^{\mathbb{N}}$, where $\mathbb{N}$ is the natural numbers with discrete topology. It can be verified that each Borel set is analytic but not conversely–this is why "nearly" is used).

In order to get a result similar to that of Theorem 2.2, or an extension of the latter, on abstract probability, we introduce a (global) restriction, following Gnedenko and Kolmogorov (1954) under the name of perfectness, which controls the ranges of random variables. This concept was analyzed in detail by Sazonov (1962). We give several other conditions analogous to this one and obtain existence results for them, and for a class of topological measure spaces employed in applications.

Definition 1. A probability space (Ω, Σ, P) is *perfect* if for each random variable $X : \Omega \to \mathbb{R}$, there is a Borel set $B \subset X(\Omega)$ of full measure, i.e., $P(X^{-1}(B)) = P(\Omega) = 1$. [Here B usually depends on X.]

To see the usefulness of this concept in applications, first we give some immediate properties of perfectness, and later present several equivalent conditions. Thus let (Ω, Σ, P) be perfect. The following properties are easy consequences of the definition.

(i) For each $A \in \Sigma$, the restriction space $(A, \Sigma(A), P(A \cap \cdot))$ is also perfect.

In fact, let $Y : A \to \mathbb{R}$ be measurable for $\Sigma(A)$. If $X = Y\chi_A$ then $X : \Omega \to \mathbb{R}$ and by the perfectness of (Ω, Σ, P), there is a Borel set $B \subset X(\Omega)$ of full measure for $P \circ X^{-1}$. Let $\tilde{B} = B$ if $0 \in X(A), = B - \{0\}$ if $0 \notin X(A)$. Then $\tilde{B} \subset X(A)$, Borel, and $P(A) = P(X^{-1}(B) \cap A)$ so that $P(A\cap\cdot)$ is perfect. [It is however possible for $(A, \Sigma(A), P|\Sigma(A))$ to be perfect even if $A \notin \Sigma$.]

(ii) For each σ-algebra $\mathcal{S} \subset \Sigma, (\Omega, \mathcal{S}, P_{\mathcal{S}})$ is also perfect.

This is clear since each random variable relative to $\mathcal{S}$ is also one for Σ.

(iii) The image triple for any $f : \Omega \to \tilde{\Omega}$, with $\tilde{E} = \{\tilde{A} \subset \tilde{\Omega} : f^{-1}(\tilde{A}) \in \Sigma\}$ and $\tilde{P}_f = P \circ f^{-1}$, denoted $(\tilde{\Omega}, \tilde{\Sigma}, \tilde{P}_f)$, is again perfect.

Indeed, if $\mathcal{S} = f^{-1}(\tilde{\Sigma}) \subset \Sigma$, and $h : \tilde{\Omega} \to \mathbb{R}$ is a $(\tilde{\Sigma})$-measurable function, then $g = h \circ f$ is a random variable for $\mathcal{S}$, and by (ii) $(\Omega, \mathcal{S}, P_{\mathcal{S}})$ is perfect. So there is a Borel set $B \subset g(\Omega) = h(f(\Omega))$ of full $\tilde{P}_f$-measure. Thus

$$1 = P(g^{-1}(B)) = P_f(h^{-1}(B)), \ B \subset g(\tilde{\Omega}) .$$

Since $P_f(\tilde{\Omega}) = 1$, this shows that $(\tilde{\Omega}, \tilde{\Sigma}, \tilde{P}_f)$ is perfect.

(iv) If (Ω, Σ, P) is perfect, then its completion $(\Omega, \hat{\Sigma}, \hat{P})$ is also perfect (and the converse holds by (ii)).

To see this, if $X : \Omega \to \mathbb{R}$ is a random variable for $\hat{\Sigma}$, then, by the standard measure theory, there is a random variable $f : \Omega \to \mathbb{R}$ such that $X = f$ a.e. So if $\Omega_0 = \{w : X(w) \neq f(w)\}$, then $\hat{P}(\Omega_0) = 0$, and $\Omega - \Omega_0 = \tilde{\Omega}$, has full measure. Hence by (i) $(\tilde{\Omega}, \Sigma(\tilde{\Omega}), P(\tilde{\Omega} \cap \cdot))$ is perfect, and there is a Borel set $B \subset f(\tilde{\Omega}) \subset X(\Omega)$ of full measure. So the completion is perfect.

The implication of this concept for regular conditioning is illuminated by the following result:

Theorem 2. *Let (Ω, Σ, P) be a perfect probability space, and let $\mathcal{S}, \mathcal{B}$ be two σ-subalgebras of Σ such that $\mathcal{S}$ is countably generated. Then there is a regular perfect version $Q(\cdot, \cdot)$ of the conditional probability function $P^{\mathcal{B}}(\cdot)(\cdot)$ on $\mathcal{S}$ in the sense that $Q(\cdot, \cdot) : \mathcal{S} \times \Omega \to \mathbb{R}^+$ satisfies the conditions:*

(a) *$Q(A, \cdot) : \Omega \to \mathbb{R}^+$ is $\mathcal{S}$-measurable*
(b) *$(\Omega, \mathcal{S}, Q(\cdot, \omega))$ is a perfect probability space for each $\omega \in \Omega$, and*
(c) *Q satisfies the functional equation,*

$$\int_B Q(A, \omega) P_{\mathcal{B}}(d\omega) = P(A \cap B), \quad A \in \mathcal{S}, \ B \in \mathcal{B} . \tag{2}$$

Proof. The result will be established using the proof of Theorem 2.2 together with some of the properties of perfectness noted above. Here are the details.

Let $\{A_n, n \geq 1\}$ be a sequence of generators of $\mathcal{S}$, and let $X : \Omega \to \mathbb{R}^+$ be the $\mathcal{S}$-measurable (Marczewski) function given in (1) for this sequence. If $P^{\mathcal{B}}$ is the conditional probability measure, define

$$Q_X(B, \omega) = P^{\mathcal{B}}(X^{-1}(B))(\omega), \quad B \in \mathcal{R}, \ \omega \in \Omega . \tag{3}$$

Then with the separability of $\mathbb{R}$, an identical argument between (5)-(8) of the preceding section shows that $Q_X : \mathcal{S} \times \Omega \to \mathbb{R}^+$ is well-defined and satisfies conclusions (a) and (c) of the theorem. Using the perfectness of P we now deduce the same property of $Q_X(\cdot, \omega)$, i.e. (b) will be established.

Since $X(\Omega) \subset \mathbb{R}^+$ need not be a Borel set, we can find, by perfectness of P, a Borel set $B_0 \subset X(\Omega)$ of full measure. Thus $Q_X(B_0, \omega) = 1$, a.a. (ω). On the exceptional P-null set here, let $Q_X(\cdot, \omega) = \tilde{Q}_1(\cdot)$, an arbitrary perfect, or Lebesgue-Stieltjes, probability on $\mathbb{R}$. If we now set $Q(A, \omega) = Q_X(B, \omega), A = X^{-1}(B), B \in \mathcal{R}$, and $\omega \in \Omega$, then $Q(\cdot, \cdot)$ is asserted to be the required perfect version.

Indeed, let $g : \Omega \to \mathbb{R}$ be $\mathcal{S}$-measurable. It is to be shown that there is a Borel set $B_1 \subset g(\Omega)$ of $Q_X(\cdot, \omega)$-full measure. Since $\mathcal{S} = X^{-1}(\mathcal{R}) \subset \Sigma$, and $(\Omega, \mathcal{S}, P_{\mathcal{S}})$ is perfect by property (ii), let $\tilde{\mathbb{R}} = X(\Omega)$

and $\tilde{\mathcal{B}} = \{\tilde{B} \subset \tilde{\mathbb{R}} : X^{-1}(\tilde{B}) \in \mathcal{S}\}$. [This contains $\mathcal{R}(\tilde{\mathbb{R}})$.] Then by the Doob-Dynkin lemma, there is a $g_1 : \tilde{\mathbb{R}} \to \mathbb{R}$, measurable for $\tilde{\mathcal{B}}$, such that $g = g_1 \circ X$. Since $B_0 \subset \tilde{\mathbb{R}}$, let $g_0 = g_1|B_0$, the restriction. Then g_0 is $\tilde{\mathcal{B}}$-measurable, and since $Q_X(\cdot, \omega)$ is a Lebesgue-Stieltjes measure on $(\mathbb{R}, \mathcal{R})$ it is perfect (in fact it is a "compact" measure and hence is automatically perfect as the definition and discussion of compactness below show this clearly). Hence there is a Borel set $B_2 \subset g_0(B_0) \subset g_1(\tilde{\mathbb{R}}) = g(X(\Omega))$ of full measure or $Q_X(B_2, \omega) = 1$. Thus

$$g^{-1}(B_2) = (g_1 \circ X)^{-1}(B_2) = X^{-1}(g_1^{-1}(B_2)) \supset X^{-1}(g_0(B_2)) \,,$$

and then

$$1 \geq Q(g^{-1}(B_2), \omega) \geq Q(g_0^{-1}(B_2), \omega) \geq Q_X(B_0, \omega) = 1 \,.$$

This implies that $Q(\cdot, \omega)$, $\omega \in \Omega$, is perfect. But by construction of $Q(\cdot, \cdot)$, it is a version of $P^{\mathcal{B}}$ on $\mathcal{S}$, exactly as in Theorem 2.2, and so (b) also holds. $\square$

In view of this result, which supplements Theorem 2.2 in some respects, we present several equivalent and (certain other) closely related conditions of perfectness. These will illuminate the intricate structure of regular conditional probability functions.

A natural question here is to find whether the hypothesis that $\mathcal{S}$ be countably generated is necessary. There are examples showing that for general σ-algebras, a regular conditional probability function, of the type asserted in the theorem, need not exist. Even if $Q(\cdot, \omega)$ is desired to be a probability (not necessarily perfect), a negative answer prevails for general $\mathcal{S}$, which is not countably generated and $\mathcal{S} \underset{\neq}{\subset} \Sigma$. For a discussion of such examples and references, one may refer to an article by D. Ramachandran (1981).

The following concepts are relevant in isolating classes of measure spaces with "nice" properties and which admit regular conditioning:

1. A probability measure P on (Ω, Σ) is called *compact* on Σ, if there is a class $\mathcal{C} \subset \Sigma$ with the property that $C_n \in \mathcal{C}$, $\bigcap_{n=1}^{\infty} C_n = \phi$ implies $\bigcap_{i=1}^{n_0} C_i = \phi$ for some $1 \leq n_0 < \infty$, ($\mathcal{C}$ is a *compact class*), then

$$P(A) = \sup\{P(C) : C \subset A, C \in \mathcal{C}\} \,, \quad A \in \Sigma \,. \tag{4}$$

This concept was introduced by E. Marczewski in 1953.

2. A probability space (Ω, Σ, P) (or simply P on Σ) is *quasi-compact* if for each $\varepsilon > 0$, and each $\{A_n, n \geq 1\} \subset \Sigma$, there is an $A_\varepsilon \in \Sigma$ such that $P(A_\varepsilon) > 1 - \varepsilon$ and $\{A_\varepsilon \cap A_n, n \geq 1\}$ becomes a compact class. This concept, also introduced in 1953, is due to C. Ryll-Nardzewski. Since when $\{A_n, n \geq 1\} \subset C, A_\varepsilon = \Omega$ is possible, this is more general than compactness.

3. If (Ω, Σ_0) is a pair with Σ_0 as an algebra of $\Omega, P : \Sigma_0 \to \mathbb{R}^+$ is additive, then a subring $\mathcal{S}_0(P)$ is called *P-pure* if (a) $A_n \in \mathcal{S}_0(P)$, disjoint, $\cup_n A_n \in \mathcal{S}_0(P)$, then $P(\underset{n \geq n_0}{\cup} A_n) = 0$ for some $1 \leq n_0 < \infty$, and (b) the Carathéodory outer measures generated by (P, Σ_0) and $(P, \mathcal{S}_0(P))$ are identical. An additive function $P : \Sigma_0 \to \mathbb{R}^+$ is termed *pure* if there is a P-pure subring $\mathcal{S}_0(P) \subset \Sigma_0$.

The above definition can be given a simpler form on noting the fact that every pure additive function is in fact σ-additive and hence a measure. We present the short proof for completeness. This will enable us to relate the concepts of purity and compactness of probability measures.

Lemma 3. *If an additive $P : \Sigma_0 \to \mathbb{R}^+$ is pure then it is σ-additive.*

Proof. Let $P : \Sigma_0 \to \mathbb{R}^+$ be additive and $\mathcal{S}_0(P) \subset \Sigma_0$ be a P-pure ring. If $A_n \in \mathcal{S}_0(P)$, disjoint, such that $A = \underset{n}{\cup} A_n \in \mathcal{S}_0(P)$, then $P(\underset{n > n_0}{\cup} A_n) = 0$ by hypothesis, $1 \leq n_0 < \infty$. Since $A_k \subset \underset{n > n_0}{\cup} A_n$, $k > n_0$, and $P(\cdot)$ is monotone, $P(A_k) = 0$. Then

$$P(A) = P(\overset{n_0}{\underset{k=1}{\cup}} A_k) + P(\underset{k > n_0}{\cup} A_k) = \overset{n_0}{\underset{k=1}{\Sigma}} P(A_k) = \overset{\infty}{\underset{k=1}{\Sigma}} P(A_k) \, .$$

If P^* is the outer measure generated by $(P, \mathcal{S}_0(P))$, then by hypothesis this is also the same as that generated by (P, Σ_0). By the standard theory (see Rao, 1987, p. 41) if $\mathcal{M}_{P^*}$ is the collection of P^*-measurable sets, then $\mathcal{S}_0(P) \subset \Sigma_0 \subset \mathcal{M}_{P^*}$, and $P^*|\mathcal{M}_{P^*}$ is σ-additive. It is Carathéodory regular (i.e., has no more σ-additive extensions by this process), and $P^*|\Sigma_0 = P$. Hence P has a unique extension to $\sigma(\Sigma_0)$ whose completion is $\mathcal{M}_{P^*}$. This shows that P is a measure on Σ_0 if it is pure. $\square$

With this fact, an alternative definition of purity of a probability can be given and it simplifies the discussion.

3′. If (Ω, Σ, P) is a probability space, then a subring Σ_0 of Σ (depending on P) is P-pure provided

$$P(A) = \inf\{ \sum_{n \geq 1} P(A_n) : A \subset \bigcup_n A_n, A_n \in \Sigma_0\}, A \in \Sigma , \qquad (5)$$

and for any $A_n \downarrow \phi, P(A_n) \downarrow 0$ implies that $P(A_n) = 0$, $n \geq n_0$ for some $1 \leq n_0 < \infty$. And P is *pure* if there exists a P-pure ring in Σ.

In view of the above lemma and the classical Carathéordory theory, we can present a proof of the following:

Proposition 4. *Let (Ω, Σ, P) be a probability space. Then P is a pure measure iff it is a compact measure in the sense of 1., above.*

Proof. Let P be a compact probability. Then, by definition, there is a compact class $\mathcal{C} \subset \Sigma$. It was shown by Marczewski (1953) that $\mathcal{C}$ can be enlarged to the class which is closed under finite unions and countable intersections. With this let $\mathcal{J} = \{A^c : A \in \mathcal{C}\} \cup \{\phi, \Omega\}$. Then $\mathcal{J}$ is a topology (called the "σ-topology") and then the compactness condition implies that P is a Radon probability in this topology on the algebra $\mathcal{A}(\mathcal{C})$ generated by $\mathcal{C}$, i.e.,

$$P(A) = \sup\{P(C) : C \subset A, C \in \mathcal{A}(\mathcal{C}), C \text{ compact for } \mathcal{J}\}.$$

Then by the known Henry extension theorem, P has an extension to be a Radon measure on the Borel σ-algebra $\mathcal{B}$, generated by the $\mathcal{J}$-open sets of Ω. Moreover, this extension is unique since $\mathcal{J}$ is generated by $\mathcal{C}$. (For these standard results, see, Rao, 1987, p. 76 and Ex. 11 on p. 80.) Since $\mathcal{B} = \sigma(\mathcal{A}(\mathcal{C}))$, and the classical Carathéodory extension procedure applied to the couple $(P, \mathcal{A}(\mathcal{C}))$ shows that P is pure relative to $\mathcal{A}(\mathcal{C})$, a P-pure algebra.

For the converse, let P be a probability relative to a P-pure ring $\mathcal{A}(P)$. We may assume that (Ω, Σ, P) is complete also, and then P is pure relative to $\mathcal{A}(P)$. Now one introduces a topology $\mathcal{J}_1$ through a lifting map in Ω (see Rao, 1987, p. 433). With this one concludes that $\mathcal{A}(P)$ is a compact class so that P is compact relative to this class. Indeed, if $\mathcal{J}_1$ is this topology, then by the above stated result the condition on $\mathcal{A}(P)$ [namely, $\tilde{\mathcal{A}} = \{\lambda(A) : A \in \mathcal{A}(P)\}$ is a compact class where $\lambda : \Sigma \to \Sigma$ is the set mapping $\lambda(A) = \rho(\chi_A)$ with ρ as a lifting on $\mathcal{L}^\infty(P)$, so that $\lambda(A) = \phi$ iff A is P-null] is precisely the compactness condition of 1., above. This is the assertion of the proposition. □

Remark. The converse part was proved by J. Pachl (1975). The direct part (without proof) and the introduction of the pure measure concept were given by the author in 1970. Some properties of pure measure were stated there somewhat in outline. That was intended only for (Carathéodory) regular measures. In view of the lifting topology used above (but its relevance was not noticed then), all those assertions are now consequences of the earlier work on compact measures.

The preceding concepts are related as follows:

Theorem 5. *Let (Ω, Σ, P) be a probability space. Then the following implications hold:*

(i) P is pure iff P is compact,

(ii) P is compact implies P is quasi-compact, and

(iii) P is quasi-compact iff P is perfect.

The first equivalence is Proposition 4, the second implication is simple as noted before, and the last equivalence is given by Ryll-Nardzewski (1953). (See also Sazonov, 1962, and Theorem 6 below.) Instead of including the details, since perfectness is a more general class for which regular conditional probabilities exist, by Theorem 2, we shall give some equivalent conditions and properties of perfectness to be used in applications. By the preceding result, regular conditional measures exist for all these classes. We include some complements to Theorem 2.

Theorem 6. (Sazonov) *For a probability space (Ω, Σ, P) the following are equivalent statements:*

(i) P is perfect,

(ii) for each measurable $f : \Omega \to \mathbb{R}$, each E in $\mathcal{S}_f = \{A \subset \mathbb{R} : f^{-1}(A) \in \Sigma\}$ and $\varepsilon > 0$, there is an open $G \supset A$ such that $P(f^{-1}(A)) > P(f^{-1}(G)) - \varepsilon,$

(iii) for each $E_n \in \Sigma, n \geq 1$, and $\varepsilon > 0$, there is an $F \in \Sigma$ such that $P(F) > 1 - \varepsilon$ and $\{F \cap E_n, n \geq 1\}$ is a compact class, and

(iv) for each countably generated σ-algebra $\mathcal{A} \subset \Sigma, (\Omega, \mathcal{A}, P_{\mathcal{A}})$ is compact.

If (Ω, Σ) is moreover a Borelian space with Ω as a separable metric space, then the following is equivalent to (i) - (iv) above:

(v) P is a Radon probability on Σ.

It is clear from these results that separability of various classes, suitably formulated, is essential for a nice behavior of P. One says that an additive $P : \Sigma \to \mathbb{R}^+$ is a τ-*smooth* function if for each decreasing net $A_\alpha \in \Sigma, A_\alpha \downarrow \phi$ we have $P(A_\alpha) \downarrow 0$. Thus a τ-smooth function is σ-additive, and so is a measure. Clearly not every probability can be τ-smooth. There exist τ-smooth probabilities which are not perfect and there are perfect probabilities that are not τ-smooth, so that these are distinct conditions on P. In general, it can happen that a τ-smooth perfect probability on a topological space is not Radon, although a Radon measure is always perfect. However, on a metric space $\Omega, (\Omega, \Sigma)$ Borelian, a probability P is Radon iff it is τ-smooth and perfect. On these matters and for a proof of Theorem 6, we refer the reader to Sazonov (1962). Part of this study was motivated by the work due to Blackwell (1955).

In view of the above discussion, it is reasonable to restrict oneself to separable metric spaces Ω, and Borelian (Ω, Σ). For such spaces the following result complements Theorem 2, and we present different proofs.

Theorem 7. *Let (Ω, Σ, P) be a probability space with Ω as a complete separable metric space, also called a Polish space, and Σ as its Borel σ-algebra. Then for each σ-algebra $\mathcal{B} \subset \Sigma, P^{\mathcal{B}}(\cdot)$ has a version $Q(\cdot, \cdot) : \Sigma \times \Omega \to \mathbb{R}^+$, that is a regular conditional probability.*

Remark. We present three proofs with and without using the preceding work. It will illuminate the subject and give a better understanding of the restrictiveness of regularity in general cases.

First proof. This demonstration is short, but it uses several "big" theorems. Since Ω is a Polish space and (Ω, Σ) is Borelian, by a classical theorem, P is both inner and outer regular, i.e., it is Radon. (For a proof of this familiar result, see Parthasarathy (1967), p. 29, Thm. 3.2.) Hence P is perfect by Theorem 5 above. [This is given in Gnedenko and Kolmogorov (1956), p. 18, Thm. 1, and is proved also in the above reference, p. 31, Thm. 4.1.] The separability of Ω implies that Σ is countably generated. [This follows by taking the generators as all open spheres with rational radii and centers at a dense set of points of Ω. Alternatively Ω is Borel isomorphic to $[0,1]$, by a classical result of C. Kuratowski, and this shows that Σ and $\mathcal{B}([0,1])$ are σ-isomorphic. But $\mathcal{B}([0,1])$ is countably generated.] So for any $\mathcal{B} \subset \Sigma$, taking $\mathcal{S} = \Sigma$ in Theorem 2, (and $X = $ identity) we deduce the existence of $Q(\cdot, \cdot) : \Sigma \times \Omega \to \mathbb{R}^+$ such that $Q(\cdot, \omega)$ is a perfect probability, $Q(A, \cdot)$ is $\mathcal{B}$-measurable,

$$\int_B Q(A, \omega) dP_{\mathcal{B}} = P(A \cap B), \ A \in \Sigma, \ B \in \mathcal{B}, \tag{6}$$

and that $Q(A, \omega) = P^{\mathcal{B}}(A)(\omega)$ for a.a. (ω). This gives more than what is asserted in the theorem. $\square$

Second proof. This time we do not need to go for the named theorems as in the preceding one, but will only use Theorem 2.2 and some simpler results, so that the argument is self-contained.

We recall the classical fact that every separable metric space has a totally bounded metric and conversely. Thus a Polish space Ω has again a totally bounded metric which is not complete (unless Ω is compact). (For a proof, see Dugundji, 1966, p. 298.) So let $C_b(\Omega_d)$ be the space of bounded uniformly continuous real functions on Ω_d, the space Ω with a totally bounded metric d. Then $C_b(\Omega_d)$ is separable (and its functions separate the points of Ω_d). Let $\{f_n, n \geq 1\}$ be a countable dense set of $C_b(\Omega_d)$. It may be assumed that these are linearly independent functions. Although $f_n(\Omega_d) \subset \mathbb{R}$ is analytic, by the earlier result of Luzin's, it is not necessarily Borel and hence Theorem 2.2 can not be applied directly. So a detour with an additional argument is needed. Thus let $\tilde{\Omega}_d$ be the completion of Ω_d under the metric d. Then $\tilde{\Omega}_d$ is compact and each f_n extends uniquely to a continuous function $\tilde{f}_n$ onto $\tilde{\Omega}_d$ (because of its uniform continuity). But then $\tilde{f}_n(\tilde{\Omega}_d)$ is compact (hence Borel) subset of $\mathbb{R}$. Let $\tilde{\Sigma}$ be the Borel σ-algebra of $\tilde{\Omega}_d$. Considering $X_n = (\tilde{f}_1, \ldots, \tilde{f}_n)$, we see that $X_n(\tilde{\Omega}_d) \subset \mathbb{R}^n$ is compact, and if $\mathcal{R}_n$ is the Borel σ-algebra of $\mathbb{R}^n$, and if $\tilde{\mathcal{S}}_n = X_n^{-1}(\mathcal{R}_n) \subset \tilde{\Sigma}$, then $\tilde{\mathcal{S}}_n \subset \tilde{\mathcal{S}}_{n+1}$; and by the linear independence of f_n's, it follows that $\tilde{\Sigma} = \sigma(\underset{n}{\cup} \tilde{\mathcal{S}}_n)$.

For each n, we have satisfied the hypothesis of Theorem 2.2 and hence there is a regular (conditional) probability function $Q_n(\cdot, \omega)$ on $\tilde{\mathcal{S}}_n$ for each $\omega \in \tilde{\Omega}_d - N_n, P(N_n) = 0, Q_n(A, \cdot)$ is $\mathcal{B}$-measurable and $Q_n(\cdot, \cdot)$ satisfies the functional equation:

$$\int_B Q_n(A, \omega) P_{\mathcal{B}}(d\omega) = P(A \cap B), \ A \in \tilde{\mathcal{S}}_n, \ B \in \mathcal{B}, \tag{7}$$

where $\mathcal{B}$ is regarded as a σ-subring of $\tilde{\Sigma}$; and the above equation holds again, since the Radon-Nikodýn theorem (on which it is based) is valid for finite measures on σ-rings, (see Berberian, 1965, p. 160. Thm. 1 or Rao (1987) p. 271, Ex. 14). If $\Pi_{mn} : \mathbb{R}^n \to \mathbb{R}^m (m < n)$ is the coordinate projection, then it follows from the definition of X_n that $\Pi_{mn}(X_n) = X_m$, and hence that

$$Q_{n+1}(\cdot,\omega)|\tilde{\mathcal{S}}_n = Q_n(\cdot,\omega), \omega \in \tilde{\Omega}_d - N ,$$

where $N = \underset{n}{\cup} N_n$, $P(N) = 0$. But this is just the compatibility condition for the regular measures $\{Q_n(\cdot,\omega), n \geq 1\}$ so that by the Kolmogorov existence theorem (actually an easy special case of it for sequences, cf., Rao, 1987, p. 366, or Theorem 8.3.2 below), there is a unique regular probability $Q(\cdot,\omega)$ such that $Q(\cdot,\omega)|\tilde{\mathcal{S}}_n = Q_n(\cdot,\omega), \omega \in \tilde{\Omega}_d - N$. From this we deduce that $Q(\cdot,\omega)$ satisfies (7) on $\underset{n}{\cup} \tilde{\mathcal{S}}_n$ and hence on $\tilde{\Sigma}$ itself by the standard Hahn extension theorem. It is now evident that (7) may be written as

$$\int_B \int_{\tilde{\Omega}_d} \tilde{f}(\omega')Q(d\omega',\omega)P(d\omega) = \int_B \tilde{f}(\omega')P(d\omega'), \tilde{f} \in C_b(\tilde{\Omega}_d) , \qquad (8)$$

for all $B \in \mathcal{B}$.

To see that $\tilde{\Omega}_d$ can be replaced by Ω_d here, let $K \subset \Omega$ be a compact set. Then [K is also compact in $\tilde{\Omega}_d$ so that] there exists a decreasing sequence $\tilde{h}_n \in C_b(\tilde{\Omega}_d)$, where $h_n \in C_b(\Omega_d)$, such that $\tilde{h}_n \downarrow \chi_K$. Hence, with the bounded convergence theorem, (8) becomes (with $\tilde{f}$ replaced by $\tilde{h}_n$)

$$\int_{\tilde{\Omega}} Q(K,\omega)P(d\omega) = P(K) . \qquad (9)$$

But P is a regular measure on $(\Omega,\Sigma), P(\Omega) = 1, (\Omega = \Omega_d)$. Consequently, for all n, there are compact sets $K_n \subset \Omega$, such that $P(K_n) > 1 - \frac{1}{n}$. It may be assumed that $K_n \subset K_{n+1}$ (replacing this by their partial finite unions). So if $\Omega_0 = \underset{n}{\cup} K_n$, then $P(\Omega_0) = 1$, and by (9), $Q(\Omega_0,\omega) = 1$ for each $\omega \in \Omega - (N \cup \tilde{N})$ with $\tilde{N} = \Omega - \Omega_0, P(\tilde{N}) = 0$. Let $Q(\cdot,\omega) = P$ for $\omega \in N \cup \tilde{N}$. Then $Q(\cdot,\cdot)$ is a regular conditional probability function on Σ for each $\omega \in \Omega$, and it satisfies (6), so that it is a version of $P^{\mathcal{B}}$. $\square$

Third proof. This time we avoid use of Theorem 2.2 as well as the Kolmogorov existence theorem in exchange for the Riesz(-Markov) theorem.

We again start with $C_b(\Omega_d)$ as in the above proof, and select a linearly independent collection $\{f_n, n \geq 1\}$ dense in $C_b(\Omega_d)$, in which $f_1 = 1$. However, the basic ideas of proof of Theorem 2.2 enter in a different form. Let $g_n = E^{\mathcal{B}}(f_n)$, the conditional expectation, so that $g_1 = 1$ a.e. Using the separability of $\mathbb{R}^n$, define

$$D_n = \{(r_1, \ldots, r_n) \in \mathbb{R}^n : \sum_{i=1}^{n} r_i f_i(\omega) \geq 0, \ r_i \text{ rational}, \ \omega \in \Omega_d\} .$$

Then D_n is countable and since $E^{\mathcal{B}}$ is order preserving, $\sum_{i=1}^{n} r_i g_i \geq 0$, a.e. Thus there is a $P_{\mathcal{B}}$-null set $N(r_1, \ldots, r_n)$ such that $\sum_{i=1}^{n} r_i g_i(\omega) \geq 0$ for all $\omega \in N^c(r_1, \ldots, r_n)$. Let $N = \bigcup_{n=1}^{\infty} \cup \{N(r_1, \ldots, r_n) : r_i \in D_n, 1 \leq i \leq n\}$. Then $P(N) = 0$. If $\omega \in \Omega - N$, and $\mathcal{L}$ is the linear span of $\{f_n, n \geq 1\}$, so that $f = \sum_{i=1}^{n} s_i f_i \in \mathcal{L}, s_i \in \mathbb{R}$, we define

$$\ell_\omega(f) = \sum_{i=1}^{n} s_i g_i(\omega) .$$

By the linear independence of f_n's and the positivity of $E^{\mathcal{B}}$, we deduce that $\ell_\omega(\cdot)$ is well-defined on $\mathcal{L}, \ell_\omega(f) \geq 0$ for $f \geq 0$ and $\ell_\omega(1) = 1$. [The positivity of ℓ_ω is clear if in the above definition s_i are rational, and in the general case the same conclusion follows by approximation since rationals are dense in $\mathbb{R}$ and $E^{\mathcal{B}}$ is a contraction.] Next using the density of $\mathcal{L}$ in $C_b(\Omega_d)$, we extend $\ell_\omega(\cdot)$ preserving the same positivity and boundedness properties first for $0 \leq f \leq 1$, and then by linearity for all of $C_b(\Omega_d)$.

To use the representation theorem, we embed $C_b(\Omega_d)$ into $C_b(\tilde{\Omega}_d)$ densely in the latter, where $\tilde{\Omega}_d$ is the completion of Ω_d (hence $\tilde{\Omega}_d$ is compact), as in the last proof, so $C_b(\tilde{\Omega}_d) = C(\tilde{\Omega}_d)$. If $\tilde{\ell}_\omega$ is the (uniquely) extended functional from $C_b(\Omega_d)$ onto $C(\tilde{\Omega}_d)$ which is positive, linear, and $\tilde{\ell}_\omega(1) = 1$, we can now invoke the classical Riesz-Markov theorem (see Rao, 1987, p. 471), to obtain a unique Radon probability $Q(\cdot, \omega)$ such that

$$\tilde{\ell}_\omega(\tilde{f}) = \int_{\tilde{\Omega}_d} \tilde{f}(\omega') Q(d\omega', \omega), \ \tilde{f} \in C(\tilde{\Omega}_d), \omega \in \tilde{\Omega}_d . \tag{10}$$

Hence in particular, for each $n \geq 1$,

$$g_n(\omega) = \ell_\omega(f_n) = \int_{\tilde{\Omega}_d} \tilde{f}_n(\omega') Q(d\omega', \omega), \ \omega \in \Omega_d . \tag{11}$$

This implies that the right side, as a function of ω is $\mathcal{B}$-measurable for all $f \in \mathcal{L}$, and then for all $\tilde{f} \in C(\tilde{\Omega}_d)$. It satisfies also

$$\int_B \int_{\tilde{\Omega}_d} \tilde{f}(\omega')Q(d\omega',\omega)P_B(d\omega) = \int_B f(\omega)P(d\omega), \; B \in \mathcal{B} \; . \qquad (12)$$

Now using the regularity of P, we can apply the same argument as in the last part of the preceding proof to deduce that $Q(\Omega_0, \omega) = 1$, a.a. (ω) with $\Omega_0 \subset \Omega$ as a σ-compact set. Then (12) again implies that $Q(\cdot,\cdot)$ is a version of $P^{\mathcal{B}}$ as desired. $\square$

Remark. The third proof is adapted from Stroock and Varadhan (1979). The Kolmogorov existence theorem used in the second proof is available in a more general form, even for arbitrary Hausdorff spaces, as given by Prokhorov and extended by Bourbaki (see Rao, 1987, p. 359).

A problem intimately related to regularity of conditional measures, which we have not emphasized thus far, should be noted. From the general definition of conditioning, presented in Chapter 2, it follows that, for each $B \in \mathcal{B}$,

$$E^{\mathcal{B}}(\chi_B) = P^{\mathcal{B}}(B) = \chi_B, a.e.$$

When does this imply that for all $\omega \in B \in \mathcal{B}$, for a regular conditional version $Q(\cdot,\cdot)$ of $P^{\mathcal{B}}$, we have $Q(B,\omega) = 1$? If this holds, then $Q(\cdot,\cdot)$ is called a *proper* regular conditional probability function. In general a proper regular version of $P^{\mathcal{B}}$ need not exist even if a regular version is available. The sharper result answering this question rather than merely obtaining regularity of $Q(\cdot,\cdot)$, is given by the following theorem due to Blackwell and Ryll-Nardzewski (1963):

Theorem 8. *Let Ω be a Polish space, Σ be its Borel σ-algebra and P be a probability on Σ. If $X : \Omega \to \mathbb{R}$ is a random variable, $\mathcal{B} = X^{-1}(\mathcal{R}) \subset \Sigma$ where $\mathcal{R}$ is the Borel σ-algebra of $\mathbb{R}$, let $Q(\cdot,\cdot)$ be a regular version of the conditional probability function, $P^{\mathcal{B}}$, guaranteed by Theorem 7. Then $Q(\cdot,\cdot)$ is proper on $\mathcal{B}$, (i.e., for each $\omega \in B \in \mathcal{B}$, $Q(B,\omega) = 1$) iff there exists a function $g : \Omega \to \Omega$ such that (i) $g^{-1}(\Sigma) \subset \mathcal{B}$, and (ii) $X(g(\omega)) = X(\omega)$ for all $\omega \in \Omega$. The existence of such a g signifies that the range $X(\Omega)$ of X is necessarily a Borel subset of $\mathbb{R}$.*

This result shows that the condition on the range assumed in Theorem 2.2 is necessary for the good behavior of a version of $P^{\mathcal{B}}$. Although we have not yet found methods of construction of these well-behaved versions to use in concrete applications, it is important to know, for theoretical developments of the subject, that such functions do exist for a large class of measure spaces. We include a proof of this result in the next section which is better suited to go with a class of propositions of similar nature for topological measure spaces. We thus turn to the work related to regular versions in the general theory of conditioning.

5.4 Disintegration of probability measures and regular conditioning

The preceding section already indicates that the existence of regular versions of a conditional probability function is a deeper problem than one might have anticipated from the results of the earlier chapters. In fact this is only a slightly disguised form of another problem in the general measure theory of considerable depth and interest. To appreciate the finer points of the subject, we need to consider the question of disintegration of a measure and then identify both the problems. This will give a better perspective.

To motivate the concept, observe that a key requirement in finding a regular version, when it exists, is one of choosing a member from the equivalence class of functions $\{P^{\mathcal{B}}(A)(\cdot),\ A \in \Sigma\}$ in such a way that the additivity and particularly the order preserving (or positivity) properties are preserved. But this is nontrivial if $\mathcal{B}$ is not generated by a simple partition. In fact, this problem is not special for the subject of conditioning. It is inherent in Lebesgue's integration theory which ignores the sets of measure zero. For instance, when we consider the real L^p-spaces, $1 \leq p \leq \infty$, their elements are equivalence classes of functions. If one has to deal with (not necessarily countable) subsets, then it is essential for many applications that we select functions from these classes in such a way as to retain the linear and order preserving relations. To appreciate the difficulty here, recall that if $\mathcal{L}^p, 1 \leq p \leq \infty$, is the class of pth power integrable real functions on (Ω, Σ, P) and $\mathcal{N}_p = \{f \in \mathcal{L}^p, \int_\Omega |f|^p d\mu = 0\}$, then $L^p = \mathcal{L}^p / \mathcal{N}_p$. So we need a mapping $\rho : L^p \to \mathcal{L}^p$ such that ρ is linear and positive. It is not easy to prove the existence of such a ρ. This mapping is called a (linear) *lifting* and it always exists if $p = +\infty$, but does not exist if $0 \leq p < \infty$ and P is diffuse. The existence problem becomes even more difficult if we demand further properties for ρ such as $\rho(\tilde{f}) = f$ for all continuous f when Ω is (say) a locally compact space. Such a ρ is then termed a *strong lifting*. A comprehensive account of the existence problem together with several theoretical applications are given in the monograph by Tulcea and Tulcea (1969). We include a relevant part in order to establish that the existence of a strong lifting on such spaces as those considered in Thorem 4 below is essentially equivalent to the existence of a proper regular conditional probability function. This will illuminate the subject and enable a better appreciation of the difficulties. More on these points will be found in Section 8.5.

Since $L^p = \mathcal{L}^p/\mathcal{N}_p$, we recognize that $\pi : \mathcal{L}^p \to L^p$ is just the quotient map from the function space $\mathcal{L}^p$ onto the factor space L^p. Then $\pi(\pi^{-1}(\tilde{f})) = \tilde{f}$, $\tilde{f} = f + \mathcal{N}_p \in L^p$. Since $\pi^{-1}(\cdot)$ is usually set valued, if there exists a (point) mapping $h : L^p \to \mathcal{L}^p$ such that $\pi(h(\tilde{f})) = \pi(f) = \tilde{f}$, then h is called a cross-section (of $\tilde{f}$). Thus $h(\cdot)$ "lifts" f from its equivalence class $\tilde{f}$. The existence of h with the linearity and order-preserving properties is sought. This is equivalent to finding a $\rho : \mathcal{L}^0 \to \mathcal{L}^0$ such that the following relations hold, (if $f = g$ a.e., we write $f \equiv g$ for simplicity): (i) $\rho(f) \equiv f$, (ii) if $f \equiv g$ then $\rho(f) = \rho(g)$, (iii) $\rho(af + bg) = a\ \rho(f) + b\ \rho(g), a, b \in \mathbb{R}$, (iv) if $f \geq 0$ a.e., then $\rho(f) \geq 0$, (v) $\rho(1) = 1$, and (vi) $\rho(fg) = \rho(f)\rho(g)$.

A mapping satisfying conditions (i) - (v) is called a *linear lifting* and that obeying (i) - (vi) is a *lifting*. If the existence of such a ρ is proved, then $h(\tilde{f}) = \rho(f) = f$ for $\tilde{f} = f + \mathcal{N}_p$, gives h unambiguously satisfying the earlier conditions, and conversely, if h exists then this equation gives ρ for which (i) - (vi) hold.

It is not immediate, but true, that for $0 \leq p < \infty$, such a lifting does not exist if the underlying P is diffuse on a set of positive measure. On the other hand, if $p = +\infty$, it is an important result, first proved by J. von Neumann in an important special case, and later by D. Maharam followed by the Ionescu Tulceas in complete generality, that a lifting ρ always exists on $\mathcal{L}^\infty$ (even when the measure space is σ-finite or more generally "strictly localizable"). This is a difficult theorem, and a relatively simple (still technical) proof in all its detail, based on T. Traynor's 1974 method, is included in the author's book (Rao, 1987, Chap. 8). We shall not repeat it here but employ the result.

Since we want to specialize the underlying space for conditional measures, it is assumed that (Ω, Σ) is a Borelian space. Let us introduce the concept of a strong lifting:

Definition 1. Let Ω be a topological space, Σ its Borel σ-algebra, and $P : \Sigma \to \mathbb{R}^+$, a Radon probability such that each nonempty open set of Ω has positive measure. Then a (linear) lifting $\rho : \mathcal{L}^\infty \to \mathcal{L}^\infty$, which exists by the preceding discussion, is (i) *strong* if $\rho(f) = f$ for all continuous f in $\mathcal{L}^\infty$. The couple (Ω, P) is also said to have the *strong lifting* property, (ii) [*strong*] *Borel lifting* if $\rho(f)$ is Borel for all Borel $f \in \mathcal{L}^\infty$ [and thus $\rho(f)$ is continuous for $f \equiv g$ with $g \in C_b(\Omega)$].

Although we introduced a lifting and a special case of it called linear lifting, it can be shown that every (strong) linear lifting extends to a (strong) lifting. Thus the existence problem is the same for both. Some equivalences for the existence of a strong lifting, when Ω is a locally compact space, are: (i) $\rho(f) = f$ for each continuous f with compact support, (ii) $\rho(\chi_U) \geq \chi_U$ for each open $U \subset \Omega$, and (iii) $\rho(\chi_F) \leq \chi_F$ for each closed $F \subset \Omega$. To give another equivalence, in cases of interest for us, we need the concept of "distintegration of a measure", which will enable one to put this on par with regular conditioning.

Definition 2. Let $(\Omega_i, \mathcal{B}_i), i = 1, 2$, be Borelian topological spaces and $\mu : \mathcal{B}_1 \to \mathbb{R}^+$ be a Radon probability with support Ω_1 so that each nonempty open set has positive measure. If $f : \Omega_1 \to \Omega_2$ is continuous [μ-measurable], $\nu = \mu \circ f^{-1} : \mathcal{B}_2 \to \mathbb{R}^+$ is the image probability, then μ is said to admit a [general] disintegration relative to f, or ν, if there is a mapping $Q : \mathcal{B}_1 \times \Omega_2 \to \mathbb{R}^+$ such that
   (i) $Q(\cdot, \omega) : \mathcal{B}_1 \to \mathbb{R}^+$ is a probability for each $\omega \in \Omega_2$,
   (ii) $Q(A, \cdot) : \Omega_2 \to \mathbb{R}^+$ is $\mathcal{B}_2$-measurable [or ν-measurable].
   (iii) $\mu(A) = \int_{\Omega_2} Q(A, \omega)\nu(d\omega), A \in \mathcal{B}_1$, and
   (iv) $\text{supp}(Q(\cdot, \omega)) \subset f^{-1}(\{\omega\})$, $\omega \in \Omega_2$, (supp = support).

Let us compare this concept with conditioning relative to a general random variable (instead of a σ-algebra). The latter definition was also discussed for Equations (2.1.7)-(2.1.10), leading to Proposition 2.1.3. Thus let (Ω, Σ, μ) be a Radon probability space and $(S, \mathcal{S})$ be a Borelian space with S as a topological space. If $h : \Omega \to S$ is measurable $(h^{-1}(\mathcal{S}) \subset \Sigma)$, then $\nu = \mu \circ h^{-1} : \mathcal{S} \to \mathbb{R}^+$ is a probability. Hence for any $f = \chi_A, A \in \Sigma, E \in \mathcal{S}$, we have

$$\int_E Q(A, s)\nu(ds) = \int_{h^{-1}(E)} \chi_A \, d\mu = \mu(A \cap h^{-1}(E)) \ , \qquad (1)$$

for a unique $\mathcal{S}$-measurable $Q(A, \cdot) : S \to \mathbb{R}^+, Q(\cdot, s)$ being σ-additive for a.a. (s). The exceptional null sets depend on the countable sequences used. This function $Q(\cdot, \cdot)$ is the conditional probability given h, also denoted $\mu(\cdot | h)$. Observe that we may regard $\mu(\cdot | h)(s)$ as an (positive) element of the Banach space $M(S, \mathcal{S})$, the space of regular σ-additive real functions on $\mathcal{S}$, with total variation as norm. But from discussion of the last paragraph, there is a lifting ρ on $\mathcal{L}^\infty(S, \mathcal{S}, \nu)$ and this determines uniquely a linear lifting ρ' on $\mathcal{L}^\infty_{\mathcal{I}}(S, \mathcal{S}, \nu)$ of $\mathcal{I}$-valued essentially bounded functions, where $\mathcal{I}$ is a Banach space, by the following correspondence:

$$\ell(\rho'(f)) = \rho(\ell(f)), \quad \text{for all} \ \ \ell \in \mathcal{I}^*, \ \ f \in \mathcal{L}^\infty_{\mathcal{I}}(S, \mathcal{S}, \nu) \ . \qquad (2)$$

Hereafter we write $f \overset{w}{\equiv} g$ if $\ell(f) = \ell(g)$, a.e., for all $\ell \in \mathcal{I}^*$. Then ρ' satisfies the obvious equations:

(i) $\rho'(f) \overset{w}{\equiv} f$, (ii) $f \overset{w}{\equiv} g$ implies $\rho'(f) = \rho'(g)$. The existence of such ρ' follows from the work on the lifting theorem (Tulcea and Tulcea, 1969, or Rao, 1987, p. 421). Now taking $\mathcal{I} = M(S, \mathcal{S})$, the lifting ρ' enables us to choose a regular version of conditioning in many cases. Using these concepts we establish the following two results on the disintegration of μ, of (1), in terms of regular (proper) conditinal probability functions on topological measure spaces of special interest in our applications. These will also be useful in Chapter 8.

Both the following results are due to Tulcea and Tulcea (1964):

Theorem 3. *Let* $(\Omega_i, \mathcal{B}_i), i = 1, 2$ *be Borelian spaces where the* Ω_i *are compact Hausdorff spaces. If* $\mu : \mathcal{B}_1 \to \mathbb{R}^+$ *is a Radon probability of full support on* Ω_1, *then the following are equivalent statements:*

(i) (Ω_1, μ) *has the strong lifting property.*

(ii) *For any continuous* $f : \Omega_1 \to \Omega_2$ *with* $f(\Omega_1) = \Omega_2, \nu_f = \mu \circ f^{-1}, \mu$ *admits a disintegration relative to* f *and* ν_f *in the sense of Definition 2.*

One half of this theorem will be deduced from the following more general and very useful result. (Here $\mathcal{I} = M(\Omega_2, \mathcal{B}_2)$ as before.)

Theorem 4. *Let Ω_i be locally compact, $\mathcal{B}_i$ the Borel σ-algebra of $\Omega_i, i = 1, 2,$ and $\mu : \mathcal{B}_1 \to \mathbb{R}^+$ be a Radon probability. If $f : \Omega_1 \to \Omega_2$ is continuous, $\nu_f = \mu \circ f^{-1} : \mathcal{B}_2 \to \mathbb{R}^+$, let $\alpha : \mathcal{B}_2 \to \bar{\mathbb{R}}^+$ be a Radon measure dominating ν_f (i.e., $\nu_f \ll \alpha$). If ρ' is a lifting on $\mathcal{L}_{\mathcal{I}}^\infty (\Omega_2, \mathcal{B}_2, \alpha)$, which exists, then μ admits a disintegration in the following sense: there is a function $Q^f : \mathcal{B}_1 \times \Omega_2 \to \mathbb{R}^+$ such that (i) $\rho'(Q^f(\cdot, \cdot)) = Q^f(\cdot, \cdot)$ in $\mathcal{L}_{\mathcal{I}}^\infty (\Omega_2, \mathcal{B}_2, \alpha)$, (ii) $Q^f(\Omega_1, s) = \frac{d\nu_f}{d\alpha}(s)$ a.e. $[\alpha]$, and (iii) we have the functional equation:*

$$\int_{\Omega_1} g_1(t)\mu(dt) = \int_{\Omega_2} \left(\int_{\Omega_1} g_1(t)Q^f(dt, s) \right) \alpha(ds), \ g_1 \in C_c(\Omega_1) . \quad (3)$$

If further ρ' is strong, then (iv) $\mathrm{supp}(Q^f(\cdot s)) \subset f^{-1}(\{s\}), \ s \in \Omega_2$. In this case the $Q(\cdot, \cdot)$ satisfying the conditions (i)-(iii) is unique outside of an α-null set. (Here and below $C_c(\Omega)$ denotes the space of real continuous compactly based functions on Ω.)

Proof. Since the $Q^f(\cdot, \cdot)$ will be analogs of conditional probability functions, their existence is proved through a suitable Radon-Nikodým theorem for certain measures. Then a (bounded) function Q^f is selected from the equivalence class through a lifting operator. These are the two central ideas in the proof. These points are detailed in steps.

(a) If $\nu_f = \mu \circ f^{-1}$, then for each $g \in C_c(\Omega_2)$, define a new measure by the formula

$$\nu_f^g : A \mapsto \int_A g(\omega)\nu_f(d\omega) = \int_{f^{-1}(A)} (g \circ f)(t)\mu(dt), \ A \in \mathcal{B}_2 . \quad (4)$$

Then the mapping $T : g \mapsto \nu_f^g$ relates $C_c(\Omega_2) \to C_c(\Omega_2)^* \cong M(\Omega_2, \mathcal{B}_2)$, (i.e. isometrically isomorphic) the space of regular bounded (real) Borel measures on $\mathcal{B}_1$, which was $\mathcal{I}$ in the statement of the theorem. Clearly T is a bounded linear operator on the space shown. [$C_c(\Omega)$ is given the uniform norm as usual.] Also it is clear that $C_c(\Omega_2) \subset \mathcal{L}^1(\Omega_2, \mathcal{B}_2, \nu_f)$ as a dense subspace since ν_f is a Radon measure. Hence T has a unique norm preserving extension, denoted by $\tilde{T}$, onto $\mathcal{L}^1(\Omega_2, \mathcal{B}_2, \nu_f)$, with range in $M(\Omega_1, \mathcal{B}_1)$. If we let $\mathbf{m}(A) = \tilde{T}(\chi_A), A \in \mathcal{B}_2$, then $\mathbf{m}(\cdot)$ is σ-additive on $\mathcal{B}_2$ with values in the (adjoint) space $M(\Omega_1, \mathcal{B}_1)$. It is of finite variation. Indeed, for each $A \in \mathcal{B}_2$, by definition of $|\mathbf{m}|(\cdot)$, one has

$$|\mathbf{m}|(A) = \sup \left\{ \sum_{i=1}^{n} \|\mathbf{m}(A_i)\| : \bigcup_{i=1}^{n} A_i = A, A_i \in \mathcal{B}_2, \text{ disjoint, } n \geq 1 \right\}$$

$$= \sup \left\{ \sum_{i=1}^{n} \mu(f^{-1}(A_i)) : \bigcup_{i=1}^{n} A_i = A, A_i \in \mathcal{B}_2, \text{ disjoint, } n \geq 1 \right\}$$

$$= \mu(f^{-1}(A)) \leq \mu(f^{-1}(\Omega_2)) = 1 , \tag{5}$$

since

$$\|\mathbf{m}(A_i)\| = \int_{f^{-1}(\Omega_2)} \chi_{f^{-1}(A_i)}\, \mu(dt) = \mu(f^{-1}(A_i)), \text{ by (4).} \tag{6}$$

It follows that $|\mathbf{m}|(\cdot) = \nu_f(\cdot)$ and hence $\mathbf{m}(\cdot)$ is a regular $M(\Omega_1, \mathcal{B}_1)$-valued measure and is α-continuous (since $\nu_f << \alpha$). At this point, we invoke the Dunford-Pettis (generalization of the Radon-Nikodým) theorem to obtain

$$\langle g, \mathbf{m}(A) \rangle = \int_A \langle g, Q_s^f \rangle\, \alpha(ds) , \quad A \in \mathcal{B}_2, \; g \in C_c(\Omega_1) . \tag{7}$$

It should be observed that the original Dunford-Pettis theorem is given only if the range of T is separable (cf. e.g., Dunford-Schwartz (1958), p. 504, VI. 8.7). But the existence of a lifting ρ' on $\mathcal{L}_T^\infty$ allows us to drop the separability assumption, so that one takes $Q_{(\cdot)}^f$ to satisfy the condition $\rho'(Q_{(\cdot)}^f) = Q_{(\cdot)}^f$ where $Q_s^f \in M(\Omega_1, \mathcal{B}_1) = \mathcal{I}$, (as shown in Tulcea and Tulcea 1969, p. 89). With this representation we establish parts (i)-(iii) of the theorem.

Here we may mention that, if Ω_1 were second countable, then the original Dunford-Pettis version suffices to prove (i) - (iii). However, the second part that Q_s^f is proper and that it is essentially unique depends on the existence of a strong lifting for the couple (Ω_2, α). Thus the existence of proper regular conditional probability functions is a deeper problem, and that is what one needs in applications.

(b) We now evaluate both sides of the quantities in the representation (7). Consider

$$\int_{\Omega_2} h(s)\nu_f(ds) = \int_{\Omega_2} h(s)\frac{d\nu_f}{d\alpha}(s)\, d\alpha(s) = \int_{\Omega_2} h(s)\|Q_s^f(\cdot)\|d\alpha(s), \tag{8}$$

for all $h \in \mathcal{L}^1(\Omega_2, \mathcal{B}, \nu_f)$, where $\|Q_s^f(\cdot)\|$ is the total variation norm in $M(\Omega_1, \mathcal{B}_1)$. Since h is arbitrary, from (8) we deduce that (i) and (ii) of the theorem hold. But the left side of (7) can be expressed as:

$$\langle g, \mathbf{m}(A) \rangle = \int_{\Omega_1} g(t) d\mu \circ (\chi_A \circ f)^{-1}$$

$$= \int_{\Omega_1} g(t)(\mu|f^{-1}(A))(dt), \text{ where } (\mu|B)(\cdot) = \mu(B \cap \cdot) ,$$

$$= \int_{f^{-1}(A)} g(t) d\mu(t), \ A \in \mathcal{B}_2 \ , g \in \mathcal{L}^1(\Omega_1, \mathcal{B}_1, \mu) .$$

Hence (7) becomes,

$$\int_{f^{-1}(A)} g(t) d\mu(t) = \int_A \langle g, Q_s^f \rangle d\alpha(s)$$

$$= \int_A \left(\int_{\Omega_1} g(t) Q_s^f(dt) \right) \alpha(ds) . \tag{9}$$

It follows from (9), on taking $g \geq 0$, that $\int_{\Omega_1} g(t) Q_s^f(dt) \geq 0$, a.a.(s). Now we use the fact that $\rho'(Q_s^f) = Q_s^f$ for all $s \in \Omega - N, \alpha(N) = 0$. If we set $Q_s^f = \mu$ for $s \in N$, then $Q_s^f \geq 0$ for all $s \in \Omega_2$. Hence (9) gives (3). So (i)-(iii) are established. Under the additional hypothesis, the last part will be proved.

(c) Let ρ be a strong lifting of $\mathcal{L}^\infty(\alpha)$. Then we assert that the support of Q_s^f is contained in $f^{-1}(\{s\})$ for each $s \in \Omega_2$. Since $Q_s^f(\cdot)$ is a Radon measure, its behavior is completely determined by its values on compact sets. But $f^{-1}(\{s\}) \subset \Omega_1$ is closed since f is continuous. So it suffices to prove that $Q_s^f(K) = 0$ for each compact set $K \subset (f^{-1}(\{s\}))^c$. If K is such a set, then $C = f(K) \subset \Omega_2$ is compact and $s \notin C$. So by a classical result there exist disjoint open sets V_1, V_2 such that $s \in V_1, C \subset V_2$. Hence $f^{-1}(\{s\}) \subset f^{-1}(V_1), \ K \subset f^{-1}(C) \subset f^{-1}(V_2)$ and $f^{-1}(V_1), \ f^{-1}(V_2)$ are disjoint open sets. Let $g \in C_c(\Omega_1)$ be chosen such that $g|K = 1, 0 \leq g \leq 1$ and $g|f^{-1}(V_1) = 0$, which is possible by Urysohn's lemma. But then

$$\int_A \chi_{V_1} < g, Q_s^f > d\alpha(s) = \int_{f^{-1}(A)} \chi_{f^{-1}(V_1)}(t) g(t) d\mu(t) = 0, A \in \mathcal{B}_2, \tag{10}$$

since $g|f^{-1}(V_1) = 0$. Hence $\langle g, Q_s^f \rangle = 0$, a.a. (s) in V_1. We also have

$$\rho(\chi_{V_1} \langle g, Q_s^f \rangle) = \rho(\chi_{V_1}) \cdot \langle g, Q_s^f \rangle = 0, \ s \in Q_2 ,$$

where ρ is the strong lifting on $\mathcal{L}^\infty(\Omega_2, \mathcal{B}_2, \alpha)$ which induces ρ'. But $\rho(\chi_V) \geq \chi_V$ so that $s \in V_1$ implies from the above equation that $\rho(\chi_{V_1}) \neq 0$ and hence $< g, Q_s^f > = 0$. Consequently

$$Q_s^f(K) = \int_{\Omega_1} \chi_K(t) Q_s^f(dt) \leq \int_{\Omega_1} g(t) Q_s^f(dt)$$
$$= \langle g, Q_s^f \rangle = 0.$$

This proves that $\text{supp}(Q_s^f) \subset f^{-1}(\{s\})$, as asserted.

(d) It remains to establish uniqueness if ρ is simply a lifting, but relations (i) - (iv) hold (even if ρ is not strong). (However under the hypothesis of Theorem 4, we shall see that ρ will be strong in that case also.) Thus let $Q_s^i(\cdot), i = 1, 2$ be two families of regular proper measures for which (i) - (iv) hold. Then we have

$$\int_{\Omega_1} g_1(t) d\mu(t) = \int_{\Omega_2} \int_{\Omega_1} g_1(t) Q_s^i(dt) \alpha(ds), \; i = 1, 2, \; g_1 \in C_c(\Omega_1) \; .$$
(11)

Since both Q_s^1, Q_s^2 are supported inside $f^{-1}(\{s\})$, $s \in \Omega_2$, for any $h \in C_c(\Omega_2)$ we get $g \circ h \in C_c(\Omega_1)$ so that replacing g_1 by $h \circ g_1$ in (11) the right side (equal) terms give

$$0 = \int_{\Omega_2} \int_{\Omega_1} g_1(t)(h \circ f)(t)[Q_s^1(dt) - Q_s^2(dt)]\alpha(ds)$$
$$= \int_{\Omega_2} h(s) \int_{\Omega_1} g_1(t)[Q_s^1(dt) - Q_s^2(dt)]\alpha(ds) \; .$$

Since h is arbitrary from $C_c(\Omega_2)$, this shows that the inner integral vanishes for a.a.(s), and all $g_1 \in C_c(\Omega_1)$. Hence $Q_s^1 \overset{w}{=} Q_s^2$, a.a. (s), and since $\rho'(Q_s^1) = Q_s^1$, $\rho'(Q_s^2) = Q_s^2$, we conclude that the left sides must agree and hence that $Q_s^1 = Q_s^2$ a.e. (α). $\square$

We now use this result to proceed with the

Proof of Theorem 3. Since Ω_1, Ω_2 are compact, the first part that (i) $\Rightarrow$ (ii) follows from Theorem 4, which in fact holds even for locally compact spaces. We thus need to prove the reverse implication.

(ii) $\Rightarrow$ (i). Let the given decomposition hold. Let $(\Omega_1, \mathcal{B}_1, \mu)$ be a compact probability space as in the hypothesis, and consider its Stone space representation $(S, \mathcal{S}, \nu)$ where S is the (compact) Stone space of $(\Omega_1, \mathcal{B}_1, \mu)$ and $\mathcal{S}$ its Borel σ-ring determined by the clopen

(= closed-open) sets and ν its image measure under $\pi : \Omega_1 \to S$, an isomorphism (see Rao, 1987, p. 509), so that $\nu = \mu \circ \pi^{-1}$. Moreover, let $U : \mathcal{L}^\infty(\Omega_1, \mathcal{B}_1, \mu) \to \mathcal{L}^\infty(S, \mathcal{S}, \nu)$ be the mapping defined by $U\chi_A = \chi_{\pi(A)}$, and extended linearly. Then it is a multiplicative linear mapping. If ρ is the lifting on $\mathcal{L}^\infty(S, \mathcal{S}, \nu)$, then $\rho(U(\tilde{f}))$ is in fact continuous, since $U(\tilde{f})$ belongs to the equivalence class of continuous functions and $\rho(U(\tilde{f}))$ denotes that unique continuous function of this class, where $\tilde{f}(\in \mathcal{L}^\infty(\Omega_1, \mathcal{B}_1, \mu))$ is the equivalence class of f.

Now by hypothesis, μ admits a disintegration satisfying (i) - (iv). If $\{Q_s^\pi, \; s \in S\}$ is the corresponding disintegration, let

$$\lambda(f)(\cdot) = \langle \rho(U(\tilde{f})), \; Q_{(\cdot)}^\pi \rangle, \; f \in \mathcal{L}^\infty(\Omega_1, \mathcal{B}_1, \mu) \; . \tag{12}$$

We claim that λ is a strong linear lifting of $\mathcal{L}^\infty(\Omega_1, \mathcal{B}_1, \mu)$ to finish the proof.

By the definition of U, properties of ρ, and of the integral, it follows that $\lambda(\cdot)$ is a positive linear mapping such that $\lambda(1) = 1$, and $f = g$ a.e. implies $\lambda(f) = \lambda(g)$. To see that $\lambda(f) = f$ a.e., for $f \in \mathcal{L}^\infty(\Omega_1, \mathcal{B}_1, \mu)$ note that $\rho(U(\tilde{f})) = g$ is a continuous function on S. Then, for each $h \in C(\Omega_1)$, from the disintegration hypothesis of ν, we have

$$\int_S (h \circ \pi) g \, d\nu = \int_{\Omega = \pi^{-1}(S)} h(t) \langle g, Q_t^\pi \rangle \mu(dt)$$

$$= \int_\Omega h(t) \lambda(f)(t) d\mu(t), \; \text{by (12)}. \tag{13}$$

However, $U\tilde{h} = \tilde{h} \circ \pi, \tilde{g} = U(\tilde{f})$ so that

$$\int_S (h \circ \pi) g \, d\nu = \int_S (U\tilde{h})(U\tilde{f}) d\nu$$

$$= \int_S U(\tilde{h}\tilde{f}) d\nu = \int_\Omega hf \, d\mu \; . \tag{14}$$

Then (13) and (14) imply $\lambda(f) = f$ a.e., since h is arbitrary in $C(\Omega_1)$. Thus λ is a linear lifting on $\mathcal{L}^\infty(\Omega_1, \mathcal{B}_1, \mu)$. It is also strong since for each $f \in C(\Omega_1)$, $U(\tilde{f}) = \tilde{f} \circ \pi$ so that $\rho(U\tilde{f}) = \rho(\tilde{f} \circ \pi) = f \circ \pi$, and $\lambda(f)(t) = \langle \rho(U\tilde{f}), Q_t^\pi \rangle = f(t), t \in \Omega_1$, since Q_t^π concentrates on $\pi^{-1}(\{t\})$. Consequently λ is strong and extends to a strong lifting. $\square$

These two results do not yet show that a disintegration (= strong lifting) exists on each couple (Ω, μ) where Ω is (locally) compact. In fact the general case is complicated (see Sec. 8.5). However, there are many useful spaces for which a useful solution obtains. We now show that this is the case if Ω is a Polish space. This will imply that Theorem 3.7 is extended by the preceding two results.

Theorem 5. *Let $(\Omega, \mathcal{B}, \mu)$ be a Radon probability space where Ω is a Polish space. Then the couple (Ω, μ) has the strong lifting property.*

Proof. Let Ω_d be the given space endowed with a totally bounded metric d, as in the second and third proofs of Theorem 3.7, and let $\tilde{\Omega}_d$ be its completion. Then $\tilde{\Omega}_d$ is a compact metric space, with Ω_d as a dense subspace. If $\tilde{\mu}$ is the unique extension of μ to the Borel σ-algebra of $\tilde{\Omega}_d$, then $\tilde{\mu}$ is again a Radon probability and if we show that $(\tilde{\Omega}_d, \tilde{\mu})$ has the strong lifting property, then its restriction to (Ω_d, μ) can be shown to have the same property. So we first establish that $\mathcal{L}^\infty(\tilde{\Omega}_d, \tilde{\mu})$ admits a strong lifting. This follows from:

Proposition 6. *Let Ω be a locally compact second countable space, and μ be a Radon probability on it with support Ω. Then the couple (Ω, μ) has the strong lifting property.*

Proof. Since Ω is second countable, $C_c(\Omega)$ is separable under the uniform norm, and let $\{f_n, n \geq 1\}$ be a dense sequence in $C_c(\Omega)$. If $\mathcal{A}$ is the linear span of $\{f_n, n \geq 1\}$ with rational coefficients, then $\mathcal{A} \subset C_c(\Omega)$ and is a dense denumerable set. But we know that $\mathcal{L}^\infty(\nu)$ admits a lifting ρ so that $\rho(f) = f$ a.e., $f \in \mathcal{L}^\infty(\nu)$. Thus there is a ν-null set N_f such that $\rho(f) = f$ on N_f^c. If $N = \cup\{N_f : f \in \mathcal{A}\}$, then $\nu(N) = 0$, and $\rho(f) = f$ on N^c for each $f \in \mathcal{A}$. To see that $\mathcal{A}$ can be replaced by $C_c(\Omega)$ here, let $g \in C_c(\Omega), g_n \in \mathcal{A}$ be chosen such that $\|g - g_n\|_\infty \to 0$, as $n \to \infty$. Since $\rho(g_n) = g_n$ on N^c, and

$$\|\rho(g_n) - \rho(g)\|_\infty = \|\rho(g_n - g)\|_\infty \leq \|g_n - g\|_\infty \to 0, \text{ as } n \to \infty , \quad (15)$$

we have

$$|\rho(g) - g|(s) \leq |\rho(g) - \rho(g_n)|(s) + |\rho(g_n) - g_n|(s) + |g_n - g|(s) \to 0$$

uniformly for $s \in N^c$. Hence $\rho(g) = g$ on N^c. For the points on N, note that for each $s \in \Omega$, there is a multiplicative linear functional ℓ_s on the Banach algebra $\mathcal{L}^\infty(\nu)$. In fact ℓ_s is an extremal point of the closed unit ball of the adjoint space $(\mathcal{L}^\infty(\nu))^*$, (see Dunford-Schwartz, 1958, p. 443). Hence $\ell_s(f) = \ell_s(g)$ for $f = g$ a.e., and $\ell_s(f) = f(s)$ if f is continuous. So if we let $\tilde{\rho}(f)(s) = \rho(f)(s), s \in N^c$ and $= \ell_s(f)$ if $s \in N$, then $\tilde{\rho}$ is a lifting on $\mathcal{L}^\infty(\nu)$ which satisfies the condition that $\tilde{\rho}(f) = f$ for each continuous f. Hence $\tilde{\rho}$ is strong, by definition. $\square$

We can now complete:

Proof of Theorem 5. By the above proposition there is a strong lifting $\tilde{\rho}$ on $\mathcal{L}^\infty(\tilde{\Omega}_d, \tilde{\nu})$. On the other hand $\Omega_d \subset \tilde{\Omega}_d$ and $\nu = \tilde{\nu}|\Omega_d$, and each continuous f on Ω_d has a unique continuous extension $\tilde{f}$ to $\tilde{\Omega}_d$. Since ν and $\tilde{\nu}$ are Radon probabilities, $\tilde{f} \in \mathcal{L}^\infty(\tilde{\nu})$, let $\rho(f) = \tilde{\rho}(\tilde{f})$, $f \in \mathcal{U}(\Omega_d)$, the space of bounded uniformly continuous functions on Ω_d. Then $\rho(1) = \tilde{\rho}(1) = 1$, and $\rho(\cdot)$ is a positive linear (multiplicative) mapping on $\mathcal{L}^\infty(\nu)$. If $g \in \mathcal{L}^\infty(\nu)$, then there is a $\tilde{g} \in \mathcal{L}^\infty(\tilde{\nu})$ such that $\tilde{g}|\Omega = g$ and so $\tilde{\rho}(\tilde{g})(s) = \tilde{g}(s) = g(s)$, a.a. $s \in \Omega_d$. But for $f \in \mathcal{U}(\Omega_d), \tilde{f} \in C(\tilde{\Omega}_d)$, so that $\rho(f) = \tilde{\rho}(\tilde{f}) = \tilde{f}$, and since $\tilde{f}|\Omega_d = f, \rho$ is also strong. $\square$

Recall that a mapping $Q : \Sigma \times \Omega \to \mathbb{R}^+$ is a regular conditional probability relative to a σ-algebra $\mathcal{B} \subset \Sigma$, if $Q(\cdot, s)$ is a regular probability for each $s \in \Omega$, and $Q(A, \cdot)$ is $\mathcal{B}$-measurable, $A \in \Sigma$, and the following functional equation holds:

$$\int_B Q(A, s)\mu_{\mathcal{B}}(ds) = \mu(A \cap B), \ A \in \Sigma, \ B \in \mathcal{B} . \tag{16}$$

Also Q is proper at s_0 if $Q(A, s_0) = 1$ for $s_0 \in A \in \mathcal{B}$. With this, as a consequence of Theorems 5 and 4 (cf. also Theorem 3.7), we have the following:

Corollary 7. *Let (Ω, Σ, μ) be a Radon probability space where Ω is a locally compact second countable, or a Polish space. Then there exists a proper regular conditional probability function $Q : \Sigma \times \Omega \to \mathbb{R}^+$, relative to any σ-algebra $\mathcal{B} \subset \Sigma$ at all points of $\Omega - N$, where N is a μ-null set.*

Proof. Let $\Omega_1 = \Omega, \mathcal{B}_1 = \Sigma$, and $(S, \mathcal{S})$ be the Stone space of $(\Omega, \mathcal{B})$, and $\pi : \Omega \to S$ be the corresponding canonical mapping. Letting $\nu = \mu_{\mathcal{B}} \circ \pi^{-1}$ where π is continuous, with $\alpha = \nu$, the hypothesis of Theorem 5 is satisfied. If ρ is a strong lifting on $\mathcal{L}^\infty(\mu)$, then there is Q satisfying (3) with $\text{supp}(Q(\cdot, s)) \subset \pi^{-1}(\{s\}), s \in S$, $\|Q(\cdot, s)\| = \frac{d\nu}{d\alpha} = 1$ a.e. (α). Thus $Q(A, s_0) = 1$ for $s_0 \in B(\in \mathcal{S})$, with $\pi^{-1}(B) \in \mathcal{B}$. The existence of such a Q is equivalent to that of the disintegration of μ relative to ν and π, which is the assertion. □

Remark. For the spaces (Ω, Σ, μ) with Ω as a σ-compact second countable metric space a somewhat stronger statement holds. Indeed, for such spaces Lloyd (1974) has shown, under the continuum hypothesis, that (Ω, μ) has the strong Borel lifting property. Thus for these spaces the disintegration property may perhaps be streamlined.

To see the technical intricacies involved in these existence results, we now present a proof of the Blackwell-Ryll-Nardzewski theorem stated in the last section, as it reveals another aspect of the structure.

Proof of Theorem 3.8. Let $Q : \Sigma \times \Omega \to \mathbb{R}^+$ be the regular conditional probability function of μ relative to $\mathcal{B}$ or X. Thus in particular (16) holds. We first give the sufficiency proof which is straightforward. Thus suppose there is a $g : \Omega \to \Omega$ such that $X(g(\omega)) = X(\omega), \omega \in \Omega$. If $\mathcal{R}$ is the Borel σ-algebra of $\mathbb{R}$, then we have

$$\mathcal{B} = X^{-1}(\mathcal{R}) = g^{-1}(X^{-1}(\mathcal{R})) = g^{-1}(\mathcal{B}) \subset g^{-1}(\Sigma) \subset \mathcal{B} . \qquad (17)$$

It follows that $g^{-1}(\mathcal{B}) = \mathcal{B}$. Moreover, the last relation implies that each B in $\mathcal{B}$ can be expressed as $B = g^{-1}(A)$ for some $A \in \Sigma$. Then (16) becomes

$$\int_C Q(B, \omega)\mu_{\mathbf{B}}(d\omega) = \int_C Q(g^{-1}(A), \omega)\mu_{\mathbf{B}}(d\omega), \quad C \in \mathcal{B} ,$$
$$= \mu(g^{-1}(A) \cap C), \text{ by (16)},$$
$$= \int_C \chi_{g^{-1}(A)}(\omega)\mu_{\mathbf{B}}(d\omega) .$$

Hence

$$\int_C [Q(g^{-1}(A), \omega) - \chi_{g^{-1}(A)}(\omega)]\mu_{\mathbf{B}}(d\omega) = 0, \quad C \in \mathcal{B} .$$

Since the integrand is $\mathcal{B}$-measurable, we deduce that

$$Q(g^{-1}(A), \omega) = \chi_{g^{-1}(A)}(\omega), \quad \text{a.e.}$$

Thus if $\tilde{Q} : \Sigma \times \Omega \to \mathbb{R}$ is defined as

$$\tilde{Q}(A, \omega) = \begin{cases} \chi_{g^{-1}(A)}(\omega), & A \in \mathcal{B}, \ \omega \in \Omega \\ Q(A, \omega), & A \in \Sigma - \mathcal{B}, \ \omega \in \Omega \end{cases}$$

then by (17) and (18) we see that $\tilde{Q}$ satisfies (16) and is a proper regular conditional probability function.

Let us also show that $X(\Omega) \in \mathcal{R}$, when $X \circ g = X$. Since $g^{-1}(\Sigma) \subset \mathcal{B} = X^{-1}(\mathcal{R})$, it follows from the Doob-Dynkin lemma, that there is a Borel function $h : \mathbb{R} \to \Omega$, (i.e., $h^{-1}(\Sigma) \subset \mathcal{R}$) such that $g = h \circ X$. Hence $X \circ h \circ X = X$ or if $y = X(\omega) \in X(\Omega)$, then $X(h(y)) = y$. Since $X \circ h : \mathbb{R} \to \mathbb{R}$, is a Borel function, we have

$$X(\Omega) = \{y : (X \circ h)(y) = y\} \in \mathcal{R} .$$

Thus the range of X is necessarily a Borel set, under the hypothesis.

For the converse, we need to find the (desired) function $g : \Omega \to \Omega$. So we suppose that $Q : \Sigma \times \Omega \to \mathbb{R}^+$ is a proper regular conditional probability function for μ and $\mathcal{B}$. Let $S = \{(\omega, \omega') : X(\omega) = X(\omega')\} \subset \Omega \times \Omega$, and $A(\omega) = \{\omega' : X(\omega') = X(\omega)\} = S(\omega)$, the ω-section of S. Since $\omega \in A(\omega)$ and $A(\omega) \in \mathcal{B}$, we have $Q(A(\omega), \omega) = 1$ by the fact that $Q(\cdot, \cdot)$ is a proper regular conditional probability. Hence if we can show that there is a function $g : \Omega \to \Omega$, with $g^{-1}(\Sigma) \subset \mathcal{B}$, such that $(\omega, g(\omega)) \in S$, then by definition of S, one gets $X(\omega) = X(g(\omega))$, $\omega \in \Omega$, and the result will follow. Now the existence of such a g is a consequence of the next somewhat more general statement.

Proposition 8. *Let Ω_1, Ω_2 be Borel subsets of a Polish space. (In the above $\Omega_1 = \Omega_2 = \Omega$.) If $\mathcal{B} = X^{-1}(\mathcal{R})$, with $X : \Omega_1 \to \mathbb{R}$ as a random variable and $(\Omega_i, \Sigma_i), i = 1, 2$, Borelian spaces, let $\mu : \Sigma_2 \times \Omega_1 \to \mathbb{R}^+$ be such that (a) $\mu(\cdot, \omega) : \Sigma_2 \to \mathbb{R}^+$ is a probability, $\omega \in \Omega_1$, (b) $\mu(A, \cdot)$ is $\mathcal{B}$-measurable, $A \in \Sigma_2$, and (c) for any $S \in \mathcal{B} \otimes \Sigma_2$, such that $\mu(S(\omega), \omega) > 0$ where $S(\omega) = \{\omega' : (\omega, \omega') \in S\}, \omega \in \Omega_1$, then there is a $(\mathcal{B}, \Sigma)$-measurable $g : \Omega_1 \to \Omega_2$ such that $(\omega, g(\omega)) \in S$ for each $\omega \in \Omega$, so that the graph of g lies in S.*

This measure theoretical result is proved on using the next lemma which is somewhat technical.

Lemma 9. *Let $\{(\Omega_i, \Sigma_i), i = 1, 2\}, \mathcal{B}, \mu$ be as in the proposition. Then for each $S \in \mathcal{B} \otimes \Sigma_2$, and $0 \leq \theta < 1$, there is an $\tilde{S} \in \mathcal{B} \otimes \Sigma_2$ such that $\tilde{S} \subset S, \tilde{S}(\omega)$ closed and $\mu(\tilde{S}(\omega), \omega) \geq \theta \mu(S(\omega), \omega)$, for all $\omega \in \Omega_1$.*

Proof of Lemma. Let S be the class of sets $S \in \mathcal{B} \times \mathcal{B}_2$ for which the conclusions of the lemma hold. If $S = A \times B, A \in \mathcal{B}$ and $B \in \mathcal{B}_2$ closed, then $\tilde{S} = S$ satisfies the conditions so that if $\tilde{\mathcal{B}}_2 \subset \mathcal{B}_2$ is the algebra generated by its closed sets, then $\mathcal{B} \times \tilde{\mathcal{B}}_2 \subset \mathcal{S}$. Since $\mathcal{B} \times \tilde{\mathcal{B}}_2$ is an algebra, it suffices to show that $\mathcal{S}$ is a monotone class which would then imply, from standard measure theoretical results, that $\sigma(\mathcal{S}) = \sigma(\mathcal{B} \times \tilde{\mathcal{B}}_2) = \mathcal{B} \otimes \mathcal{B}_2$ and the lemma will follow. So let $S_n \in \mathcal{S}$, $\{S_n, n \geq 1\}$ a monotone sequence with limit S. We show that $S \in \mathcal{S}$.

Let $S_n \uparrow S$. Since $S_n \in \mathcal{S}$, we may choose $\tilde{S}_n \subset S_n$, $\tilde{S}_n(\omega)$ closed and $\mu(\tilde{S}_n(\omega), \omega) > \theta^{\frac{1}{2}} \mu(S(\omega), \omega)$, all $\omega \in \Omega_1$. Replacing $\tilde{S}_n$ by $\bigcup_{k=1}^{n} \tilde{S}_k \subset S_n$ if necessary, we may also assume that $\tilde{S}_n \uparrow$. If T_n is defined as

$$T_n = \{\omega \in \Omega_1 : \mu(\tilde{S}_n(\omega), \omega) \geq \theta \mu(S(\omega), \omega)\}, \tag{19}$$

then $T_n \uparrow$ since $\tilde{S}_n \uparrow$. From the fact that $\mu(\cdot, \omega)$ is a measure, it is easily deduced that $\lim_n T_n = \Omega_1$, so that $\tilde{S} \subset S$ and $S \in \mathcal{S}$.

Now let $S_n \downarrow S$. Define a new function $\lambda : \mathcal{B}_2 \times \Omega_1 \rightarrow \mathbb{R}^+$ as

$$\lambda(B, \omega) = \begin{cases} \frac{\mu(S(\omega) \cap B, \omega)}{\mu(S(\omega), \omega)}, & \text{if } \mu(S(\omega), \omega) > 0 \\ 0, & \text{otherwise.} \end{cases}$$

Let $\theta_n = 1 - (1-\theta)2^{-n}$, and choose $\tilde{S}_n \subset S_n$, with $\tilde{S}_n(\omega)$ closed, such that $\lambda(\tilde{S}_n(\omega), \omega) \geq \theta_n \lambda(S_n(\omega), \omega)$, all $\omega \in \Omega_1, n \geq 1$. This is possible by the given conditions on μ. In fact, by hypothesis we can choose $\tilde{S}_n \subset S_n$ with the stated properties for $\mu(\cdot, \cdot)$ to satisfy

$$\mu(\tilde{S}_n(\omega), \omega) \geq \theta_n \mu(S_n(\omega), \omega), \omega \in \Omega_1. \tag{20}$$

Since $\tilde{S}_n(\omega), S_n(\omega)$ are in $\mathcal{B}_2$ for all ω, the inequality (20) holds on the trace $\mathcal{B}_2(S(\omega))$ for each $\omega \in \Omega_1$. Thus for each $S_n \in \mathcal{S}(S)$, $S_n(\omega) \in \mathcal{B}_2(S(\omega))$, and (20) holds if $\tilde{S}_n \subset S_n$, with $S_n(\omega) \in \mathcal{B}_2(S(\omega))$, $\omega \in \Omega_1$. This means we can replace $\tilde{S}_n$ and S_n by $(\tilde{S}_n \cap S)(\omega)$ and $(S_n \cap S)(\omega)$ in (20). Dividing the resulting inequality by the positive quantity $\mu(S(\omega), \omega)$, we get the inequality with the λ-measure. Let $\tilde{S} = \cap_n \tilde{S}_n$. Then

$$\lambda(\tilde{S}(\omega), \omega) = 1 - \lambda(\bigcap_n \tilde{S}_n^c(\omega), \omega)$$

$$\geq 1 - \sum_n \lambda(S_n(\omega), \omega)$$

$$= 1 - \sum_n (1 - \lambda(\tilde{S}_n(\omega), \omega))$$

$$\geq 1 - \sum_n (1 - \theta_n) = \theta .$$

Since $\tilde{S}(\omega) \subset \tilde{S}_n(\omega)$, this inequality gives

$$\mu(\tilde{S}, (\omega), \omega) \geq \theta\mu(S, \omega), \omega) .$$

In case $\mu(S(\omega), \omega) = 0$, this is always valid, so that in all cases $S \in \mathcal{S}$ and $\mathcal{S}$ is thus shown to be a monotone class. $\square$

With this result we can now present:

Proof of Proposition 8. Let $\tilde{\Omega}_2$ be the completion of the separable Ω so that $\tilde{\Omega}_2$ is Polish. If $S \in \mathcal{B} \otimes \mathcal{B}_2$, choose $S_1 \subset S$ such that $S(\omega)$ is closed and $\mu(S_1(\omega), \omega) > 0$, for all $\omega \in \Omega_1$, which is possible by hypothesis. For $\varepsilon > 0$, let $F_1^\varepsilon, F_2^\varepsilon, \ldots$ be a countable cover of closed sets of $\tilde{\Omega}_2$ of diameter less than ε. Let

$$n(\omega, \varepsilon) = \inf\{k > 0 : \mu(S_1(\omega) \cap F_k^\varepsilon, \omega) > 0\} .$$

Then $n(\cdot, \varepsilon)$ is $\mathcal{B}$-measurable. Let $S_2 \subset S_1$ be defined by

$$S_2(\omega) = \{\omega \in S_1(\omega) \cap F_k^1 : n(\omega, 1) = k\},$$

and if $S_{m-2} \subset S_{m-1}$ is chosen, inductively, let

$$S_m(\omega) = \{\omega \in S_{m-1}(\omega) \cap F_k^{1/(m-1)} : n(\omega, \frac{1}{m-1}) = k\} .$$

Thus $S_1 \supset S_2 \supset \ldots$ with diameter of $S_m < (m-1)^{-1}, \mu(S_m(\omega), \omega) > 0$, and each $S_m(\omega)$ is closed. If $S' = \cap_n S_m$, then $S'(\omega) = \{\omega_2\}(\subset \Omega_2)$, a singleton. Thus, for each $\omega \in \Omega_1, S'(\omega)$ reduces to a single point and so S' is the graph of a function $g : \Omega_1 \to \Omega_2$. But $S' \in \mathcal{B} \otimes \mathcal{B}_2$, and by standard results in measure theory g is $(\mathcal{B}, \mathcal{B}_2)$-measurable. (See Rao, 1987, p. 324. There $\Omega_2 = \mathbb{R}$, but the same result holds in the present case.) This is the conclusion of the proposition and this establishes Theorem 3.8 also. $\square$

The proof of this result can be simplified considerably if the conditioning σ-algebra $\mathcal{B}$ is partition generated. For the general case which is needed in applications, the detailed and involved arguments of the preceding theory seem unavoidable. For comparison, we also present the special case:

Corollary 10. *Let (Ω, Σ, μ) be a Radon probability space as in the theorem above. If $\mathcal{B} \subset \Sigma$ is a σ-algebra generated by a (countable) partition, then there exists a proper regular conditional measure Q, given $\mathcal{B}$.*

Proof. By definition of Q, it satisfies (16). So if $B \in \mathcal{B}$ we get

$$Q(B, \omega) = \chi_B(\omega), \text{ a.a.}(\omega) . \tag{21}$$

If $\mathcal{P}$ is the countable partition generating $\mathcal{B}$, then (21) holds for each $B \in \mathcal{P}$ or $B \in \mathcal{B}_0$, the algebra generated by $\mathcal{P}$, which is still countable. Thus if N_B is the exceptional null set, and $N = \cup\{N_B : B \in \mathcal{B}_0\}$, then $\mu(N) = 0$. So (21) holds for all $B \in \mathcal{B}_0$ and $\omega \in N^c$. Since $Q(\cdot, \omega)$ and $\chi_{(\cdot)}(\omega)$ are σ-additive and bounded, by the standard Hahn extension theorem the same result holds for all $B \in \sigma(\mathcal{B}_0) = \mathcal{B}$, and $\omega \in N^c$. Thus if $A(\omega) = \cap\{A : \omega \in A \in \mathcal{B}\}$, then $A(\omega) \in \mathcal{B}$, and by the above result (note that $Q(\cdot, \cdot)$ is regular in this case by our theory in Sect. 3 already) we get

$$Q(A(\omega), \omega) = \chi_{A(\omega)}(\omega) = 1, \text{ all } \omega \in N^c .$$

It follows that Q is proper if we set $Q(A(\omega), \omega) = 1$ for $\omega \in N$, so that it holds for all $\omega \in \Omega$. $\square$

We include some complements to the above work in the next section to conclude this discussion, and then present some methods of evaluating regular conditional probabilities for a special class of processes.

5.5 Further results on disintegration

In view of the work of the last section, it is natural to consider some related results on general probability spaces to admit regular conditioning. The theory of Section 3 shows that the desirable spaces for this purpose are the perfect spaces. So we include here some useful complements to that work. Let us first recall a concept for the ensuing discussion.

Definition 1. Let $S \subset \mathbb{R}$, and $(S, \mathcal{S})$ be the Borelian space, and $\mathcal{F}$ be the set of all Radon (= Lebesgue-Stieltjes) probability measures on S. If $\hat{\mathcal{S}}_F$ is the completion of $\mathcal{S}$ for $F \in \mathcal{F}$, then the σ-algebra

$\hat{S} = \cap\{\hat{S}_F : F \in \mathcal{F}\}$ is termed *universally* (or absolutely) *measurable*, and each $A \in \hat{S}$, a *universally* measurable subset of S.

This concept is also meaningful if S is a separable metric space. In either case, $\tilde{S}$ contains all Borel as well as analytic (also called Souslin) sets of S. However, these properties need not hold for all topological (even locally compact) spaces S.

The next result, due to V.V. Sazonov (1962), slightly sharpens Theorem 3.2.

Theorem 2. *Let (Ω, Σ) be a measurable space and $(S, \tilde{S})$ be a universally measurable space with $S = \mathbb{R}$. If every random variable $X : \Omega \to \mathbb{R}$ satisfies the condition that $X(\Omega) \in \tilde{S}$, then (Ω, Σ, μ) is a perfect probability space for every probability $\mu : \Sigma \to \mathbb{R}^+$. Conversely, if for each countably generated σ-algebra $\mathcal{A} \subset \Sigma$ the triple $(\Omega, \mathcal{A}, \mu_{\mathcal{A}})$ is perfect, then for each random variable X on Ω, $X(\Omega) \in \tilde{S}$. When either of these conditions is satisfied and $\mathcal{B} \subset \Sigma$ is any σ-algebra, then μ admits a disintegration relative to $\mathcal{B}$ and $\mu_{\mathcal{B}}$, so that there is a $Q : \Sigma \times \Omega \to \mathbb{R}^+$ such that it is a perfect regular conditional probability function in the sense of Theorem 3.2.*

Remark. Here the basic measure space is more restricted than the earlier result, but $\mathcal{B}$ is not. On the other hand we are not able to assert that $Q(\cdot, \omega)$ is also proper. Thus the result complements the previous work only in some respects. [In the assertion Q may be chosen to satisfy $\rho'(Q_{(\cdot)}) = Q_{(\cdot)}$ where ρ' is a lifting on $M(\Omega, \Sigma)$-valued bounded functions, as in the last section.]

The proof of the above theorem depends on the following:

Proposition 3. *A probability space (Ω, Σ, μ) is perfect iff for each random variable $X : \Omega \to \mathbb{R}$, $X(\Omega)$ is universally measurable, or equivalently, there is a set $\Omega_0 \in \Sigma$ such that $\mu(\Omega_0) = 1$ and $X(\Omega_0)$ is a Borel set.*

Proof. When μ is perfect then, by definition, for each random variable X, there is a Borel set $B \subset X(\Omega)$ such that $\mu(X^{-1}(B)) = \mu(\Omega) = 1$. Let $\nu = \mu \circ X^{-1}$. If ν^* is the outer measure generated by $(\nu, \mathcal{R}), \mathcal{R}$ being the Borel algebra of $\mathbb{R}$, then ν^* is a Carathéodory regular outer measure. Hence by the classical measure theory there is a $B_1 \in \mathcal{R}$ such that $B_1 \supset X(\Omega)$ and $\nu^*(B_1) = \nu^*(X(\Omega))$. ($B_1$ is a measurable cover of $X(\Omega)$, see Rao, 1987 p. 41.) But $B_1 \supset X(\Omega) \supset B$ and B, B_1 are ν^*-measurable, so that $\nu^*(B_1 - B) = 0$, whence $X(\Omega) \in \mathcal{M}_{\nu^*}$, the ν^*-measurable sets. Since μ and hence $\nu (= \mu \circ X^{-1})$ are arbitrary, $X(\Omega) \in \bigcap_{\mu} \mathcal{M}_{(\mu \circ X^{-1})} = \tilde{S}$.

Conversely, let $X(\Omega) \in \tilde{S}$, where X is a random variable $X : \Omega \to \mathbb{R}$, and μ be a probability on Σ. Then $\tilde{S} \subset \mathcal{M}_{\mu \circ X^{-1}}$, the class of $\mu \circ X^{-1}$ measurable sets in the sense of Carathéodory. But then for each $A \in \mathcal{M}_{\mu \circ X^{-1}}$ there is a Borel set $B \subset A$, such that $\mu \circ X^{-1}(A - B) = 0$, (cf., again e.g., Rao (1987), p. 95). Taking $A = X(\Omega) \in \tilde{S} \subset \mathcal{M}_{\mu \circ X^{-1}}$, we get a $B \subset A$ as above and

$$1 \geq \mu(X^{-1}(B)) = \mu \circ X^{-1}(X(\Omega)) \geq \mu(\Omega) = 1 .$$

Hence μ is perfect. $\square$

Proof of Theorem 2. Let μ be a probability measure on Σ, and let $X : \Omega \to \mathbb{R}$ be a random variable, $\nu = \mu \circ X^{-1}$. Then $\mathcal{M}_\nu \supset \tilde{S}$, and since ν is a Lebesgue-Stieltjes measure on $\mathcal{R}$ and $S(\Omega) \in \tilde{S}$, by Proposition 3 above, there is a Borel set $B \subset X(\Omega)$ such that $\nu(B) = \nu(X(\Omega)) = 1$. So μ is perfect.

For the converse, let $\mathcal{A} \subset \Sigma$ be countably generated and $(\Omega, \mathcal{A}, \mu)$ be perfect for any $\mu : \mathcal{A} \to \mathbb{R}^+$, a probability. Since $\mathcal{A}$ is countably generated, there is a random variable $f : \Omega \to \mathbb{R}$ such that $f^{-1}(\mathcal{R}) = \mathcal{A}$ (see Eq. (1) of Sect. 3). Let $\nu : \mathcal{R} \to \mathbb{R}^+$ be a Radon probability measure. If ν^* is the outer measure generated by $(\nu, \mathcal{R})$, then by the classical theory (used in Proposition 3) there exists a $B \in \mathcal{R}$ such that it is a measurable cover of $f(\Omega)$. If $\nu^*(f(\Omega)) = 0$, then $f(\Omega)$ is already ν-measurable. So let $\nu^*(f(\Omega)) > 0$. We assert that there is a Borel set $B_1 \subset f(\Omega)$ such that $\nu(B) = \nu(B_1)$ and then deduce that $f(\Omega)$ is universally measurable.

Since $\nu(B) > 0$, define a probability measure ξ on $\mathcal{R}(B)$ as follows. Although $\mathcal{A} = f^{-1}(\mathcal{R})$, a set $A \in \mathcal{A}$, may be covered by several Borel sets, e.g., $A = f^{-1}(C_1) = f^{-1}(C_2), C_i \in \mathcal{R}$, since f need not be onto. If

$$\xi(A) = \nu(C_1 \cap B)/\nu(B) , \tag{1}$$

we assert that the formula (1) gives ξ unambiguously. This is because $C_1 \cap C_2 \cap B \supset f(\Omega)$, and the Borel set $C_1 \cap C_2 \cap B$ is also a measurable cover of $f(\Omega)$. This implies (by definition of such a cover)

$$\nu(B) \geq \nu(C_i \cap B) \geq \nu(C_1 \cap C_2 \cap B) \geq \nu^*(f(\Omega)) = \nu(B) . \qquad (2)$$

Hence there is equality, in (2), which shows that (1) is unaltered if C_1 is replaced by C_2. So $\xi(\cdot)$ is well-defined and $\xi(\Omega) = 1$. By hypothesis $(\Omega, \mathcal{A}, \xi)$ is a perfect probability space so that there is a Borel set $B_0 \subset f(\Omega)$ of full ξ-measure. Since $B_0 \subset f(\Omega) \subset B$, (1) gives

$$1 = \xi(f^{-1}(B_0)) = \nu(B \cap B_0)/\nu(B) = \nu(B_0)/\nu(B) . \qquad (3)$$

Thus $\nu(B - B_0) = 0$, and $f(\Omega)$ is ν-measurable. Since ν is arbitrary we deduce that $f(\Omega) \in \tilde{\mathcal{S}}$. But $\mathcal{A} \subset \Sigma$ is any countably generated σ-algebra. If $X : \Omega \to \mathbb{R}$ is any random variable, then using the separability of $\mathbb{R}$, we conclude that $X^{-1}(\mathcal{R})(\subset \Sigma)$ is countably generated (using, for instance, open intervals with rational end points) we get by the preceding result that $X(\Omega)$ is universally measurable.

Finally, the disintegration assertion now follows immediately from the first part of the proof of Theorem 4.4 in which $\Omega_1 = \Omega, \mathcal{B}_2 = \mathcal{B}$, and no topology is needed for (i) - (iii) there. $\qquad \square$

It should be noted that there is a certain asymmetry in the theorem, since from the second half we cannot conclude that (Ω, Σ, μ) is itself perfect for all probability measures μ. If the latter is perfect, then $(\Omega, \mathcal{A}, \mu_\mathcal{A})$ is perfect for every σ-subalgebra $\mathcal{A}$ of Σ since any random variable on $(\Omega, \mathcal{A})$ is also one on (Ω, Σ). The converse of the latter holds under a strengthening, however. Thus if for each countably generated $\mathcal{A} \subset \Sigma$, the restriction $\mu_\mathcal{A}$ is compact (or pure) then the given $\mu :$ $\Sigma \to \mathbb{R}^+$ will be perfect and conversely. But if Σ itself is countably generated, then this difficulty disappears. We state the latter case for reference as follows.

Corollary 4. *Let (Ω, Σ) be a measurable space, Σ countably generated. Then for any probability $\mu : \Sigma \to \mathbb{R}^+, (\Omega, \Sigma, \mu)$ is perfect iff for each random variable $X : \Omega \to \mathbb{R}$, $X(\Omega)$ is universally measurable. When this holds, for each σ-algebra $\mathcal{B}(\subset \Sigma)$, the measure μ admits a disintegration into regular conditional probabilities $Q : \Sigma \times \Omega \to \mathbb{R}^+$ relative to $\mathcal{B}$.*

Next we briefly consider the transitivity of regular conditional measures on different spaces through disintegration formulas. This will be useful in applications such as Markov processes, discussed later in Chapter 9. The problem is motivated by our previous Propositions 2.2.1 and 2.3.2. The first one states that for σ-algebras $\mathcal{B}_1 \subset \mathcal{B}_2 \subset \Sigma$, we have

$$E^{\mathcal{B}_1}(P^{\mathcal{B}_2}(A)) = P^{\mathcal{B}_1}(A), \text{ a.e. } (\mu_B), \quad A \in \Sigma . \tag{4}$$

The second result says, in the current terminology, that if $Q(\cdot, s) : \Sigma \to \mathbb{R}^+, s \in S$, is a family of probability measures, indexed by S, and $Q(A, \cdot)$ is S-measurable, where $(S, \mathcal{S}, \mu)$ is a probability space, then there is a probability P on the product measurable space $(\Omega \times S, \Sigma \otimes \mathcal{S})$, such that we have

$$\int_{\Omega \times S} X dP = \int_S \int_\Omega X(\omega, s) Q(d\omega, s) \mu(ds) , \tag{5}$$

for all random variables $X : \Omega \times S \to \mathbb{R}^+$. Since (4) also implies that

$$E(E^{\mathcal{B}_1}(P^{\mathcal{B}_2}(A))) = E(P^{\mathcal{B}_1}(A)) = E(\chi_A) , \tag{6}$$

both (4) and (5) can be combined into a common generalization of successive (or product) disintegration formulas. Such an extension has already been considered by L. Schwartz (1973). Here we discuss a simple case that suffices for our later applications.

The inclusion relations in (6) for $\mathcal{B}_0 \subset \mathcal{B}_1 \subset \mathcal{B}_2 \subset \Sigma$ extend the following types. Let $\Omega = \mathbb{R}^{m_1} \times \mathbb{R}^{m_2} \times \mathbb{R}^{m_3}, \pi_{m_1} : \Omega \to \mathbb{R}^{m_1}$, and $\pi_{m_1+m_2} : \Omega \to \mathbb{R}^{m_1} \times \mathbb{R}^{m_2} = \mathbb{R}^{m_1+m_2}$, be coordinate projections. If $\mathcal{R}_{m_1}, \mathcal{R}_{m_1+m_2}$ denote the Borel σ-algebras of $\mathbb{R}^{m_1}$ and $\mathbb{R}^{m_1+m_2}$, then $\mathcal{B}_1 = \pi_{m_1}^{-1}(\mathcal{R}_{m_1})$ and $\mathcal{B}_2 = \pi_{m_1+m_2}^{-1}(\mathcal{R}_{m_1+m_2})$ are the cylindrical σ-algebras contained in Σ, the Borel σ-algebra of Ω, and these satisfy the inclusion relations for (4) and (6). With this motivation, the general case can be given as follows.

Let $(\Omega_i, \Sigma_i), i = 1, 2, 3$, be measurable spaces, $p_{12} : \Omega_1 \to \Omega_2, p_{23} : \Omega_2 \to \Omega_2$ be measurable functions and let $\mu_1 : \Sigma \to \mathbb{R}^+$ be a probability. Let $\mu_2 = \mu_1 \circ p_{23}^{-1}, \mu_3 = \mu_2 \circ p_{23}^{-1}$ be the respective image measures. We can find conditions for the disintegration of $(\mu_1, \mu_2), (\mu_2, \mu_3)$ and (μ_1, μ_3). In general, how will they relate to each other? Although $\mu_3 = \mu_1 \circ (p_{23} \circ p_{12})^{-1}$, if $p_{13} : \Omega_1 \to \Omega_3$ is another measurable mapping and $\tilde{\mu}_3 = \mu_1 \circ p_{13}^{-1}$, then μ_3 and $\tilde{\mu}_3$ are not necessarily the same probability measures; and μ_1 may have two disintegrations relative to μ_3 and $\tilde{\mu}_3$.

Recalling that the existence of a disintegration is equivalent to the existence of a regular conditioning, the above situation can be stated precisely as follows:

Proposition 5. *Let* (Ω, Σ, μ) *be a probability space,* $(S, \mathcal{S})$ *and* $(T, \mathcal{T})$ *be measurable spaces. If* $p : \Omega \to S$ *and* $q : S \to T$ *are measurable mappings and* $\nu = \mu \circ p^{-1}, \xi = \nu \circ q^{-1}$ *are the image measures on* S *and* $\mathcal{T}$, *suppose that* $Q : \Sigma \times \mathbb{R} \to \mathbb{R}^{+}$ *is a disintegration of* μ *relative to* ξ *and* $p \circ q$. *Then* $\tilde{Q} : (A, s) \mapsto Q(A, q(s))$, $A \in \Sigma$, $s \in S$, *defines a disintegration of* μ *relative to* ν *and* p.

This result follows immediately from the formulas,

$$\mu(A \cap p^{-1}(B)) = \int_{B} \tilde{Q}(A, s) d\nu(s)), \ A \in \Sigma, B \in \mathcal{S}, \tag{7}$$

and

$$\mu(A \cap (p \circ q)^{-1}(C)) = \int_{C} Q(A, t) d\xi(t), \ A \in \Sigma, \ C \in \mathcal{T}, \tag{8}$$

together with the facts that $Q(\cdot, t)$ is regular, $p^{-1}(q^{-1}(\mathcal{T})) \subset p^{-1}(\mathcal{S}) \subset \Sigma$.

We now present some methods of evaluation of conditional expectations for certain classes of processes having considerable practical interest.

5.6 Evaluation of conditional expectations by Fourier analysis

In the preceding sections we have presented results on regularity of conditional probability functions so that conditional expectations can be represented as ordinary (Stieltjes) integrals. However, we still do not have a recipe to evaluate these integrals. A formula will now be presented on calculating some conditional expectations in terms of ordinary expectations, using certain classical Fourier analytic methods. We essentially utilize some ideas of Yeh's papers (1974 and 1975).

The procedure is basically to employ the disintegration formula [see Eq. (5.5)] which connects the absolute and the regular conditional measures. Then the Lévy inversion theorem is employed. The precise result is as follows.

Theorem 1. *Let* (Ω, Σ, P) *be a probability space,* $X : \Omega \to \mathbb{R}^k$ *and* $Y : \Omega \to \mathbb{R}$ *be random variables such that* $E(|Y|) < \infty$, *and* $Y(\Omega) \subset \mathbb{R}$ *is a Borel set. If* $\varphi_Y(t) = E(Y e^{i(t,X)})$, *and* F_X *is the distribution function of* X, *then* $\varphi_Y(\cdot)$ *is a Fourier-Stieltjes transform of a signed measure and the conditional expectation* $E(Y|X)$ *is obtainable from the relation* $(P_X(A) = P[X \in A])$:

$$(E(Y|X))(a) \; P_X(A) = \lim_{T \to \infty} \frac{1}{(2\pi)^k} \int_{-T}^{T} \cdots \int_{-T}^{T} \left[\prod_{j=1}^{k} \left(\frac{1 - e^{it_j h_j}}{it_j} \right) e^{-it_j a_j} \right] \times$$

$$\varphi_Y(t_1, \ldots, t_k) dt_1 \ldots dt_k, \quad (1)$$

for each $a = (a_1, \ldots, a_k) \in \mathbb{R}^k$, *with* $A = \underset{j=1}{\overset{k}{\times}} [a_j, a_j + h_j)$ *as a continuity set of* F_X.

Proof. The argument is based on a careful use of Eq. (5.5) and the Lévy inversion formula, as already noted. Here are the details.

Let $D = Y(\Omega)$, $\mathcal{S} = \sigma(X)(\subset \Sigma)$ and $\mathcal{B} = \sigma(Y)(= Y^{-1}(\mathcal{R}(D)), \mathcal{R}(D)$ being the Borel σ-algebra of D). Then $P^{\mathcal{S}} : \mathcal{B} \times \mathbb{R}^k \to \mathbb{R}^+$ is a regular conditional probability which we denote by $Q(\cdot, \cdot)$ to use the notation of Theorem 2.2. Hence for each measurable mapping $Z : \Omega \times \mathbb{R}^k \to \mathbb{R}^+$, we have

$$\int_{\Omega \times \mathbb{R}^k} Z(\omega, r) d\tilde{P} = \int_{\mathbb{R}^k} \int_{\Omega} Z(\omega, r) Q(d\omega, r) \mu(dr), \quad (2)$$

where $\mu = P \circ X^{-1}(= P_X)$ and $\tilde{P}$ is the probability measure on $\mathcal{B} \otimes \mathcal{R}^k$ determined by $Q(\cdot, \cdot)$ and μ. Note that in (2) Z may be replaced by any $\tilde{P}$ integrable function. Also as a consequence of the regularity of the conditional measure $P^{\mathcal{S}}(\cdot)$ we have

$$E^{\mathcal{S}}(Y)(r) = E(Y|X)(r) = \int_{\Omega} Y(\omega) Q(d\omega, r), \quad (3)$$

which holds first for simple functions Y, and then the general case follows by the Lebesgue dominated convergence criterion since $Q(\cdot, r)$ is now a measure. Set $(u, X) = \sum_{i=1}^{k} u_i X_i$.

Let $Z = e^{i(u,X)} Y$ in (2). Since Z is $\tilde{P}$-integrable, we get

$$\varphi_Y(u) = E(e^{i(u,X)} Y) = \int_{\mathbb{R}^k} \int_{\Omega} e^{i(u,r)} Y(\omega)Q(d\omega, r)d\mu(r)$$

$$= \int_{\mathbb{R}^k} e^{i(u,r)} E(Y|X)(r)d\mu(r), \quad \text{by (3),}$$

$$= \int_{\mathbb{R}^k} e^{i(u,r)} d\nu(r), \tag{4}$$

where $\nu : A \mapsto \int_A E(Y|X)(r)d\mu(r)$ is a signed measure on $\mathcal{R}^k$, the Borel σ-algebra of $\mathbb{R}^k$. Thus (4) shows that $\varphi_Y(\cdot)$ is a Fourier-Stieltjes transform of ν.

Since ν can be expressed (by the Jordan decomposition) as the difference of a pair of finite positive measures on $\mathcal{R}^k$, we can apply the classical multidimensional Lévy inversion formula (see Rao, 1984, p. 257) which gives (1) immediately. $\square$

It may be of interest to note that when $\varphi_Y(\cdot)$ is Lebesgue integrable, then we can divide (1) by $(h_1 h_2, \ldots h_k)$ and take limits as $h_j \to 0$ so that P_X or F_X has a bounded continuous density. If this is denoted by f_X, we get (1) simplified as:

$$E(Y|X)(r)f_X(r) = \frac{1}{(2\pi)^k} \int_{\mathbb{R}^k} e^{-i(u,r)} \varphi_Y(u)du_1 \ldots du_k . \tag{5}$$

We state this result for a convenient reference as:

Corollary 2. *Let X, Y be as in the theorem and suppose that $\varphi_Y :$ $t \mapsto E(Ye^{i(t,X)})$ is Lebesgue integrable. Then the image measure P_X of X has a bounded continuous density $f_X(\cdot)$ and the conditional expectation $E(Y|X)$ is given by (5). [If X, Y are independent, the $\varphi_Y(u) = E(Y)\varphi_X(u)$, and (5) reduces to the familiar identity.]*

In the above assertions, the requirement that $Y(\Omega)$ be a Borel set is unpleasant. But without some additional condition, the regularity of conditional measures cannot be demanded, as we know by counterexamples. However, this condition can be exchanged for several others if the basic probability space is restricted. For instance, with Theorem 3.2 the above becomes:

Corollary 3. *If (Ω, Σ, P) is a perfect probability space, Y is any P-integrable random variable and X is a k-random vector, then again* *the conditional expectation is given by* (1), *and, if* $\varphi_Y(\cdot)$ *is integrable, formula* (5) *holds.*

Using Theorem 3.7 in place of 3.2 we can assert the following which is more useful in applications:

Corollary 4. *Let* Ω *be a Polish space,* Σ *be its Borel* σ*-algebra and* $P : \Sigma \to \mathbb{R}^+$ *be a probability measure. Then for* X, Y *as in Corollary 3, we get* $E(Y|X)$ *given by* (1) *or* (5) *accordingly as the corresponding hypothesis holds.*

As an example of such an Ω we can consider the space of bounded continuous $\mathbb{R}^k$-valued functions $C(\mathbb{R}^+, \mathbb{R}^k)$ with the uniform metric. Such spaces appear in the function space representations of stochastic processes with random vectors as evaluation functionals. We shall also illustrate this point by some explicit calculations of conditional expectations.

In (5) we demanded the integrability of $\varphi_Y(\cdot)$. However, this may be weakened with the classical $(c, 1)$-summability methods of Fourier analysis if we assume that the image measure $\mu(= P_X)$ of X is diffuse. The precise statement is as follows.

Corollary 5. *Let* X, Y *be as in Theorem 1 and* μ *be diffuse. Then we have*

$$E(Y|X)(r)f_X(r) = \lim_{a \to \infty} \frac{1}{(2\pi)^k} \int_{-a}^{a} \cdots \int_{-a}^{a}$$
$$\prod_{i=1}^{k} \left(1 - \frac{t_i}{a}\right) \cdot \varphi_Y(t_1, \ldots, t_k) dt_1 \ldots dt_k,$$

for a.a. $r \in \mathbb{R}^k$, *where* $f_X(\cdot)$ *is the density of* P_X *for* μ.

This is a consequence of the fact that if $f : \mathbb{R}^k \to \mathbb{R}$ is Lebesgue integrable with $\hat{f}$ as its Fourier transform, then we have for a.a.(u)

$$f(u) = \lim_{T \to \infty} \frac{1}{(2\pi)^k} \int_{-T}^{T} \cdots \int_{-T}^{T} \prod_{j=1}^{k} [(1 - \frac{|t_i|}{T}) e^{-iu_j t_j}]$$
$$\hat{f}(t_1, \ldots, t_k) dt_1, \ldots, dt_k .$$

In the next section we use this Corollary to get an extension of the conditional Feynman-Kac formula. To consider such an application, however, we need to recall some facts and the function space representation of the (abstract) Wiener space.

Observe that if $\{X_t, t \in I\}$ is a process on $(\Omega, \Sigma, P), I \subset \mathbb{R}$, and $t_1 < t_2 < \cdots < t_n$ are points from I, then the distributions

$$F_{t_1,\ldots,t_n}(x_1,\ldots,x_n) = P[X_{t_1} < x_1,\ldots,X_{t_n} < x_n], \; x_i \in \mathbb{R},$$

satisfy the *compatibility relations*:

(i) $F_{t_{i_1},\ldots,t_{i_n}}(x_{i_1},\ldots,x_{i_n}) = F_{t_1,\ldots,t_n}(x_1,\ldots,x_n)$

(ii) $\lim\limits_{x_n \to \infty} F_{t_1,\ldots,t_n}(x_1,\ldots,x_n) = F_{t_1,\ldots,t_{n-1}}(x_1,\ldots,x_{n-1}),$

where $(i_1,\ldots,i_n)$ is a permutation of $(1,\ldots,n)$. Conversely if an indexed family $\{F_{t_1,\ldots,t_n}, \; t_i \in \mathbb{R}, \; n \geq 1\}$ of distributions satisfying the above pair of compatibility conditions is given, then by Theorem 8.4.2 (proved later) of Kolmogorov says that we can construct a probability space (Ω, Σ, P) and a stochastic process $\{X_t, t \in I\}$ on it, having the given family as its finite dimensional distributions. Here $\Omega = \mathbb{R}^I$, the space of real functions on I, with Σ as the σ-algebra generated by all the cylinder sets $\{\omega \in \Omega : \omega(t_1) < x_1,\ldots,\omega(t_n) < x_n, t_i \in I, n \geq 1\}$, $X_t : \Omega \to \mathbb{R}$ is the coordinate projection, $X_t(\omega) = \omega(t), t \in I$, and

$$P[\omega : X_{t_i}(\omega) < x_i, i = 1,\ldots,u] = \int\limits_{-\infty}^{x_1} \cdots \int\limits_{-\infty}^{x_n} dF_{t_1,\ldots,t_n}(u_1,\ldots,u_n) .$$

With this set up, if we choose $I = \mathbb{R}^+, 0 = t_0 < t_1 < \cdots < t_n$, and

$$F_{t_1,\ldots,t_n}(x_1,\ldots,x_n) = [(2\pi)^n \Pi_{j=1}^n (t_j - t_{j-1})]^{-\frac{1}{2}} \int\limits_{-\infty}^{x_1} \cdots \int\limits_{-\infty}^{x_n}$$

$$\exp[-\frac{1}{2} \sum_{j=1}^n \frac{(u_j - u_{j-1})^2}{t_j - t_{j-1}}]du_n \ldots du_1, \quad (6)$$

then the resulting process $\{X_t, t \geq 0\}$ with $X_0 = 0$, is called the *Wiener* or *Brownian motion process* starting at 0. Moreover, it has almost all of its sample paths $t \mapsto X_t(\omega)$ continuous and the process has independent Gaussian increments, i.e., $X_{t_j} - X_{t_{j-1}}$ is $N(0, t_j - t_{j-1})$. The set $C_0(\mathbb{R}^+) \subset \Omega = \mathbb{R}^{\mathbb{R}^+}$ of continuous functions vanishing at 0, is the space on which P is supported. Consequently, one may replace (Ω, Σ, P) with $(C_0(\mathbb{R}^+), \hat{\mathcal{B}}, P_1)$ where $\hat{\mathcal{B}}$ is the completed σ-algebra of the cylinder sets of $C_0(\mathbb{R}^+)$, and $P_1 = P|\hat{\mathcal{B}}$. (For direct proof of these assertions one may also consult, Rao, 1979, §III.2 and especially p. 189.) An important consequence of this representation is that each X_t satisfies $X_t(\Omega) = \mathbb{R}$ which is thus a Borel set. Hence, *the conditional probability function of X_t given a random variable on such Ω is automatically regular* by Theorem 2.2. A similar statement holds for many function space representations. If $I \subset \mathbb{R}^+$ is a compact interval, then $C_0(I)$ is a Polish space, under the uniform norm, and Theorem 3.7 is applicable to obtain regular conditional probabilities.

We now illustrate, with the above facts, the exact evaluation of conditional expectations of some functionals of Brownian motion.

Example 6. Let $\{X_t, t \geq 0\}$ be a Brownian motion, $a > 0$, and Z be a functional defined as $Z_a = \int\limits_0^a X_s^2 \, ds$, the pathwise stochastic integral. It is desired to find the conditional expectation $E(Z_a | X_a)$. We apply Corollary 4. Note that by (2) and (3) we can express (5) as:

$$E(g(X)Y) = \int_{\mathbb{R}^k} g(v) E(Y|X)(v) dP_X(v), \tag{7}$$

for any integrable random variable Y and a bounded Borel function g. Taking $g(u) = \frac{1}{2h}\chi_{[u-h,u+h]}$ for $h > 0$, and noting that P_X is a diffuse measure, we get for (7), on setting $k = 1$,

$$\lim_{h \to 0} E(\chi_{[u-h,u+h]}(X)Y) = \lim_{h \to 0} \frac{1}{2h} \int\limits_{u-h}^{u+h} E(Y|X)(v) dP_X(v)$$

$$= E(Y|X)(u) f_X(u), \quad a.a. \ (u), \tag{8}$$

by the Lebesgue differentiation theorem (see Rao, 1987, p. 237). Since $f_{X_a}(u) = (2\pi a)^{-\frac{1}{2}} \exp(-u^2/2a)$, it suffices to evaluate the left side of (8) when $Y = Z_a$ and $X = X_a$. This is obtained as follows. Consider:

$$E(\chi_{(u-h,u+h)}(X_a)Z_a) = \int_{\Omega} \chi_{(u-h,u+h)}(X_a)\left(\int\limits_0^a X_s^2 \, ds\right) dP$$

$$= \int\limits_0^a \int\limits_{-\infty}^{\infty} \int\limits_{-\infty}^{\infty} \chi_{(u-h,u+h)}(v_2) v_1^2 [(2\pi)^2 s(a-s)]^{-\frac{1}{2}} \times$$

$$\exp\{-\frac{1}{2}\frac{v_1^2}{s} + \frac{(v_2 - v_1)^2}{a - s}\} dv_1 \, dv_2 \, ds,$$

by (6) and the image probability law, (see Rao, 1984, p. 19),

$$
= \int_0^a 4\pi^2 s(a-s) \Bigg[\int_{-\infty}^\infty dv_2 \int_{u-h}^{u+h} [(v_1 - \frac{s}{a}v_2)^2
$$

$$
+ \frac{2s}{a}v_2(v_1 - \frac{s}{a}v_2) + (\frac{s}{a}v_2)^2] \times
$$

$$
\exp\{-\frac{1}{2}\frac{a}{s(a-s)}(v_1 - \frac{s}{a}v_2)^2 - \frac{1}{2}\frac{v_2^2}{a}\}d\, v_1 \Bigg] ds
$$

$$
= \frac{1}{\sqrt{2\pi a}\cdot 2h} \int_{u-h}^{u+h} \Bigg[\int_0^a (s - \frac{s^2}{a} + \frac{s^2 v_2^2}{a^2})ds \Bigg] \times
$$

$$
\exp(-\frac{v_2^2}{2a})dv_2,
$$

using the fact that the odd moments of X_t vanish, the even moments are given by simple formulas, and the Fubini-Tonelli theorem applies,

$$
= \frac{1}{2a} \int_{u-h}^{u+h} (\frac{a^2}{6} + \frac{av_2^2}{3})\frac{1}{\sqrt{2\pi a}} \exp(-v_2^2/2a)dv_2
$$

$$
\rightarrow (\frac{a^2}{6} + \frac{a}{3}u^2) \frac{1}{\sqrt{2\pi a}} \exp(-u^2/2a), \text{ as } h \rightarrow 0.
$$
(9)

It follows from (8) and (9) that for a.a. (u)

$$
E(Z_a|X_a)(u) = \frac{a^2}{6} + \frac{a}{3}u^2,
$$
$$
(= E(Z_a|X_a = u) \text{ in the old notation}).
$$
(10)

Note that from (10) one gets the known result that

$$
E(Z_a) = E(E(Z_a|X_a))
$$

$$
= \int_{-\infty}^\infty (\frac{a^2}{6} + \frac{a}{3}u^2)\cdot \frac{1}{\sqrt{2\pi a}} \exp(-u^2/2a)du
$$

$$
= \frac{a^2}{6} + \frac{a}{3}\cdot a = \frac{a^2}{2} = E\left(\int_0^a X_s^2 ds \right)
$$

$$
= \int_0^a E(X_s^2)\, ds = \int_0^a s\, ds \ .
$$
(11)

Remark. This computation does not use the L'Hôpital rule because the event $[X_a = u]$ has probability zero. The latter method can yield other values than the correct result given by (10), depending on the approximation used, besides being quite difficult to calculate.

Next let us consider the Feynman-Kac formula, evaluating again a conditional expectation of a suitable functional on a Wiener space as in the above example. We first set up the problem.

Let $V : \mathbb{R} \to \mathbb{R}^+$ be a bounded continuous function and $\{X_t, t \geq 0\}$ be the Brownian motion. Then the function $u(\cdot, \cdot) : \mathbb{R} \times \mathbb{R}^+ \to \mathbb{R}$ defined by the equation

$$u(x, t) = E[\exp(- \int_0^t V(X_s)ds) \cdot \delta(X_t - x)] \tag{12}$$

($\delta(\cdot)$ being the Dirac delta function) is shown, by Feynman and Kac, to satisfy the integral equation for $t > 0$:

$$u(x, t) = f_{X_t}(x) - \frac{1}{\sqrt{2\pi}} \int_0^t ds \int_{-\infty}^{\infty} \frac{1}{\sqrt{t-s}}$$

$$\exp\left\{-\frac{1}{2}\frac{(x-y)^2}{t-s}\right\} \cdot V(y)u(y, s) \, dy \tag{13}$$

subject to the boundary condition

$$\lim_{t \to 0} \int_{-\varepsilon}^{\varepsilon} u(x, t)dx = 1, \ \varepsilon > 0 . \tag{14}$$

The solution $u(\cdot, \cdot)$ of (13) which satisfies (12) is called the *Feynman-Kac formula*.

Taking advantage of the existence of regular conditional probability measures in this case, with the function space representation, there is a related but distinct solution of (13). One notes that $u(\cdot, \cdot)$ of (13) also satisfies the partial differential equation with (14) holding:

$$\frac{\partial u}{\partial t}(x, t) = \frac{1}{2}\frac{\partial^2 u}{\partial x^2}(x, t) - V(x)u(x, t), \ (x, t) \in \mathbb{R} \times \mathbb{R}^+ . \tag{15}$$

Example 7. Let $0 < a < \infty$ be fixed and for the Brownian motion process $\{X_t, t \geq 0\}$, consider $Y_t = \exp(-\int_0^t V(X_s) \, ds)$, where $V \geq 0$ also satisfies $E(V(X_t)) < \infty$, $t > 0$. Let $u(x, t) = E(Y_t|X_t)(x)f_{X_t}(x)$. We assert that $u(\cdot, \cdot)$ satisfies (13) subject to (14) and (15).

The idea of proof is again to use Corollary 5 to calculate $u(\cdot, \cdot)$ by converting the conditional expectation into an absolute integral with an appropriate Fourier transform. Thus u is expressible as:

$$u(x, t) = \lim_{T \to \infty} \frac{1}{2\pi} \int_{-T}^{T} (1 - \frac{|v|}{T}) e^{ivx} E(e^{ivX_t} Y_t) \, du \ . \tag{16}$$

Observe that a differntiation and pathwise integration of $\exp(\int_0^t g(s)ds)$ for an appropriate $g(\cdot)$ gives:

$$Y_t = 1 - \int_0^t V(X_s) \exp(-\int_0^s V(X_r)dr)ds \ . \tag{17}$$

From (16) and (17) we have

$$u(x, t) = \lim_{T \to \infty} \frac{1}{2\pi} \int_{-T}^{T} (1 - \frac{|v|}{T}) e^{ivx} [E(e^{ivX_t}) - J_t(v)]dv$$

$$= I_1(x, t) - I_2(x, t), \text{ (say)},$$

$$\tag{18}$$

where

$$J_t(v) = E\left[e^{ivX_t} \int_0^t V(X_s) \exp(-\int_0^s V(X_r)dr)ds\right] \ . \tag{19}$$

Remembering the expression for the characteristic function of the Gaussian random variable X_t, $N(0, t)$, and with a standard justification for a change of the order of integration, one gets

$$I_1(x, t) = (2\pi t)^{-\frac{1}{2}} \exp(-x^2/2t) \ .$$

To simplify $J_t(v)$, recall that for $0 < s < t, X_s$ and $X_t - X_s$ are independent Gaussian random variables so that

$$J_u(v) = \int_0^t \exp(-v^2/2(t-s)) \left[\int_{\mathbb{R}} e^{ivy} V(y) E(Y_s|X_s)(y) P_{X_s}(dy) \right] ds$$

$$= \int_0^t \exp(-v^2/2(t-s)) \left[\int_{\mathbb{R}} e^{ivy} V(y) u(y,s) dy \right] ds . \tag{20}$$

With this in $I_2(x,t)$ and a change of the order of integration (standard justification again) we get from Corollary 5 the following:

$$I_2(x,t) = \int_0^t \left[\int_{\mathbb{R}} V(y) u(y,s) [2\pi(t-s)]^{-\frac{1}{2}} \exp(-(x-y)^2/2(t-s)) \, dy \right] ds .$$

$$\tag{21}$$

Substituting (20) and (21) in (18), one obtains the desired integral equation (13).

5.7 Further evaluations of conditional expectations

Although the Fourier analytic methods give sharp results when they are applicable, they are not easily employed for nonlinear and infinite dimensional functionals. We now present a useful extension to abstract Wiener spaces and functionals on them. This again is ultimately based on the existence of regular conditional probabilities guaranteed by Theorem 3.7. In this application we use some computations given in D. M. Chang and S.J. Kang (1989).

As in the last section, we need to recall some facts on abstract Wiener spaces and discuss conditioning on them. Thus if $\mathcal{H}$ is a Hilbert space with $\langle \cdot, \cdot \rangle$ as its inner product and $\|\cdot\|$ the norm, let $P(\cdot)$ be the measure defined on the cylinder sets $A = \Pi^{-1}(F)$ based in finite dimensional spaces $\mathbb{R}^n$, where $\Pi : \mathcal{H} \to \mathbb{R}^n$ is a projection, given by

$$P(A) = (2\pi)^{-\frac{n}{2}} \int_F \exp\left(-\frac{1}{2}\|x\|^2\right) dx . \tag{1}$$

Here we may and do identify each n-dimensional subspace of $\mathcal{H}$ with $\mathbb{R}^n$. Then the class $\mathcal{C}$ of all cylinders of $\mathcal{H}$ forms a semi-algebra on which

P is finitely additive. Suppose now there is a positive homogeneous subadditive functional $q(\cdot)$–a seminorm–given on $\mathcal{H}$ such that for each $\varepsilon > 0$ there is a projection $\Pi_0(= \Pi_0(\varepsilon))$ with finite dimensional range satisfying

$$P[x : q(\Pi(x)) > \varepsilon] < \varepsilon \qquad (2)$$

for each finite dimensional projection $\Pi \perp \Pi_0$. Such a $q(\cdot)$ is called a *measurable seminorm*, the existence of which implies that $\mathcal{H}$ is also separable and $q(\cdot)$ is weaker than $\|\cdot\|$. Let $B = \overline{sp}(\mathcal{H})$ for $q(\cdot)$. Then, P has a countably additive extension to the Borel σ-algebra $\mathcal{B}$ of B, and that the support of this extended measure, again denoted by P, is all of B. This is an important result due to L. Gross. For details of these statements, one may refer to Kuo (1975) and Rao (1979, pp 30-35). Thus if B^* is the dual of the Banach space B, then $B^* \subset \mathcal{H}^* \cong \mathcal{H} \subset B$ and these inclusions are continuous. The triple $(B, \mathcal{B}, P)$ or if $i : \mathcal{H} \hookrightarrow B$, the $(i, \mathcal{H}, B)$ is termed an *abstract Wiener space*. The following are some concrete examples of these spaces:

(i) Let $B = \{f \in C[0, a] : f(0) = 0\}, \mathcal{H} = \{f \in B : \int_0^a |f'(t)|^2 \, dt = \langle f', f' \rangle < \infty\}$ the classical Wiener space with $q(f) = \|f\|_\infty$.

(ii) Let $\mathcal{H}$ be as in (i) and for $0 < \alpha < \frac{1}{2}$, let $q_\alpha(f) = \sup\{|f(s) - f(t)|/|t - s|^\alpha : 0 \le s \ne t \le a\}$ and $B_\alpha = \overline{sp}\{\mathcal{H}, q_\alpha\}$.

(iii) Let $\{\mathcal{H}, \langle \cdot, \cdot \rangle\}$ be a separable Hilbert space, $A : \mathcal{H} \to \mathcal{H}$ be a positive definite nuclear (= trace class) operator so that $\langle Ax, x \rangle = \langle x, Ax \rangle > 0$ for $0 \ne x \in \mathcal{H}$, and $\sum_{n=1}^{\infty} \|Ae_n\| < \infty$ for some (hence all) complete orthonormal sequence $\{e_n, n \ge 1\}$. Here $q(x) = \sqrt{\langle Ax, x \rangle}$ and $B = \overline{sp}\{\mathcal{H}, q(\cdot)\}$.

Of course, not every Hilbert space need admit a measurable seminorm. The following properties are of interest here. For each complete orthonormal sequence $\{e_j, j \ge 1\}$ of $\mathcal{H}$, let $f \in \mathcal{H}$ be defined as a functional on $B = \overline{sp}\{\mathcal{H}, q(\cdot)\}$ by the equation:

$$f(x) = \lim_{n \to \infty} \sum_{j=1}^{n} \langle f, e_j \rangle (e_j, x), \quad x \in B, \qquad (3)$$

if this limit exists and set $f(x) = 0$ otherwise. Here $(\cdot, \cdot)$ is the duality pairing of B and B^*. Since $e_j \in \mathcal{H} \subset B$, and $B^* \subset \mathcal{H}$, the expression (3) is meaningful. It may be verified that f is a random variable on $(B, \mathcal{B}, P)$ and is $N(0, \|f\|^2)$. Moreover, if $f_1, \ldots, f_n$ are orthogonal in $\mathcal{H}$, then $\{f_i(\cdot), 1 \le i \le n\}$ forms a mutually independent set on $(B, \mathcal{B}, P)$. These statements have obvious counterparts for complex spaces.

We now give a form of Theorem 6.1 in the infinite dimensional case:

Theorem 1. *Let $(B, \mathcal{B}, P)$ be an abstract Wiener space so that B is already a Polish space. Let $g \in \mathcal{H}$ and define the random variables X_g and Y_h as $X_g(x) = (g, x)$, and*

$$Y_h(x) = \exp\{i(Ah, x) - \alpha(x, Ax)\}, \quad x \in B, \ h \in \mathcal{H}, \ \alpha > -\frac{1}{2\lambda_1} \quad (4)$$

where λ_1 is the maximum eigenvalue of the positive definite nuclear operator A on $\mathcal{H}$ such that $(I + 2\alpha A)^{-1}$ exists. Then the conditional expection $E(Y_h | X_g)$ is given by the equation

$$E(Y_h|X_g)(y)f_{X_g}(y) = [2\pi \det(I + 2\alpha A)^{-1}]^{-\frac{1}{2}} \langle g, (I + 2\alpha A)^{-1} g \rangle^{\frac{1}{2}} \times$$
$$\exp\{\frac{[\langle Ah, (I + 2\alpha A)^{-1} g \rangle + iy]^2}{2\langle g, (I + 2\alpha A)^{-1} g \rangle} -$$
$$\frac{1}{2}\langle Ah, (I + 2\alpha A)^{-1} Ah \rangle\}, \quad (5)$$

for a.a.(y) in $\mathbb{R}$ (Leb.) and $f_{X_g}(\cdot)$ is the Gaussian density of X_g.

Proof. Using the properties of A and of the abstract Wiener space, we reduce this result to that of Theorem 6.1 (or its Corollary 6.2). Thus let λ_j be the eigenvalues and e_j the corresponding normalized eigenvectors (forming a complete orthonormal set) of the positive definite nuclear (hence compact) A so that $\lambda_n \to 0$, as $n \to \infty$. By relabelling if necessary, we may assume that $\lambda_n \ge \lambda_{n+1}$. Let $g_j = \langle g, e_j \rangle$ and $h_j = \langle h, e_j \rangle$ so that $g = \sum_{j=1}^{\infty} g_j e_j$, $h = \sum_{j=1}^{\infty} h_j e_j$ the series converging in norm. If we define X_j by $X_j(x) = e_j(x)$, then the X_j are mutually independent. Also

$$X_g = \sum_{j=1}^{\infty} g_j X_j, \ Ah = \sum_{i=1}^{\infty} h_j(Ae_j) = \sum_{j=1}^{\infty} \lambda_j h_j e_j,$$

and

$$(Ah, x) = \sum_{j=1}^{\infty} \lambda_j h_j(e_j, x) = \sum_{j=1}^{\infty} \lambda_j h_j X_j(x).$$

With these representations, we use Corollary 6.2 for which one needs to simplify $\varphi_{Y_h}(\cdot)$ there. So consider

$$\varphi_{Y_h}(u) = E(e^{uX_g} Y_h)$$

$$= \int_B \exp\left\{iu \sum_{i=1}^{\infty} g_j X_j\right\}. \exp\{i(Ah, \cdot) - \alpha(\cdot, A)\} \, dP_{X_g}$$

$$= \int_B \prod_{j=1}^{\infty} \exp[i\, ug_j X_j + i\lambda_j h_j X_j - \alpha\lambda_j X_j^2] \, dP_{X_g},$$

since

$$Ax = \sum_{j=1}^{\infty} e_j(x)Ae_j = \sum_{j=1}^{\infty} X_j(x)\lambda_j e_j,$$

and

$$(Ax, x) = \sum_{j=1}^{\infty} X_j^2(x)\lambda_j, \quad x \in \mathcal{H},$$

$$= \prod_{j=1}^{\infty} \int_B \exp\{i(ug_j + \lambda_j h_j)X_j - \alpha\lambda_j X_j^2\} \, dP_X,$$

by the independence of X_j,

$$= \prod_{j=1}^{\infty} \int_{\mathbb{R}} \exp\left\{i(ug_j + \lambda_j h_j)v - (\alpha\lambda_j + \frac{1}{2})v^2\right\} \frac{dv}{\sqrt{2\pi}},$$

by the image law theorem,

$$= \left\{\prod_{j=1}^{\infty}(1 + 2\alpha\lambda_j)^{-\frac{1}{2}}\right\} \exp\left\{-\frac{1}{2}\sum_{j=1}^{\infty} \frac{(ug_j + \lambda_j h_j)^2}{1 + 2\alpha\lambda_j}\right\}, \tag{6}$$

using the results for the moment generating and characteristic functions of normal random variables and the fact that $\sum_{j=1}^{\infty} \lambda_j < \infty$.

Since $E(|Y_h|) < \infty$, we can now use Corollary 6.2 to get

$$E(Y_h|X_g)(y)f_{X_g}(y) = \frac{1}{2\pi} \int_{\mathbb{R}} e^{-iuy} \varphi_{Y_h}(u) \, du$$

$$= \frac{1}{2\pi}[\det(P + 2\alpha A)]^{-\frac{1}{2}} \int_{\mathbb{R}} \exp\left\{-iuy - \frac{1}{2} \times\right.$$

$$\left. \sum_{j=1}^{\infty} \frac{(ug_j + \lambda_j - h_j)^2}{1 + 2\alpha\lambda_j}\right\} \, du,$$

since the first factor of (6) is the Fredholm determinant given above,

$$
= \frac{1}{\sqrt{2\pi}} [\det(I + 2\alpha A)]^{-\frac{1}{2}} \left(\sum_{j=1}^{\infty} \frac{g_j^2}{1 + 2\alpha\lambda_j} \right)^{-\frac{1}{2}} \times
$$

$$
\exp \left\{ \sum_{j=1}^{\infty} \frac{(g_j h_j \lambda_j + iy(1 + 2\alpha\lambda_j)^2 / (1 + 2\alpha\lambda_j)^2}{2 \sum_{j=1}^{\infty} g_j^2 (1 + 2\alpha\lambda_j)^{-1}} - \right.
$$

$$
\left. \frac{1}{2} \sum_{j=1}^{\infty} h_j^2 \lambda_j^2 (1 + 2\alpha\lambda_j) \right\} .
$$

This is just (5) in a different form. $\square$

As an application, we can give a simplified proof of an extension of the Feynman-Kac formula, established by Donsker and Lions (1962).

Let $B = C_0[0, t], \mathcal{H} = \{ f \in B : \int_0^t |f'(u)|^2 du = \|f'\|_2^2 < \infty \}$ be as in example (i) and P be the Wiener probability measure on $\mathcal{B}$. If $h \in \mathcal{H}$, define

$$
u(x, t, h) = E[\exp(i \int_0^t h(s) X_s \, ds - \frac{1}{2} \int_0^t X_s^2 ds) \cdot \delta(X_t + x)], \qquad (7)
$$

where $X_t(x) = x(t)$, $x \in B$ and $\delta(\cdot)$ is the Dirac delta function. Then

$$
\frac{\partial u}{\partial t} - \frac{1}{2} \frac{\partial^2 u}{\partial x^2} = -\frac{1}{2} x^2 u + \frac{1}{2} x \, h(t) \, u, \ 0 < t < \infty , \qquad (8)
$$

with the boundary condition

$$
\lim_{t \downarrow 0} u(x, t, h) = \delta(x) .
$$

Note that if $h = 0$ and $V(x) = x/\sqrt{2}$, (8) is analogous to the earlier case, although u is defined somewhat differently there. Extending the Feynman-Kac result, Donsker and Lions have shown that a solution of (8) satisfies the following integral equation (however, this is slightly different in from, hence does not reduce to the previous case):

$$u(x,t,h) = (2\pi \sinh\ t)^{-\frac{1}{2}} \exp\ \{-\frac{x^2}{2\tanh\ t}$$

$$-\frac{ix}{\tanh\ t} \int_0^t r(t,v;-1)h(v)\ dv\} \times$$

$$\exp\ \{\frac{1}{2} \int_0^t \int_0^t r(v,s;-1)h(v)h(s)dvds\ \} \times$$

$$\exp\ \{\frac{1}{2\tanh\ t} \int_0^t \int_0^t r\ (t,v;-1)r(t,s;-1)h(v)h(s)dvds\}, \tag{9}$$

where the kernel $r(\cdot,\cdot;-1)$ is given by

$$r(x,y,-1) = -\ \frac{\cosh(x-y)\sinh(\min(x,y))}{\cosh\ y},\ 0 \leq x,y\ . \tag{10}$$

We now show that, by a suitable specialization of the preceding theorem, the Donsker-Lions solution can be obtained by an evaluation of (5). The analogy with the last section will then be clear.

Note that the kernel (10) may be obtained as the reciprocal kernel of $\rho(s,t) = \min(s,t)$, $0 \leq s,t \leq a$, $0 < a < \infty$. This means, if $\rho_1 = \rho$ and for $k > 1$, if we set $\rho_k(s,t) = \int_0^a \rho(s,v)\rho_{k-1}(v,t)dv$, then

$$r(s,t;\lambda) = -\ \sum_{k=1}^{\infty} \lambda^{k-1} \rho_k(s,t),$$

which converges absolutely if $|\lambda|$ max $|\rho(s,t)| < 1$. Then r satisfies the equation

$$r(s,t;\lambda) = -\rho(s,t) + \lambda \int_0^a \rho(s,t)r(v,t;\lambda)dv. \tag{11}$$

It can be verified that $r(\cdot,\cdot;\cdot)$ given by (10) satisfies (11) with the above value of ρ and $\lambda = -1$, $(0 < a < 1)$.

Define the operator $A : \mathcal{H} \to \mathcal{H}$ by the equation, for fixed $t > 0$,

$$(Ax)(\tau) = \int_0^t \rho(\tau,s)x(s)\ ds,\quad x \in \mathcal{H}\ . \tag{12}$$

Then converting this into a differential equation, it is seen that A is a nuclear operator with eigenvalues $\lambda_n(t) = t^2(n - \frac{1}{2})^{-2}\pi^{-2}$ and the corresponding eigen functions $e_{n,t}(\tau) = \sqrt{2t}[(n - \frac{1}{2})\pi]^{-1}\sin(\frac{(n-\frac{1}{2})\pi}{t}\tau)$. Then the integral equation

$$X(\tau) = y(\tau) - \int_0^t \rho(\tau, s)x(s)\,ds = y(\tau) - (Ax)(\tau) \qquad (13)$$

can be solved to obtain $x(\tau) = (I + T)y(\tau)$ where

$$(Ty)(\tau) = \int_0^t r(\tau, s; -1)y(s)\,ds. \qquad (14)$$

Now simplifying the right side of (5) for the A given by (12), with $\alpha = \frac{1}{2}$ and $g(\tau) = \tau$, we get

$$\det(I + A) = \prod_{j=1}^{\infty}\left(1 + \frac{t^2}{(j - \frac{1}{2})\pi^2}\right) = \cosh t,$$

$$g_j = \langle g, e_j\rangle = \frac{2t}{(j - \frac{1}{2})\pi},$$

and hence

$$\langle g, (I + A)^{-1}g\rangle = \sum_{j=1}^{\infty}\frac{g_j^2}{1 + \lambda_j}$$

$$= \sum_{j=1}^{\infty}\frac{2t}{(j - \frac{1}{2})^2\pi^2}\left(1 - \frac{1}{(j - \frac{1}{2})^2\pi^2}\right)^{-1}$$

$$= \tan t. \qquad (15)$$

Since $(I + A)^{-1}g = (I + T)g$, we have

$$\langle Ah, (I + A)^{-1}g\rangle = \int_0^t h(\tau)(I + T)g(\tau)\,d\tau$$

$$= \int_0^t h(\tau)[\tau + \int_0^t r(\tau, s; -1)s\,ds]\,d\tau$$

$$= \int_0^t h(\tau)\rho(\tau; t)\,d\tau + \int_0^t\int_0^t \rho(s, t)r(\tau, s; -1)h(\tau)ds\,d\tau$$

$$= -\int_0^t h(\tau)r(t, \tau; -1)d\tau\ ,\ \text{by (11).} \qquad (16)$$

Also

$$\langle Ah, (I + A)^{-1} Ah \rangle = \int_0^t h(\tau)[(Ah)(\tau) + \int_0^t r(\tau, s; -1)(Ah)(s) \ ds] \ d\tau$$

$$= - \int_0^t \int_0^t h(\tau)h(s)r(\tau, s; -1)ds d\tau, \text{ by (11) and}$$

$$\text{a simplification}. \tag{17}$$

Substituting the expressions between (14)-(17), in the right side of (5), we get (9).

Remark. In addition to the regularity of conditional probabilities, we need to use additional, often nontrivial and new, techniques to evaluate conditional expectations as defined by Kolmogrov. Methods in non Wiener measure spaces are still desirable for similar computations.

5.8 Bibliographical notes

One of the key problems raised in the preceding chapters is to find hypotheses for conditional probability functions to be used as ordinary measures in applications. This is a nontrivial matter which demands that we find classes of measure spaces admitting such conditioning. For abstract spaces this can be done if the measures are perfect, pure or compact. The first one was introduced by Gnedenko and Kolmogorov in 1949 and the third one by Marczewski in 1953 and a related concept, called quasi-compactness of measures by Ryll-Nardzewski the same year which was later found to be equivalent to perfectness. The concept of pure measures for a similar purpose was introduced by the author in 1970 and some years later was found to be equivalent to compactness. A thorough analysis of some of these concepts (especially perfectness) is to be found in Sazonov (1962). We have included a somewhat detailed treatment of these concepts in order to explain and later use the regular conditioning more thoroughly than that available in previous publications. Actually the early investigations on this subject seem to have been pursued by Blackwell since 1942. The sufficient condition that the range of a conditioning random variable be a Borel set was given by Doob (1953) refining the previous attempts, but which is closely related

is closely related to perfectness. Its significance and the necessity of
the condition for proper regular conditional probability functions were
obtained only in 1963 by Blackwell and Ryll-Nardzewski. One of the
alternative characterizations of perfectness, and hence its employment
for perfect regular conditioning, was established by Sazonov (1962),
(see also Kallianpur (1959)), given here as Theorem 5.2. Most of these
results are for abstract probability spaces, although from time to time
topology has played a (supplementary) role.

The second part of our work, also intimately related to regular conditioning, is to disintegrate the given probability by its image measure.
Refined results here are possible for topological measure spaces, especially if they are locally compact second countable or Polish spaces in
general. One can obtain the existence of regular conditional probabilities on these spaces, but for the more useful *proper* regular conditional
measures further restrictions are needed. The results here are closely
related to, (and in fact equivalent with) the existence of strong lifting
operators on these measure spaces. Major contributions to this problem are due to Tulcea and Tulcea (1964). A detailed discussion of the
latter topic is found in their monograph (1969). See also Dinculeanu
(1966). Another important memoir on this set of problems is that of
Schwartz's (1973). Although nonfinite Radon measures are considered
in these references, the theory has a smooth flow especially for probability measures. We have included several results here since the explicit
connection of regular conditioning with disintegration illuminates the
subject, as introduced originally by Kolmogorov, better than viewing
it in isolation. Moreover, in all but simple cases, the very connection
with lifting shows that we can only prove the existence results, but the
actual construction of these regular conditional probabilities is really a
difficult problem. Since a constructive method must necessarily avoid
Zorn's lemma (a key step in the existence proof of lifting), it follows
that we have to impose some separability hypotheses at almost every
turn. This point has been largely unrecognized in the literature.

The several examples of Chapter 3 show that there are no "easy"
methods of construction that can be employed in applications. Any
reasonable procedure appears to need a differentiation basis, and much
remains to be done on this problem. This is why we consider here
primarily the *existence* problem, but the last two sections show how for

certain classes, especially for Wiener spaces, conditional expectations can be gainfully calculated and used. The results here are abstracted from Yeh (1974) and (1975) as well as additional applications from Chang and Kang (1989). The treatment here follows the recent paper by the author (Rao, 1992). The special case contained in Corollary 6.2 has been generalized in Zabell (1979) to the LCA groups. His method does not include Theorem 6.1 or Corollary 6.5 easily. It is useful to note that $E(X|Y)$ given in Corollaries 6.2–6.5 is the only correct expression that satisfies the Kolmogorov definition.

Since regular conditioning is part of the definition of Rényi's (1955) axiomatic approach to conditional probability, the above difficulties are disallowed there. However, his theory also excludes problems with conditioning events having probability zero. But the above elaborate set up was needed to solve precisely these problems which have great practical significance. The extension indicated by Rényi to solve the Borel paradox, for instance, was not satisfactory as the solution depends on the method used. We shall see in Chapter 7, that an abstract version of Kolmogorov's theory works also for σ-finite measures and hence can include Rényi's conditional measures. In this sense we shall eventually have a unified account of both conditional measures. The work in the present chapter will thus form an important foundation of this abstract formulation and will be a key component for several applications including such areas as Markov processes and martingales, to be considered in Chapter 9.

We next turn to another application, the "sufficiency" concept, introduced by R.A. Fisher, of great interest in statistical inference theory. Its very definition depends crucially on the conditioning concept as formulated by Kolmogorov (and is distinct from Rényi's model).

CHAPTER 6: SUFFICIENCY

The theory of conditioning so far developed is elaborated, extended, and given an important application in this chapter, to a study of sufficient statistics. The concept of sufficiency plays an important role in a branch of statistical inference. Here the study consists of two parts: the dominated and the undominated cases. Both aspects are considered below in detail. The applications include statistical experiments and estimation. These results establish the principal role played by Kolmogorov's general concept of conditioning in statistical theory, especially in the area of sufficiency, and also elsewhere. An abstract formulation of the latter concept is discussed in the last section.

6.1 Introduction

To motivate the concept, consider a Gaussian probability measure P_α on a measurable space (Ω, Σ) which has mean α and unit variance so that for the corresponding random variable $X : \Omega \to \mathbb{R}$ one has:

$$P_\alpha(X(\omega) < x) = \frac{1}{\sqrt{2\pi}} \int_{-\infty}^{x} \exp(-\frac{1}{2}(u - \alpha)^2)du, \quad -\infty < \alpha < \infty, \ x \in \mathbb{R}.$$

$$(1)$$

A Borel function $\varphi(X)$ of X is called an *estimator* of α, and is said to

180

be *optimal* or *best* provided

$$E_\alpha(\psi(\varphi(X) - \alpha)) = \int_\Omega \psi(\varphi(X) - \alpha)dP_\alpha \qquad (2)$$

is a minimum, as α varies, relative to a given nonnegative function $\psi(\cdot)$, termed a *loss function*. If moreover the conditional loss or distribution of X, given $\varphi(X)$, does not involve α then $\varphi(X)$ is said to be *sufficient* for X, since such a conditional distribution does not play any part in describing the properties of α. This concept was first introduced by R.A. Fisher in the early 1920s into statistical inference theory. However, the full potential of this idea was realized only in the 1940s when the axiomatic formulation and the general theory of conditioning, due to Kolmogorov, were disseminated and appreciated. This chapter is thus devoted to this topic and its ramifications while other applications of conditioning will occupy the later parts of this book.

The following is the traditional (i.e., essentially Fisherian) concept using the contemporary terminology.

Definition 1. Let $\{(\Omega, \Sigma, P_\alpha) : \alpha \in I\}$ be a family of probability spaces describing an experiment. Then a σ-subalgebra $\mathcal{B} \subset \Sigma$ is said to be *sufficient* for the collection $\mathcal{P} = \{P_\alpha, \alpha \in I\}$ of probability measures (cf. (1)), if for each bounded real function measurable relative to Σ, or $f \in B(\Omega, \Sigma)$, there is a $\mathcal{B}$-measurable function $\tilde{f}$ such that the following equations hold:

$$\int_\Omega f g dP_\alpha = \int_\Omega \tilde{f} g dP_\alpha, \quad \alpha \in I, \quad g \in B(\Omega, \mathcal{B}). \qquad (3)$$

Note that if $I = \{\alpha_0\}$ is a singleton, then $\tilde{f} = E^{\mathcal{B}}_{\alpha_0}(f)$, the conditional expectation of f relative to $\mathcal{B}$, which always exists and is $(P_{\alpha_0}|\mathcal{B})$-unique. But (3) imposes a stronger restriction that the same $\tilde{f}$, or its "equivalence class", should hold for *all* members of $\mathcal{P}$ and this becomes more stringent if I is uncountable, as in the example of (1) above. If $\mathcal{B} = \Sigma$ then $\tilde{f} = f$, so that Σ is trivially sufficient for $\mathcal{P}$. Hence the existence of $\tilde{f}$ arises when $\mathcal{B} \neq \Sigma$. Now (3) can be restated, in the notation of conditional expectations, as follows. A σ-subalgebra $\mathcal{B} \subset \Sigma$ is sufficient for $\mathcal{P}$ iff for each bounded measurable f (relative to Σ), there is a $\mathcal{B}$-measurable F_f such that

$$E^{\mathcal{B}}_\alpha(f) = F_f, \quad a.e.[P_\alpha], \quad \alpha \in I. \qquad (4)$$

Here F_f equals each element $E_\alpha^\mathcal{B}(f)$ on $(\Omega, \Sigma, P_\alpha)$, but does not itself depend on α. In (4) we may take f in $\cap\{L^1(\Omega, \Sigma, P_\alpha); \ \alpha \in I\}$ and then F_f is required to be in $\cap\{L^1(\Omega, \mathcal{B}, P_\alpha): \ \alpha \in I\}$. The class of "$\mathcal{P}$-null" sets has a key role to play in the technical analysis and complicates the ensuing work.

An analogous concept for random functions can be given as:

Definition 2. Let $\{(\Omega, \Sigma, P_\alpha), \alpha \in I\}$ and $\mathcal{B} \subset \Sigma$ be as in Definition 1. If $(\tilde{\Omega}, \tilde{\Sigma})$ is a measurable space, $T : \Omega \to \tilde{\Omega}$ is a measurable mapping then T is called a *sufficient statistic* for the family $\mathcal{P} = \{P_\alpha, \ \alpha \in I\}$ if the generated σ-algebra $\mathcal{B}_T = T^{-1}(\tilde{\Sigma})$ is sufficient in the sense of Definition 1, i.e., (3) or (4) should hold with $\mathcal{B}_T$ in place of $\mathcal{B}$ there.

These abstract definitions are less simple to consider for specific problems. Under further restrictions on $\mathcal{P}$, they can be simplified. But first we present some discussion which clarifies the concepts. The main contributions to the general (not necessarily dominated) sufficiency theory are due to Bahadur (1954), Pitcher (1957, 1965) and Burkholder (1961, 1962). Most of the (negative) results here show the difficulties in the general case contrasting it with the (positive) results in the dominated case, originally detailed by Halmos and Savage (1949).

Just as in the case $\mathcal{B} = \Sigma$, the other extreme $\mathcal{B} = \{\phi, \Omega\}$ is also trivial as seen from the following:

Lemma 3. *The σ-algebra $\mathcal{B} = \{\phi, \Omega\}$ satisfies (3) iff $I = \{\alpha\}$, a singleton.*

Proof. If α_1, α_2 are two element of $I, f = \chi_A$ for $A \in \Sigma$, then taking $g = 1$, (3) becomes

$$P_{\alpha_1}(A) = \int_\Omega f dP_{\alpha_1} = \int_\Omega E^\mathcal{B}(f) dP_{\alpha_1}$$

$$= \int_\Omega E^\mathcal{B}(f) dP_{\alpha_2} = \int_\Omega f dP_{\alpha_2} = P_{\alpha_2}(A). \tag{4}$$

Hence $P_{\alpha_1} = P_{\alpha_2}$ so that $\alpha_1 = \alpha_2$. The converse is obvious. $\square$

The preceding result motivates the following weaker concept:

Definition 4. Let $\{(\Omega, \Sigma, P_\alpha), \ \alpha \in P\}$ be a family of probability spaces. Then a σ-subalgebra $\mathcal{B} \subset \Sigma$ is termed *pairwise sufficient* for

$\mathcal{P} = \{P_\alpha, \alpha \in I\}$, if for each couple $(P_{\alpha_1}, P_{\alpha_2})$ of $\mathcal{P}$ and each f in $B(\Omega, \Sigma)$, there is a $\mathcal{B}$-measurable $\tilde{f}_{\alpha_1, \alpha_2}$ such that

$$\int_\Omega fg dP_{\alpha_i} = \int_\Omega \tilde{f}_{\alpha_1, \alpha_2} g dP_{\alpha_i}, \quad i = 1, 2, \quad g \in B(\Omega, \mathcal{B}). \tag{5}$$

It is evident that, if $\mathcal{B}$ is sufficient for $\mathcal{P}$, then it is also pairwise sufficient. But the converse does not hold, as the following example shows. Later, in Section 3 below, it will be shown that both concepts coincide for the dominated families and hence this is also a useful weakening.

Example 5. (Pairwise sufficiency $\not\Rightarrow$ Sufficiency) Let $\Omega = [0, 1] \times \{0, 1\} \subset \mathbb{R}^2$ and $\Sigma = \mathcal{B}([0, 1]) \otimes 2^{\{0,1\}}$. For each $\alpha \in [0, 1]$, let $P_\alpha : \Sigma \to [0, 1]$ be given by

$$P_\alpha(E) = \frac{1}{2}\{\mu(A_1) + \delta_\alpha(A_2)\} \tag{6}$$

where $E = (A_1 \times \{0\}) \cup (A_2 \times \{1\}) \in \Sigma$ with $A_i \in \mathcal{B}([0, 1])$. Here $\mathcal{B}([0, 1])$ is the Borel σ-algebra of $[0, 1], 2^{\{0,1\}}$ is the power set of the two point set $\{0, 1\}, \mu$ the Lebesgue measure, and $\delta_\alpha(\cdot)$ the point measure so that $\delta_\alpha(A) = 1$ if $\alpha \in A$ and $= 0$ if $\alpha \notin A$. Then $P_\alpha(\cdot)$ is a probability measure on Σ for each α. Let $\tilde{\mathcal{B}} = \mathcal{B}([0, 1]) \otimes \mathcal{A}_0$ where $\mathcal{A}_0 = \{\phi, \{0, 1\}\}$ the trivial algebra of $\{0, 1\}$. Then $\tilde{\mathcal{B}} \subsetneqq \Sigma$, and it is *claimed* that $\tilde{\mathcal{B}}$ is pairwise sufficient but not sufficient for $\mathcal{P} = \{P_\alpha, 0 \le \alpha \le 1\}$.

To see that $\tilde{\mathcal{B}}$ is pairwise sufficient, let A be a generator of Σ so that it is of the form E in (6). Let $g \in B(\Omega, \tilde{\mathcal{B}})$ and consider (4) for $f = \chi_A$. Then

$$\int_\Omega \chi_A g dP_{\alpha_i} = \int_\Omega \tilde{f}_{\alpha_1, \alpha_2} g dP_{\alpha_i}, \quad i = 1, 2.$$

Since we may take $g = g_1 \chi_{\{0,1\}}$, $g_1 \in B([0, 1], \mathcal{B}([0, 1]))$, the above pair of equations will hold if $\tilde{f}_{\alpha_1, \alpha_2} = (\chi_{A_1 - \{\alpha_1, \alpha_2\}} + \chi_{A_2 \cap \{\alpha_1, \alpha_2\}})\chi_{[0,1]}$. This shows that $\tilde{\mathcal{B}}$ is pairwise sufficient for $\mathcal{P}$.

For the sufficiency case, suppose for any $f \in B(\Omega, \Sigma)$ there exists an $\tilde{f}$ satisfying (4). So $\tilde{f} = \tilde{f}_1 \chi_{\{0,1\}}$, and (4) becomes with $g = \chi_B, B$ in $\tilde{\mathcal{B}}$ of the form $B = \{\alpha\} \times \{0, 1\}$, and $f = \chi_{[0,1]} \chi_{\{0\}}$,

$$\int_\Omega \tilde{f} g dP_\alpha = \int_B \tilde{f} dP_\alpha = \int_\Omega fg dP_\alpha = \int_B \chi_{[0,1]} \chi_{\{0\}} dP_\alpha, \quad 0 \le \alpha \le 1. \tag{7}$$

Then the right side of (7), with (6), gives $\frac{1}{2}\delta_\alpha([0,1]) = \frac{1}{2}$. But the left side is

$$\text{LHS of (7)} = \frac{1}{2}\left[\int_B \tilde{f}_1 d\mu + \int_B \tilde{f}_1 d\delta_\alpha\right] = \frac{1}{2}(0 + \tilde{f}_1(\alpha)),$$

since a finite set has Lebesgue measure zero. Hence $\tilde{f}_1(\alpha) = 0$ for $0 < \alpha < 1$, so that $\tilde{f}_1 = 0$ a.e., and then $\tilde{f} = 0$ a.e. $[P_\alpha]$ whereas the right side $= \frac{1}{2}$. This contradiction shows that $\tilde{f}$ satisfying (4) does not exist.

This implies that, in the nontrivial cases, sufficient σ-algebras do not always exist. Consequently, we need to answer the following three questions at a minimum: (i) When is a given σ-algebra $\mathcal{B} \subsetneq \Sigma$ sufficient for a family $\mathcal{P} = \{P_\alpha, \alpha \in I\}$, if I has at least two elements? (ii) Is there a minimal sufficient σ-algebra when there is a positive answer to (i)? (iii) When do sufficiency and pairwise sufficiency coincide? We devote this chapter to solving these and related problems.

6.2 Conditioning relative to families of measures

In Definition 1.1, one may evidently restrict $f = \chi_A, A \in \Sigma$. Thus $\mathcal{B} \subset \Sigma$ is a sufficient σ-algebra for $\mathcal{P} = \{P_\alpha, \alpha \in I\}$ if there is an $\tilde{f}_A \in B(\Omega, \mathcal{B})$ satisfying

$$\int_\Omega \chi_A g \, dP_\alpha = \int_\Omega \tilde{f}_A g \, dP_\alpha, \quad \alpha \in I, \ g \in B(\Omega, \mathcal{B}). \tag{1}$$

Letting $g = \chi_B, B \in \mathcal{B}$, this becomes

$$\int_B \tilde{f}_A dP_\alpha = P_\alpha(A \cap B), \quad A \in \Sigma, \ B \in \mathcal{B}, \ \alpha \in I. \tag{2}$$

Thus form the work in Chapter 2, we have $\tilde{f}_A = P_\alpha^{\mathcal{B}}(A)$, a.e. $[P_\alpha|\mathcal{B}], \alpha \in I$, where $P_\alpha^{\mathcal{B}}(\cdot)$ is the conditional probability function on $(\Omega, \Sigma, P_\alpha)$ relative to $\mathcal{B}$ and $0 \le \tilde{f}_A \le 1$, a.e. $[P_\alpha|\mathcal{B}]$, as an element of $B(\Omega, \mathcal{B})$. Since $P_\alpha^{\mathcal{B}}(\cdot)$ is σ-additive for each α in I, it is natural to ask whether $\tilde{f}_{(\cdot)}$ is σ-additive in the vector space $B(\Omega, \mathcal{B})$. It is evident that $\tilde{f}_{(\cdot)}$ is finitely additive, and for the σ-additivity the role of the family $\mathcal{P} = \{P_\alpha, \alpha \in I\}$ should be clarified.

Consider the space $B(I)$ of real bounded functions on I. It becomes a Banach lattice under uniform norm and pointwise ordering. The given family $\mathcal{P}$ can be regarded as a mapping $\mathcal{P} : \Sigma \to B(I)$ which is additive. If $\ell_\alpha : B(I) \to \mathbb{R}$ is an evaluation functional, defined as $\ell_\alpha(f) = f(\alpha)$ for $f \in B(I)$, then ℓ_α is a continuous linear functional and $\ell_\alpha \circ P$, defined as $\ell_\sigma(\mathcal{P}(E)) = \mathcal{P}(E)_\alpha = P_\alpha(E), E \in \Sigma$, is a (probability) measure. Although $(\ell_\alpha \circ \mathcal{P})(\cdot)$ is thus a measure for each $\alpha \in I$, one needs to show that $(x^* \circ \mathcal{P})(\cdot)$ is a signed measure for *all* x^* in $(B(I))^*$ to conclude that $\mathcal{P}(\cdot)$ is a weakly and hence, by a classical result due to B. J. Pettis also strongly σ-additive. In fact we have the following general statement.

Proposition 1. *If I is infinite, $\mathcal{P} : \Sigma \to B(I)$ need not be σ-additive in the weak or (equivalently) strong topology of the Banach space $B(I)$, and it does not have finite variation.*

Proof. To verify the first statement, suppose that $\mathcal{P}$ is in fact always σ-additive. We now construct an example to contradict this supposition.

Let $\Omega = [0,1], \Sigma = $ Borel σ-algebra of Ω and P_α be the Bernouli measure assigning probability $\frac{1}{2}$ to α and $\frac{1}{2}$ to $\alpha + \frac{1}{2}$ where $\alpha \in I = [0, \frac{1}{2}]$. Then

$$\|\mathcal{P}(E)\| = \sup_{\alpha \in I} |P_\alpha(E)| \geq \frac{1}{2}, \quad \phi \neq E \in \Sigma. \tag{3}$$

Let $\tilde{\mu}(A) = \sup\{\|\mathcal{P}(E)\| : E \subset A, E \in \Sigma\}$, $A \in \Sigma$. Then $\tilde{\mu}(A) \geq \frac{1}{2}$ for $\phi \neq A \in \Sigma$. Since $\mathcal{P}(\cdot)$ is a vector measure by assumption, it follows by a result of Gould (1965, Cor. 3.6) that $\tilde{\mu}(\cdot)$ is subadditive and, what is more important, for any $A_n \searrow \phi$, $A_n \in \Sigma$, one has with (3),

$$\frac{1}{2} \leq \lim_n \tilde{\mu}(A_n) = \tilde{\mu}(\lim_n A_n) = \tilde{\mu}(\phi) = 0 \tag{4}$$

This contradiction shows that the above $\mathcal{P}(\cdot)$ cannot be σ-additive.

Regarding the *variation* of $\mathcal{P}$ on any open set $E \subset \Omega$, by definition,

$$|\mathcal{P}|(E) = \sup\{\sum_{i=1}^{n} \|\mathcal{P}(A_i)\| : A_i \text{ disjoint, } \overset{n}{\underset{i=1}{\cup}} A_i, A_i \in \Sigma(E)\}, \tag{5}$$

and for the above $\mathcal{P}$, each term $\|\mathcal{P}(A_i)\| \geq \frac{1}{2}$. Hence $|\mathcal{P}|(E) = +\infty$. Any nonempty measurable set will have a similar property. $\square$

A natural question now is to utilize the lattice properties of the range space $B(I)$ of $\mathcal{P}(\cdot)$, and to find the behavior of $\mathcal{P}(\cdot)$ although it is not σ-additive in the norm topology. Indeed $\mathcal{P} : \Sigma \to B(I)$ has the following properties: (i) $\mathcal{P}(\phi) = 0$, (ii) $\mathcal{P}(A) \geq 0$ for each $A \in \Sigma$, where one uses the pointwise ordering in $B(I)$, and (iii) if $A_n \in \Sigma$, disjoint, $A = \underset{n}{\cup} A_n = \lim_{n} \overset{n}{\underset{i=1}{\cup}} A_i \in \Sigma$, one has $\mathcal{P}(A) = \bigvee_n \overset{n}{\underset{i=1}{\Sigma}} \mathcal{P}(A_i) = \bigvee_n \mathcal{P}(\overset{n}{\underset{i=1}{\cup}} A_i)$, (by definition of supremum). The last equality is clear from

$$
\begin{aligned}
\sup_n \mathcal{P}(\tilde{A}_n) &= \sup_n \{P_\alpha(\tilde{A}_n), \ \alpha \in I\}, \ \tilde{A}_n = \overset{n}{\underset{i=1}{\cup}} A_i, \\
&= \{P_\alpha(A), \ \alpha \in I\}, \text{ since each } P_\alpha(\cdot) \text{ is a measure}, \\
&= \mathcal{P}(A), \text{ by definition}.
\end{aligned}
\tag{6}
$$

It follows from (5) and (6) that $\mathcal{P}(\cdot)$ does not, in general, have finite variation, but it is σ-additive in the order topology but not necessarily so in norm. Also from standard results in Abstract Analysis (the Stone-Gel'fand theorem), it follows that $B(I)$ is isometrically lattice isomorphic to the space of real continuous functions $C(S)$ on a compact Hausdorff space. [Here S is the "Stone space", the space whose topology is extremally disconnected, so that the closure of each open set is open. See, e.g., Dunford-Schwartz (1958), Ch. IV, Sec. 5.] If $J : B(I) \to C(S)$ is this isomorphism and $\tilde{\mathcal{P}} = J \circ \mathcal{P}$, then $(\Omega, \Sigma, \tilde{\mathcal{P}})$ has values in $\mathcal{E} = C(S)$. In this case, integrals $\int_\Omega f d\tilde{\mathcal{P}}$ can be defined for each $f \in B(\Omega, \Sigma)$, and they have many (although *not* all) properties of the Lebesgue-Stieltjes type, including the monotone convergence theorem. These integrals have been developed by J.D.M. Wright (1969) who also established a Radon-Nikodým theorem and then the existence of an operator analogous to conditional expectations. To understand these results in our context, it is necessary to study sufficiency in the dominated case, and some pathological properties of $\mathcal{P}$ in the undominated case. After these problems are analyzed, we shall treat the abstract case in Section 5 below where, using Wright's integrals, it will be possible to answer the first question stated at the end of the preceding section, essentially completely.

6.3 Sufficiency: the dominated case

Let us recall a result from the vector measure theory. If $\mathcal{P} : \Sigma \to B(I)$ is σ-additive, then by a classical result of Bartle-Dunford-Schwartz there is a positive finite measure λ on Σ such that $\lim_{\lambda(E) \to 0} \|\mathcal{P}\|(E) = 0$ where $\|\mathcal{P}\|(\cdot)$ is the *semivariation*, defined as:

$$\|\mathcal{P}\|(A) = \sup\{\| \sum_{i=1}^{n} a_i \mathcal{P}(A_i)\| : |a_i| \leq 1, \quad A_i \subset A, \text{ disjoint, } A_i \in \Sigma\}. \tag{1}$$

It can be shown that $\|\mathcal{P}\|(A)$ is always finite, $A \in \Sigma$, and $\|\mathcal{P}\|(A) \leq |\mathcal{P}|(A)$, where $|\mathcal{P}|(\cdot)$ is the variation of $\mathcal{P}$ (cf., (5) of Sec. 2). The above inequality implies that $\lambda(E) \leq \|\mathcal{P}(E)\|$ and $\lim_{\lambda(E) \to 0} \|\mathcal{P}(E)\| = 0$, with $\|\cdot\|$ denoting the (uniform) norm of $B(I)$. We now suppose merely the conclusion of the above result, and use it as a definition of domination:

Definition 1. Let $\mathcal{M} = \{\mu_\alpha, \alpha \in I\}$ be a set of scalar measures on a measurable space (Ω, Σ). Then it is said to be *dominated* if there is a σ-finite measure λ on Σ such that $\mu_\alpha << \lambda, \alpha \in I$. This will be denoted simply as $\mathcal{M} << \lambda$. [Thus $\lambda(A) = 0 \Rightarrow \mu_\alpha(A) = 0$, uniformly in $\alpha \in I$.]

It is clear that $\mathcal{P} = \{P_\alpha, \alpha \in I\} : \Sigma \to B(I)$ being additive, if it is dominated, then $\lim_{\lambda(A) \to 0} \|\mathcal{P}(A)\| = 0$ since λ is equivalent to a finite measure (see Lemma 3 below). Hence $\mathcal{P}$ will be σ-additive in $B(I)$. The notion of domination in statistical literature predates the BDS theorem recalled above. In many applications $\mu_\alpha = P_\alpha$ will be a probability and λ will be σ-finite. However, the most general λ that can be admitted here is what is called a "localizable measure". We state this concept now, since it is then possible to discuss many of the results in a more general form and this is also important for some applications of sufficiency, especially if Ω is a locally compact group and λ can be a Haar measure which is not σ-finite but is localizable.

Definition 2. Let $(\Omega, \Sigma, \lambda)$ be a measure space satisfying the (nonrestrictive) condition that $\Sigma = \sigma(\Sigma_0)$ where $\Sigma_0 = \{A \in \Sigma : \lambda(A) < \infty\}$. Then $\lambda(\cdot)$ is said to be *localizable* if every (possibly uncountable) collection $\mathcal{C} \subset \Sigma$ has a supremum, say B, in Σ. This means (i) for each $C \in \mathcal{C}$, $\lambda(C - B) = 0$ and (ii) if $\tilde{B} \in \Sigma$ satisfies (i), then $\lambda(B - \tilde{B}) = 0$. One says that λ is *strictly localizable* (or *decomposable* or has *the direct* *sum property*) if there is a disjoint (perhaps uncountable) collection $\{A_i, i \in I\} \subset \Sigma$ such that (a) $0 < \lambda(A_i) < \infty$, (b) $\Omega - \underset{i \in I}{\cup} A_i \subset N$ with $\lambda(N) = 0$, and (c) $A \in \Sigma, \mu(A) < \infty$ implies $\{i \in I : \lambda(A \cap A_i) > 0\}$ is at most countable.

It is clear that σ-finiteness implies decomposability which in turn implies localizability. Also the latter is stronger than the *finite subset property*, i.e., $A \in \Sigma, \lambda(A) > 0$ implies the existence of a $B \in \Sigma_0$, $B \subset A$ with $\lambda(B) > 0$. These relations are all strict. The concept of localizability was introduced by I.E. Segal in 1954 who also showed that any measure $\nu : \Sigma \to \bar{\mathbb{R}}^+$ with $\nu << \mu$ for some measure $\mu : \Sigma \to \bar{\mathbb{R}}^+$, there exists a measurable μ-unique $f_0 : \Omega \to \bar{\mathbb{R}}^+$ such that

$$\int_\Omega f d\nu = \int_\Omega f f_0 d\mu, \quad f \in L^1(\nu), \tag{1}$$

iff μ is localizable. Thus the Radon Nikodým theorem holds for (ν, μ) iff $\nu << \mu$ and μ is localizable. (See for a proof, Rao, 1987, p. 276.) The f_0 is the density of ν relative to μ, and if ν is σ-finite, then f_0 is finite a.e. $[\mu]$. This generality is needed in some applications if ν is a probability measure, but μ is not necessarily σ-finite. (A Haar measure is known to be localizable but not necessarily σ-finite on a locally compact group.) It will also be seen later that a weaker condition than domination uses localizability. Thus this concept is relevant for our study.

Let us start with a useful technical result which is further explained later (see the discussion following Definition 4.8 below):

Lemma 3. *Let $\mathcal{M} = \{\mu_i, i \in I\}$ be a set of scalar measures on (Ω, Σ) dominated by some measure $\lambda : \Sigma \to \bar{\mathbb{R}}^+$ which is localizable. Then there is a (not necessarily localizable) measure $\tilde{\lambda} : \Sigma \to \bar{\mathbb{R}}^+$ such that $\tilde{\lambda} << \lambda$, and each $\mu_i \equiv \tilde{\lambda}$, for all $i \in I$. [This is written as $\mathcal{M} \equiv \tilde{\lambda} << \lambda$.] If, further, λ is σ-finite then $\tilde{\lambda}$ can be chosen to be a linear combination of at most a countable number of members of the set $\{|\mu| : \mu \in \mathcal{M}\}$ and $\tilde{\lambda}(\Omega) < \infty$. [$\mu_i \equiv \tilde{\lambda}$ means $\mu_i << \tilde{\lambda}$ and $\tilde{\lambda} << \mu_i$.]*

Proof. By hypothesis $\mu_i << \lambda$, and by the (earlier) result (1), $f_i = \frac{d\mu_i}{d\lambda}$ exists, $i \in I$. Let $C_i = \{\omega : |f_i|(\omega) > 0\}$. Then $C_i \in \Sigma$ and it is the support of μ_i and $\lambda(C_i) > 0$. Let $\mathcal{C} = \{C_i : i \in I\} \subset \Sigma$. By the localizability of $\lambda, \mathcal{C}$ has a supremum, say B in Σ. Define $\tilde{\lambda} : \Sigma \to \bar{\mathbb{R}}^+$ as:

$$\tilde{\lambda}(A) = \lambda(B \cap A), \quad A \in \Sigma. \tag{2}$$

Then $\tilde{\lambda} : \Sigma(B) \to \bar{\mathbb{R}}^+$ is a measure, and $\mathcal{M} << \tilde{\lambda}$. Moreover, if $A \in \Sigma$, $|\mu_i|(A) = 0$, for all i in I, then $\lambda(A \cap C_i) = 0$, $i \in I$, since

$$0 = |\mu_i|(A \cap C_i) = \int_{A \cap C_i} |f_i| d\lambda \qquad (3)$$

and $|f_i| > 0$ on C_i. Thus $\tilde{\lambda}(A) = 0$ and $\mathcal{M} \equiv \tilde{\lambda}$ ($<< \lambda$ by construction).

Now suppose λ is σ-finite. Then $\Omega = \underset{n}{\cup} \Omega_n$, $\Omega_n \in \Sigma$, disjoint and $0 < \lambda(\Omega_n) < \infty$. Define $\lambda_0 : \Sigma \to \bar{\mathbb{R}}^+$ as:

$$\lambda_0(\cdot) = \sum_{n=1}^{\infty} 2^{-n} \lambda(\Omega_n \cap \cdot)/\lambda(\Omega_n) . \qquad (3)$$

Clearly $\lambda_0(\Omega) = 1$ and $\lambda \equiv \lambda_0$. Hence $\mathcal{M} << \lambda_0$. Let $\tilde{f}_i = \frac{d\mu_i}{d\lambda_0}$ and $\tilde{C}_i = \{\omega : |\tilde{f}_i|(\omega) > 0\}$. Since a finite measure is always localizable, it has a supremum B_0 and $\lambda_0(B_0) < \infty$, and $\{i \in I : \lambda_0(C_i) = \lambda_0(B_0 \cap C_i) > 0\}$ is at most countable, say $= \{i_n, n \geq 1\} \subset I$. By definition of B_0 the set $\tilde{B}_0 = \overset{\infty}{\underset{n=1}{\cup}} C_{i_n}$ satisfies $\lambda_0(B \Delta B_0) = 0$ so that B_0 is also a supremum of $\mathcal{C}$. Define $\tilde{\lambda}_0$ as in (3), but with the modification:

$$\tilde{\lambda}_0(\cdot) = \sum_{n=1}^{\infty} 2^{-n} |\mu_{i_n}|(\cdot)/|\mu_{i_n}|(C_{i_n})$$

$$= \sum_{n=1}^{\infty} 2^{-n} |\mu_{i_n}|(\cdot)/|\mu_{i_n}|(\Omega) = \sum_{n=1}^{\infty} a_n |\tilde{\mu}_{i_n}|(\cdot). \qquad (4)$$

It follows that $\mathcal{M} \equiv \tilde{\lambda}_0 << \lambda_0$, and $\tilde{\lambda}_0$ is a linear combination of a countable set of $|\mu_i|(\cdot)$, $\tilde{\lambda}(\Omega) = 1$, $a_n = 2^{-n}$ and $|\tilde{\mu}_i|(\cdot) = |\mu_i|(\cdot)/|\mu_i|(\Omega)$. This is the desired result. $\square$

We can now present a technical characterization of sufficient σ-algebras, essentially following Halmos and Savage (1949) as follows:

Theorem 4. *Let $\mathcal{M} = \{P_\alpha, \alpha \in I\}$ be a family of probability measures on (Ω, Σ) dominated by a measure $\lambda : \Sigma \to \bar{\mathbb{R}}^+$. If $B \subset \Sigma$ is a σ-algebra such that $\lambda|B$ is σ-finite (or localizable) and each $\frac{dP_\alpha}{d\lambda}$ is B-measurable for all $\alpha \in I$, then B is sufficient for $\mathcal{M}$.*

Conversely, if $B \subset \Sigma$ is a sufficient σ-algebra for $\mathcal{M}$ which is dominated by a σ-finite measure λ, and λ_0 is the associated finite measure (given by Lemma 3 above) so that $\mathcal{M} \equiv \lambda_0$, then each $\frac{dP_\alpha}{d\lambda_0}$ is B-measurable, $\alpha \in I$.

Remark. In the converse part the σ-finiteness of λ assures that λ_0 is a linear combination of a countable number of P_{α_i} from $\mathcal{M}$, and this essential property will be lost if λ is localizable but not σ-finite.

Proof. For the direct part, let $\mathcal{B}$ be such that $\lambda|\mathcal{B}$ is σ-finite (or localizable) and consider $f_\alpha = \frac{dP_\alpha}{d\lambda}, \alpha \in I$, which exists since λ has the same property on Σ. By hypothesis f_α is $\mathcal{B}$-measurable. Now for each $A \in \Sigma$, $B \in \mathcal{B}$, consider the following:

$$\int_B E^{\mathcal{B}}_\lambda(\chi_A)dP_\alpha = \int_B E^{\mathcal{B}}_\lambda(\chi_A)f_\alpha \ d\lambda, \text{because of the hypothesis on } \lambda|\mathcal{B} ,$$

$$= \int_B E^{\mathcal{B}}_\lambda(f_\alpha\chi_A)d\lambda, \text{ since } f_\alpha \text{ is } \mathcal{B}\text{-measurable,}$$

$$= \int_B f_\alpha\chi_A \ d\lambda, \text{ by definition of } E^{\mathcal{B}}_\lambda(\cdot),$$

$$= \int_B \chi_A dP_\alpha . \tag{5}$$

By the linearity of $E^{\mathcal{B}}_\lambda(\cdot)$ this holds if χ_A is replaced by a simple (Σ-measurable) function f. Hence (5) becomes

$$\int_\Omega \chi_B f \ dP_\alpha = \int_\Omega \chi_B E^{\mathcal{B}}_\lambda(f) \ d \ P_\alpha . \tag{6}$$

The same reasoning implies that χ_B can be replaced by a simple ($\mathcal{B}$-measurable) function g in (6). Since simple functions are dense in $B(\Omega, \Sigma)$ and in $B(\Omega, \mathcal{B})$, it follows that

$$\int_\Omega fgdP_\alpha = \int_\Omega \tilde{f}gdP_\alpha, \ \alpha \in I, \ f \in B(\Omega, \Sigma), g \in B(\Omega, \mathcal{B}). \tag{7}$$

Hence by Definition 1.1, $\mathcal{B}$ is sufficient for the dominated $\mathcal{M}$.

For the oppose direction, let λ and λ_0 be as in the statement. Then λ_0 of Lemma 3 is obtained by taking $|\mu_{\alpha_n}|(\cdot) = P_{\alpha_n}(\cdot)$ there, so that

$$\lambda_0(\cdot) = \sum_{n=1}^\infty a_n P_{\alpha_n}(\cdot), \ a_n \geq 0 . \tag{8}$$

Hence for any $A \in \Sigma$, $B \in \mathcal{B}$, and sufficiency of $\mathcal{B}$ for $\mathcal{M}(\equiv \lambda_0)$ one has

$$\int_B \chi_A dP_\alpha = \int_B \tilde{\chi}_A dP_\alpha, \ \alpha \in I, \tag{9}$$

where $\tilde{\chi}_A$ is some $\mathcal{B}$-measurable function, not depending on α. But

$$\int_B \chi_A \, d\lambda_0 = \sum_{n=1}^{\infty} a_n \int_B \chi_A \, dP_{\alpha_n}, \text{ by (8) and monotone convergence},$$

$$= \sum_{n=1}^{\infty} a_n \int_B \tilde{\chi}_A \, dP_{\alpha_n}, \text{ by (7)},$$

$$= \int_B \tilde{\chi}_A \, d\lambda_0, \ B \in \mathcal{B}, \ A \in \Sigma. \tag{10}$$

This implies (since $\lambda_0(\Omega) < \infty$) that $\tilde{\chi}_A = E_{\lambda_0}^{\mathcal{B}}(\chi_A)$, a.e. $(\lambda_0|\mathcal{B})$. Now consider $f_\alpha = \frac{dP_\alpha}{d\lambda_0}$, $\alpha \in I$, and $A \in \Sigma$:

$$\int_A f_\alpha d\lambda_0 = \int_A dP_\alpha = \int_\Omega \chi_A \, dP_\alpha$$

$$= \int_\Omega \tilde{\chi}_A \, dP_\alpha \text{ , by (9)},$$

$$= \int_\Omega E_{\lambda_0}^{\mathcal{B}}(\chi_A) dP_\alpha, \text{ by a consequence of (10)},$$

$$= \int_\Omega E_{\lambda_0}^{\mathcal{B}}(\chi_A) \cdot E_{\lambda_0}^{\mathcal{B}}(f_\alpha) d\lambda_0, \text{ since } P_\alpha|\mathcal{B} << \lambda_0|\mathcal{B}, \text{ and}$$

$$(P_\alpha|\mathcal{B})(B) = \int_B E_{\lambda_0}^{\mathcal{B}}(f_\alpha) d(\lambda_0|\mathcal{B}),$$

$$= \int_\Omega E_{\lambda_0}^{\mathcal{B}}(\chi_A E^{\mathcal{B}}(f_\alpha)) d\lambda_0, \text{ by the averaging property of}$$

conditional expectations,

$$= \int_A E_{\lambda_0}^{\mathcal{B}}(f_\alpha) \, d\lambda_0, \text{ by definition}.$$

Since A in Σ is arbitrary and the extreme integrands are both measurable for Σ, they can be identified, i.e., $E_{\lambda_0}^{\mathcal{B}}(f_\alpha) = f_\alpha$ a.e. $[\lambda_0]$. But since the left side is $\mathcal{B}$-measurable, so must f_α be, as desired. $\square$

The above result takes a more attractive form for applications, if $\mathcal{B}$ is generated by a statistic T. This one, in an important special case, was first considered by R. A. Fisher and J. Neyman in different ways. It says that each density f_α above, may be factorized into a density of T and a positive function. A precise generalized result is given by:

Corollary 5. (Factorization criterion) *Let $\mathcal{M} = \{P_\alpha, \alpha \in I\}$ be a dominated family of probability measures on (Ω, Σ) by a σ-finite measure λ and let T be a statistic (i.e., $T : \Omega \to \mathbb{R}^n$ is a measurable function, or if $(\tilde{\Omega}, \tilde{\Sigma})$ is a measurable space $T : \Omega \to \tilde{\Omega}$ satisfies $T^{-1}(\tilde{\Sigma}) \subset \Sigma$). Then T is sufficient for $\mathcal{M}$, in the sense of Definition 2, iff we have the factorization:*

$$f_\alpha = \frac{dP_\alpha}{d\lambda} = (g_\alpha \circ T) \cdot h \quad , \quad a.e. \ [\lambda] \tag{11}$$

where $g_\alpha : \tilde{\Omega} \to \mathbb{R}^+$ is (Borel, if $\tilde{\Omega} = \mathbb{R}^n$) measurable and $h : \Omega \to \mathbb{R}^+$ is a measurable function (for Σ), independent of α.

Proof. Let $\mathcal{B}_T$ be the σ-algebra generated by T. Then we can replace λ by an equivalent finite measure $\tilde{\lambda}$ on Σ. [E.g., let $\Omega_n \uparrow \Omega, \Omega_n \in \Sigma, \lambda(\Omega_n) < \infty$. Take $\tilde{\lambda}(\cdot) = \sum_n 2^{-n} \lambda(\Omega_n, \cap \cdot) / \lambda(\Omega_n)$.] Thus there is a probability measure λ_0 such that $\mathcal{M} \equiv \lambda_0$ by Lemma 3 and $\lambda_0 \ll \tilde{\lambda}$.

If T is sufficient, or equivalently if $\mathcal{B}_T$ is sufficient, (cf., Definition 2) then the functions $\tilde{f}_\alpha = \frac{dP_\alpha}{d\lambda_0}$ are $\mathcal{B}_T$-measurable for all $\alpha \in I$. Hence by the Doob-Dynkin lemma there is a (Borel) measurable function $g_\alpha : \tilde{\Omega} \to \mathbb{R}^+$ such that $\tilde{f}_\alpha : \Omega \to \mathbb{R}^+$ is composed of $\Omega \xrightarrow{T} \tilde{\Omega} \xrightarrow{g_\alpha} \mathbb{R}^+$, i.e., $\tilde{f}_\alpha = g_\alpha \circ T$. Since $\lambda | \mathcal{B}_T$ is σ-finite, $h = \frac{d\lambda_0}{d\lambda}$ exists as a measurable function for Σ, and by the Radon-Nikodým form of the chain rule

$$f_\alpha = \frac{dP_\alpha}{d\lambda} = \frac{dP_\alpha}{d\lambda_0} \cdot \frac{d\lambda_0}{d\lambda} = (g_\alpha \circ T) \cdot h, \quad a.e. \ [\lambda],$$

so that (11) holds.

Conversely, if (11) holds, then, still under the domination hypothesis (using Lemma 3), there is a $\tilde{\lambda}_0$ of the above type such that $\mathcal{M} \equiv \tilde{\lambda}_0 \ll \tilde{\lambda}$ and for each α in I, $f_\alpha = \tilde{g}_\alpha \cdot h$ a.e. $[\tilde{\lambda}]$, where $\tilde{g}_\alpha$ is $\mathcal{B}_T$-measurable and h is Σ-measurable, both being nonnegative. However $\tilde{\lambda}_0$ is determined by a countable collection $\{P_{\alpha_n}, n \geq 1\}$ so that $\tilde{\lambda}_0 = \sum_{n=1}^{\infty} a_n P_{\alpha_n}(\cdot)$ with $0 \leq a_n, \sum_{n=1}^{\infty} a_n = 1$. This gives on using the fact that $dP_\alpha = \tilde{g}_\alpha \cdot h \, d\lambda$,

$$d\tilde{\lambda}_0 = \left(\sum_{n=1}^{\infty} a_n \tilde{g}_{\alpha_n} \right) \cdot h \, d\lambda = p \cdot h \, d\lambda, \tag{12}$$

where $p(\cdot)$, defined by the sum, is a nonnegative $\mathcal{B}_T$-measurable function, and $\lambda_0(\Omega) = 1$. Hence $P_\alpha << \tilde{\lambda}_0$, and one has:

$$f_\alpha = \frac{dP_\alpha}{d\lambda} = \frac{dP_\alpha}{d\tilde{\lambda}_0} \cdot \frac{d\tilde{\lambda}_0}{d\lambda}, \quad \text{a.e. } [\tilde{\lambda}]$$

$$\tilde{g}_\alpha \cdot h = \frac{dP_\alpha}{d\tilde{\lambda}_0} \cdot p.h, \quad \text{by (12)} . \tag{13}$$

Consequently $\frac{dP_\alpha}{d\lambda} = \tilde{g}_\alpha/p$ a.e. $[\tilde{\lambda}_0]$, $\alpha \in I$. This is well-defined (set it $= 0$ where $p = 0$) and $\mathcal{B}_T$-measurable. By the direct part of the theorem then $\mathcal{B}_T$, hence T, is sufficient. $\square$

Remark. If $\lambda : \Sigma \to \bar{\mathbb{R}}^+$ is σ-finite (and nonfinite), and $\mathcal{B}_T \subsetneq \Sigma$ then $\lambda|\mathcal{B}_T$ need not be σ-finite. So, for the above argument, we need to replace λ by the *equivalent* measure $\tilde{\lambda}$ on Σ. If $\lambda(\Omega) < \infty$, then the additional replacement is, of course, unnecessary. Thus the generality afforded by the σ-finiteness condition here is illusory, since only their null sets play any role.

If a statistic T is sufficient for a family $\mathcal{M}$, so that $\mathcal{B}_T$ is a sufficient σ-algebra, then both $\mathcal{B}_T$ and Σ are sufficient and $\mathcal{B}_T \subset \Sigma$. But is there a smallest sufficient σ-algebra? This question has a negative answer in general, but has a positive solution in the dominated case. First we state the concept precisely and then present the positive assertion.

Definition 6. For a given family $\mathcal{M} = \{P_\alpha, \alpha \in I\}$ of probability measures on (Ω, Σ) a σ-algebra $\mathcal{B} \subset \Sigma$ is *minimal sufficient* if $\mathcal{B}$ is sufficient and for any other sufficient σ-algebra $\tilde{\mathcal{B}} \subset \Sigma$ one has: for each $A \in \Sigma$ there is $\tilde{A} \in \tilde{\mathcal{B}}$ such that $P_\alpha(A\Delta\tilde{A}) = 0$ for all $\alpha \in I$. This is written symbolically as $\mathcal{B} \subset \tilde{\mathcal{B}}$ a.e. $[\mathcal{M}]$.

The desired solution of the above problem is as follows:

Corollary 7. *If a family $\mathcal{M}$ of probability measures on (Ω, Σ) is dominated by a σ-finite measure λ on Σ, then there exists a minimal sufficient σ-algebra $\mathcal{B}_0 \subset \Sigma$.*

Proof. By Lemma 3, there is a probability measure λ_0 such that $\mathcal{M} \equiv \lambda_0 << \lambda$. Let $f_\alpha = \frac{dP_\alpha}{d\lambda}, \{P_\alpha, \alpha \in I\} = \mathcal{M}$, and let $\mathcal{B}_0$ be the σ-algebra generated by $\{f_\alpha, \alpha \in I\}$ and completed for λ_0 (or equivalently for $\mathcal{M}$). Then by the preceding theorem $\mathcal{B}_0$ is sufficient and every other sufficient σ-algebra contains $\mathcal{B}_0$ except for null sets. This last qualification cannot be omitted since $\lambda_0(\equiv \mathcal{M})$ is not unique. $\square$

The main force of domination is the following result, due also to Halmos and Savage (1949), illuminating the structure of sufficiency.

Theorem 8. *If $\mathcal{M}$ is a set of probability measures on (Ω, Σ) dominated by a σ-finite measure, then a σ-algebra $\mathcal{B} \subset \Sigma$ is sufficient for $\mathcal{M}$ iff it is pairwise sufficient (cf. Definition 1.4).*

Proof. Since sufficiency obviously implies the weaker pairwise sufficiency whether or not the family is dominated, only the converse has to be established.

Thus let $\mathcal{B}$ be a pairwise sufficient σ-algebra and $\lambda_0 (\equiv \mathcal{M})$ be given by Lemma 3 or by (8), so $\lambda_0(\Omega) = 1$, $\lambda_0 << \lambda$, a dominating σ-finite measure for $\mathcal{M}$. Consider the collection $\{P_{\alpha_n}, n \geq 1\}$ of (8) defining λ_0, and let P_α be an arbitrary element of $\mathcal{M}$. By hypothesis $\mathcal{B}$ is sufficient for the pair $\{P_\alpha, P_{\alpha_n}\}$, P_{α_n} being a fixed element of the above collection. Then by Definition 1.4, for $f = \chi_A$, $A \in \Sigma$, there is a $\mathcal{B}$-measurable $\tilde{f}_{\alpha,\alpha_n}$ such that

$$E_\alpha^{\mathcal{B}}(\chi_A) = \tilde{f}_{\alpha,\alpha_n} \quad \text{a.e. } [P_\alpha], \quad E_{\alpha_n}^{\mathcal{B}}(\chi_A) = \tilde{f}_{\alpha,\alpha_n} \quad \text{a.e. } [P_{\alpha_n}]. \qquad (14)$$

Note that, if $\tilde{\lambda}_0 = \lambda_0 | \mathcal{B}$ and $\tilde{P}_\alpha = P_\alpha | \mathcal{B}$, then $\tilde{P}_\alpha \equiv \tilde{\lambda}_0$ implies the existence of a $\mathcal{B}$-measurable $\tilde{h}_{\alpha_n} (= \frac{d\tilde{P}_{\alpha_n}}{d\tilde{\lambda}_0})$. Define

$$\tilde{f}_{\alpha,A} = \sum_{n=1}^{\infty} a_n \tilde{f}_{\alpha,\alpha_n} h_{\alpha_n}. \qquad (15)$$

Then $\tilde{f}_{\alpha,A}$ is $\mathcal{B}$-measurable, and we assert that this is independent of α, and satisfies Definition 1.1 because of the two equations of (14).

Indeed consider a substitution of $\tilde{f}_{\alpha,\alpha_n}$ form (14) in (15) to obtain:

$$\int_B \tilde{f}_{\alpha,A} d\tilde{P}_\alpha = \sum_{n=1}^{\infty} a_n \int_B E_\alpha^{\mathcal{B}}(\chi_A) \tilde{h}_{\alpha_n} d\tilde{P}_\alpha, \quad B \in \mathcal{B},$$

$$= \sum_{n=1}^{\infty} a_n \int_B E_\alpha^{\mathcal{B}}(\chi_A \tilde{h}_{\alpha_n}) d\tilde{P}_\alpha, \quad \text{since } \tilde{h}_{\alpha_n} \text{ is } \mathcal{B}\text{-measurable},$$

$$= \int_{A \cap B} \left(\sum_{n=1}^{\infty} a_n \tilde{h}_{\alpha_n} \right) dP_\alpha,$$

by definition of conditional expectation and the monotone convergence theorem,

$$= \int_{A \cap B} dP_\alpha,$$

the integrand being unity because $\tilde{\lambda}_0(B) = \lambda_0(B) = \int_B (\sum_n a_n \tilde{h}_{\alpha_n}) d\lambda_0$,

$$= \int_B \chi_A dP_\alpha = \int_B E_\alpha^B(\chi_A) d\tilde{P}_\alpha . \tag{16}$$

Hence (16) gives

$$\tilde{f}_{\alpha,A} = E_\alpha^B(\chi_A), \quad \text{a.e.}[\tilde{P}_\alpha], \quad \alpha \in I. \tag{17}$$

Similarly consider, with the second equation of (14) in(15),

$$\int_B \tilde{f}_{\alpha,A} d\tilde{\lambda}_0 = \sum_{n=1}^\infty a_n \int_B E_{\alpha_n}^B(\chi_A) \tilde{h}_{\alpha_n} d\tilde{\lambda}_0 ,$$

$$= \sum_{n=1}^\infty a_n \int_B E_{\alpha_n}^B(\chi_A) d\tilde{P}_{\alpha_n} , \quad \text{by definition of } \tilde{h}_{\alpha_n} ,$$

$$= \sum_{n=1}^\infty a_n \int_B \chi_A \, dP_{\alpha_n}$$

$$= \int_{A \cap B} \sum_{n=1}^\infty a_n \, dP_{\alpha_n}$$

$$= \int_{A \cap B} d\lambda_0 = \int_B E_{\lambda_0}^B(\chi_A) d\tilde{\lambda}_0 . \tag{18}$$

Since $B \in \mathcal{B}$ is arbitrary, (18) gives

$$\tilde{f}_{\alpha,A} = E_{\lambda_0}^B(\chi_A), \quad \text{a.e. } [\tilde{\lambda}_0], \quad \alpha \in I . \tag{19}$$

Comparison of (17) and (19) shows that the following holds:

$$\int_B E_{\lambda_0}^B(\chi_A) dP_\alpha = \int_B \chi_A \, dP_\alpha, \quad B \in \mathcal{B}, \, \alpha \in I, \, A \in \Sigma . \tag{20}$$

By linearity of the integral, this implies for all $f \in B(\Omega, \Sigma), g \in B(\Omega, \mathcal{B})$,

$$\int_\Omega f g dP_\alpha = \int_\Omega \tilde{f} g \, d\, P_\alpha, \quad \alpha \in I,$$

where $\tilde{f} = E_{\lambda_0}^B(f)$ a.e., $\alpha \in I$. Hence $\mathcal{B}$ is sufficient for $\mathcal{M}$. $\square$

After seeing the pleasant nature of the above result in the presence of a dominating σ-finite measure λ, it is natural to ask about the restrictions imposed by this assumption on the family $\mathcal{M}$. A subtle nature of the condition is exemplified by the following example where a sufficient σ-algebra $\mathcal{B} \subsetneq \Sigma$ exists for an undominated family, but there is no minimal sufficient one. It is due to Pitcher (1957).

Example 9. Let (I, Σ, P) be the Lebesgue unit interval and $A_1 \subset [0, \frac{1}{2}]$ be a Borel set, $A_2 \subset [\frac{1}{2}, 1]$ be a set to be specified shortly. Let $\varphi : A_1 \to A_2$ be a one-to-one, onto, mapping that is not (Borel) measurable relative to the (Borel) σ-algebras of the two intervals. [If it is Borel measurable then A_2 must also be a Borel set by a classical theorem of Kuratowski, (see Parthasarathy, 1967, p. 21).] Thus A_2 is not a Borel set, (may be analytic) and hence A_1 must be infinite. For each $x \in A_1$, let m_x be a probability measure assigning measure $\frac{1}{2}$ to x and $\frac{1}{2}$ to $\varphi(x)$. This is well-defined since points are Borel measurable. Let $\mathcal{M}$ be the collection of all m_x for x in $A_1 \cup A_2$ and the Dirac measures δ_x, $x \notin A_1 \cup A_2$. Then $\mathcal{M}$ is not dominated. If $\mathcal{B}_x$ is the σ-algebra generated by the set $\{B \in \mathcal{B} : B \supset \{x, \varphi(x)\} \text{ or } B^c \supset \{x, \varphi(x)\}\}$, $x \in I$, ($\mathcal{B}$ is the Borel σ-algebra of I) then for each x, $\mathcal{B}_x$ has the set $\{x, \varphi(x)\}$ as an atom and $m_x | \mathcal{B}_x$ is a point measure. Consequently $\mathcal{B}_x$ will be sufficient for $\mathcal{M}$ for each x. However, there is no minimal sufficient σ-algebra $\mathcal{B}_0 \subset \underset{x \in I}{\cap} \mathcal{B}_x$ for $\mathcal{M}$. For otherwise, we may compute $E^{\mathcal{B}_0}(f)(y)$ and find it to be $= \frac{1}{2}$ if $y \in \{x, \varphi(x)\}$, $x \in A_1$ and $= 0$ otherwise. Taking $f = \chi_{A_1}$, this gives $E^{\mathcal{B}_0}(\chi_{A_1}) = \frac{1}{2}(\chi_{A_1} + \chi_{A_2})$ so that the right side is $\mathcal{B}_0$ measurable. But by choice, it is not $\mathcal{B}_x$ measurable for any $x \in A_1$. Thus such a minimal $\mathcal{B}_0$ can not exist.

Motivated by this and other examples of a similar nature, we analyze the undominated case in some detail, without which the mysterious nature of this useful concept will not be properly understood.

6.4 Sufficiency: the undominated case

Let us start with a simple isomorphism result to indicate that such transformations are not sharp enough for the present study.

Proposition 1. *Let $\mathcal{M} = \{P_\alpha, \alpha \in I\}$ be a family of probability measures on (Ω, Σ). Then there is a topological measurable space $(\tilde{\Omega}, \tilde{\Sigma})$ and a family $\mathcal{M}^* = \{P_\alpha^*, \alpha \in I\}$, indexed by the same set I, and a (set) isomorphism $\tau : \Sigma \to \tilde{\Sigma}$ such that $\mathcal{M}^* \circ \tau = \mathcal{M}$ (i.e., $P_\alpha(\cdot) = P_\alpha^* \circ \tau(\cdot)$) and there is a minimal sufficient σ-algebra $\tilde{\mathcal{B}} \subset \tilde{\Sigma}$ for $\mathcal{M}^*$.*

Proof. Let $\mathcal{M} : \Sigma \to B(I)$ be the given mapping considered as a vector valued additive set function where $B(I)$ is the Banach space, under the uniform norm, of real bounded functions on I. For the Borelian space (Ω, Σ), let $(\tilde{\Omega}, \tilde{\Sigma})$ be the Stone representation space. Here $\tilde{\Omega}$ is an extremally disconnected compact space representing the Boolean algebra Σ, and if $\hat{\Sigma}$ is the algebra of all clopen (= closed-open) sets of $\tilde{\Omega}$, then $\tilde{\Sigma}$ is the σ-algebra generated by $\hat{\Sigma}$. Let $\tau : \Sigma \to \hat{\Sigma}$ be the (set) isomorphism given by this representation (see Dunford-Schwartz, 1958, IV.9.10, or Rao, 1987, Chap. 10). Let $\mathcal{M}^* = \{P_\alpha^*, \alpha \in I\}$ with $P_\alpha^* \circ \tau = P_\alpha$. Then $\mathcal{M}^*$ is additive and regular. It is *strongly* σ-bounded on $\hat{\Sigma}$ with its values in the positive cone of $B(I)$, [i.e., if $A_n \in \hat{\Sigma}$, disjoint, $\underset{n}{\cup} A_n \in \hat{\Sigma}$. Then $\|\mathcal{M}^*(A_n)\| \to 0$ as $n \to \infty$, which is true since the union contains only a finite number of nonempty terms, being a clopen subset of the compact $\tilde{\Omega}$]. Hence $\mathcal{M}^*$ has a unique σ-additive extension to $\tilde{\Sigma} = \sigma(\hat{\Sigma})$, as observed by Kupka (1978), and it will be denoted by the same symbol. Thus $\mathcal{M}^*$ is a vector measure and therefore there is a finite positive measure λ dominating $\mathcal{M}^*$ by the classical result of Bartle-Dunford-Schwatz. Hence by Corollary 3.7 there is a minimal sufficient σ-algebra $\tilde{\mathcal{B}} \subset \tilde{\Sigma}$ for $\mathcal{M}^*$, as asserted. $\square$

The point of this result is that every family of probability measures describing an "experiment" (Ω, Σ), is isomorphic to some "nicer" experiment having a minimal sufficient σ-algebra for its family of probability measures indexed by the same set. The weakness is that the latter cannot be constructed [it uses the axiom of choice] and is not available to the experimenter. Also, in the above result $\tau^{-1}(\tilde{\mathcal{B}})$ can be Σ itself. Thus the utility of such a result is limited. So we need to analyze the structure of the original $(\Omega, \Sigma, \mathcal{M})$ itself in more detail.

Let us start with a simple question. If $\mathcal{B} \subsetneq \Sigma$ is a sufficient σ-algebra for $\mathcal{M}$, is every intermediate σ-algebra $\mathcal{B}_1$ (i.e., $\mathcal{B} \subset \mathcal{B}_1 \subset \Sigma$) also sufficient? A negative answer is provided by the following example. This as well as the next two results are adapted from Burkholder (1961).

Example 2. Let $(\mathbb{R}, \mathcal{B})$ be the Borelian line and $\mathcal{B}_0 = \{A \in \mathcal{B}, A = -A\}$, the symmetric σ-subalgebra of $\mathcal{B}$ where $-A = \{x \in A : -x \in A\}$. If $\mathcal{M} = \{P_\alpha, \alpha \in I\}$ is a family of symmetric probability measures on $\mathcal{B}_1$ (i.e., $P_\alpha(A) = P_\alpha(-A)$, $A \in \mathcal{B}$), then we claim that $\mathcal{B}_0$ is sufficient for $\mathcal{M}$, but there is a nonsufficient σ-algebra $\mathcal{B}_1$ satisfying $\mathcal{B}_0 \subset \mathcal{B}_1 \subset \mathcal{B}$, for $\mathcal{M}$.

Indeed, the σ-subalgebra $\mathcal{B}_0$ of $\mathcal{B}$ satisfies, for each bounded $\mathcal{B}$-

measurable f, and $\tilde{f} : x \mapsto \frac{1}{2}(f(x) + f(-x))$ which is $\mathcal{B}_0$-measurable,

$$\int_{\mathbb{R}} \chi_B f dP_\alpha = \int_B f dP_\alpha = \int_B \tilde{f} dP_\alpha = \int_{\mathbb{R}} \tilde{f} \chi_B dP_\alpha, \ B \in \mathcal{B}_0, \ \alpha \in I. \quad (1)$$

The general case where χ_B is replaced by a $\mathcal{B}_0$-simple and then general bounded $\mathcal{B}_0$-measurable function follows by linearity and bounded convergence. Hence by Definition 1.1, $\mathcal{B}_0$ is sufficient for $\mathcal{M}$. Next let

$$\mathcal{B}_1 = \{A \cup A_0 : \ A_0 \in \mathcal{B}_0, \ A \in \mathcal{B}(S)\} \quad (2)$$

where S is a nonempty symmetric subset of $\mathbb{R}$, and $\mathcal{B}(S) = \{A \cap S : A \in \mathcal{B}\}$ the *trace* of $\mathcal{B}$ on S. Taking $A = \emptyset$, we have $\mathcal{B}_0 \subset \mathcal{B}_1$ and $\mathcal{B}_1 \subset \mathcal{B}$ is clear. To see that $\mathcal{B}_1$ is a σ-algebra, since it is evidently closed under countable unions, it suffices to show that, for each $B \in \mathcal{B}_1$, $B^c \in \mathcal{B}_1$. Now by definition $B = A \cup A_0$, $A \in \mathcal{B}(S)$, $A_0 \in \mathcal{B}_0$. So $A \subset S$ and hence $-A \subset -S = S$, and if $C_0 = (-A) \cup A$, $C = C_0 - A$, then $C_0 \in \mathcal{B}(S)$ and $C \in \mathcal{B}(S)$. Also $A = C_0 - C$. Thus

$$B^c = A^c \cap A_0^c = (C \cup C_0)^c \cap A_0^c = (C \cap A_0^c) \cup (C_0^c \cap A_0^c). \quad (3)$$

Since $\{C - A_0, \ A_0^c \cap C_0^c\} \subset \mathcal{B}(S)$ the latter being a σ-algebra, (3) implies $B^c \in \mathcal{B}_1$ so that $\mathcal{B}_1$ is a σ-algebra.

However, $\mathcal{B}_1$ is not sufficient for $\mathcal{M}$. For otherwise, for each bounded $\mathcal{B}$-measurable f, there must exist a $\mathcal{B}_1$-measurable $\tilde{f}$ such that

$$\int_B f dP_\alpha = \int_B \tilde{f} \, dP_\alpha, \ \alpha \in I, \ B \in \mathcal{B}_1 . \quad (4)$$

If $B = \{x\}, x \in S$, then (4) gives $f(x) = \tilde{f}(x)$ when $P_\alpha(\{x\}) > 0$. In particular, let $S = S_0 \subset \mathbb{R}$ be a non-Borel set containing $0, S_0 = -S_0$, (e.g. S_0 is a symmetric analytic set). If $P_\alpha(\cdot) = \frac{1}{2}(\delta_\alpha + \delta_{-\alpha})(\cdot), \alpha \in S_0$, ($\delta_\alpha$ is the Dirac measure), $f(x) = 0$ for $x > 0$ and $= 0$ for $x \leq 0$, then $f(x) = \tilde{f}(x)$ for $x \in S_0$, and for $x \in S_0^c$, $\{x_1, -x_1\} \in \mathcal{B}$. So (4) gives for these $P_\alpha, \alpha \in S_0, \tilde{f}(x) = \frac{1}{2}$. Since $\tilde{f}$ must be $\mathcal{B}_1$-measurable, we get

$$\{x : \tilde{f}(x) = 1\} \cup \{x : \tilde{f}(x) = 0\} = S_0 \in \mathcal{B}_1 \subset \mathcal{B} . \quad (5)$$

But this contradicts the choice of S_0 which is not a Borel set. Thus $\mathcal{B}_1$ is not sufficient for $\mathcal{M}$, as asserted.

In the positive direction we have the following result:

Theorem 3. *Let* $\mathcal{M} = \{P_\alpha, \alpha \in I\}$ *be a family of probabilities on* (Ω, Σ) *and* $\{\mathcal{B}_n, n \geq 1\}$ *be a monotone sequence of sufficient σ-algebras in Σ for $\mathcal{M}$. Then* $\lim_n \mathcal{B}_n = \mathcal{B}$ *is also sufficient for $\mathcal{M}$. Moreover if $\mathcal{A}$ is a countably generated σ-subalgebra of Σ, then* $\mathcal{C}_n = \sigma(\mathcal{B}_n, \mathcal{A}) \subset \Sigma$ *is also sufficient for $\mathcal{M}, n \geq 1$. [Thus each sufficient σ-subalgebra of Σ, can be countably augmented.]*

Proof. The first assertion is a direct consequences of a standard martingale convergence theorem, to be discussed in Section 9.3. It states that, in the present context, if f is a bounded measurable function for Σ, then $g_n = E_\alpha^{\mathcal{B}_n}(f) \to g = E_\alpha^{\mathcal{B}}(f)$ a.e. and in $L^1(P_\alpha)$ where $\mathcal{B} = \lim_n \mathcal{B}_n$ $(= \cap_n \mathcal{B}_n$ in the decreasing case and $= \sigma(\cup_n \mathcal{B}_n)$ in the increasing case). But by hypothesis of sufficiency g_n is $\mathcal{B}_n$-measurable for all $\alpha \in I$, and hence the limit g is $\mathcal{B}$-measurable for all $\alpha \in I$. Thus $\mathcal{B}$ is sufficient for $\mathcal{M}$ by Definition 1.1.

For the second part, let $A_1, A_2, \dots$ be a sequence of generators of $\mathcal{A}$. It is enough to consider one $\mathcal{B}_n$, denoted $\tilde{\mathcal{B}}$, and let $\tilde{\mathcal{C}}_m = \sigma(\tilde{\mathcal{B}}, A_1, \dots, A_m)$ so that $\tilde{\mathcal{C}} = \lim_m \tilde{\mathcal{C}}_m$. Because of the first part, we only need to show that $\tilde{\mathcal{C}}_m$ is sufficient for each m. By induction this will follow if we show $\mathcal{D} = \sigma(\tilde{\mathcal{B}}, A_1)$ is sufficient for $\mathcal{M}$, and this will now be verified.

It is well-known that $\mathcal{D}$ can be alternately described as (let $A = A_1$):

$$\mathcal{D} = \{(A \cap B_1) \cup (A^c \cap B_2): \ B_i \in \mathcal{B}, \ i = 1, 2\}. \tag{6}$$

To verify the sufficiency of $\mathcal{D}$ for $\mathcal{M}$, let $0 \leq f \leq 1$ be a measurable function for Σ, and consider $\mathcal{D}$-measurable functions g_1, g_2 where

$$g_1 = E_\alpha^{\mathcal{B}}(\chi_A f / E_\alpha^{\mathcal{B}}(\chi_A)), \quad g_2 = E_\alpha^{\mathcal{B}}(\chi_{A^c} f / E_\alpha^{\mathcal{B}}(\chi_{A^c})), \tag{7}$$

where $\frac{0}{0}$ is taken to be zero. Since $\mathcal{B}$ is sufficient for $\mathcal{M}$, it follows that g_1, g_2 do not vary with α. We now assert that $g = \chi_A g_1 + \chi_{A^c} g_2$ is identifiable with $E_\alpha^{\mathcal{B}}(f)$, $\alpha \in I$, so that $\mathcal{D}$ will be sufficient for $\mathcal{M}$.

Consider a typical element of $\mathcal{D}$, given by (6). One has

$$\int_{B_1 \cap A} f dP_\alpha = \int_{B_1} (\chi_A f) dP_\alpha = \int_{B_1} E^{\mathcal{B}}(\chi_A f) dP_\alpha$$

$$= \int_{B_1} g_1 E_\alpha^{\mathcal{B}}(\chi_A) dP_\alpha, \text{ by (7)},$$

$$= \int_{B_1} E_\alpha^{\mathcal{B}}(g_1 \chi_A) dP_\alpha = \int_{B_1} g_1 \chi_A dP_\alpha = \int_{B_1 \cap A} g_1 dP_\alpha .$$

$$\tag{8}$$

In the last but one equality the $\mathcal{B}$-measurability of g_1 is used. Similarly

$$\int_{B_2 \cap A^c} f \, dP_\alpha = \int_{B_2 \cap A^c} g_2 \, dP_\alpha. \tag{9}$$

Adding (8) and (9) one gets

$$\int_D f \, dP_\alpha = \int_D g \, dP_\alpha, \quad D \in \mathcal{D}, \ \alpha \in I .$$

Hence $g = \tilde{f}$, and $\mathcal{D}$ is sufficient for $\mathcal{M}$. □

In general one cannot expect for arbitrary sufficient σ-algebras $\mathcal{B}_1, \mathcal{B}_2$ the same property for $\mathcal{B}_1 \cap \mathcal{B}_2$ or $\sigma(\mathcal{B}_1 \cup \mathcal{B}_2)$, and counterexamples can be constructed. However, the following restricted statement can be established.

Proposition 4. *If $\mathcal{M} = \{P_\alpha, \alpha \in I\}$ is a probability family on (Ω, Σ) where Σ is countably generated, then the sufficiency of any sequence of σ-subalgebras $\mathcal{B}_n$ of Σ, $n \geq 1$, implies the same for $\sigma(\underset{n}{\cup} \mathcal{B}_n)$ [but that property need not hold for $\mathcal{B}_i \cap \mathcal{B}_j$, $i \neq j$].*

Proof. The first assertion follows from the preceding result, using induction, if it is established for $\mathcal{B}_1$ and $\mathcal{B}_2$. Although Σ is countably generated $\mathcal{B}_i$ need not be. [For instance, if $\mathcal{B}$ is the Borel σ-algebra of $\mathbb{R}$ and $\mathcal{B}_1 \subset \mathcal{B}$ is the σ-algebra of countable and co-countable sets, then $\mathcal{B}_1$ is not countably generated, even though $\mathcal{B}$ is.] Here the trick is to replace $\mathcal{B}_i$ by a countably generated sufficient σ-algebra for $\mathcal{M}$.

Thus let $\mathcal{B}_1$ be sufficient for $\mathcal{M}$ and let $A_n, n \geq 1$, be a countable set of generators of Σ which exist by hypothesis. Let $\mathcal{A} = \{A_n, n \geq 1\}$

so that $\Sigma = \sigma(\mathcal{A})$. Consider for $A \in \mathcal{A}$,

$$\int_B \chi_A dP_\alpha = \int_B E^{\mathcal{B}}(\chi_A) dP_\alpha \ , \ \alpha \in I \ , \tag{10}$$

where $g_A = E^{\mathcal{B}}(\chi_A)$ does not vary with α by the sufficiency hypothesis. Let $\mathcal{B}_0$ be the σ-algebra generated by $\{g_A, \ A \in \mathcal{A}\}$. Hence $\mathcal{B}_0$ is countably generated. Also by linearity (10) holds for all $\mathcal{A}$-simple and then for all $(\sigma(\mathcal{A}) =) \Sigma$-measurable bounded functions f in place of χ_A. Thus $\mathcal{B}_0$ is sufficient for $\mathcal{M}$. But it is immediately seen that $\mathcal{B} \subset \sigma(\mathcal{B}_0, \mathcal{N})$ where $\mathcal{N}$ is the σ-ideal of all $\mathcal{M}$-null sets of Σ. Since in the sufficiency definition all the σ-algebras may be taken to be complete, the above result shows that $\sigma(\mathcal{B}_0, \mathcal{N})$ is sufficient for $\mathcal{M}$.

To complete the proof, note that since the given $\mathcal{B}_1, \mathcal{B}_2$ are sufficient for $\mathcal{M}$, by the preceding argument there exist countably generated sufficient σ-algebras $\tilde{\mathcal{B}}_i \subset \mathcal{B}_i \subset \sigma(\tilde{\mathcal{B}}_i, \mathcal{N})$, $i = 1, 2$. Then by the preceding proposition $\sigma(\tilde{\mathcal{B}}_1, \tilde{\mathcal{B}}_2)$ is sufficient for $\mathcal{M}$. Since

$$\sigma(\tilde{\mathcal{B}}_1, \tilde{\mathcal{B}}_2) \subset \sigma(\mathcal{B}_1, \mathcal{B}_2) \subset \sigma(\tilde{\mathcal{B}}_1, \tilde{\mathcal{B}}_2, \mathcal{N})$$

and the extremes are sufficient for $\mathcal{M}$, and they differ only by $\mathcal{M}$-null sets, it follows that $\sigma(\mathcal{B}_1, \mathcal{B}_2)$ is sufficient for $\mathcal{M}$. This, by the initial reduction, proves the proposition. $\square$

Remark. These results imply that, in the absence of monotonicity conditions between sufficient σ-algebras, their order structure is little related to the lattice structure of the algebras they determine. These difficulties disappear when the family $\mathcal{M}$ is dominated. Also the positive statements indicate that there must be some other useful conditions in between domination and complete generality, and we now explore this idea.

A condition more general than domination, but not too general was introduced by Pitcher (1965) and further extended by Rosenberg (1968). The earlier results mostly depended on the σ-finiteness of the dominating measure, although a few statements are seen to be true (in Sect. 3 above) for localizable dominating measures. The new condition introduced by Pitcher is called "compactness", and we shall explore this and later connect it with the localizability property of a dominating measure. These considerations are appropriately given with some elementary facts from function space aspects of functional analysis.

Let $p \geq 1$ and consider the linear span $\mathcal{E}_p(\mathcal{M})$ of some continuous linear functionals on $L^p(P_\alpha), \alpha \in I$, to define a topology on $L^p(\mathcal{M})$, the space of scalar functions f on Ω which belong to all $L^p(P_\alpha)$, $\alpha \in I$, i.e., $L^p(\mathcal{M}) = \cap\{L^p(P_\alpha) : \alpha \in I\}$. More precisely $\mathcal{E}_p(\mathcal{M})$ is as follows. Consider the vector space of (continuous) linear functionals

$$\mathcal{E}_p(\mathcal{M}) = sp\{\ell_{h,\alpha} : \ell_{h,\alpha}(f) = \int_\Omega fh \, dP_\alpha, \ f \in L^p(P_\alpha), \ \alpha \in I,$$

$$h \text{ is bounded and measurable for } \Sigma \}. \qquad (11)$$

Then $\mathcal{E}_p(\mathcal{M})$ is total for the space $L^p(\mathcal{M}) = \underset{\alpha \in I}{\cap} L^p(P_\alpha)$, in the sense that $\ell_{h,\alpha}(f) = 0$ for all $\ell_{h,\alpha} \in \mathcal{E}_p(\mathcal{M})$ implies $f = 0$, a.e $[P_\alpha]$ for all $\alpha \in I$. Now we can define the $\mathcal{E}_p(\mathcal{M})$-topology of $L^p(\mathcal{M})$ by the neighborhood system:

$$N(f, F, \varepsilon) = \{g \in L^p(\mathcal{M}) : |\ell(f) - \ell(g)| < \varepsilon, \ \ell \in F\}, \ f \in L^p(\mathcal{M}),$$
$$(12)$$

where F is a finite subset of $\mathcal{E}_p(\mathcal{M})$ and $\varepsilon > 0$. This system defines a locally convex Hausdorff topology on $L^p(\mathcal{M})$ as the classical results in linear analysis (see Dunford-Schwartz, 1958, V. 3.9) show. We denote the "unit ball" in this space by $U_p(\mathcal{M})$, i.e.,

$$U_p(\mathcal{M}) = \{f \in L^p(\mathcal{M}) : \sup_\alpha \|f\|_{p,\alpha} \leq 1\} . \qquad (13)$$

With these concepts, we now can present the desired condition as:

Definition 5. The family $\mathcal{M} = \{P_\alpha, \alpha \in I\}$ on (Ω, Σ) is *compact* if $U_p(\mathcal{M})$ is compact in the $\mathcal{E}_p(\mathcal{M})$-topology for some real $p \geq 1$.

This definition can and will be extended by replacing the L^p-spaces with more flexible Orlicz spaces, L^φ, later. Let us first present an example to show that this concept is indeed more general than domination of $\mathcal{M}$, by a finite (or σ-finite) measure λ[i.e., $P_\alpha << \lambda$ for all $\alpha \in I$]. Several of the following results (ideas) are adapted from Pitcher (1965).

Example 6. Let $\{(\Omega_\alpha, \Sigma_\alpha, P_\alpha), \alpha \in I\}$ be a family of probability spaces and consider the set sum $\Omega = \cup\{\{\alpha\} \times \Omega_\alpha : \alpha \in I\}$ for an uncountable index set I. Let $\Sigma = \{A \subset \Omega : A \cap (\{\alpha\} \times \Omega_\alpha) \in \Sigma_\alpha, \ \alpha \in I\}$ where we identify the spaces Ω_α and $\{\alpha\} \times \Omega_\alpha$. Let $\tilde{P}_\alpha(A) = P_\alpha(A \cap \Omega_\alpha)$ for $A \in \Sigma$. Then the set $\mathcal{M} = \{\tilde{P}_\alpha, \alpha \in I\}$ is not dominated, but it is

$\mathcal{E}_p(\mathcal{M})$-compact for each $p \geq 1$. On the other hand if $\mathcal{M} = \{\delta_x, \ x \in I = [0,1]\}$ are point measures, Σ is the Borel σ-algebra of I, and $A \subset I$ is a non-Borel set, then $\mathcal{M}^* = \{\delta_x, \ x \in A\}$ is not $\mathcal{E}_p(\mathcal{M}^*)$-compact for any $p \geq 1$, so that the compactness condition of Definition 5 is not totally unrestrictive.

These assertions are not obvious and need an elaboration. However, they follow from a characterization of compactness to be given below. We then also present a generalization of the result given in Corollary 3.7 on the existence of a minimal sufficient σ-algebra for compact $\mathcal{M}$.

The statements of the preceding example can be verified from the next result which admits a far reaching generalization.

Theorem 7. *Let $\mathcal{M}$ be a family of measures on (Ω, Σ) as before, and $U_p(\mathcal{M})$ be the unit ball introduced in (13). Let $\mathcal{M}' = \{P_\alpha, \alpha \in I'\}$ where $I' \subset I$. Then the compactness of $\mathcal{M}$ in the sense of Definition 5 implies the same property of $\mathcal{M}'$ as well as of $\mathcal{M}''$ where $\mathcal{M}''$ consists of all probability measures ν on Σ each of which is dominated by some countable collection of elements of $\mathcal{M}$. In particular, $\mathcal{M}$ is compact if it is of a countable collection. Moreover, $U_p(\mathcal{M})$ is $\mathcal{E}_p(\mathcal{M})$-compact for some $1 < p < \infty$ iff $U_\infty(\mathcal{M})$ is $\mathcal{E}_1(\mathcal{M})$-compact. In fact on $U_\infty(\mathcal{M})$, the $\mathcal{E}_1(\mathcal{M})$ and $\mathcal{E}_p(\mathcal{M}), p > 1$, topologies coincide.*

This result as well as the compactness concept will be better understood if the role of "sup" in (13) is clarified with its relation to $p \geq 1$. In the Lebesgue space context $\varphi_p(x) = |x|^p, p \geq 1$, a convex function, plays a key role, but "sup" is also a convex function and this property has to be illuminated in relation to φ_p and the family $\mathcal{M}$. This and the consequent generalization of the above theorem will now be discussed after recalling certain function spaces, called Orlicz spaces.

Let Φ and Θ be Young's functions, i.e., nonnegative symmetric convex functions on $\mathbb{R}$ vanishing at the origin and not identically zero. Then an Orlicz space $L^\Phi(\mu)$ on (Ω, Σ, μ) is the set of equivalence classes of scalar measurable functions f such that $\|f\|_{\Phi,\mu} < \infty$ where

$$\|f\|_{\Phi,\mu} = \inf\{k > 0 : \int_\Omega \Phi\left(\frac{|f|}{k}\right) d\mu \leq 1\} . \tag{14}$$

It can be verified that $\{L^\Phi(\mu), \|\cdot\|_{\Phi,\mu}\}$ is a Banach space (i.e. a complete normed linear space). If $\mathcal{M} = \{P_\alpha, \alpha \in I\}$ is a family of probability measures on (Ω, Σ), indexed by a set I, and $G : I \to (0, \infty)$ is a (weight) function, consider the associated measure $\lambda_G(\cdot)$ on the power set $2^I = \mathcal{Q}$ of I defined as

$$\lambda_G(A) = \sum_{i \in A} G(i) \quad , \quad A \in 2^I = \mathcal{Q} . \tag{15}$$

Then $\lambda_G : 2^I \to \bar{\mathbb{R}}^+$ is σ-additive, and $(I, \mathcal{Q}, \lambda_G)$ becomes a measure space. Now a generalization of the space $\mathcal{E}_p(\mathcal{M})$ is definable as:

$$\mathcal{E}_\Phi(\Theta, \mathcal{M}) = \{f \in \underset{\alpha \in I}{\cap} L^\Phi(P_\alpha) : \quad F_f \in \ell^\Theta(\lambda_G)\} \tag{16}$$

where $F_f(\alpha) = \|f\|_{\Phi, P_\alpha}$, so that $F_f : I \to \bar{\mathbb{R}}^+$ is defined, and

$$\ell^\Theta(\lambda_G) = \{h : I \to \mathbb{R} \big| \ N_{\Theta, G}(h) < \infty\} \tag{17}$$

with

$$N_{\Theta, G}(h) = \inf\{k > 0 : \ \sum_{\alpha \in I} \Theta\left(\frac{|h|}{k}\right)\lambda_G(\alpha) \le 1\} . \tag{18}$$

It follows that $\{\ell^\Theta(\lambda_G), \ N_{\Theta, G}(\cdot)\}$ is a Banach space and $F_\alpha \in \ell^\Theta(\lambda_G)$ implies $F_f(\alpha) < \infty$ for all $\alpha \in I$. Further, it can be verified that if we set $N^G_{\Theta, \Phi} : f \mapsto N_{\Theta, G}(F_f)$, then $\{\mathcal{E}_\Phi(\Theta, \mathcal{M}), \ N^G_{\Theta, \Phi}(\cdot)\}$ becomes a Banach space. If $\Theta : \mathbb{R} \to \bar{\mathbb{R}}^+$ is the special (two-valued) Young function given by

$$\Theta(x) = \begin{cases} 0, & |x| \le 1 \\ +\infty, & |x| > 1, \end{cases} \tag{19}$$

then $\ell^\Theta(\lambda_G)$ contains bounded functions and $f \in \mathcal{E}_\Phi(\Theta, \mathcal{M})$ iff one has $\sup\{\|f\|_{\Phi, P_\alpha} : \quad \alpha \in I\} < \infty$ which does not depend on G. In this case, we denote this set by $\mathcal{E}_\Phi(\mathcal{M})$ and note that when $\Phi(x) = |x|^p$, $1 \le p < \infty$, this space reduces to $\mathcal{E}_p(\mathcal{M})$. The fact that $\mathcal{E}_\Phi(\Theta, \mathcal{M})$ is a Banach space need some, not entirely simple but standard, computation which may be found in Rosenberg (1970). The theory of Orlicz spaces based on abstract measure spaces that is used in the present context may be referred to Rao and Ren (1991), and for the classical treatment based on a bounded Lebesgue domain in $\mathbb{R}^n$ to the book, Krasnoselskii and Rutickii (1958). A few standard results from these sources will be utilized here without special elaboration. We now present a generalization of the compactness condition of Definition 5 for $\mathcal{M}$.

For each Young function Φ, there is a complementary function Ψ, (in the sense of Young) having similar properties, defined by:

$$\Psi(x) = \sup\{|x|y - \Phi(y) : y \geq 0\}. \tag{20}$$

The couple (Φ, Ψ) is termed a complementary Young pair. Then $L^\Phi(P_\alpha)$ and $L^\Psi(P_\alpha)$ on (Ω, Σ) define a duality pair. More particularly, for each $\alpha \in I$, each $h \in L^\Psi(P_\alpha)$ the functional $x^*_{h,\alpha} : \mathcal{E}_\Phi(\Theta, G) \to \mathbb{R}$ defined by

$$x^*_{h,\alpha}(f) = \int_\Omega fh \ dP_\alpha \ ,$$

is linear and continuous. In fact, using a Hölder inequality one finds

$$|x^*_{h,\alpha}(f)| \leq 2\|f\|_{\Phi,P_\alpha} \|h\|_{\Psi,P_\alpha}$$

$$\leq 2[\Theta^{-1}\left(\frac{1}{G(\alpha)}\right)\|h\|_{\Psi,P_\alpha}]N^G_{\Theta,\Phi}(f). \tag{21}$$

Taking supremum over f with $N^G_{\Theta,\Phi}(f) \leq 1$, one sees that $x^*_{h,\alpha}(\cdot)$ is bounded. Hence $\mathcal{E}_\Psi = sp\{x^*_{h,\alpha} : \alpha \in I, \ h \in L^\Psi(\mu) \cap L^\infty(\mu)\}$ is a total subspace of $(\mathcal{E}_\Phi(\Theta, G))^*$, so that the $\mathcal{E}_\Psi$-topology, with the neighborhood system as in (12), is Hausdorff. With this observation, Definition 5 extends to:

Definition 8. If $\mathcal{M} = \{P_\alpha, \alpha \in I\}$ is a family of probability measures on (Ω, Σ) and Θ, G are the Young and weight functions, then $\mathcal{M}$ is (Θ, G)-compact if for some Young function Φ the unit ball $U^G_{\Theta,\Phi}$ of $\mathcal{E}_\Phi(\Theta, G)$ is $\mathcal{E}_\Psi$-compact (or compact in the [weak] $\mathcal{E}_\Psi$-topology).

It will be verified that, under some natural conditions, every dominated family $\mathcal{M}$ is (Θ, G)-compact as it subsumes Definition 5. More explicitly, this holds if the complementary function Ψ of Φ is continuous, $\Psi(x) = 0$ only for $x = 0$, and $\{\Theta^{-1}(1/G(\alpha)) : \alpha \in I\}$ is a bounded subset of $\mathbb{R}$. We shall not give details of this result here. If Θ is a two valued convex function as in (19), so that the above set is contained in the unit interval of $\mathbb{R}$, when $G(\alpha) \equiv 1$, the classical case considered by Pitcher is then covered with $\Phi(x) = |x|^p$, for some real $p > 1$. On the other hand, it was noted in the literature that the compactness of $\mathcal{M}$ (of Definition 5) is equivalent to the existence of a localizable measure μ on (Ω, Σ) such that P_α is equivalent to μ for all

$\alpha \in I$, (cf. Ghosh, Morimoto and Yamada (1981), Ramamoorthi and Yamada (1983)). [Regarding localizability concept, see Definition 3.2, and compare the above statement with Lemma 3.3 which is slightly weaker than the present one.] This indicates that (Θ, G)-compactness is weaker than compactness of Definition 5, since in the current case one may not conclude that there is such a μ. Several results on (Θ, G)-compactness concept were given by Rosenberg (1970), where the reader will also find details of some of the statements following Definition 8. In fact a complete generalization of Theorem 7 for (Θ, G)-compactness holds true. These are not crucial for what follows and the details may be found in Rosenberg (1970). Here we intend to discuss only their application to sufficiency.

If $\mathcal{M}$ is a set of probability measures, as before, let $\mathcal{M}_c$ be its convex hull, so that if $\mathcal{M}$ has two distinct elements then $\mathcal{M}_c$ is infinite. For any σ-algebra $\mathcal{B} \subset \Sigma$ of $(\Omega, \Sigma, \mathcal{M})$, let $\hat{\mathcal{B}}$ be $\mathcal{M}_c$ complete, i.e.,

$$\hat{\mathcal{B}} = \{A \subset \Omega : \quad \text{there is } A_\mu \in \mathcal{B} \text{ such that } \mu^*(A \Delta A_\mu) = 0\}, \quad (22)$$

where μ^* is the outer measure generated by $(\mu, \mathcal{B})$ in the sense of Carathéodory. The following properties of $\hat{\mathcal{B}}$ are easily established.

(1) $\mathcal{B} \subset \hat{\mathcal{B}}$,   (2) $\hat{\mathcal{B}} = \hat{\hat{\mathcal{B}}}$

(3) $\hat{\mathcal{B}}$ is a σ-algebra, (4) $f : \Omega \to \mathbb{R}$ is $\hat{\mathcal{B}}$ measurable iff for each $\mu \in \mathcal{M}_c$, there is a $\mathcal{B}$-measurable f_μ such that $f = f_\mu$ a.e. $[\mu]$,

(5) if $\mathcal{B}$ is sufficient for $\mathcal{M}_c$, then $\mathcal{B} = \hat{\mathcal{B}}$.

(The fourth property may be verified first for simple functions and then the general case follows by the structure theorem for measurable functions. The last relation is then proved easily.) In terms of this "enlargement" of $\mathcal{B}$ into $\hat{\mathcal{B}}$, we can present some relations between compactness and sufficiency concepts, as well as the existence of (minimal) sufficient σ-subalgebras, together with a comparison of pairwise sufficiency and intermediate algebras between sufficient ones. The results are based on Pitcher's (1965) original work, as extended by Rosenberg (1968). The details are too technical and so we only present here a general discussion of them.

In this extended formulation the unit ball $U_\infty(\Theta, G)$ having a strictly positive element plays a significant role in addition to the compactness hypothesis of Definition 8. Here is the main statement:

Theorem 9. *Let* $(\Omega, \Sigma, \mathcal{M})$ *be* (Θ, G) *compact and* $U_\infty(\Theta, G)$ *have a strictly positive element. Then the following conclusions hold: (i)* $\hat{\Sigma} = \Sigma$, *(ii)* $\mathcal{B} \subset \Sigma$ *is a* σ-*algebra, implies* $(\Omega, \hat{\mathcal{B}}, \mathcal{M})$ *is also* (Θ, G)-*compact, (iii) there exists a unique minimal sufficient (for* $\mathcal{M}$*)* σ-*algebra* $\mathcal{B} \subset \Sigma$, *(iv)* $\mathcal{B}_1 \subset \Sigma$ *is a pairwise sufficient* σ-*algebra implies* $\hat{\mathcal{B}}_1$ *is actually sufficient; and finally (v)* $\mathcal{B} \subset \mathcal{B}_1 \subset \Sigma$ *and* $\mathcal{B}$ *is sufficient for* $\mathcal{M}$, *implies* $\hat{\mathcal{B}}_1$ *is sufficient.*

The last two parts should be compared with Theorem 3.8 and with Counterexample 2 above. Under compactness and existence of a positive element, sufficiency and pairwise sufficiency are essentially equivalent (i.e., after an enlargement). The last part extends Theorem 3 in a certain sense. The existence problem is a form of a new Radon-Nikodým theorem which we state it for reference as follows:

Theorem 10. *Let* (Ω, Σ) *be a measurable space and* $\mathcal{M} = \{P_\alpha \in I\}$ *be an uncountable and undominated set of probability measures (so* $\mathcal{M}$: $\Sigma \to B(I)$ *is an additive vector function which does not have finite variation). If* $\mathcal{M}$ *is* (Θ, G)-*compact and* $U_\infty(\Theta, G)$ *has an* $f_0 > 0$, *then there exists a proper* σ-*subalgebra* $\mathcal{B}$ *of* Σ *such that*

$$\int\limits_A f \, d\mathcal{M} = \int\limits_A g_f d\bar{\mathcal{M}} \quad , \quad A \in \mathcal{B}, \; f \in B(\Omega, \Sigma), \tag{23}$$

for a unique $\mathcal{B}$-*measurable* g_f *(depending only on* f*), where* $\bar{\mathcal{M}} = \mathcal{M}|\mathcal{B}$. *The integral here is defined using the order topology of the Banach lattice* $B(\Omega, \Sigma)$. *(See the next section on this where it is shown that* $\mathcal{M}$ *is* σ-*additive in the order topology.)*

The existence of a minimal sufficient $\mathcal{B}$ in (iii) above is proved by considering the set of $\frac{d\mu_i}{d(\mu_1+\mu_2)}, i = 1, 2$ for each pair (μ_1, μ_2) of $\mathcal{M}$, and let $\mathcal{B}_0$ be the smallest σ-algebra relative to which all the above functions are measurable. Then as the argument of Corollary 3.7 shows that $\mathcal{B}_0$ is contained in each sufficient σ-algebra if the latter exists. Using Theorem 7 above one shows that $\mathcal{B}_1 = \hat{\mathcal{B}}_0$ is indeed the minimal sufficient σ-algebra. We shall not include the details.

As a consequence of the last part of Theorem 9, and Theorem 3, one has:

Corollary 11. *Let* $\mathcal{B}_n \subset \mathcal{B}_{n+1} \subset \Sigma$ *be* σ-*subalgebras of* $(\Omega, \Sigma, \mathcal{M})$ *which are* (Θ, G) *compact, with* $U_\infty(\Theta, G)$ *having a strictly positive* *element. Then the sufficiency of $\mathcal{B}_1$ implies that $\mathcal{B}_1 = \hat{\mathcal{B}}_1 \subset \hat{\mathcal{B}}_2 \subset \cdots \subset \hat{\Sigma} = \Sigma$ each $\hat{\mathcal{B}}_n$ is sufficient and $\tilde{\mathcal{B}}_\infty = \sigma(\underset{n}{\cup}\hat{\mathcal{B}}_n) \subset \Sigma$ is sufficient. Moreover, $\hat{\mathcal{B}}_\infty = \tilde{\mathcal{B}}_\infty (\supset \mathcal{B}_\infty = \underset{n}{\lim}\mathcal{B}_n = \sigma(\underset{n}{\cup}\mathcal{B}_n))$, but $\mathcal{B}_\infty$ need not be sufficient.*

We now consider an alternative treatment of results based on, and motivated by, the above Theorem 10 in what follows.

6.5 Sufficiency: another approach to the undominated case

If $\mathcal{M} = \{P_\alpha, \alpha \in I\}$ is an uncountable set of (undominated) probability measures on Σ, then it may be regarded as an additive function on Σ into $B(I)$, the Banach space of bounded real functions on I with the uniform (or supremum) norm. Since $B(I)$ is also a (vector) lattice under pointwise ordering (i.e., $f \leq g$ iff $f(x) \leq g(x)$, $x \in I$), and since each bounded increasing sequence has a supremum in $B(I)$ it is boundedly σ-complete. Using this ordering we observe that $\mathcal{M} : \Sigma \to B(I)$ has the following properties: (i) $\mathcal{M}(A) \geq 0$, $A \in \Sigma$, and $\mathcal{M}(\emptyset) = 0$, and (ii) if $A_n \in \Sigma$, disjoint, then it possesses the (weaker) σ-additivity, property so that

$$\mathcal{M}(\overset{\infty}{\underset{n=1}{\cup}} A_n) = \sup_{n \geq 1}(\overset{n}{\underset{k=1}{\Sigma}} \mathcal{M}(A_k)), \tag{1}$$

since the right side is an increasing bounded sequence of $B(I)$. This "order σ-additivity" of $\mathcal{M}$ can be used to define an integral of a scalar function relative to $\mathcal{M}$ for which the dominated convergence theorem holds, and for which "good" conditions can be found in order that a Radon-Nikodým theorem is valid. One can use it to define sufficiency (reversing the role of Theorem 4.10 into 4.9). The necessary background analysis is available from J. D. M. Wright (1969), and a slight improvement of it is given by R. Haydon (1977). We outline these results (see Sec. 2 above) in a suitable form here.

As usual, if $f = \overset{n}{\underset{i=1}{\Sigma}} a_i \chi_{A_i} \in B(I)$, is a simple function, $A_i \in \Sigma$, disjoint, then define

$$\int_\Omega f \, d\mathcal{M} = \overset{n}{\underset{i=1}{\Sigma}} a_i \mathcal{M}(A_i), (\in B(I)) \tag{2}$$

which is seen to be unambiguously defined. If $0 \leq f$ is any measurable function (for Σ), then there exist $0 \leq f_n \uparrow f$ pointwise, f_n simple, so that one defines

$$\int_\Omega f \, d\mathcal{M} = \sup_n \int_\Omega f_n \, d\mathcal{M}, \quad \text{or } = +\infty, \tag{3}$$

accordingly as the right side is in $B(I)$ or not. Next for $f : \Omega \to \bar{\mathbb{R}}$, measurable ($\Sigma$), we let $f = f^+ - f^-$ and put

$$\int_\Omega f \, d\mathcal{M} = \int_\Omega f^+ \, d\mathcal{M} - \int_\Omega f^- \, d\mathcal{M} \tag{4}$$

if at least one of the right side terms is not infinity. Now one verifies that the mapping $f \mapsto \int_\Omega f \, d\mathcal{M}$ is well-defined, linear, and the monotone and dominated convergence statements hold. Since $\mathcal{M}$ does not have finite variation (see Prop. 2.1), this integral cannot, in general, be reduced to the Lebesgue-Daniell type integral and the necessary reworking of the proofs has been presented by Wright (1969). Indeed further restrictions are necessary on the (vector) measure $\mathcal{M}$ for a Radon-Nikodým type result which we now present. The desired form is based on a slightly improved version of Wright's result, due to R. Haydon (1977), as noted already.

A motivation for the above condition will be given before the actual result is presented, since this can indicate a "rationale" for the search. Definition 1.1 and Lemma 1.3 together show for a (nonscalar) vector measure $\mathcal{M}$ of the type considered with $\mathcal{N} = \mathcal{M}|\mathcal{B}$ where $\mathcal{B} \subset \Sigma$ is a σ-subalgebra, no $\mathcal{B}$-measurable function $f : \Omega \to \mathbb{R}$ exists to satisfy

$$\int_A d\mathcal{M} = \mathcal{M}(A) = \int_A f \, d\mathcal{N}, \quad A \in \mathcal{B}, \tag{5}$$

if $\mathcal{B}$ is just $\{\emptyset, \Omega\}$. On the other hand if $\mathcal{B} \subsetneq \Sigma$ is "too rich" then also such an f need not exist; but if $\mathcal{M}$ satisfies a certain compactness restriction, then Theorem 4.10 shows that there are certain σ-algebras $\mathcal{B} \subsetneq \Sigma$ and a $\mathcal{B}$-measurable f such that (5) holds. As noted by Wright (1969), difficulties can arise even in simple measurable spaces. For instance, let $\Omega = \{\omega_1, \omega_2\}, \Sigma =$ power set of $\Omega, B(I) = \mathbb{R}^2, \mathcal{M}, \mathcal{N}$ be defined as, $\mathcal{M}(\{\omega_1\}) = \mathcal{N}(\{\omega_2\}) = (1,0), \mathcal{M}(\{\omega_2\}) = \mathcal{N}(\{\omega_1\}) = (0,1)$. Then evidently $\mathcal{N}(A) = 0$ implies $\mathcal{M}(A) = 0$ for $A \in \Sigma$, and if there exists an $f : \Omega \to \mathbb{R}$, one must have for $A = (\{\omega_1\})$,

$$(1,0) = \mathcal{M}(\{\omega_1\}) = \int_A f \, d\mathcal{N} = f(\omega_1)(0,1),$$

which is impossible for any scalar $f(\omega_1)$. Thus the condition for a positive solution must involve both the σ-algebra $\mathcal{B}$ and $\mathcal{M}$ (or its range space). For us the range is $B(I)$ and $\mathcal{B} \subsetneq \Sigma$ and the restriction to be found should automatically include the domination property. These competing properties indicate some subtle problems to be met. The condition we consider is again functional analytic in nature, in line with the compactness of the last section but is distinct from it. In order to state, we introduce the $L^p(\mathcal{M})$ spaces of scalar functions on (Ω, Σ) or $(\Omega, \mathcal{B})$ suitably involving both the algebra Σ, or $\mathcal{B}$ and the range space of $\mathcal{M}$ and specialize to suit the present needs.

Since $\mathcal{M}(A) \geq 0$, if $B \subset A$ are measurable, $\mathcal{M}(A) = 0$, then $\mathcal{M}(B) = 0$ also. If $\mathcal{P}$ is the collection of all subsets of sets on which $\mathcal{M}$ vanishes, then it is a ring [in fact an ideal], and one can extend $\mathcal{M}$ onto $\Sigma \cup \mathcal{P}$ by defining $\mathcal{M}^*(\tilde{A}) = 0$ if $\tilde{A} = A \cup N$, $A \in \Sigma$ and $N \in \mathcal{P}$. If Σ^* is the ring generated by $\Sigma \cup \mathcal{P}$, then the classical arguments imply that $\mathcal{M}^* : \Sigma \to B(I)$ is σ-additive (in the order topology) and we can say $\mathcal{M}^*$ is complete. Thus f and g are termed equivalent, written $f \sim g$, if $\{\omega : f(\omega) \neq g(\omega)\}$ is $\mathcal{M}^*$-null. Hereafter we assume $\mathcal{M}$ is complete so $\mathcal{M}^* = \mathcal{M}$, as we can complete it. Defining $\int_\Omega f \, d\mathcal{M}$ as before, for simple f, and extending it for all $\mathcal{M}^*$-measurable $f \geq 0$, let $\mathcal{L}^1(\mathcal{M})$ be the space of all $\mathcal{M}$-integrable functions. Further $f \sim g$ implies $\int_\Omega f \, d\mathcal{M}$ and $\int_\Omega g \, d\mathcal{M}$ are $\mathcal{M}$-equivalent, and each is in $B(I)$. We say $f \in \mathcal{L}^p(\mathcal{M})$ iff $|f|^p \in \mathcal{L}^1(\mathcal{M})$. These are linear spaces and identifying the equivalence classes one obtains the vector spaces $L^p(\Omega, \Sigma^*, \mathcal{M})$, or $L^p(\mathcal{M})$, for which a *norm* $\| \cdot \|$ is definable on the equivalence classes $[f]$ of f as:

$$\|[f]\|_p = \| \int_\Omega |f|^p d\, \mathcal{M}\|_\infty^{\frac{1}{p}} \, , \quad \text{if } 1 \leq p < \infty$$

$$= \text{ ess. sup} |f|, \text{ if } p = \infty \, , \ f \in \mathcal{L}^p(\mathcal{M}). \tag{6}$$

Then one can verify in the usual way that $\{L^p(\mathcal{M}), \| \cdot \|_p\}$, $p \geq 1$, is a Banach lattice. In (6), the outer symbol is the uniform norm in $B(I)$. The spaces $L^p(\mathcal{M})$ have the property that for each increasing sequence $\{f_n, n \geq 1\}$ in the unit ball, the supremum is also in the same ball, but this need not hold for generalized sequences. (The latter property is usually termed *Dedekind completeness*.) For our $\mathcal{M} = \{P_\alpha, \alpha \in I\}$, the elements of $B(I)$ act as constants in integration. In fact if each

$b \in B(I)$ is identified as $\tilde{b}$ in $L^p(\mathcal{M})$, then one has

$$\int_\Omega \tilde{b} f \, d\mathcal{M} = b \int_\Omega f d\mathcal{M}. \qquad (7)$$

and the mapping $\pi : b \to \tilde{b}$ is an algebra homomorphism in the sense that $\pi(b_1 b_2) = \pi(b_1)\pi(b_2)$, $\pi(1) = \tilde{1}$ a.e., and it is positivity preserving. If $\mathcal{M} = \{P(\cdot|x), x \in I\}$ is a family of regular conditional measures (cf. Chapter 5), then also (7) will hold. This is called *modularity* property of $\mathcal{M}$, but for general measures $\mathcal{M}' : \Sigma \to B(I)$ it need not be true. We now introduce the following:

Definition 1. If $\mathcal{M} : \Sigma \to B(I)$ is a vector measure which is (order) σ-additive, as above, then it has *property D* (for Dedekind) provided the Banach lattice $L^1(\Sigma, \mathcal{M})$ has that property, i.e., each upward directed set in its unit ball has a supremum in that ball.

This condition is a restriction on I [hence on $B(I)$], if Σ is general and a restriction on Σ if I [hence $B(I)$] is general. It is shown by Haydon (1977, page 38) that this is equivalent to Dedekind completeness of $L^\infty(\Sigma, \mathcal{M})$ for positive finite $\mathcal{M}$, as in our case. The above L^∞-space has the last property in case $\mathcal{M} = \{P_\alpha, \alpha \in I\}$ is dominated by a localizable measure. Thus condition (D) can be somewhat more general than this type of domination.

We now present analogs of Theorem 4.10 in this context and then relate them to sufficiency without assuming domination.

Theorem 2. *Let (Ω, Σ) be a measurable space, $\mathcal{M} = \{P_\alpha, \alpha \in I\}$ be a collection of (possibly) undominated probability measures on Σ, so that $\mathcal{M} : \Sigma \to B(I)$ is an (order) σ-additive vector measure. If $\mathcal{B} \subset \Sigma$ is a σ-algebra such that $\mathcal{M} : \mathcal{B} \to B(I)$ has property (D), then for each $f \in B(\Omega, \Sigma)$, there exists an $\mathcal{M}$-unique ($\mathcal{B}$-measurable) $g_f : \Omega \to \mathbb{R}$ such that*

$$\int_A f \, d\mathcal{M} = \int_A g_f d\bar{\mathcal{M}}, \quad A \in \mathcal{B}, \qquad (8)$$

where $\bar{\mathcal{M}} = \mathcal{M}|\mathcal{B}$. The integral in (8) is defined as in (4).

Proof. The main point of the argument is to show that the $\mathcal{M}$ here satisfies an improved version (due to Haydon, 1977, Thm. 6.G) of Wright's (1969) Radon-Nikodým Theorem. This verification will now be sketched.

Since $B(I)$ is a function algebra, by a classical result, it is isometrically lattice isomorphic to $C(S)$, the space of real continuous functions on a compact Stone space [the space of maximal ideals of $B(I)$]. If $\tau : B(I) \to C(S)$ is this isomorphism and $\pi_1 = \pi \circ \tau^{-1} : C(S) \to L^1(\mathcal{M})$ is a modular map (π being defined after (7)), then $\tau \circ \mathcal{M} : \Sigma \to C(S)$ is a finite "π_1-modular" $C(S)$-valued measure. Also $\tau \circ \bar{\mathcal{M}} : \mathcal{B} \to C(S)$ has the same properties and in addition verifies the critical condition (D) by hypothesis, so that $\bar{\mathcal{M}}$ is "ample" in Wright's terminology. Let $\nu : A \mapsto \tau(\int_A f \, d\mathcal{M})$, $A \in \mathcal{B}$. Then the above properties imply that ν is a π_1-modular $C(S)$-valued measure which vanishes on $\tau \circ \bar{\mathcal{M}}$-null sets. Hence by the improved Radon-Nikodým theorem, there is a $g \in L^1(\mathcal{B}, \bar{\mathcal{M}})$ such that (8) holds, when τ^{-1} is applied suitably. $\square$

This following complements the above result in some cases:

Proposition 3. *Let $(\Omega, \Sigma, \mathcal{M})$ be a vector measure space where $\mathcal{M} = \{P_\alpha, \alpha \in I\}$ is as before. Suppose that (in lieu of condition (D)) the representation space $C(S)$ of $B(I)$ satisfies the countable chain condition in the sense that each bounded subset of $C(S)$ has a countable subset having the same least upper bound as the former. If $\mathcal{B} \subsetneq \Sigma$ is a σ-subalgebra and $\pi_1(C(S)) \subset L^\infty(\mathcal{B}, \bar{\mathcal{M}})$ where $\bar{\mathcal{M}} = \mathcal{M}|\mathcal{B}$ (and π_1 is the modular map), then for each $f \in L^1(\Sigma, \mathcal{M})$ there is a unique $g_f \in L^1(\mathcal{B}, \bar{\mathcal{M}})$ such that*

$$\int_A f \, d\mathcal{M} = \int_A g_f d\, \bar{\mathcal{M}}, \quad A \in \mathcal{B} . \tag{9}$$

Remark. In contrast to Theorem 2, here the restriction is (indirectly) on the index set I, or equivalently on its representation space S; but $\mathcal{B}$ should not be too small, since $\pi_1(C(S))$ should lie in $L^\infty(\mathcal{B}, \bar{\mathcal{M}})$.

Proof. As shown by Wright (1969) (Prop. 3.3), for a finite $\mathcal{M}$ as here, $L^\infty(\Sigma, \mathcal{M})$ is also a Stone algebra satisfying the countable chain condition, and $L^\infty(\mathcal{B}, \bar{\mathcal{M}})$ is its Stone subalgebra which clearly obeys the same condition. But then by Haydon's version noted in the preceding proof, $\bar{\mathcal{M}}$ has property (D). Hence the result follows from the preceding one. $\square$

The foregoing two statements can be given the following form in the context of sufficiency that is of immediate concern here.

Theorem 4. *Let $\mathcal{M} = \{P_\alpha, \alpha \in I\}$ be a family of probability measures on (Ω, Σ), Σ completed for $\mathcal{M}$. If $\mathcal{B} \subset \Sigma$ is a σ-algebra such that either (i) $\mathcal{M}$ has property (D) when restricted to $\mathcal{B}$, or (ii) the representation space $C(S)$ of $B(I)$ satisfies the countable chain condition and $\pi_1(C(S)) \subset L^1(\mathcal{B}, \mathcal{M})$, then $\mathcal{B}$ is sufficient for $\mathcal{M}$.*

The result is immediate from the preceding ones since then

$$\int_A f \, d\mathcal{M} = \int_A g_f \, d\mathcal{M}, \quad A \in \mathcal{B}, \ f \in B(I),$$

and if $T : B(I) \to B(I)$ is a positive linear mapping which preserves monotone limits and intertwines with π, then T commutes with the integral as seen from definition contained in (4). In particular taking T as an evaluation functional on $B(I)$, one gets from the above

$$\int_A f \, dP_\alpha = \int_A g_f \, dP_\alpha, \quad A \in \mathcal{B}, \tag{10}$$

and g_f does not depend on α. Hence $\mathcal{B}$ is sufficient as asserted.

It should be observed that these conditions are "good" sufficient conditions, but are not the best in that they are not necessary since $\mathcal{B} = \Sigma$ is always a sufficient σ-algebra without any restrictions. On the other hand, we have no simple condition asserting the existence of a minimal sufficient σ-algebra of $\{P_\alpha, \alpha \in I\}$. In the case of the alternative hypothesis (of Prop. 3 above), it is easy to see that the monotone sequence of σ-algebras $\mathcal{B}_n$ all of which satisfy the condition that $\pi_1(C(S)) \subset L^\infty(\mathcal{B}_n, \mathcal{M})$, when $C(S)$ obeys the countable chain condition, have their limits which are again sufficient. However, this type of extension will not be pursued now since these hypotheses are also harder to verify in applications, and demand more prerequisites.

The above considerations show that the vector integration with measures which are not of bounded variation need further analysis, since they should answer problems arising from sufficiency theory when no domination is assumed. We thus leave the topic here, and turn to an examination of Kolmogorov's formulation of conditioning from a general point of view in the following chapter, and show how Rényi's new axiomatic approach can be considered as part of that frame work.

6.6 Bibliographical notes

The concept of sufficiency, introduced by R.A. Fisher in early 1920s into the statistical inference theory, plays a vital role. The classical theory of this is detailed in Lehmann (1959, 1983). The general undominated case is usually not covered in most books, perhaps because of the abstract mathematical level it demands and also because the results are still being refined and updated. After the basic theory by Halmos-Savage (1949) and Bahadur (1954) which we included in Section 3, the general case has been primarily considered by Burkholder (1961) and in several papers by Pitcher (1958–65). Some of these results are given in Section 4 where a general formulation of Pitcher's compactness condition, as further extended by Rosenberg (1968, 1970), are discussed. It is clear how functional analysis (especially Orlicz spaces) naturally appears in this case. We included the discussion without too many details, since that will be a great digression here. Some of this has also been considered by Heyer (1982), but we included several recent developments from a broader perspective.

There is an alternative view of considering the set of probability measures as an infinite vector which is σ-additive in the order topology. This, based on Wright's (1969) theory of integration adapted to our case, is discussed in Section 5. It clearly shows how the existing vector measure theory is not adequate for the statistical sufficiency requirements. We therefore used further refinements with Haydon's (1977) discussion on extension using vector lattice methods. It is noted in the work of Luschgy and Mussmann (1985) (see also their references to earlier contributions) that in some ways the compactness condition in a general case is related to domination of $\mathcal{M}$, by a localizable measure. A similar relation with conditions (D) exists between our vector measure $\mathcal{M}$ and a "simultaneous localizability" of $\{P_\alpha, \alpha \in I\}$, as formulated by McShane (1962). It is not yet clear how all these concepts are interlaced together, because of the different contexts in which they arose.

It is possible to concentrate on weakening the concept of domination of the family $\{P_\alpha, \alpha \in I\}$ and consider the structure of the underlying measurable space (Ω, Σ) that supports these measures. Then one can study the solid subspace of the abstract L-space generated by $\{P_\alpha, \alpha \in I\}$ in the space of all signed measures on (Ω, Σ) which becomes a Banach space under variation norm, and define an invariant as the dimension of this subspace. These concepts lead to set theoretical studies of the problems and to the existence (or otherwise) of nonmeasurable cardinals and their equivalences with the problems at hand. In our study we have concentrated on the functional analytic aspects since then the general structure of the problem becomes simpler to understand. But the latter study did not fit into our framework, and we therefore concluded the discussion at this point. It is, however, clear that a considerable amount of research is needed to fully understand the sufficiency theory in the undominated case.

CHAPTER 7: ABSTRACTION OF KOLMOGOROV'S FORMULATION

The formulation of Kolmogorov's conditional probability model is abstractly discussed in this chapter viewing it as a subclass of vector measures on function spaces and the conditional expectation as a projection operator on the same spaces. Characterizations of both these classes are presented for a family set of function spaces, and their structure is thereby illuminated. In this context the Rényi (new) model is compared with, and shown to be a particular case of, Kolmogorov's extended formulation. Finally a discussion of the relations between conditioning and differentiation, complementing the work of Section 3.4, is given and a general result on exact evaluations is also included.

7.1 Introduction

In the last two chapters we considered the regularity properties of conditional measures and applications of conditioning to sufficiency. Here the general case, without regard to regularity, will be analyzed as originally formulated where the concept is based on the abstract Radon-Nikodým theorem. This will clarify the underlying structure of the functional operation involved and then we characterize it after recalling the (stronger) Dunford-Schwartz integral as applied to conditional measures. This work enables us to study Rényi's axiomatic formulation as a specialization of Kolmogorov's concept. (It was originally indicated earlier, Rao, 1981, p. 93.) The point will be elaborated in Section 4 below. The fifth section contains a (further) analysis of the basic relation between conditioning and differentiation theory which is particularly useful in exact evaluations of conditional expectations in the Kolmogorov model for a wide class of problems.

With a view to several applications, we consider a nonfinite measure space (Ω, Σ, μ) in this chapter and reformulate some results from Chapter 2 in a general form which often need new proofs. If $\mathcal{B} \subset \Sigma$ is a σ-algebra, we suppose that the restriction of μ to $\mathcal{B}$, denoted $\mu_{\mathcal{B}}$, is σ-finite or more generally localizable. If $f : \Omega \to \bar{\mathbb{R}}$ is measurable and μ-integrable on Ω, consider $\nu_f : A \mapsto \int_A f d\mu$, $A \in \mathcal{B}$. Then $\nu_f : \mathcal{B} \to \mathbb{R}$ is a signed measure and is $\mu_{\mathcal{B}}$-continuous (i.e., $\nu_f \ll \mu_{\mathcal{B}}$). Then by the Radon-Nikodým theorem (see Rao, 1987, Sec. 5.4 for the general localizable case) there is a unique $\mathcal{B}$-measurable $\tilde{f}$ such that

$$\int_A f d\mu = \nu_f(A) = \int_A \tilde{f} \, d\mu_{\mathcal{B}}, \ A \in \mathcal{B} \ . \tag{1}$$

The mapping $E^{\mathcal{B}} : f \mapsto \tilde{f}$ is well-defined, is a positive contractive linear operator on $L^1(\mu)$ into $L^1(\mu)$. Note that the same argument is applicable if $f : \Omega \to \bar{\mathbb{R}}^+$, which is measurable but not necessarily μ-integrable. Then for a general $f : \Omega \to \bar{\mathbb{R}}$, μ-measurable, one can define $E^{\mathcal{B}}(f^{\pm})$ where $f = f^+ - f^-$ and if f^+ or f^- is μ-integrable, then one can again define $E^{\mathcal{B}}(f) = E^{\mathcal{B}}(f^+) - E^{\mathcal{B}}(f^-)$, and the thus obtained $E^{\mathcal{B}}$ is a positive linear operator which is a contraction on $L^1(\mu)$ into $L^1(\mu)$. Moreover, $E^{\mathcal{B}}(1) = 1$, a.e., always holds.

Let $\Sigma_0 = \{A \in \Sigma : \mu(A) < \infty\}$, which is a δ-ring. If we set $P^{\mathcal{B}}(A) = E^{\mathcal{B}}(\chi_A)$, $A \in \Sigma_0$ where $\mu_{\mathcal{B}}$ is localizable, then by the last paragraph $P^{\mathcal{B}}(\Omega) = 1$ a.e., $0 \le P^{\mathcal{B}}(A) \le 1$ a.e., $A \in \Sigma$, and it is a.e. σ-additive on Σ. It is also σ-additive on Σ_0 in the L^p-norm as the following simple computation shows: Let $\{A_n, n \ge 1\} \subset \Sigma_0$ be disjoint such that $A = \bigcup_{n=1}^{\infty} A_n \in \Sigma_0$. Then

$$\|P^{\mathcal{B}}(A) - \sum_{k=1}^{n} P^{\mathcal{B}}(A_k)\|_p^p = \int_{\Omega} [P^{\mathcal{B}}(\bigcup_{k>n} A_k)]^p d\mu_{\mathcal{B}}$$

$$= \int_{\Omega} [E^{\mathcal{B}}(\chi_{\bigcup_{k>n} A_k})]^p d\mu_{\mathcal{B}}$$

$$\leq \int_\Omega E^B(\chi_{\underset{k>n}{\cup} A_k})d\mu_B,$$

by the conditional Jensen inequality
(see Sec. 2.2, and also Prop. 2, next section),

$$= \mu(\underset{k>n}{\cup} A_k) \to 0, \text{ as } n \to \infty, \tag{2}$$

by the dominated convergence theorem since $\mu(A) < \infty$. This will be false for $p = \infty$, but the result holds in the order topology, as noted in Section 6.5. Moreover $P^B : \Sigma_0 \to L^p(\mu)$ has finite variation if $p = 1$ and finite semi-variation (on Σ_0) if $1 < p < \infty$ as seen from

$$\text{var } (P^B)(A) = |P^B|(A) = \sup\{\sum_{i=1}^{n} \|E^B(\chi_{A_i})\|_1 : A_i \in \Sigma(A), \text{ disjoint}\}$$

$$= \sup\{\sum_{i=1}^{n} \int_\Omega \chi_{A_i} d\mu : A_i \in \Sigma(A), \text{ disjoint}\}$$

$$= \mu(A) < \infty, \quad A \in \Sigma_0, \tag{3}$$

where $\Sigma(A) = \{A \cap B : B \in \Sigma\}$ is the trace σ-algebra of Σ on A, and

semivar $(P^B)(A) = \|P^B\|_p(A)$

$$= \sup\{\|\sum_{i=1}^{n} \alpha_i P^B(A_i)\|_p : A_i \in \Sigma(A), \text{ disjoint } |\alpha_i| \leq 1\}$$

$$= \sup\{\|P^B(\underset{i=1}{\overset{n}{\cup}} A_i)\|_p : A_i \in \Sigma(A), \text{ disjoint}\}$$

$$\leq \|E^B(\chi_A)\|_p \leq [\mu(A)]^{\frac{1}{p}} < \infty, \quad A \in \Sigma_0. \tag{4}$$

Thus μ acts as a "control measure" for $P^B(\cdot)$, so that the Dunford-Schwartz integration of scalar functions relative to a vector measure can be developed, and that $\int_\Omega f dP^B$ is definable for all bounded μ-measurable functions $f : \Omega \to \mathbb{R}$. This can be given a simpler proof than the original general case, and it will be used, (see Prop. 2.1 below).

7.2 Integration relative to conditional measures and function spaces

The vector integral, to be presented here, is in many respects similar to that discussed in Section 6.5 using order properties. The modularity property considered there reappears in disguise.

If $\mathcal{B} \subset \Sigma$ is a σ-algebra in (Ω, Σ, μ) with $\mu_{\mathcal{B}}$ localizable and $P^{\mathcal{B}}$ is the conditional measure having μ as its control measure, so that $P^{\mathcal{B}} : \Sigma \to L^{\infty}(\mathcal{B}) \cap L^{p}(\mathcal{B})$, and is σ-additive in L^{p}-mean, consider, for each simple function $f = \sum_{i=1}^{n} a_i \chi_{A_i}, A_i \in \Sigma_0 = \{A \in \Sigma, \mu(A) < \infty\}$, the integral defined as:

$$\int_{A} f dP^{\mathcal{B}} = \sum_{i=1}^{n} a_i P^{\mathcal{B}}(A \cap A_i), \ A \in \Sigma \ . \tag{1}$$

The mapping $f \mapsto \int_A f dP^{\mathcal{B}}$ is well-defined [i.e., if $f = g$ a.e. (μ) then the integrals agree a.e. (μ)], linear, and order preserving. Moreover,

$$\| \int_{A} f dP^{\mathcal{B}} \|_p = \| \sum_{i=1}^{n} a_i E^{\mathcal{B}}(\chi_{A \cap A_i}) \|_p, \ \text{by (1)},$$

$$= \|E^{\mathcal{B}}(f)\|_p \leq \|f\|_p, \tag{2}$$

since $E^{\mathcal{B}}$ is a contraction on $L^{p}(\mu), 1 \leq p \leq \infty$. (This is clear for $p = 1$ and $p = +\infty$, and the general case follows from Prop. 2 below.) Since the set of simple functions is norm dense in $L^{p}(\mu), 1 \leq p < \infty$, and the range space $L^{p}(\mathcal{B})$ is complete, the mapping has a unique extension to all of $L^{p}(\mu)$, denoted by the same symbol. For the L^{∞}-case one needs to use the order continuity, and the result holds in that case also, at least when $\mu(\Omega) < \infty$, as seen in Section 6.5. This discussion can be summarized in the following:

Proposition 1. *Let* (Ω, Σ, μ) *be a measure space* $\mathcal{B} \subset \Sigma$, *a* σ-*algebra such that* $\mu_{\mathcal{B}}$ *is localizable. Then* $P^{\mathcal{B}} : \Sigma \to L^{p}(\mathcal{B}, \mu_{\mathcal{B}})$ *is a vector measure, and the mapping* $f \mapsto \int_{\Omega} f dP^{\mathcal{B}}, f \in L^{p}(\mu), 1 \leq p < \infty$ *[and for* $p = \infty$ *with* $\mu(\Omega) < \infty$*], is well-defined relative to the control measure* $\mu_{\mathcal{B}}$, *linear, and the dominated and monotone convergence statements hold for it. Moreover, one has the representation*

$$E^{\mathcal{B}}(f) = \int_{\Omega} f \ dP^{\mathcal{B}}, \ f \in L^{p}(\mu), \tag{3}$$

as a function space integral, $1 \leq p < \infty$, *in norm, or in order topology for* $p = \infty$. *This is also order preserving and faithful in that* $f \geq 0$, *and* $\int_{\Omega} f \ dP^{\mathcal{B}} = 0$ *implies* $f = 0$ *a.e.* (μ).

Remark. The integral for the case $1 \leq p < \infty$ is the same as that of Dunford and Schwartz and for $p = +\infty$ that of Wright's. The definition becomes simpler than the original sources for the conditional measures.

Let us now present a general form of the conditional Jensen inequality (used in (2) above), for nonfinite measure spaces. This extends the result of Section 2.2, and it is due to Chow (1960).

Proposition 2. *Let (Ω, Σ, μ) be a measure space, $\mathcal{B} \subset \Sigma$ a σ-algebra and $\mu_\mathcal{B}$ be σ-finite. Let $X : \Omega \to \mathbb{R}$ be a random variable such that X^+ or X^- is μ-integrable. If $\varphi : \mathbb{R} \to \mathbb{R}$ is a (measurable) convex function such that $\varphi^+(X)$ is μ-integrable, then the conditional expectation $E^\mathcal{B}(X)$, which exists, satisfies*

$$\varphi(E^\mathcal{B}(X)) \leq E^\mathcal{B}(\varphi(X)), \quad a.e. \tag{4}$$

Moreover, if $\varphi(\cdot)$ is strictly convex then there is equality in (4) iff X is $\mathcal{B}$-measurable.

Proof. Since $\varphi(\cdot)$ is convex (measurability implies continuity on open sets) its right and left derivatives exist everywhere and are distinct at most for a countable set of points (see Rao, 1987, p. 242). Let φ' be the right derivative which is a nondecreasing (hence Borel) function so that $\varphi'(E^\mathcal{B}(X))$ is $\mathcal{B}$-measurable. By convexity one has

$$\varphi(x) - \varphi(y) \geq \varphi'(y)(x - y), \tag{5}$$

for any x, y in $\mathbb{R}$. Since X^+ and $\varphi^+(X)$ are integrable and $\mu_\mathcal{B}$ is σ-finite, there exist $\Omega_n \in \mathcal{B}$, $\Omega_n \uparrow \Omega$, such that on each Ω_n, $X, \varphi(X)$ and $\tilde{X} = E^\mathcal{B}(X)$ have well-defined integrals relative to μ. Thus for each $B \in \mathcal{B}(\Omega_n)$, the trace of $\mathcal{B}$ on Ω_n, if $B_k = B \cap [|\varphi'(\tilde{X})| \leq k]$ then $\varphi'(\tilde{X})(X - \tilde{X})$ has a well-defined integral on B_k and one has

$$\int_{B_k} \varphi'(\tilde{X})(X - \tilde{X})d\mu = \int_{B_k} E^\mathcal{B}(\varphi'(\tilde{X})(X - \tilde{X}))d\mu$$

$$= \int_{B_k} \varphi'(\tilde{X})E^\mathcal{B}(X - \tilde{X})d\mu = 0. \tag{6}$$

It follows from (5) and (6) that

$$\int_{B_k} [\varphi(X) - \varphi(\tilde{X})]d\mu \geq 0. \tag{7}$$

Let $B = [\varphi(\tilde{X}) > 0] \cap \Omega_n (\in \mathcal{B}(\Omega_n))$ in the above. Then (7) gives

$$\int_{\Omega_n \cap [\varphi(\tilde{X}) > 0, \, |\varphi'(\tilde{X})| \leq k]} (\varphi^+(X) - \varphi^+(\tilde{X})) d\mu$$

$$\geq \int_{\Omega_n \cap [\varphi(\tilde{X} > 0, |\varphi'(\tilde{X})| \leq k]} (\varphi(X) - \varphi(\tilde{X})) d\mu$$

$$\geq 0.$$

This inequality, under the hypothesis, yields

$$\infty > \int_{\Omega_n} \varphi^+(X) d\mu \geq \int_{\Omega_n \cap [|\varphi'(\tilde{X})| \leq k]} \varphi^+(\tilde{X}) d\mu.$$

Now let $k \uparrow \infty$ in this to infer that $\int_{\Omega_n} \varphi^+(\tilde{X}) d\mu < \infty$. Then (7) implies

$$\int_{B_k} \varphi(X) d\mu \geq \int_{B_k} \varphi(\tilde{X}) d\mu, \quad B_k \in \mathcal{B}(\Omega_n). \tag{8}$$

Letting $k \uparrow \infty$ in (7) one gets the same result with B in place of B_k. Since n is arbitrary, we conclude that $E^{\mathcal{B}}(\varphi(X)) \geq \varphi(\tilde{X}) = \varphi(E^{\mathcal{B}}(X))$, a.e. proving (4). Refering back to the inequalities (5)–(8) above we can deduce that there is strict inequality if $\varphi(\cdot)$ is strictly convex (i.e., $\varphi'(\cdot)$ is strictly increasing) and X is not $\mathcal{B}$-measurable. It is trivial that, conversely, if X is $\mathcal{B}$-measurable there is always equality in (4) whether or not $\varphi(\cdot)$ is strictly convex. $\square$

An interesting consequence of this result is:

Corollary 3. *If the convex function φ is also nondecreasing (in the proposition) and $Y \leq E^{\mathcal{B}}(X)$ a.e. where $\mathcal{B} \subset \Sigma$ is such that $\mu_{\mathcal{B}}$ is σ-finite, then $\varphi(Y) \leq E^{\mathcal{B}}(\varphi(X))$ a.e. holds.*

This is because by the nondecreasing property of $\varphi(\cdot)$ and the proposition one has

$$\varphi(Y) \leq \varphi(E^{\mathcal{B}}(X)) \leq E^{\mathcal{B}}(\varphi(K)), \quad a.e.$$

Remarks. 1. In Proposition 1, we have taken $\varphi(x) = |x|^p, p \geq 1$, and

$X \in L^p(\mu)$ with $\mu_{\mathcal{B}}$ localizable so that $E^{\mathcal{B}}(X)$ exists and

$$\|E^{\mathcal{B}}(X)\|_p^p = \int_\Omega \varphi(E^{\mathcal{B}}(X))d\mu$$

$$\leq \int_\Omega E^{\mathcal{B}}(\varphi(X))d\mu, \text{ by Proposition 2 and the fact that } X$$

vanishes outside a σ-finite set,

$$= \int_\Omega \varphi(X)d\mu = \|X\|_p^p.$$

Thus $E^{\mathcal{B}}(\cdot)$ is a contraction on $L^p(\mu)$.

2. A similar statement holds also on Orlicz spaces which were considered in Chapter 6. In fact, if $X \in L^\varphi(\mu)$ and $k_0 = \|X\|_\varphi$ so that $k_0 = \inf\{k > 0 : \int_\Omega \varphi(\frac{|X|}{k})d\mu \leq 1\}$ where $\varphi(\cdot)$ is a Young function and $\mu_{\mathcal{B}}$ is localizable, one can verify that

$$\int_\Omega \varphi\left(\frac{E^{\mathcal{B}}(X)}{k_0}\right)d\mu \leq \int_\Omega E^{\mathcal{B}}(\varphi(\frac{X}{k_0}))d\mu$$

$$= \int_\Omega \varphi(\frac{X}{k_0})d\mu \leq 1 .$$

Hence $\|E^{\mathcal{B}}(X)\|_\varphi \leq k_0$, as desired.

Special properties of the vector integral, considered in Proposition 1, include its modularity condition, of importance, in this work. This follows from the definition of $E^{\mathcal{B}}$-itself. More precisely,

Proposition 4. *If $(\Omega, \Sigma, \mathcal{B}, \mu)$ is as in Proposition 2, and $P^{\mathcal{B}}$ is the conditional measure on $\Sigma \rightarrow L^p(\mathcal{B})$, $p \geq 1$, then it is a modular measure, so that for all $f \in L^p(\mu)$ and $g \in L^\infty(\mathcal{B}, \mu)$,*

$$\int_\Omega f[g]dP^{\mathcal{B}} = g\int_\Omega f \, dP^{\mathcal{B}}, \quad a.e.[\mu], \tag{9}$$

where $[g]$ is the equivalence class of all elements of g. Thus the elements of $L^\infty(\mathcal{B}, \mu)$ act as constants in this integral and $\int_\Omega gdP^{\mathcal{B}} = g$, a.e.

Proof. By Proposition 1, one has

$$E^{\mathcal{B}}(f) = \int_\Omega f \, dP^{\mathcal{B}}, \quad f \in L^p(\mu),$$

and it is an element of $L^p(\mathcal{B}, \mu)$. But by the properties of $E^{\mathcal{B}}(\cdot)$ [or by considering simple functions and approximating the integral as usual] (9) follows and since $E^{\mathcal{B}}(1) = 1$ a.e., the last statement holds. Thus if $L^p(P^{\mathcal{B}})$ is the space of scalar f for which $|f|^p$ is $P^{\mathcal{B}}$-integrable, then $L^p(P^{\mathcal{B}})$ is a module over $L^\infty(\mathcal{B}, \mu)$. $\square$

Remark. Note that in the above discussion, no regularity of $P^{\mathcal{B}}$ is involved and that property is neither available nor needed for this work.

We also recall that if $\mathcal{B}_1 \subset \mathcal{B}_2 \subset \Sigma$ are σ-algebras such that $\mu_{\mathcal{B}_i}$ is localizable, and X is μ-integrable then

$$E^{\mathcal{B}_1}(E^{\mathcal{B}_2}(X)) = E^{\mathcal{B}_2}(E^{\mathcal{B}_1}(X)) = E^{\mathcal{B}_1}(X), a.e.$$

This property may be iterated to get the following product conditional integral formula as a consequence of Propositions 1 and 4:

Corollary 5. *If $\{\Omega, \Sigma, \mathcal{B}_i \subset \mathcal{B}_{i+1}, \mu, i \geq 0\}$ is a measure space with $\mu_{\mathcal{B}_i}$ localizable, and X is μ-integrable, then*

$$\int_\Omega X dP^{\mathcal{B}_0} = \int_\Omega \int_\Omega \cdots \int_\Omega X \, dP^{\mathcal{B}_{n+1}} \, dP^{\mathcal{B}_n} \ldots dP^{\mathcal{B}_0}, \quad a.e. \qquad (10)$$

In particular, if $\mu(\Omega) = 1$ and $\mathcal{B}_0 = \{\emptyset, \Omega\}$, then $P^{\mathcal{B}_0} = \mu$ in (10).

This result motivates a general analysis of product conditional measures, to be studied in the following chapter. Such product integrals, extending (10), are of interest in the existence theory of Markov processes which will be considered in Chapter 9. However, we now establish some general properties of conditional expectations and probabilities which illuminate the structure of these objects.

To include diverse applications of these operators and measures we introduce a class of function spaces that include both the Lebesgue and Orlicz spaces, and show that the conditioning concept extends optimally to this class only. Thus if (Ω, Σ, μ) is a measure space and M is the class of real μ-measurable functions, let $\rho : M \to \bar{\mathbb{R}}^+$ be a *function norm* in the sense that: (i) $\rho(f) = \rho(|f|)$, (ii) $0 \leq f_1 \leq f_2$ a.e. $\implies \rho(f_1) \leq \rho(f_2)$, and (iii) $\rho(\cdot)$ is a positively homogeneous subadditive functional vanishing only on μ-null elements so that it is a norm on M. Identifying the a.e. equal functions as usual, we let $L^\rho(\Sigma, \mu) = \{f \in M : \rho(f) < \infty\}$. Suppose also that $\rho(\cdot)$ has the *Fatou*

property i.e., $0 \le f_n \uparrow f \Rightarrow \rho(f_n) \uparrow \rho(f) \le \infty$. Then one can verify
that $L^\rho(\Sigma, \mu)$ is a Banach space, and if $\rho(f) = \|f\|_\varphi$, the gauge norm
for a Young function $\varphi : \mathbb{R} \to \bar{\mathbb{R}}^+$, one gets the Orlicz spaces $L^\varphi(\Sigma, \mu)$
and then the Lebesgue spaces $L^p(\Sigma, \mu)$ are included when $\varphi(x) = |x|^p$
or $\rho(\cdot) = \|\cdot\|_p$. To avoid some uninteresting pathology, we always
assume that $\rho(\cdot)$ has the finite subset property, i.e., for each set $A \in \Sigma$
with $\rho(\chi_A) > 0$, there is a $B \in \Sigma$, $B \subset A, \mu(B) > 0$ and $\rho(\chi_B) < \infty$.
This is sometimes called the *saturation property* of ρ which is present
in the $L^\varphi(\mu)$ spaces when μ has the finite subset property. We describe
further conditions that are assumed through out.

   An element f in $L^\rho(\mu)$ has *absolutely continuous norm* (a.c.n.) iff for
each $A_n \downarrow \emptyset, A_n \in \Sigma, \rho(f\chi_{A_n}) \downarrow 0$, and ρ is an a.c.n. if every element of
$L^\rho(\mu)$ has a.c.n. In what follows we associate a signed measure ν_f with
each $f \in L^\rho(\mu)$ by setting $\nu_f : A \mapsto \int_A f \, d\mu, A \in \Sigma_0 = \{B \in \Sigma : \mu(B) <$
$\infty\}$. Since Σ_0 is a δ-ring, this definition is useful only if ν_f is σ-additive.
For this purpose, we assume from now on that $L^\rho(\mu) \subset L^1_{loc}(\mu)$, the
space of scalar μ-measurable $g : \Omega \to \mathbb{R}$ such that $g\chi_A$ is μ-integrable
for each $A \in \Sigma_0$, and that the inclusion map is locally bounded. So for
each $A \in \Sigma_0, L^\rho(\mu_A) \subset L^1(\mu_A)$ the inclusion being continuous where
$\mu_A(\cdot) = \mu(A \cap \cdot)$. The function norm $\rho(\cdot)$ has the (strict) localizable
property if (i) $\Sigma_1 = \{A \in \Sigma_0, \rho(\chi_A) < \infty\} \Rightarrow \Sigma = \sigma(\Sigma_1)$ and (ii)
$f \in L^\rho(\mu) \Rightarrow$ the support of f is (strictly) localizable relative to μ on
Σ. Note that if $\rho(\cdot) = N_\varphi(\cdot)$, the gauge norm of an Orlicz space (or
$= \|\cdot\|_p, p \ge 1$ for the Lebesgue case), then these are automatic if μ is
localizable and $\varphi(\cdot)$ is a continuous Young function vanishing only at
0. Let $\rho'(\cdot)$ be defined by

$$\rho'(f) = \sup\{|\int_\Omega fg d\mu| : \quad \rho(g) \le 1\}. \tag{11}$$

Then $\rho'(\cdot)$ is a function norm. Under the Fatou property of $\rho(\cdot)$ it
follows that on iterating $\rho'' = (\rho')' = \rho$ itself. It should be remarked
that the definition of $\rho(\cdot)$ can be taken so obliquely that it need not
relate well with the underlying measure space, unlike the Orlicz and
Lebesgue cases. This is the reason why we hasten to impose several
of the above conditions to remedy this defect. Here $\rho'(\cdot)$ is called the
(first) *associate norm* of ρ. (For an account of these spaces, one may
consult Zaanen, 1967, Chap. 15. Here we merely state the desired

properties which are simple for the Orlicz and Lebesgue spaces.) From (11) one has for $A \in \Sigma_0$,

$$\rho'(\chi_A) = \sup\{|\int_\Omega \chi_A g \, d\mu| : \rho(g) \le 1\}$$

$$= \sup\{\|\tau_A g\|_1 : \rho(g\chi_A) \le 1\}, \quad \tau : L^\rho(\mu_A) \to L^1(\mu_A),$$

$$\le \|\tau_A\| < \infty, \text{ since the embedding is locally bounded.}$$

Replacing ρ' by ρ'' shows that $\rho'(\chi_A) < \infty \Rightarrow \rho''(\chi_A) = \rho(\chi_A) < \infty$ so that $\Sigma_0 = \{A \in \Sigma : \rho'(\chi_A) < \infty\}$ also holds.

With these preliminaries, we can establish the following:

Theorem 6. *Let $L^\rho(\Sigma)$ be the space defined above on (Ω, Σ, μ) where ρ has the localizable property in addition. If $\mathcal{B} \subset \Sigma$ is a μ-completed σ-algebra such that $\mu_\mathcal{B}$ is localizable, and $L^\rho(\mathcal{B})$ is the subspace of $L^\rho(\Sigma)$ consisting of $\mathcal{B}$-measurable functions, then there is a contractive idempotent operator $E^\mathcal{B}$ on $L^\rho(\Sigma)$ into $L^\rho(\mathcal{B})$, (i.e., $E^\mathcal{B} \in B(L^\rho(\Sigma), L^\rho(\mathcal{B}))$ and $E^\mathcal{B} E^\mathcal{B} = E^\mathcal{B}$ where $B(\mathcal{X}, \mathcal{Y})$ is the space of bounded linear operators between the normed vector spaces $\mathcal{X}$ and $\mathcal{Y}$) such that*

$$\int_A f \, d\mu = \int_A E^\mathcal{B}(f) d\mu_\mathcal{B}, f \in L^\rho(\Sigma), \ A \in \mathcal{B}, \ \mu(A) < \infty, \qquad (12)$$

and for any $\mathcal{B} \subset \mathcal{B}_1 \subset \Sigma$, σ-algebras as above, one has
  (i) $E^\mathcal{B}(f E^\mathcal{B}(g)) = E^\mathcal{B}(f) E^\mathcal{B}(g)$, a.e., $f, g \in L^\rho(\Sigma) \cap L^\infty(\Sigma)$,
  (ii) $E^\mathcal{B}(E^{\mathcal{B}_1}(f)) = E^{\mathcal{B}_1}(E^\mathcal{B}(f)) = E^\mathcal{B}(f)$, a.e., $f \in L^\rho(\Sigma)$.
  In other words, $E^\mathcal{B}$ is a "generalized" conditional expectation on $L^\rho(\Sigma)$ into $L^\rho(\mathcal{B})$.

Proof. Let $\mathcal{B}_0 = \{A \in \mathcal{B} : \rho(\chi_A) < \infty\}$ and $\tilde{\mathcal{B}}$ be the σ-ring generated by $\mathcal{B}_0$ and completed for μ. Then $L^\rho(\mathcal{B}) = L^\rho(\tilde{\mathcal{B}})$. Indeed since both are vector lattices, $\rho(f) = \rho(|f|)$, and $L^\rho(\mathcal{B}) \supset L^\rho(\tilde{\mathcal{B}})$, it suffices to consider, for the opposite inclusion, $0 \le f \in L^\rho(\mathcal{B})$ and show that $f \in L^\rho(\tilde{\mathcal{B}})$. But by the structure theorem there exist $0 \le f_n \uparrow f$ pointwise where $f_n = \sum_{k=1}^n a_k^n \chi_{A_k^n}$ and hence $f_n \in L^\rho(\mathcal{B})$, so that $A_k^n \in \mathcal{B}_0$ or $f_n \in L^\rho(\tilde{\mathcal{B}})$. By the Fatou property of $\rho, \rho(f_n) \uparrow \rho(f) < \infty$. Hence $f \in L^\rho(\tilde{\mathcal{B}})$ so that $L^\rho(\mathcal{B}) = L^\rho(\tilde{\mathcal{B}})$. Also $L^\rho(\Sigma)$ is continuously embedded in $L^1_{loc}(\mu)$, so $0 \le f \in L^\rho(\Sigma)$ is locally integrable. If $\nu_f : A \mapsto \int_A f \, d\mu$, it is σ-additive on Σ and $\mu_\mathcal{B}$-continuous. Let $\tilde{\nu}_f = \nu_f | \mathcal{B}$. We now define

E^B through $\tilde{\nu}_f$ and show that it has the desired properties. Since $\tilde{\mathcal{B}}$ need not be a σ-algebra, we shall reduce the problem for a σ-subalgebra of $\tilde{\mathcal{B}}$.

For this, note that if S_0 is the support of $\tilde{\nu}_f$, then $S_0 \in \tilde{\mathcal{B}}$. [Recall that S_0 is the smallest set such that on every measurable subset of S_0^c (the complement), $\tilde{\nu}_f$ vanishes.] Since ρ is a localizable norm and $\mu_{\mathcal{B}}$ is localizable, $S_0 \in \mathcal{B}$. Moreover, the associate norms ρ', ρ'' exist and (Fatou's property) $\rho'' = \rho$, by the properties recalled earlier from Zaanen's book. Hence

$$\infty > \rho(f) = \rho''(f) = \sup\{|\int_\Omega g \, d\nu_f| : \rho'(g) \le 1\}$$

$$= \sup\{|\int_\Omega g \, f d\mu| : \rho'(g) \le 1\}. \qquad (13)$$

Since $L^{\rho'}(\mathcal{B}) \subset L^{\rho'}(\Sigma)$, and $\rho(\nu_f) = \rho(f)$, consider

$$\rho''(\nu_f) \ge \rho''(\tilde{\nu}_f) = \sup\{|\int_\Omega g \, d\tilde{\nu}_f| : \rho'(g) \le 1, \, g \in L^{\rho'}(\tilde{\mathcal{B}})\}$$

$$= \lim_{n\to\infty} \int_\Omega g_n \, d\tilde{\nu}_f, \rho'(g_n) \le 1, 0 \le g_n \in L^{\rho'}(\tilde{\mathcal{B}}). \qquad (14)$$

If $S_n = \text{supp}\,(g_n)$, then $S_n \in \tilde{\mathcal{B}}$ and $\tilde{S} = \overset{\infty}{\underset{n=1}{\cup}} S_n \in \tilde{\mathcal{B}}$, since the latter is a σ-ring. Also $S_0 \supset S_n$ so that $\tilde{S} \subseteq S_0$. If there is strict inclusion here, since $S_0 - \tilde{S} \in \mathcal{B}$, there exists some $A \in \mathcal{B}_0, \nu_f(A) > 0$ and $A \subset S_0 - \tilde{S}$. Hence by the saturated property of ρ and ρ', there is a $0 < g_0 \in L^{\rho'}(\tilde{\mathcal{B}})$ such that $\alpha = \int_A g_0 d\tilde{\nu}_f > 0$, and $A \cap \tilde{S} = \emptyset$. So

$$\rho''(\tilde{\nu}_f) = \sup\{\int_\Omega g \, d\tilde{\nu}_f : \rho'(g) \le 1, 0 \le g \in L^{\rho'}(\tilde{\mathcal{B}})\}$$

$$\ge \sup\{\int_A g \, d\tilde{\nu}_f + \int_{\tilde{S}} d \, \tilde{\nu}_f : 0 \le g \in L^\rho(\tilde{\mathcal{B}}), \rho'(g) \le 1\}$$

$$\ge \alpha + \lim_{n\to P} \int_{\tilde{S}} g_n d\tilde{\nu}_f = \alpha + \rho''(\tilde{\nu}_f), \text{ by (14)}.$$

This contradiction shows that A is μ-null and $S_0 = \tilde{S}$ a.e., so $S_0 \in \tilde{\mathcal{B}}$.

Next consider $\tilde{\mathcal{B}}(S_0)$, the trace of $\tilde{\mathcal{B}}$ on S_0 and it is a σ-algebra. By the hypothesis on ρ (excluding the trivial case that $L^\rho(\mathcal{B}) = \{0\}$),

$\mu_1 = \mu|\tilde{\mathcal{B}}(S_0)$ is localizable and $\tilde{\nu}_f << \mu_1$ on $\tilde{\mathcal{B}}(S_0)$. Hence by the Radon-Nikodým theorem, there is a μ_1-unique $\tilde{f} \in L^1_{loc}(\tilde{\mathcal{B}}(S_0))$, such that $\tilde{\nu}_f(A) = \int_A \tilde{f} d\mu_1 : A \in \tilde{\mathcal{B}}(S_0)$. Since by (14) $\rho(\tilde{\nu}_f) < \infty$, ($\rho = \rho''$ here) it follows that $\rho(\tilde{f}) < \infty$ and then $\tilde{f} \in L^\rho(\tilde{\mathcal{B}})$. By the μ_1-(hence $\mu_{\tilde{\mathcal{B}}}$-) uniqueness of $\tilde{f}$, the operator $E^{\mathcal{B}} : f \mapsto \tilde{f}$ is well-defined, linear, and positivity preserving. (In case $L^\rho(\mathcal{B}) = \{0\}$ set $E^{\mathcal{B}} = 0$.) So $E^{\mathcal{B}} : L^\rho(\Sigma) \to L^\rho(\mathcal{B})$, and using $\rho = \rho''$ again,

$$\infty > \rho(f) = \rho''(f) = \rho''(\nu_f)$$
$$= \sup\{|\int_\Omega g d\nu_f| : \rho'(g) \le 1, \ g \in L^{\rho'}(\Sigma)\}$$
$$\ge \sup\{|\int_\Omega g d\nu_f| : \rho'(g) \le 1, g \in L^{\rho'}(\mathcal{B})\}$$
$$= \sup\{|\int_\Omega g \, E^{\mathcal{B}}(f) d\mu_{\mathcal{B}}| : \rho'(g) \le 1, \ g \in L^{\rho'}(\mathcal{B})\}.$$

Hence $E^{\mathcal{B}} \in B(L^\rho(\Sigma), \ L^\rho(\mathcal{B}))$, with operator bound 1, establishing the first (main) part.

Finally, (i) and (ii) should be verified for only locally integrable bounded f, g. But they follow immediately from Proposition 2.2.1, and the same argument extends to the case here. $\square$

Since $\rho(\cdot) = N_\varphi(\cdot)$, the gauge norm, the result includes Orlicz spaces. It was seen there that $N_\varphi(E^{\mathcal{B}}(f)) \le N_\varphi(f)$, as a consequence of the Jensen inequality. This however need not be true for the $\rho(\cdot)$ norm without this additional property, as the following trivial example shows. [However, $\rho'(\cdot)$ is not $N_\psi(\cdot)$, but it is another equivalent (called the Orlicz) norm in these spaces; see Rao and Ren, 1991.]

Example 7. Let $\Omega = \{1,2\}, \Sigma =$ power set of $\Omega, \mathcal{B} = \{\phi, \Omega\}$ and $\mu : \Sigma \to \mathbb{R}^+$ be defined as $\mu(\{1\}) = \frac{1}{3}, \mu(\{2\}) = \frac{2}{3}$. Let $F = \{1\}$, and $G = \{2\}$, and for any $f : \Omega \to \mathbb{R}$, let $\rho(\cdot)$ be given by

$$\rho(f) = \int_F |f| d\mu + [\int_G |f|^2 d\mu]^{\frac{1}{2}}.$$

One finds that $E^{\mathcal{B}}(f) = \frac{1}{2}\{f(1) + f(2)\}$. Taking $f = 3\chi_F + 2\chi_G, f \in L^\rho(\mu)$, it is verified that

$$\rho(f) = 1 + 2(\frac{2}{3})^{\frac{1}{2}} < \rho(E^{\mathcal{B}}(f)) = \frac{7}{3}(1 + (\frac{2}{3})^{\frac{1}{2}}).$$

Thus $E^B(\cdot)$ is not a contraction on $L^\rho(\Sigma)$ into $L^\rho(\Sigma)$.

To avoid such pathological behavior of ρ, we impose the Jensen condition in the following precise form to be used in applications.

Definition 8. A function norm $\rho(\cdot)$ has the *Jensen* or (J-) *property* if for each finite disjoint collection $\mathcal{A} = \{A_1, \ldots, A_n\}$ of sets from $\Sigma, 0 < \mu(A_i) < \infty, i = 1, \ldots, n$, and $f \in L^\rho(\Sigma)$, the averaged function $\tilde{f}$ given by

$$\tilde{f} = \sum_{i=1}^{n} \left(\frac{1}{\mu(A_i)} \int_{A_i} f d\mu \right) \chi_{A_i}, \tag{15}$$

satisfies $\rho(\tilde{f}) \le \rho(f)$.

This condition makes $\rho(\cdot)$ well-behaved and several of the properties assumed on ρ for the above theorem are automatic as shown below. More importantly, the difficulty appearing in Example 7 is eliminated by the (J)-property and $E^B : L^\rho(\Sigma) \to L^\rho(\Sigma)$ is a contraction. This condition and the resulting analysis on the $L^\rho(\Sigma)$-spaces has been studied by Gretsky (1968). The following result is adapted from his work with minor alterations.

Proposition 9. *Let $\rho(\cdot)$ be a function norm with property (J). Then the following statements hold:*

(i) *For any A, B from Σ of finite positive μ-measure, one has*

$$\rho(\chi_A) \ge \frac{\mu(A)}{\mu(A \cup B)} \rho(\chi_B),$$

(ii) $0 < \mu(A) < \infty \Rightarrow 0 < \rho(\chi_A) < \infty,$

(iii) $\mu(A) < \infty \Rightarrow \mu(A) = \rho(\chi_A)\rho'(\chi_A),$ *where $\rho'(\cdot)$ is the associate norm of $\rho(\cdot)$,*

(iv) $\{A \in \Sigma : \mu(A) < \infty, \rho(\chi_A) < \infty\} = \{A \in \Sigma : \mu(A) < \infty, \rho'(\chi_A) < \infty\},$

(v) $\rho'(\cdot)$ *has property (J).*

Proof. (i) Let $A_1 = A \cup B$ and $f = \chi_A$. If $\mathcal{A} = \{A_1\}$ is the trivial one element "partition," then by definition

$$\tilde{f} = \frac{1}{\mu(A_1)} \left(\int_{A_1} f d\mu \right) \chi_{A_1} = \frac{\mu(A)}{\mu(A_1)} \chi_{A_1}.$$

Hence by condition (J) one gets

$$\rho(\chi_A) = \rho(f) \geq \rho(\tilde{f}) = \frac{\mu(A)}{\mu(A_1)}\rho(\chi_{A_1}), \text{ since } \rho(\cdot) \text{ is a norm,}$$

$$\geq \frac{\mu(A)}{\mu(A \cup B)}\rho(\chi_B) \text{ since } B \subset A_1 \text{ and } \rho \text{ is monotone.}$$

Thus (i) holds for any sets of finite positive measure.

(ii) Since $\rho(\chi_A) = 0$ iff $\mu(A) = 0$, let $0 < \mu(A_0) < \infty$ so that $\rho(\chi_{A_0}) > 0$. By hypothesis of saturatedness, there is a $B_0 \in \Sigma$, such that $\rho(\chi_{B_0}) < \infty$ and $0 < \mu(B_0) < \infty$. Hence by (i), taking A for B_0 and B for A_0 there, one has

$$\infty > \rho(\chi_{B_0}) \geq \frac{\mu(B_0)}{\mu(A_0 \cup B_0)}\rho(\chi_{A_0}),$$

so that $\rho(\chi_{A_0}) < \infty$, which is (ii).

(iii) Again we only need to consider $A \in \Sigma, 0 < \mu(A) < \infty$. If $f = \sum_{i=1}^{n} a_i\chi_{A_i} \in L^\rho(\Sigma)$, [so $\mu(\bigcup_{i=1}^{n} A_i) < \infty$, see (ii)] consider $B = A \cap \bigcup_{i=1}^{n} A_i$ and $\mathcal{A} = \{B\}$, a one element "partition." Letting $\tilde{f}$ be the averaged element of $f \in L^\rho(\Sigma)$ based on $\mathcal{A}$, one has

$$\int_A \tilde{f}d\mu = \int_B fd\mu, \text{ since } \tilde{f} = \frac{1}{\mu(B)}(\int_B fd\mu)\chi_B, \tag{16}$$

and

$$\int_A fd\mu = \int_{A\cap(\bigcup_{i=1}^{n} A_i)} fd\mu = \int_B fd\mu$$

$$= \int_B \tilde{f}d\mu, \text{ by (16).} \tag{17}$$

Hence by definition of ρ' and the fact that $\rho(\tilde{f}) \leq \rho(f)$, one has the following simplication.

$$\rho'(\chi_A) = \sup\{|\int_\Omega \chi_A f d\mu| : \ \rho(f) \leq 1, \ f \text{ simple}\}$$

$$= \sup\{|\int_A \tilde{f} d\mu| : \ \rho(f) \leq 1, \ f \text{ simple and supported in } A\}$$

$$= \sup\{\int_A k\chi_B d\mu : \ \rho(k\chi_B) \leq 1, \ B \subset A, B \in \Sigma\},$$

since $\tilde{f}$ is constant on A and by (17)

one can restrict it to step functions $k\chi_B$,

$$= \sup\{k\mu(A) : \ \rho(k\chi_A) \leq 1\}, \text{ since } \widetilde{k\chi_B} = k\tilde{\chi}_B \text{ and use (17)},$$

$$= \mu(A) \sup\{k : k \leq \frac{1}{\rho(\chi_A)}\}$$

$$= \mu(A)\rho(\chi_A)^{-1}.$$

This means $\rho(\chi_A)\rho'(\chi_A = \mu(A)$ which also holds for $\mu(A) = 0$. Thus (iii) follows.

(iv) This is a direct consequence of (iii).

(v) First note that for any $0 \leq f \in L^\rho(\Sigma), 0 \leq g \in L^{\rho'}(\Sigma)$ and $\mathcal{A} = \{A_1, \ldots, A_n\}, 0 < \mu(A_i) < \infty$, disjoint, one has the following relation between the averaged functions $\tilde{f}$ and $\tilde{g}$ derived from f and g respectively:

$$\int_\Omega f\tilde{g} d\mu = \int_\Omega f \left(\sum_{i=1}^n \left(\int_{A_i} g \, d\mu\right) \frac{\chi_{A_i}}{\mu(A_i)}\right) d\mu$$

$$= \sum_{i=1}^n \left(\int_{A_i} g \, d\mu\right) \left(\int_\Omega f \chi_{A_i} \, d\mu\right) \frac{1}{\mu(A_i)}$$

$$= \int_\Omega \left(\sum_{i=1}^n \frac{\chi_{A_i}}{\mu(A_i)} \int_{A_i} g \, d\mu\right) \left(\sum_{i=1}^n \int_{A_i} f \, d\mu\right) \frac{\chi_{A_i}}{\mu(A_i)} d\mu$$

$$= \int_\Omega \tilde{f}\tilde{g} d\mu$$

$$= \int_\Omega g\tilde{f} d\mu, \text{ by symmetry.} \qquad (18)$$

Hence

$$\rho'(\tilde{g}) = \sup\{\int_\Omega \tilde{g}h \, d\mu : 0 \le h \in L^\rho(\Sigma), \rho(h) \le 1\}$$

$$= \sup\{\int_\Omega \tilde{g}\tilde{h}d\mu : \ 0 \le h \in L^\rho(\Sigma), \rho(h) \le 1\}, \text{ by (18)},$$

$$\le \sup\{\int_\Omega \tilde{g}\tilde{h}d\mu : \ 0 \le h \in L^\rho(\Sigma), \rho(\tilde{h}) \le 1\}, (J)\text{-property of } \rho,$$

$$\le \sup\{\int_\Omega fh \, d\mu : \ 0 \le h \in L^\rho(\Sigma), \rho(h) \le 1\},$$

since $\tilde{h} \in L^\rho(\Sigma)$ and is in its unit ball,

$$= \rho'(h).$$

Hence (v) holds, and thus all properties are established. □

Remarks. 1. In the case of Orlicz spaces with $\rho(\cdot) = N_\varphi(\cdot)$, the adjoint norm $\rho'(\cdot) = \|\cdot\|_\psi$ is not the gauge norm $N_\psi(\cdot)$ in general. For the case of Lebesgue spaces, one has $\rho(\cdot) = \|\cdot\|_p$ and $\rho'(\cdot) = \|\cdot\|_q$, with $\frac{1}{p} + \frac{1}{q} = 1$, $p \ge 1$. In this context, if one uses $\rho(\cdot) = N_\varphi(\cdot)$, and $\rho'(\cdot) = N_\psi(\cdot)$ then in lieu of (iii) the following will be true:

$$C_1 \rho(\chi_E)\rho'(\chi_E) \le \mu(E) \le C_2 \rho(\chi_E)\rho'(\chi_E)$$

for some absolute constants C_i with $0 < C_1, C_2 < \infty$.

2. If $\rho(\cdot)$ does not satisfy the condition (J), then (iii) is not true. For example, let (Ω, Σ, μ) be the Lebesgue unit interval and $\rho_0(f) = \int_\Omega |f(x)|\frac{d\mu(x)}{x}$. If $\rho(\cdot) = \rho_0'(\cdot)$, which has the Fatou property we have $\rho' = \rho_0'' = \rho_0$. However $\rho'(\chi_\Omega) = +\infty$, and

$$\rho(\chi_\Omega) = \rho''(\chi_\Omega) = \sup\{\int_\Omega |f(x)|d\mu(x) : \ \rho_0(f) \le 1\} = 1,$$

since $|f(x)/x| \ge |f(x)|$ on Ω. Hence

$$1 = \mu(\Omega) \ne \rho'(\chi_\Omega)\rho(\chi_\Omega), \ (= +\infty).$$

Here $\rho(\cdot)[= \rho_0'(\cdot)]$ does not have the (J)-property, and $L^\rho(\Sigma), L^{\rho'}(\Sigma)$ are *not* continuously imbeddable in $L^1_{\text{loc}}(\mu)$. Thus the preceding conditions are essentially optimal for our work.

3. It is not hard to verify that if $\rho(E^\mathcal{B}(f)) \le \rho(f)$ for all $f \in L^\rho(\Sigma)$ ($\subset L^1_{\text{loc}}(\mu)$ continuously) and $\mathcal{B} \subset \Sigma$ is any σ-algebra then $\rho(\cdot)$

satisfies condition (J), as observed by Gretsky in 1973. Thus this is a desirable property to demand of function spaces that appear in analysis with conditional measures and operators. *From now on in this book a function norm $\rho(\cdot)$ of the $L^\rho(\mu)$-spaces will be assumed to possess the (J)-property.*

Now that the basic frame work for conditional measures is set up, we turn to some characterizations of $E^{\mathcal{B}}(\cdot)$ and $P^{\mathcal{B}}(\cdot)$ on such function spaces, and consider their dependence on the conditioning σ-subalgebras $\mathcal{B}$ in the next section.

7.3 Functional characterizations of conditioning

Since $E^{\mathcal{B}}(\cdot)$ and $P^{\mathcal{B}}(\cdot)$ are functional operators and measures unless $\mathcal{B} = \{\phi, \Omega\}$, it is important to learn as clearly and completely as possible about their structure. Let us first discuss these objects as the σ-algebras $\mathcal{B}$ expand or contract, or move generally, as $\{\mathcal{B}_\alpha, \alpha \in I\}$ has a limit in a suitable sense. This has a reasonably simple answer in the context of Orlicz spaces on probability measures. Its extension for $L^\rho(\mu)$-spaces will be omitted. We thus present:

Theorem 1. *Let (Ω, Σ, μ) be a probability space, $\{\mathcal{B}_\alpha, \mathcal{B}, \alpha \in I\}$ be a net of σ-subalgebras of Σ where I is a directed (upward) set, and $\mathcal{B} \doteq \sigma(\cup_\alpha \mathcal{B}_\alpha)$ which is completed for μ. Let φ be a continuous Young function, $\varphi(x) = 0$ iff $x = 0$, and $L^\varphi(\Sigma)$ be the Orlicz space on (Ω, Σ, μ). For $f \in L^\varphi(\Sigma)$, let $g_\alpha = E^{\mathcal{B}_\alpha}(f)$ and $g = E^{\mathcal{B}}(g)$. Then $g_\alpha \to g$ in φ-mean in the sense that $\int_\Omega \varphi(\frac{|g_\alpha - g|}{\beta})d\mu \to 0$ for some $\beta \geq 1$ as $\alpha \uparrow$, iff $g_\alpha \to g$ in μ-measure. If either f is bounded or φ is Δ_2 (i.e. $\varphi(2x) \leq C\varphi(x), x \geq x_0, C > 0$) then φ-mean convergence can be replaced by strong or norm convergence here.*

Proof. First let $I = \mathbb{N}$, the natural numbers. Let $N_\varphi(\cdot)$ be the gauge norm in $L^\varphi(\Sigma)$ so that

$$N_\varphi(f) = \inf\{k > 0 : \int_\Omega \varphi(\frac{|f|}{k})d\mu \leq 1\}, \tag{1}$$

and $N_\varphi(E^{\mathcal{B}_0}(f)) \leq N_\varphi(f)$ for any σ-algebra $\mathcal{B}_0 \subset \Sigma$, since $N_\varphi(\cdot)$ has the (J)-property. Also if $h_\alpha = \varphi(E^{\mathcal{B}_\alpha}(f))$, (taking $N_\varphi(f) = 1$ for convenience, excluding the trivial case that $f = 0$ a.e.) one gets $\{h_\alpha, \alpha \in I\}$

to be a uniformly integrable class, since

$$0 \le h_\alpha = \varphi(E^{\mathcal{B}_\alpha}(f)) \le E^{\mathcal{B}_\alpha}(\varphi(f)), \text{ a.e., by Prop. 2.2,} \qquad (2)$$

and then

$$\int_{[h_\alpha > \lambda]} h_\alpha \, d\mu \le \int_{[h_\alpha \ge \lambda]} E^{\mathcal{B}_\alpha}(\varphi(f)) d\mu, \text{ by (2),}$$

$$= \int_{[h_\alpha \ge \lambda]} \varphi(f) d\mu, \text{ since } [h_\alpha \ge \alpha] \in \mathcal{B}_\alpha,$$

$$\to 0, \text{ as } \alpha \uparrow, \qquad (3)$$

since $\mu([h_\alpha \ge \lambda]) \le \frac{1}{\lambda} \int_\Omega h_\alpha \, d\mu \le \frac{1}{\lambda} \int_\Omega \varphi(f) d\mu \to 0$ as $\lambda \to \infty$, uniformly in α, by (2), and the fact that $\varphi(f) \in L^1(\mu)$. Thus $\{\varphi(g_\alpha), \alpha \in I\}$ is a uniformly integrable set in $L^1(\mu)$ where $g_\alpha = E^{\mathcal{B}_\alpha}(f)$.

Suppose $g_n \to g$ is μ-measure. Since $f \in L^\varphi(\mu)$, $N_\varphi(f) = 1, \varphi(f) \in L^1(\mu)$, and (2) implies that $\{\varphi(g_n), n \ge 1\}$ is a bounded set in $L^1(\mu)$. But by a classical theorem, there exists a subsequence $g_{n_k} \to g$, a.e., and hence (by the continuity of φ), $\varphi(g_{n_k}) \to \varphi(g)$, a.e. It follows that (by Fatou's lemma) $\varphi(f) \in L^1(\mu)$ so that $g \in L^\varphi(\mu)$. Replacing f by $\frac{f}{\beta}$ for a larger $\beta \ge 1$, we may assume that $N_\varphi(f) = \frac{1}{2}$ so that $N_\varphi(g_n) \le \frac{1}{2}$ and $N_\varphi(g) \le \frac{1}{2}$. Now for any $\varepsilon > 0$ one can choose n_0 such that $n \ge n_0 \Rightarrow$ the set $F_n = \{\omega : |g_n - g|(\omega) > \varepsilon\}$ satisfies $\mu(F_n) < \varepsilon$ (by the convergence in measure condition). Hence

$$\int_\Omega \varphi(g_n - g) d\mu = \int_{F_n} \varphi\left(\frac{2(g_n - g)}{2}\right) d\mu + \int_{\Omega - F_n} \varphi(g_n - g) d\mu$$

$$\le \frac{1}{2}\left(\int_{F_n} \varphi(2g_n) d\mu + \int_{F_n} \varphi(2g) d\mu\right) + \varphi(\varepsilon)\mu(\Omega). \qquad (4)$$

By the uniform integrability of $\{\varphi(2g_n), n \ge 1\}$ shown above, since $\mu(F_n) < \varepsilon$, each of the first two terms on the right of (4) can be made $< \frac{\eta}{2}$ if n is large, so that the right side $\le \varphi(\varepsilon)\mu(\Omega) + \eta$. This implies that $g_n \to g$ in $\varphi(\cdot)$-mean, since $\eta > 0$ and $\varepsilon > 0$ are arbitrary.

Conversely, if $g_n \to g$ in φ-mean, then for an $\varepsilon > 0$, by Markov's inequality one has

$$\mu(|g_n - g| > \varepsilon) \le \mu(\varphi(|g_n - g|) \ge \varphi(\varepsilon)) \le \frac{1}{\varphi(\varepsilon)} \int_\Omega \varphi(g_n - g) d\mu.$$

The right side $\to 0$ as $n \to \infty$ by hypothesis so that $g_n \to g$ in measure.

Now suppose that I is merely a directed set, which need not be countable. Suppose the assertion is false in this case. To derive a contradiction, let $d(\cdot, \cdot)$ be the metric of convergence in measure, so that $d(f, g) = \int_\Omega \frac{|f-g|}{1+|f-g|} d\mu$, and $f_n \to f$ in measure iff $d(f_n, f) \to 0$, as $n \to \infty$. By supposition $\int_\Omega \varphi(\frac{|g_n - g|}{c}) d\mu \not\to 0$. Hence there is an $\varepsilon > 0$, and for each $\alpha \in I$ there is a $\gamma_\alpha \geq \alpha$ in I such that $\int_\Omega \varphi(\frac{|g_{\gamma_\alpha} - g|}{c}) d\mu \geq \varepsilon, (*)$. The new sequence $\{g_{\gamma_\alpha}, \gamma_\alpha \in I\}$ still converges in measure to g. For each $n \geq 1$, if we consider the sphere $B_n = \{h : d(g, h) \leq \frac{1}{n}\}$ of radius $\frac{1}{n}$, then there is some g_{γ_n} of our subnet in B_n. So $g_{\gamma_n} \to g$ in measure. But then by the special case considered above, applied to the sequence $\{g_{\gamma_n}, n \geq 1\}$, we conclude that $g_{\gamma_n} \to g$ in φ-mean. This is the desired contradiction of $(*)$, and hence the result must hold as stated. The converse is obtained at once by use of the Markov inequality as before.

Finally if φ is Δ_2 then it is well-known that the φ-mean and norm convergence are equivalent (see for instance, Rao and Ren, 1991, p. 83). Also if f is bounded, then the g_n and g are also bounded so that they all belong to the subspace M^φ of $L^\varphi(\mu)$ determined by bounded functions in which the above equivalence is known to hold. This proves all the assertions. $\square$

The result of de la Vallée Poussin's on uniform integrability in $L^1(\mu)$ has the following interesting consequence for statistical applications.

Proposition 2. *Let φ be a strictly convex Young function and $L^\varphi(\Sigma)$ be the Orlicz space on a measure space (Ω, Σ, μ). If $f, g \in L^\varphi(\Sigma)$ then $E(f|g)$ and $E(g|f)$ exist. Now if $E(f|g) = g$ and $E(g|f) = f$ a.e., then $f = g$, a.e. If $\mu(\Omega) < \infty$, and $f, g \in L^1(\mu)$ satisfying the same pair of conditional equations, then also the conclusion that $f = g$, a.e., holds, although $\varphi(x) = |x|$ is not strictly convex.*

Proof. Since φ is a Young function $\varphi(0) = 0$ and being strictly convex vanishes only at that point. Further, by Markov's inequality,

$$\mu\{\omega : |f(\omega)| > \frac{1}{n}\} = \mu\{\omega : \varphi(\frac{f}{\alpha}(\omega)) > \varphi(\frac{1}{\alpha n})\}$$

$$\leq \frac{1}{\varphi(\frac{1}{\alpha n})} \int_\Omega \varphi(\frac{f}{\alpha}) d\mu < \infty.$$

Hence $\{\omega : |f(\omega)| > 0\}$ is σ-finite and the same is true of g. Thus the set $\{\omega : |f(\omega)| > 0, |g(\omega)| > 0\}$ is σ-finite. So for this proof μ can be assumed σ-finite. Since $L^\varphi(\mu) \subset L^1_{\text{loc}}(\mu)$, with μ σ-finite the desired conditional expectations exist. Now for the first part, using Proposition 2.2, we get

$$\varphi(g) = \varphi(E(f|g)) \leq E(\varphi(f)|g) \quad \text{a.e.,} \tag{5}$$

and

$$\varphi(f) = \varphi(E(g|f)) \leq E(\varphi(g)|f) \quad \text{a.e.,} \tag{6}$$

with strict inequality on a set of positive measure unless f is $\sigma(g)$ and g is $\sigma(f)$ measurable. Hence integrating (5) and (6) there results

$$\int_\Omega \varphi(g)d\mu < \int_\Omega E(\varphi(f)|g)d\mu = \int_\Omega \varphi(f)d\mu < \int_\Omega E(\varphi(g)|f)d\mu$$
$$= \int_\Omega \varphi(g)d\mu. \tag{7}$$

This is impossible, unless there is equality a.e. in (5) and (6) which implies $f = g$ a.e.

For the last part, $\mu(\Omega) < \infty$ so that the finite set $\{f, g\}$ is uniformly integrable. Hence by the de la Vallée Poussin theorem, there exists a φ_0 which may be assumed to be a strictly convex Young function ($\frac{\varphi_0(x)}{x} \uparrow \infty$ as $x \uparrow \infty$ also) such that $f, g \in L^{\varphi_0}(\mu)$. But then the work of the preceding paragraph is immediately applicable, and so $f = g$, a.e., holds again. $\square$

We now turn to the characterization question which is the main purpose of this section. First let us restate the existence theorem for the conditional operators in an abstract form for reference as follows:

Theorem 3. *Let* (Ω, Σ, μ) *be a measure space* $\mathcal{B} \subset \Sigma$, *a* σ-*algebra such that* $\mu_{\mathcal{B}}$ *is localizable, and* $L^\rho(\Sigma)$ *be the function space introduced in the last section with the function norm having the* (J)-*property. Then the conditional expectation operator* $Q = E^{\mathcal{B}} : L^\rho(\Sigma) \to L^\rho(\Sigma)$ *exists as a positive linear contraction such that* $Q^2 = Q, Q(L^\rho(\Sigma)) = L^\rho(\mathcal{B}), Q1 = 1$, *a.e., and* $Q(L^\rho \cap L^\infty(\Sigma)) \subset L^\rho \cap L^\infty(\Sigma)$. *If* $\mathcal{B}_1 \subset \mathcal{B}_2 \subset \Sigma$ *are* σ-*algebras such that* $\mu_{\mathcal{B}_i}$ *is localizable,* $Q_i = E^{\mathcal{B}_i}, i = 1, 2$, *then one has:*

(i) (*commutativity*)

$$Q_1 Q_2 = Q_2 Q_1 = Q_1 \tag{8}$$

(ii) (averaging) $f, g \in L^\rho(\Sigma)$ *with f or g bounded $\Rightarrow$*

$$Q(fQg) = (Qf)(Qg), \text{ a.e.} \tag{9}$$

(iii) (Šidák's identity) $f, g \in L^\rho(\Sigma) \Rightarrow (\vee = \max)$

$$Q((Qf) \vee (Qg)) = (Qf) \vee (Qg), \text{ a.e.} \tag{10}$$

Since for every $\mathcal{B} \subset \Sigma$ with $\mu_{\mathcal{B}}$ localizable, there is a projection operator (namely conditional expectation) satisfying (8) - (10) in $L^\rho(\Sigma)$, it is desirable to characterize those operators which have the above properties but which are (related to) conditional expectations. This will give us a better understanding of the structure of these operators on $L^\rho(\Sigma)$. The first problem here is to describe precisely the range of such an operator and show it is the subspace $L^\rho(\mathcal{B})$ for some σ-algebra $\mathcal{B} \subset \Sigma$, called a *measurable subspace*, and then show that every such projection is essentially a conditional expectation. Thus the converse direction is nontrivial, reflecting the intricate aspect of conditioning and of its functional operational character.

The following concept of a weak unit plays a key role:

Definition 4. Let $\mathcal{M} \subset L^\rho(\Sigma)$ be a subspace. Then a set $\Omega_0 (\in \Sigma)$ is termed the *support* of $\mathcal{M}$ if $f \in \mathcal{M} \Rightarrow \text{supp}(f) \subset \Omega_0$ and $\Omega_1 \in \Sigma$ having the same property implies $\Omega_0 \subset \Omega_1$, except for a μ-null set. A family $F = \{f_\alpha, \alpha \in I\} \subset \mathcal{M} \cap L^\infty(\Sigma)$ is called a *generalized weak unit* (g.w.u.) for $\mathcal{M}$ if $\text{supp}(f_\alpha) \cap \text{supp}(f_{\alpha'})$ is a μ-null set for $\alpha \neq \alpha', 0 < \mu(\text{supp}(f)) < \infty$, and $\bigcup_\alpha \text{supp}(f_\alpha)$ differs from Ω by a μ-null set.

If μ is σ-finite, then $I = \mathbb{N}$ and $f_n = \chi_{A_n}, n \in \mathbb{N}$ can be chosen with additional properties $\rho(\chi_{A_n}) < \infty, \rho'(\chi_{A_n}) < \infty$. If μ is only localizable, I may be uncountable. The existence of a g.w.u. implies that μ is (strictly) localizable on $\mathcal{B} = \sigma(A_\alpha, \alpha \in I)$, a σ-subalgebra of Σ. Following the case of the Orlicz (or Lebesgue) spaces $L^\varphi(\Sigma)(L^\rho(\Sigma))$, we say that the function norm $\rho(\cdot)$ is *strictly monotone* if $f, g \in L^\rho(\Sigma), 0 \leq f \leq g$ and $\rho(f) = \rho(g)$ implies, $f = g$, a.e.

The following characterization of measurable subspaces of $L^\rho(\Sigma)$ plays a basic role in the present analysis. We usually take $F = \{\chi_{A_\alpha}, \alpha \in I\}$ for a g.w.u. for simplicity.

Theorem 5. *Let $\mathcal{M}$ be a closed subspace of $L^p(\Sigma)$ introduced above. Then $\mathcal{M}$ is a measurable subspace (i.e., $\mathcal{M} = L^p(\mathcal{B})$ for a unique σ-algebra $\mathcal{B} \subset \Sigma$ with $\mu_{\mathcal{B}}$ strictly localizable) iff the following (alternative) sets of conditions hold:*

(a) $f \in \mathcal{M} \Rightarrow$ the complex conjugate $\bar{f} \in \mathcal{M}$,

(b) the real functions of $\mathcal{M}$ form a lattice and there is a g.w.u. $F_0 \subset \mathcal{M}$,

(c) $0 \leq f_n \uparrow f$, a.e., $f_n \in \mathcal{M}, f \in L^p(\Sigma) \Rightarrow f \in \mathcal{M}$.

or, (a) holds and (b), (c) are replaced by (b'), (c') where

(b') bounded functions of $\mathcal{M}$ form a (norm) dense algebra in $\mathcal{M}$, containing a g.w.u. F_0,

(c') $f_n \in \mathcal{M}, f_n \to f$, a.e., $|f_n| \leq g \in L^p(\Sigma) \Rightarrow f \in \mathcal{M}$.

Before proving this result, we apply it in characterizing conditional expectations so that its significance and utility can be better appreciated. One also should note that if ρ is an a.c.n., then both (c) and (c') are automatic and hence may be omitted. Our first substantial application is given by the following result.

Theorem 6. *Let $T : L^p(\Sigma) \to L^p(\Sigma)$ be a positive contractive projection. Suppose $\rho(\cdot)$ is a strictly monotone a.c.n. with the (J)-property, and there is a g.w.u. $F_0 = \{\chi_{A_\alpha}, \alpha \in I\}$ in $L^p(\Sigma)$. If $TF_0 = F_0$ (i.e., $T\chi_{A_\alpha} = \chi_{A_\alpha}, \alpha \in I)$ then $T = E^{\mathcal{B}}$ for a unique σ-algebra $\mathcal{B} \subset \Sigma$.*

Proof. Let $\mathcal{M}$ be the range of T in $L^p(\Sigma)$. Then $\mathcal{M}$ is a vector space containing the g.w.u. F_0 and is closed. Only the last statement could possibly be not obvious. For this let the sequence $\{f_n, n \geq 1\} \subset \mathcal{M}$ be Cauchy. Then it has the same property in $L^p(\Sigma)$ and by the completeness of the latter, there exists an $f \in L^p(\Sigma)$ such that $\rho(f_n - f) \to 0$. However T being a projection,

$$\rho(f - Tf) \leq \rho(f - f_n) + \rho(Tf_n - Tf), \text{ since } Tf_n = f_n,$$
$$\leq 2\rho(f - f_n), \text{ since } T \text{ is a contraction}.$$

But the right side tends to zero as $n \to \infty$, and the left side is independent of n. So it must vanish, and hence $f = Tf \in \mathcal{M}$. Now we note that the real elements of $\mathcal{M}$ form a lattice. In fact, if $f \in \mathcal{M}$ is real, $f = f^+ - f^-$ so that $f^{\pm} \leq |f|$ and the positivity of T implies

$$|f| = |Tf| \leq Tf^+ + Tf^- = T(|f|).$$

Since $\rho(f) = \rho(|f|) \leq \rho(T(|f|)) \leq \rho(|f|)$, by the contractivity of T, with the strict monotonicity of ρ this implies that $|f| = T(|f|)$ a.e. Hence $|f| \in \mathcal{M}$ and this gives the lattice property immediately. Thus (a) and (b) of the preceding theorem are satisfied. Regarding (c), let $0 \leq f_n \uparrow f$, $f_n \in \mathcal{M}$, and $f \in L^\rho(\Sigma)$. But then $f - f_n \downarrow 0$ and since $0 \leq f - f_n \leq f$, $\rho(f) < \infty$, the a.c.n. property of ρ implies $\rho(f - f_n) \downarrow 0$. By completeness of $\mathcal{M}$, $f \in \mathcal{M}$. Hence (c) also holds, and by Theorem 5, $\mathcal{M} = L^\rho(\mathcal{B})$ for a unique σ-algebra $\mathcal{B} \subset \Sigma$, relative to which F_0 is measurable and $\mu_\mathcal{B}$ is localizable. It remains to show that $T = E^\mathcal{B}$.

The above statement, however, is an immediate consequence of the following result which we present separately because of its use in other cases and also of its independent interest.

Theorem 7. *Let $L^\rho(\Sigma)$ be a function space, ρ with the (J)-property, F_0 being a g.w.u. in it, and $T : L^\rho(\Sigma) \to L^\rho(\Sigma)$ a bounded projection with range $L^\rho(\mathcal{B})$ for a σ-algebra $\mathcal{B} \subset \Sigma$, and $\mu_\mathcal{B}$ localizable and F_0 in the range of T. Then $Tf = E^\mathcal{B}(fg)$, a.e., $E^\mathcal{B}(g) = 1$, a.e., for a locally integrable g. In case ρ is also a.c.n., then $g = 1$, a.e.*

Proof. Various properties of $L^\rho(\Sigma)$ discussed in the preceding section including those in Proposition 2.9, will be used without comment. Thus let $A \in \mathcal{B}, \mu(A) < \infty$ and $\rho(\chi_A) < \infty$ (hence $\rho'(A) < \infty$ as well). Consider x_A^* defined on $L^\rho(\Sigma)$ by the equation

$$x_A^*(f) = \int_\Omega \chi_A(Tf) d\mu_A, \quad f \in L^\rho(\Sigma). \tag{11}$$

Then x_A^* is a bounded linear functional since (linearity being clear) $\|x_A^*\| = \sup\{|x_A^*(f)| : \rho(f) \leq 1\} \leq \rho'(\chi_A) < \infty$. Hence $x_A^* \in (L^\rho(\Sigma(A)))^*$ where $\Sigma(A)$ is the trace σ-algebra of Σ on A and $\mu_A(\cdot) = \mu(A \cap \cdot)$ is a finite measure on $\Sigma(A)$. Then x_A^* admits an integral representation (see Zaanen, 1967, p. 467; Gretsky, 1968, p. 52; and Rao, 1970$_c$, I.1.14) uniquely as follows:

$$x_A^*(f) = \int_A f\, g_A d\mu_A + \int_A \tilde{f} dG_A, \quad f \in L^\rho(\Sigma), \tag{12}$$

where $g_A \in L^{\rho'}(\Sigma(A))$, $G_A(\cdot)$ is a purely finitely additive set function on $\Sigma(A)$, vanishing on μ_A-null sets, and $\tilde{f}$ is a certain canonical image of f, and the representation is unique. Equating (11) and (12) one finds that the last integal of (12), which is purely finitely additive, is equal to σ-additive integrals and hence it must vanish. Thus (12) becomes:

$$\int_A Tf\, d\mu_B = \int_A Tf\, d\mu_A = \int_A E^B(fg_A)d\mu_B, \quad A \in \mathcal{B}_0, \ f \in L^\rho(\Sigma),$$
$$(13)$$

where $\mathcal{B}_0 = \{B \in \mathcal{B} : \mu(B) < \infty\}$. But the class $\{g_A, \ A \in \mathcal{B}_0\}$ satisfies (by uniqueness) on $A \cap B$, $g_A = g_B = g_{A \cap B}$, and the localizability of μ_B then implies the existence of a locally integrable g such that $g_A = g\chi_A$, $A \in \mathcal{B}_0$. From this it follows that ($E^B(fg)$ being $\mathcal{B}$-measurable)

$$Tf = E^B(fg), \text{ a.e.}, \quad f \in L^\rho(\Sigma). \tag{14}$$

Now replacing f by f_α of the g.w.u. in (14) one gets

$$f_\alpha = Tf_\alpha = E^B(f_\alpha g) = f_\alpha E^B(g), \text{ a.e.}, \quad \alpha \in I. \tag{15}$$

But $\cup_\alpha \text{supp}(f_\alpha) \doteq \Omega$ and hence (15) implies $E^B(g) = 1$ a.e. establishing the first part of the theorem.

Suppose now ρ is an a.c.n. We then assert that $g = 1$ a.e. Under the present hypothesis, $(L^\rho(\Sigma))^* = L^{\rho'}(\Sigma)$ and since $T : L^\rho(\Sigma) \to L^\rho(\Sigma)$ is a continuous projection, its adjoint $T^* : L^{\rho'}(\Sigma) \to L^{\rho'}(\Sigma)$ also has similar properties, which is a well-known fact from abstract analysis, so that

$$\int_\Omega T^*(h)f d\mu = \int_\Omega h(Tf)d\mu = \int_\Omega h\, E^B(fg)d\mu, \ f \in L^\rho(\Sigma), \ h \in L^{\rho'}(\Sigma),$$

$$= \int_\Omega E^B(hE^B(fg))d\mu, \quad \text{by definition of } E^B,$$

$$= \int_\Omega E^B(h)E^B(fg)d\mu, \quad \text{by (9)},$$

$$= \int_\Omega E^B(h)(Tf)d\mu, \quad \text{by the first part},$$

$$= \int_\Omega T^*(E^B(h))f\, d\mu, \quad \text{by definition of } T^*. \tag{16}$$

Hence $T^* = T^* E^B$. On the other hand $E^B(L^{\rho'}(\Sigma)) = L^{\rho'}(\mathcal{B})$, and for any $f \in L^p(\mathcal{B}) = \mathcal{M} = T(L^p(\Sigma))$, we have

$$\int_\Omega E^{\mathcal{B}}(T^*h)f \, d\mu_{\mathcal{B}} = \int_\Omega E^{\mathcal{B}}((T^*h)\cdot f)d\mu_{\mathcal{B}}, h \in L^{\rho'}(\Sigma),$$

$$= \int_\Omega (T^*h)\cdot f \cdot d\mu$$

$$= \int_\Omega hTf d\mu = \int_\Omega hf d\mu, \quad \text{since } f \in \mathcal{M},$$

$$= \int_\Omega E^{\mathcal{B}}(h)f d\mu_{\mathcal{B}}. \tag{17}$$

Since f is arbitrary in $L^p(\mathcal{B})$, (17) implies $E^{\mathcal{B}}T^* = E^{\mathcal{B}}$. Using the commutativity of $E^{\mathcal{B}}$ and T^* following from (16) and (17) we now determine T^*. [The (possible) nonreflexivity of $L^p(\Sigma)$ demands additional work.]

The averaging property of $E^{\mathcal{B}}$ implies that the operator is "self adjoint" in the following sense. Let $f \in L^p(\Sigma)$, $g \in L^{\rho'}(\Sigma)$ and be bounded. Then

$$\int_\Omega f E^{\mathcal{B}}(g)d\mu = \int_\Omega E^{\mathcal{B}}(f E^{\mathcal{B}}(g))d\mu_{\mathcal{B}}, \text{ by definition},$$

$$= \int_\Omega E^{\mathcal{B}}(f)E^{\mathcal{B}}(g)d\mu_{\mathcal{B}}, \text{ by the averaging property},$$

$$= \int_\Omega E^{\mathcal{B}}(gE^{\mathcal{B}}(f))d\mu_{\mathcal{B}}$$

$$= \int_\Omega g \, E^{\mathcal{B}}(f)d\mu.$$

Thus $(E^{\mathcal{B}})^* = E^{\mathcal{B}}$ on bounded elements of these spaces, and then the result holds, by a standard argument, on all of $L^p(\Sigma)$ and $L^{\rho'}(\Sigma)$. Also $E^{\mathcal{B}}(L^p(\Sigma)) = L^p(\mathcal{B})$ and $E^{\mathcal{B}}(L^{\rho'}(\Sigma)) = L^{\rho'}(\mathcal{B})$. We now use some elementary relations between a pair of continuous projections (of the type $E^{\mathcal{B}}$ and T^*) from a general Banach space and use them in the present context to show that the ranges of T^* and $E^{\mathcal{B}}$ in $L^{\rho'}(\Sigma)$ agree. This will imply the desired conclusion.

Let $M = $ range (T), $N = $ null space (T), and similarly let (M_1, N_1) and $(\bar{M}, \bar{N})$ be the corresponding range, null space pairs of T^* and $E^{\mathcal{B}}$ in $L^{\rho'}(\Sigma)$. By hypothesis $M = L^p(\mathcal{B})$, and by the elementary properties of $E^{\mathcal{B}}$ in $L^{\rho'}(\Sigma)$, we have $\bar{M} = L^{\rho'}(\mathcal{B})$. Also $E^{\mathcal{B}}T^* = E^{\mathcal{B}}$ implies the relation $N_1 \subset \bar{N}$, and $T^*E^{\mathcal{B}} = T^*$ similarly implies $\bar{N} \subset N_1$

(see Dunford-Schwartz, 1958, p. 514; Ex. 23 (i)), so that $N_1 = \bar{N}$.
Also, since $L^\rho(\Sigma) = M \oplus N$ we get $L^\rho(\Sigma))^* = L^{\rho'}(\Sigma) = N^\perp \oplus M^\perp$
using the a.c.n. property of ρ and the localizability of μ (and also
Ex. 22 of the preceding reference on p. 514). Here $N^\perp = $ range
$(T^*) = M_1$ and $M^\perp = $ null space $T^* = N_1(= \bar{N})$, where $N^\perp = \{x^* \in$
$(L^\rho(\Sigma))^* : x^*(N) = 0\}$. However, the g.w.u. F_0 is in $L^\rho(\mathcal{B})$ so that $\mu_\mathcal{B}$
is localizable, and $T|L^\rho(\mathcal{B}) = $ identity $= I_1$. So I_1^* on $(L^\rho(\mathcal{B}))^* = L^{\rho'}(\mathcal{B})$
is also the identity (this being its range), and $T^*|(L^\rho(\mathcal{B}))^* = I_1^*$. Hence
range $T^* = M_1 \supset L^{\rho'}(\mathcal{B})$. [Alternatively, $M^* \cong (L^\rho(\Sigma))^*/M^\perp \cong M_1$
and there is equality with $L^{\rho'}(\mathcal{B})$, with the quotient properties.] Since
$T^*|L^{\rho'}(\mathcal{B}) = $ identity, and since the g.w.u. also belongs to $L^{\rho'}(\mathcal{B})$, by
definition, one gets

$$T^*\chi_{A_\alpha} = \chi_{A_\alpha} \quad \text{or} \quad T^*F_0 = F_0. \tag{18}$$

We can now complete the argument quickly. Let $f \in L^\rho(\Sigma)$, $h \in$
$L^{\rho'}(\Sigma)$. Then

$$\int_\Omega (T^*h)f d\mu = \int_\Omega h(Tf)d\mu = \int_\Omega hE^\mathcal{B}(fg)d\mu, \quad \text{by (14)},$$

$$= \int_\Omega E^\mathcal{B}(h)E^\mathcal{B}(fg)d\mu_\mathcal{B}, \quad \text{by the averaging identity},$$

$$= \int_\Omega E^\mathcal{B}(fgE^\mathcal{B}(h))d\mu_\mathcal{B} = \int_\Omega f \cdot (gE^\mathcal{B}(h))d\mu. \tag{19}$$

Since this identity holds for all $f \in L^\rho(\Sigma)$, we conclude that $T^*h = gE^\mathcal{B}(h)$ a.e. Taking $h = \chi_{A_\alpha} \in F_0$, we get $\chi_{A_\alpha} = g\chi_{A_\alpha}$, by (18), and
$A_\alpha \in \mathcal{B}$, $\alpha \in I$. But $\bigcup_\alpha A_\alpha \doteq \Omega$ so that $g = 1$, a.e., as asserted. In this
case $Tf = E^\mathcal{B}(f)$ and hence Theorem 6 as well as the present theorem
are established. $\square$

Remarks. 1. If ρ is not a.c.n. but the rest of the hypothesis holds, then
$(L^\rho(\Sigma))^* \supsetneq L^{\rho'}(\Sigma)$ and the range of T^* may go beyond the subspace
$L^{\rho'}(\Sigma)$. Hence the above calculations will not hold, and considering
$\rho(\cdot) = N_\varphi(\cdot)$, for a φ, an exponentially growing Young function, one
sees that the result itself is not true. The role of g.w.u. will be discussed
in Section 10.5.

2. The contractivity of T in Theorem 6 was used in establishing its
range to be $L^\rho(\mathcal{B})$ for some σ-algebra $\mathcal{B} \subset \Sigma$. If the range is given to

be of this form, then we only used the continuity of T for most of the rest of that analysis.

The importance of Theorem 7 lies in the fact that several other properties of operators satisfying different identities can be identified with this class. Let us verify this statement for Šidák and averaging identities.

Theorem 8. *Let $T : L^\rho(\Sigma) \to L^\rho(\Sigma)$ be a contractive linear operator and ρ be a localizable and a.c.n. functional having the (J)-property. If there is a g.w.u. F_0 in $L^\rho(\Sigma)$ such that $TF_0 = F_0$ and either (i) T satisfies the averaging identity, or (ii) ρ is strictly monotone, T positive, and the Šidák identity holds for T, then $T = E^{\mathcal{B}}$ for a unique σ-algebra $\mathcal{B} \subset \Sigma$. [For (ii) if ρ is not an a.c.n. but T is positive, then $Tf = E^{\mathcal{B}}(fg)$ as in Theorem 7.]*

Proof. First let us show that T is a projection in both cases. For (i) let $f \in L^\rho(\Sigma)$ and $h = Tf$. Since $T\chi_{A_\alpha} = \chi_{A_\alpha}, \alpha \in I$, where $\chi_{A_\alpha} \in F_0 = \{\chi_{A_\alpha}, \alpha \in I\}$, the g.w.u., we have

$$\chi_{A_\alpha} T(f) = (T\chi_{A_\alpha})(Tf) = T(\chi_{A_\alpha} h), \quad \text{by the averaging identity,}$$
$$= T(hT\chi_{A_\alpha}) = T(h)T(\chi_{A_\alpha}) = \chi_{A_\alpha} T^2(f), \alpha \in I. \qquad (20)$$

Since $\bigcup_\alpha A_\alpha \doteq \Omega$, (20) implies $T^2 f = Tf$, so that T is a projection.

Next let T be a Šidák operator. Then by the identity (10)

$$T((Tf)^+) = T((Tf) \vee 0) = (Tf) \vee 0 = (Tf)^+ \qquad (21)$$
$$T((Tf)^-) = T((-Tf)^+) = T((T(-f))^+)$$
$$= (T(-f))^+ = (Tf)^-, \quad \text{by (21).} \qquad (22)$$

Here we used $f^- = (-f)^+$ and so $(Tf)^- = (-Tf)^+ = (T(-f))^+$. Thus for any $f \in L^\rho(\Sigma)$ one has

$$T^2(f) = T((Tf)^+ - (Tf)^-)$$
$$= T((Tf)^+) - T((Tf)^-)$$
$$= (Tf)^+ - (Tf)^- = (Tf)$$

Hence T is a projection operator in either case. It is to be shown that T is also positive which is the nontrivial part. We use different methods in both cases to establish this fact.

Let us consider the case of Šidák operator for which the representation can be obtained directly. If $\mathcal{M} = T(L^\rho(\Sigma))$, then we assert that for $f \in \mathcal{M}$, real, $|f| \in \mathcal{M}$, whence its real elements form a (vector) lattice. In fact, by (21) and (22), $|f| = |Tf| = (Tf)^+ + (Tf)^-$ and

$$T(|f|) = T(f^+) + T(f^-)$$
$$= (Tf)^+ + (Tf)^-, \quad \text{since } T(f^+) = T((Tf) \vee 0) = (Tf) \vee 0,$$
$$\text{and similarly for } T(f^-), \text{ by (21)},$$
$$= |(Tf)| = |f|. \tag{23}$$

Thus $|f| \in \mathcal{M}$, and the real elements of $\mathcal{M}$ form a lattice.

Suppose now that ρ is strictly monotone and T is also positive. [Note that (23) implies easily the positivity of T on $\mathcal{M}$, but we need it on *all* of $L^\rho(\Sigma)$, and this stronger property is the assumption.] To verify (c) of the hypothesis of Theorem 5, let $0 \leq f_n \uparrow f$, $f_n \in \mathcal{M}$, $f \in L^\rho(\Sigma)$. Then $f_n = Tf_n \leq Tf$ so that $f \leq Tf$ and one has

$$\rho(f) \leq \rho(Tf) \leq \rho(f)$$

since T is a contraction. The strict monotonicity of ρ implies $f = Tf$ a.e. so that $f \in \mathcal{M}$. Thus by Theorem 5, $\mathcal{M} = L^\rho(\mathcal{B})$ for some σ-algebra $\mathcal{B} \subset \Sigma$, and Theorem 7 implies that $Tf = E^\mathcal{B}(fg)$ for some g as in the statement and that $g = 1$ a.e. when ρ is an a.c.n.

For the remaining part, let T be an averaging operator. We assert that T is a contraction, not only on $L^\rho(\Sigma)$ as given but on $L^\infty(\Sigma)$ as well, and then it is positive.

Let $h \in L^\rho(\Sigma)$, $|h| \leq 1$ a.e. and $g = Th$. We claim that $|g| \leq 1$, a.e. Indeed, since T is a projection,

$$T(g^2) = T(gTh) = (Tg)(Th) = (T^2h)g = (Th)g = g^2.$$

By induction, we get $T(hg^{n-1}) = T(h)T(g^{n-1}) = g^n$. Then

$$\rho(g^n) = \rho''(g^n) = \sup\{|\int_\Omega g^n f d\mu| : \quad \rho'(h) \leq 1\}$$
$$= \sup\{|\int_\Omega T(hg^{n-1}) f d\mu| : \rho'(h) \leq 1\}$$
$$\leq \sup\{|\int_\Omega (hg^{n-1}) f d\mu| : \rho'(h) \leq 1\},$$
$$\text{since } T \text{ is a contraction},$$
$$= \rho(hg^{n-1}) \leq \rho(g^{n-1}), \text{ since } |h| \leq 1, \text{ a.e.} \tag{24}$$

This shows that $\rho(g^n) \leq \cdots \leq \rho(g) \leq \rho(h)$. If $A = \{\omega : |g(\omega)| > 1\}$, then (24) gives (since $(|g|\chi_A)^n \leq |g|^n$) $\rho(g^n\chi_A) \leq \rho(h)$ for all n. Now let $n \to \infty$ and use the Fatou property of $\rho(\cdot)$ to conclude that $\mu(A) = 0$. Hence $|g| \leq 1$ a.e. and $T : L^\infty(\Sigma) \to L^\infty(\Sigma)$ and is a contraction. Thus $T(L^\rho(\Sigma) \cap L^\infty(\Sigma)) \subset L^\rho(\Sigma) \cap L^\infty(\Sigma)$.

Using the a.c.n. property of $\rho(\cdot)$ at this point, we can conclude that (b') and (c') of Theorem 5 hold. If $f \in \mathcal{M} \subset L^\rho(\Sigma)$, and $\varepsilon > 0$, since $f = Th$ for some $h \in L^\rho(\Sigma)$, there is a bounded $h_\varepsilon \in L^\rho(\Sigma)$ (by density of bounded functions) satisfying $\rho(h - h_\varepsilon) < \varepsilon$. If $f_\varepsilon = Th_\varepsilon$, then f_ε is bounded and $\rho(f - f_\varepsilon) = \rho(T(h - h_\varepsilon)) \leq \rho(h - h_\varepsilon) < \varepsilon$. Thus bounded functions are dense in $\mathcal{M}$. Clearly $f_i \in \mathcal{M} \Rightarrow f_i = Tg_i$ and $f_1 f_2 = (Tg_1)(Tg_2) = T(g_1 Tg_2) \in \mathcal{M}$, so bounded elements form an algebra. If $f_n \in \mathcal{M}, f_n \to f$ a.e. and $|f_n| \leq h \in L^\rho(\Sigma)$, then by the a.c.n. again, $\rho(f - f_n) \to 0$ as $n \to \infty$, so that $f \in \mathcal{M}$. By Theorem 5, there is a σ-algebra $\mathcal{B} \subset \Sigma$ such that $\mathcal{M} = L^\rho(\mathcal{B})$. The conclusion now follows from Theorem 7 in both cases. $\quad\square$

Remark. The positivity of T on $L^\infty(\Sigma)$ is proved quickly as follows. If it were false, then there exist an f such that $0 < f < \chi_A$, $\chi_A \in F_0$, the g.w.u. and $Tf \leq -\alpha < 0$ on a set $B \subset A$ of positive measure for some $\alpha > 0$. Since $T\chi_A = \chi_A$, this implies $T(\chi_A - f) \geq \chi_A + \alpha$ on B. Hence we have

$$1 \geq \|\chi_A - f\|_\infty \geq \|T(\chi_A - f)\|_\infty \geq 1 + \alpha > 1.$$

This contradiction shows that T is positive on bounded (and then on all) elements of M^ρ since bounded elements are dense in it.

The preceding result can be simplified and refined for the L^p-spaces, i.e., if $\rho(\cdot) = \|\cdot\|_p$, $p \geq 1$, which we present for comparison.

Proposition 9. *Let* $T : L^p(\Sigma) \to L^p(\Sigma)$ *be a contractive averaging operator with* $TF_0 = F_0$ *relative to a g.w.u.* F_0 *in* $L^p(\Sigma)$. *Then* T *is a conditional expectation relative to a* σ-*algebra* $\mathcal{B} \subset \Sigma$ *for* $1 \leq p < \infty$, *and for* $p = \infty$ *it is representable as* $Tf = E^{\mathcal{B}}(fg)$ *whenever, T, which is positive, is order continuous. Here g is a locally integrable function with* $E^{\mathcal{B}}(g) = 1$, *a.e., and when* $\mu(\Omega) < \infty, g = 1$, *a.e., iff* $T^*\mu = \mu, T^*$ *being the adjoint operator of T acting on* $(L^\infty(\mu))^*$.

Proof. The part for $1 \leq p < \infty$, is a specialization of the preceding theorem since $\rho(\cdot) = \|\cdot\|_p$ is an a.c.n. So consider the case $p = +\infty$. If

$\mathcal{M} = T(L^{\infty}(\Sigma))$, then it is a closed subspace (since T is a continuous projection) and the averaging identity implies that it is also an algebra. The above remark shows that T is positive on $L^{\infty}(\Sigma)$. Thus conditions (a), (b) and (b') of Theorem 5 hold for $\mathcal{M}$. Let $0 \leq f_n \in \mathcal{M}$ such that $f_n \uparrow f, f \in L^{\infty}(\Sigma)$. Then the order continuity of T implies $f = \lim_n f_n = \lim_n T f_n = T f$ so that $f \in \mathcal{M}$. Hence condition (c) [as well as (c')] also holds so that by Theorem 5, $\mathcal{M} = L^{\infty}(\mathcal{B})$ for a σ-algebra $\mathcal{B} \subset \Sigma$. Since $F_0 \subset \mathcal{M}, \mu_{\mathcal{B}}$ is localizable, and by the first part of Theorem 7, $Tf = E^{\mathcal{B}}(fg)$ for a g of the statement.

Finally, let $\mu(\Omega) < \infty$, and T^* be the adjoint of T on $(L^{\infty}(\Sigma))^*$. This space is identifiable with $ba(\Omega, \Sigma, \mu)$, the space of bounded additive set functions vanishing on μ-null sets. When $\mu(\Omega) < \infty$, we can identify it also with an element of the latter space. Suppose $T^*\mu = \mu$. Then

$$\int_{\Omega} fg d\mu = \int_{\Omega} E^{\mathcal{B}}(fg) d\mu = \int_{\Omega} T f d\mu, \text{ since } Tf = E^{\mathcal{B}}(fg),$$
$$= \int_{\Omega} f d(T^*\mu) = \int_{\Omega} f d\mu, \quad f \in L^1(\Sigma). \tag{25}$$

Hence $g = 1$, a.e., holds. Conversely, if $g = 1$ a.e., then taking $f = \chi_A, A \in \Sigma$ in (25) one gets, on using $Tf = E^{\mathcal{B}}(f)$,

$$\int_A d\mu = \int_{\Omega} E^{\mathcal{B}}(\chi_A) d\mu = \int_{\Omega} T\chi_A d\mu = \int_{\Omega} \chi_A d(T^*\mu).$$

Hence $T^*\mu$ is σ-additive and $(T^*\mu)(A) = \mu(A), A \in \Sigma$, so that $T^*\mu = \mu$. Note that $T^*\mu$ will be undefined if $\mu(\Omega) < \infty$ is not assumed. Thus all the assertions are established. $\square$

Remark. The existence of a g.w.u. plays a key role in all the above work, especially that it should belong to the range of the operator T. Without such an assumption, there will be some "distortion" of the range being a measurable subspace. We shall discuss this case in Chapter 10 since that situation occurs in applications to analysis but it is infrequent in probability contexts.

Now that we have seen the usefulness of Theorem 5 in these characterization problems, we shall turn to its proof. An auxiliary result, of Stone-Weierstrass type, is needed. Several kinds of such approximations are available, but we present the following (taken from Rao, 1969; see also Rota, 1960, p. 55) for immediate application.

Theorem 10. *Let S be a subset of $L^\rho(\Sigma) \cap L^\infty(\Sigma)$ which is either an algebra or a vector space whose real elements form a lattice. Suppose in both cases the following three conditions hold:*

(i) S is self adjoint, i.e., $f \in S \Rightarrow \bar{f} \in S$,

(ii) there is a g.w.u. $F_0 = \{f_\alpha, \alpha \in I\} \subset S$, so $\bigcup_\alpha supp(f_\alpha) \doteq \Omega$,

(iii) for any disjoint E_1, E_2 in Σ of positive μ-measure there is a $g \in S$ such that $g > 0$, on E_1, and $g \leq 0$, on E_2, a.e. so S separates "points" of Σ.

Then the σ-algebra $\tilde{\Sigma}$ generated by S and completed for μ has the property that $L^\rho(\tilde{\Sigma}) = L^\rho(\Sigma)$. Moreover, the norm closure of S is $L^\rho(\Sigma)$ if ρ is an a.c.n., and in general is the subspace determined by all its bounded elements.

Proof. Let $S_1 = L^\rho(\Sigma) \cap L^\infty(\Sigma)$ and Σ_1 be the σ-algebra generated by S_1 and completed for μ. If $L^\rho(\Sigma_1) = \{f \in L^\rho(\Sigma) : f$ is Σ_1-measurable$\}$ then we observe that $L^\rho(\Sigma_1) = L^\rho(\Sigma)$. Indeed, since $L^\rho(\Sigma_1) \subset L^\rho(\Sigma)$, let $f \in L^\rho(\Sigma)$ and $f_n = f\chi_{[|f| \leq n]}$. Then $f_n \in S_1$ so that it is Σ_1-adapted, and since $f_n \to f$, a.e., as $n \to \infty$, f is Σ_1-adapted and hence $f \in L^\rho(\Sigma_1)$ so that $L^\rho(\Sigma_1) = L^\rho(\Sigma)$. Similarly, since $S \subset S_1$ it is clear that $L^\rho(\tilde{\Sigma}) \subset L^\rho(\Sigma_1)$. To see there is equality here, it suffices to show that $\tilde{\Sigma} = \Sigma_1$ or, equivalently, the L^∞-closures of S and S_1, denoted $\tilde{S}$ and $\tilde{S}_1$, are the same.

Consider the case that the spaces are real. Then by (ii) and (iii) [(i) is vacuous now], $\tilde{S}$ and $\tilde{S}_1$ are abstract (M)-spaces in the sense of Kakutani. By a classical representation theorem, both these spaces are isometrically (lattice) isomorphic to subspaces C_1 and C_2 of $C(S)$, the Banach space (under uniform norm) of real continuous functions on a compact Hausdorff space S. Conditions (ii) and (iii) now imply that $C_1(\subset C_2)$ possesses the properties: a) there is an $\hat{f}_0 > 0$ on S such that $\hat{f}_0 \in C_1 \subset C_2$ and b) C_1 (hence C_2) separates points of S (i.e., for any distinct s_1, s_2 of S, there is $\hat{f} \in C_1$ such that $\hat{f}(s_1) > 0$ and $\hat{f}(s_2) \leq 0$). Hence by the fundamental Stone-Weierstrass theorem $\bar{C}_1 = \bar{C}_2 = C(S)$. By the isometry, therefore, $\tilde{S} = \tilde{S}_1$ so that $\tilde{\Sigma} = \Sigma_1$ as desired.

In the complex case $L^\rho_{\mathbb{C}}(\Sigma) = L^\rho_{\mathbb{R}}(\Sigma) + iL^\rho_{\mathbb{R}}(\Sigma)$, in obvious notation, and applying the above case, using (i), one gets the stated result. $\square$

Another proof using the Gel'fand theory of commutative B-algebras, together with applications, can be found in the above reference. We now turn finally to the proof that was postponed.

Proof of Theorem 5. As in the above result, it suffices to treat the real case. So let $L^p(\Sigma)$ be a real function space. If $\mathcal{M} = L^p(\mathcal{B})$ with $\mu_{\mathcal{B}}$ localizable, then (a), (b), (c) and (a), (b′), (c′) hold for this space. Thus only the converse is nontrivial and let us prove it.

Let $\mathcal{M} = T(L^p(\Sigma))$ and F_0 be a g.w.u. in it, where $F_0 = \{\chi_{A_\alpha}, \alpha \in I\}$. If $\mathcal{A}$ is the collection of sets from Σ such that for each disjoint pair of its members, of positive μ-measure, there is a bounded function of $\mathcal{M}$ that distinguishes them, then $A_\alpha \in \mathcal{A}$, $\alpha \in I$. Let $\mathcal{B}$ be the σ-algebra generated by $\mathcal{A}$ and completed for μ. Then $F_0 \subset L^p(\mathcal{B})$, the subspace of $L^p(\Sigma)$, of $\mathcal{B}$-measurable elements. If $\mathcal{S} \subset \mathcal{M}$ is the set of bounded functions, then $F_0 \subset \mathcal{S}$. The hypotheses (a), (b), (c) imply the conditions of Theorem 10 and so the σ-algebra generated by $\mathcal{S}$ and completed for μ is $\mathcal{B}$ since $\mathcal{S}$ separates the "points" of $\mathcal{B}$. We now assert that $L^p(\mathcal{B}) \subset \mathcal{M}$. Indeed, if $0 \le f \in L^p(\mathcal{B})$, and $f_n = f \wedge n$, and if M^p is the ρ-closure of $L^p(\mathcal{B}) \cap L^\infty(\mathcal{B})$, then $F_0 \subset M^p$ and, by Theorem 10, $M^p = \rho$-closure of $\mathcal{S}$. Since $\mathcal{S} \subset \mathcal{M} \cap L^p(\mathcal{B})$, we deduce that $M^p \subset \mathcal{M}$ and hence $f_n \in M^p \subset \mathcal{M}$. So by (c) $f \in \mathcal{M}$. Both spaces being lattices this shows that $L^p(\mathcal{B}) \subset \mathcal{M}$. (Here $\wedge = \min$.)

To establish the opposite inclusion we again can consider $0 \le f_0 \in \mathcal{M}$ where f_0 is an elementary function in $\mathcal{M}$ determined by a subset of $\{A_\alpha, \alpha \in I\}$ of the g.w.u., so that $f_0 = \sum\limits_{i=1}^{\infty} a_i \chi_{A_i}$, $a_i \ge 0$. [If I is a singleton $\{\chi_\Omega\}$, then $f_0 = \chi_\Omega$.] Let $0 \le f \in \mathcal{M}$ such that $\text{supp}(f) \subset \bigcup\limits_{i=1}^{\infty} A_{\alpha_i}$. Then for any $\beta > 0$, and $n \ge 1$, consider $f_n = [n(f - \beta f_0)^+] \wedge f_0 \le f_0$. Now the lattice property of $\mathcal{M}$ implies that $f_n \in \mathcal{M}$. Also $f_n \uparrow h \le f_0$ a.e. so that by (c), $h \in \mathcal{M}$. But if $A = \{\omega : f(\omega) - \beta f_0(\omega) > 0\}$, then $h = f_0 \chi_A$ a.e. Moreover, f_0 is $\mathcal{B}$-measurable and bounded so that $h \in \mathcal{S} \subset \mathcal{M}$. Hence h is $\mathcal{B}$-measurable ($\mathcal{B}$ is the μ completion of $\sigma\{g, g \in \mathcal{S}\}$), implying that $\chi_A = \frac{h}{f_0}$ is also, and then $A(= [\frac{f}{f_0} > \beta]) \in \mathcal{B}$. Since $\beta > 0$ is arbitrary, $\text{supp}(f) \subset \text{supp}(f_0)$, and f_0 is $\mathcal{B}$-measurable, this shows that f is also. If $0 \le f \in \mathcal{M}$ is arbitrary, the above argument implies that $f \chi_{A_\alpha}$ is $\mathcal{B}$-measurable for each $\alpha \in I$. Since $\bigcup\limits_{\alpha} A_\alpha \doteq \Omega$, and $\mu_{\mathcal{B}}$ is localizable, this implies that f itself is $\mathcal{B}$-measurable. But $\rho(f) < \infty$, so that $f \in L^p(\mathcal{B})$, and $\mathcal{M} \subset L^p(\mathcal{B})$, as desired.

It remains to consider the second part (algebra conditions). If $\mathcal{S} \subset$

$\mathcal{M} = T(L^\rho(\Sigma))$ is the algebra of bounded functions, then $F_0 \subset \mathcal{S}$. Let $\mathcal{B}$ be the σ-algebra generated by $\mathcal{S}$ and completed for μ. We again have $\mathcal{S} \subset \mathcal{M} \cap L^\rho(\mathcal{B})$. If $f \in L^\rho(\mathcal{B})$, and $f_n = f\chi_{[|f|\leq n]}$, then $f_n \in L^\rho(\mathcal{B}) \cap L^\infty \subset \mathcal{M}$. Since $f_n \to f$, a.e., and $|f_n| \leq |f|$, $\rho(f) < \infty$, it follows from the completeness of $\mathcal{M}$ and (c') that $f \in \mathcal{M}$ so $L^\rho(\mathcal{B}) \subset \mathcal{M}$. For the opposite inclusion, by (b'), if $f \in \mathcal{M}$ and $\varepsilon > 0$ is given then there is f_ε in $\mathcal{S}$ such that $\rho(f - f_\varepsilon) < \varepsilon$. Since $\mathcal{S} \subset L^\rho(\mathcal{B})$, $f_\varepsilon \in L^\rho(\mathcal{B})$ and by the completeness of the latter space, $f \in L^\rho(\mathcal{B})$ so that $\mathcal{M} \subset L^\rho(\mathcal{B})$. Then $\mathcal{M} = L^\rho(\mathcal{B})$. The uniqueness of $\mathcal{B}$ is immediate so that all assertions are established. $\square$

The preceding work did not fully utilize the linearity of $\mathcal{M}$. It admits extensions when it is, for instance, the positive cone in the spaces. We shall discuss later in Chapter 11, some of these relaxations.

Employing the above results, it is now possible to characterize conditional probabilities from a class of vector measures into L^ρ-spaces. This will give a better understanding of their structures. Recall that, if $\nu : \Sigma \to \mathcal{X} = L^\rho(\Sigma)$, is a vector measure, ρ with the (J)-property and (Ω, Σ, μ) a localizable measure space, then the ρ'-variation of ν [ν vanishes on μ-null sets], is defined on $A \in \Sigma_0 = \{B \in \Sigma, \ \mu(B) < \infty\}$ as:

$$\rho'(\nu_A) = \sup\{\sup[| < x^*, \int_A f d\nu > | : \rho(f) \leq 1, \ f \text{ simple and}$$
$$\text{supported by } A, \ x^* \in \mathcal{X}^* : \|x^*\| \leq 1\}, \quad (26)$$

where $\nu_A(\cdot) = \nu(A \cap \cdot)$ and $< x^*, x >$ is the duality pairing of $\mathcal{X}^*$ and $\mathcal{X}$. In (26) x^*'s can be restricted to any norm determining subset of $\mathcal{X}^*$, e.g., to simple functions of $L^{\rho'}(\Sigma)$. If $\sup\{\rho'(\nu_A) : A \in \Sigma_0\} = \rho'(\nu) < \infty$, then ν has ρ'-*semivariation* finite. Here ρ' is, as usual, the (first) associate norm of ρ, and if $\rho = \|\cdot\|_\infty$ then ρ'-semivariation reduces to the classical semivariation given in Dunford and Schwartz (1958, p.320), as can be easily verified. If in (26) $f = \sum_{i=1}^{n} a_i\chi_{A_i}$, $A_i \in \Sigma_0$, and the right side is replaced by

$$\rho'[\nu] = \sup[\sup\{\sum_{i=1}^{n} |a_i| \ \|\nu(A_i)\|_\mathcal{X} : \rho(f) \leq 1, A_i \subset A\} : A \in \Sigma_0].$$

Then one has the ρ'-*variation*. However, this will not be of much use here since it is infinite in most interesting cases, and so will not be discussed further. The case that $\nu = P^\mathcal{B}$, a conditional measure, is of interest below. It inherits the contraction property of E^B in the following form, to be employed.

Lemma 11. *Under the localizable hypothesis of μ_B and the (J)-property of $\rho(\cdot)$ one has $\rho'(P^B) \leq 1$.*

Proof. Let $f = \overset{n}{\underset{i=1}{\Sigma}} a_i \chi_{A_i}$, $A_i \in \Sigma_0$, disjoint, $\rho(f) \leq 1$, and similarly $x^* = \overset{n}{\underset{j=1}{\Sigma}} b_j \chi_{B_j}$, $B_j \in \mathcal{B} \cap \Sigma_0$, disjoint, and $\rho'(x^*) \leq 1$. One can take the same n by letting some a_i's or b_j's be zero. With $\nu = P^B$, consider

$$< x^*, \int_\Omega f dP^B > = \overset{n}{\underset{i=1}{\Sigma}} \overset{n}{\underset{j=1}{\Sigma}} a_i \bar{b}_j \int_\Omega \chi_{B_j} E^B(\chi_{A_i}) d\mu_B,$$

$$\text{since} < h, k >= \int_\Omega h\bar{k} d\mu,$$

$$= \overset{n}{\underset{i=1}{\Sigma}} a_i \int_{A_i} x^* \, d\mu = \overset{n}{\underset{i=1}{\Sigma}} a_i G(A_i), \qquad (27)$$

where $G : A \mapsto \int_A x^* dP$, is a scalar measure on Σ_0, and $\rho'(G) = \rho'(x^*) \leq 1$. Thus

$$\sup\{|\overset{n}{\underset{i=1}{\Sigma}} a_i G(A_i)| : f = \overset{n}{\underset{i=1}{\Sigma}} a_i \chi_{A_i}, \rho(f) \leq 1\} = \rho'(G)$$

and since $\rho = \rho''$ and x^*'s of the above form determine the norm $\rho''(= \rho)$ [the (J)-property of ρ implies the same for ρ', cf. Prop. 2.9] we get

$$\rho'(P^B) = \sup\{\sup[| < x^*, \int_\Omega f dP^B > | : \rho(f) \leq 1, f \text{ simple}] : \rho'(x^*) \leq 1\}$$

$$= \sup\{\rho'(G) : \rho'(x^*) \leq 1\}$$

$$= \sup\{\rho'(x^*) : \rho'(x^*) \leq 1\} \leq 1, \text{ since } \rho'(x^*) = \rho'(G).$$

This is the assertion. $\square$

Since $P^B(A) = E^B(\chi_A)$, $A \in \Sigma_0$, the conditional measure P^B satisfies the functional equation:

$$\int_B E^B(\chi_A) d\mu_B = \int_B P^B(A) d\mu_B = \mu(A \cap B), \quad A \in \Sigma, B \in \mathcal{B}. \quad (27)$$

Hence the earlier properties of E^B discussed in Theorem 3 translate into the following result on P^B, and this motivates characterizations of conditional measures analogous to the case of E^B:

Proposition 12. *Let* (Ω, Σ, μ) *be a measure space* $\mathcal{B} \subset \Sigma$ *a* σ-*algebra and* $\mu_{\mathcal{B}}(= \mu|\mathcal{B})$ *be localizable. Then the conditional measure* $P^{\mathcal{B}}$: $\Sigma \to L^{\rho}(\mathcal{B})$ *has the properties: (i)* $0 \leq P^{\mathcal{B}}(A) \leq 1$, *a.e.,* $A \in \Sigma$, *(ii)* $\mu(A) = 0$ *implies* $P^{\mathcal{B}}(A) = 0$ *a.e., and* $\mu(A) > 0 \Rightarrow P^{\mathcal{B}}(A) > 0$, *a.e.,* $P^{\mathcal{B}}(\Omega) = 1$, *a.e., and (iii)* $A_n \in \Sigma$, *disjoint,* $A = \underset{n}{\cup} A_n \Rightarrow P^{\mathcal{B}}(A) =$

$$\lim_n P^{\mathcal{B}}(\overset{n}{\underset{k=1}{\cup}} A_k) = \overset{\infty}{\underset{n=1}{\Sigma}} P^{\mathcal{B}}(A_n), \text{ a.e.}$$

We now characterize vector measures on Σ into $L^{\rho}(\Sigma)$ satisfying some functional integral equations analogous to averaging and Šidák type relations, and some of those of the above proposition.

Theorem 13. *Let* $L^{\rho}(\Sigma)$ *on* (Ω, Σ, μ) *be the function space, as before, having a g.w.u.* $F_0 = \{\chi_{A_\alpha}, \alpha \in I\} \subset L^{\rho}(\Sigma) \cap L^{\rho'}(\Sigma)$ *and* ρ *with the strict monotonicity, the* (J)-, *and the a.c.n. properties. Let* ν : $\Sigma_0 \to L^{\rho}(\Sigma)$ *be a vector measure on* $\Sigma_0 = \{A \in \Sigma : \mu(A) < \infty\}$. *Then* ν *is a conditional measure relative to a unique* σ-*algebra* $\mathcal{B} \subset \Sigma, A_\alpha \in \mathcal{B}, \alpha \in I$, *iff (i)* $0 \leq \nu(A)$, $A \in \Sigma_0$, *(ii)* $\rho'(\nu) \leq 1$, *(iii)* $\nu(A_\alpha) = \chi_{A_\alpha} \in F_0, \alpha \in I$, *and either of the following integral equations holds:*

(iv) for any A, B, C *in* Σ_0,

$$\int_{\Omega} \chi_A(\nu(B) \vee \nu(C))d\mu = \int_{\Omega} \nu(A)(\nu(B) \vee \nu(C))d\mu, \tag{28}$$

(iv ') for any $A \in \Sigma_0$ *(the following is a Dunford-Schwartz integral),*

$$\nu(A) = \int_{\Omega} \nu(A)(\omega)\nu(d\omega). \tag{29}$$

When the above conditions are met, $\mathcal{B}$ *is generated by* $\{A \in \Sigma : \nu(A) = \chi_A, a.e.\}$ *and completed for* μ.

Proof. If $\nu = P^{\mathcal{B}}$ for a σ-algebra $\mathcal{B} \subset \Sigma$, with $A_\alpha \in \mathcal{B}$, $\alpha \in I$, then by Lemma 11 and Proposition 12, (i) – (iii) are true. Note that $\{A_\alpha, \alpha \in I\} \subset \mathcal{B} \Rightarrow \mu_{\mathcal{B}}$ is localizable so that $P^{\mathcal{B}}$ and $E^{\mathcal{B}}$ exist. Since $P^{\mathcal{B}}(A) = E^{\mathcal{B}}(\chi_A)$ and $E^{\mathcal{B}}$ = identity on $L^{\rho}(\mathcal{B})$, one has

$$\int_{\Omega} \chi_A(\nu(B) \vee \nu(C))d\mu = \int_{\Omega} E^{\mathcal{B}}(\chi_A)(\nu(B) \vee \nu(C))d\mu_{\mathcal{B}}.$$

so that (28) holds. Regarding (29), observe that

$$E^{\mathcal{B}}(f) = E^{\mathcal{B}}(E^{\mathcal{B}}(f)) = \int_{\Omega} E^{\mathcal{B}}(f)dP^{\mathcal{B}}, \quad f \in L^{\rho}(\Sigma).$$

as a Dunford-Schwartz integral, by Proposition 2.1. Then $\nu(B) = \chi_B$, $B \in \mathcal{B}$. Thus only the converse needs a detailed analysis.

If ν is a vector measure as in the statement, consider the linear operator $T : L^p(\Sigma) \to L^p(\Sigma)$ defined by $Tf = \int_\Omega f d\nu, f \in L^p(\Sigma)$, where the integral is in the Dunford-Schwartz sense. Now (i) implies that T is positive and (ii), with a standard property of this vector integral, gives

$$\rho(Tf) \le \rho(f)\rho'(\nu) \le \rho(f), \quad \text{by Lemma 11.} \tag{30}$$

Hence T is a contraction. By (iii), $T\chi_{A_\alpha} = \chi_{A_\alpha}, \alpha \in I$, and we use (iv). Since $\nu(A) = T\chi_A$, (and $\chi_A \in L^p(\Sigma) \Rightarrow \rho(\chi_A) < \infty$) we have by the identity (28),

$$\int_\Omega \chi_A(T\chi_B \vee T\chi_C)d\mu = \int_\Omega \chi_A T^*(T\chi_B \vee T\chi_C)d\mu, \tag{31}$$

where T^* is the adjoint of T acting on $L^{\rho'}(\Sigma)$ by the a.c.n. property. Also (i) and (iii) imply, $0 \le \nu(A) \le 1$, a.e., and $T : L^p(\Sigma) \cap L^\infty(\Sigma) \to L^p(\Sigma) \cap L^\infty(\Sigma)$ which follows for simple functions and the general case is then a consequence. But $A \in \Sigma_0$ is arbitrary and simple functions are dense in $L^p(\Sigma)$. So (31) implies the key relation:

$$T\chi_B \vee T\chi_C = T^*(T\chi_B \vee T\chi_C), \quad \text{a.e.} \tag{32}$$

Taking $C = \emptyset$ and using the linearity of the integral, it follows that (28) holds if χ_A, χ_B are replaced by f, g in $L^p \cap L^{\rho'} \cap L^\infty(\Sigma)$ and one obtains

$$\int_\Omega fTg d\mu = \int_\Omega (Tf)(Tg)d\mu. \tag{33}$$

If we let $f = \chi_A$, $A \in \Sigma(A_\alpha) = \{B \cap A_\alpha : B \in \Sigma\}$ in (33), we have

$$\int_A Tg d\mu = \int_\Omega (T\chi_A)(Tg)d\mu = \int_\Omega g\, T\chi_A d\mu, \quad \text{by (33),}$$

$$= \int_\Omega (T^*g)\chi_A d\mu = \int_A T^*g\, d\mu. \tag{34}$$

Since $\bigcup_\alpha A_\alpha \doteq \Omega$, and $A \in \Sigma(A_\alpha)$ is arbitrary in (34) it follows that the same relation holds for all $A \in \Sigma_0$, and hence $Tg = T^*g$, a.e. Substituting this in (32), one gets for all simple f, g in $L^p(\Sigma)$,

$$(Tf) \vee (Tg) = T((Tf) \vee (Tg)), a.e. \tag{35}$$

By the linearity and continuity of T this implies that T is a Šidák operator satisfying the hypothesis of Theorem 8 so that $T = E^{\mathcal{B}}$ for a unique σ-algebra $\mathcal{B} \subset \Sigma$ with the stated properties. Since $\nu(A) = T(\chi_A) = E^{\mathcal{B}}(\chi_A) = P^{\mathcal{B}}(A)$, $A \in \Sigma_0$, the result holds in this case.

Next suppose (iv') holds. Since $\rho'(\nu) \leq 1$, $Tf = \int_\Omega f d\nu$ exists again as a Dunford-Schwartz integral, and $\rho(Tf) \leq \rho(f)$, by (i) and (ii). Moreover $T : L^\rho(\Sigma) \to L^\rho(\Sigma)$ is a positive contractive linear operator. The result will then follow from Theorem 8 if we show $T^2 = T$. To see that this is indeed true, let $f = \sum_{i=1}^{n} a_i \chi_{A_i}$, $A_i \in \Sigma_0$, disjoint. By relation (29) we get

$$T^2 f = T(Tf) = \int_\Omega (Tf)(\omega)d\nu(\omega)$$

$$= \int_\Omega (\sum_{i=1}^{n} a_i \nu(A_i))(\omega)\nu(d\omega)$$

$$= \sum_{i=1}^{n} a_i \int_\Omega \nu(A_i)(\omega)\nu(d\omega)$$

$$= \sum_{i=1}^{n} a_i \nu(A_i) = Tf, \text{ by (29).} \qquad (36)$$

By linearity and density of simple functions in $L^\rho(\Sigma)$ (because of the a.c.n.), we can conclude that (36) implies $T^2 = T$, as asserted. $\square$

This result raises the possibility of characterizing $P^{\mathcal{B}}$ through another equation which might lead to the averaging identity. Indeed it is possible, and the following is such a statement.

Theorem 14. *Let $L^\rho(\Sigma)$ be as in the above theorem having a g.w.u. $F_0 = \{\chi_{A_\alpha}, \alpha \in I\}$ and $\nu : \Sigma_0 \to L^\rho(\Sigma)$ be a vector measure. Then ν is a conditional measure relative to a σ-algebra $\mathcal{B} \subset \Sigma$, iff*

(i) $0 \leq \nu(A) \leq 1$ a.e., $A \in \Sigma_0$, (ii) $\nu(A_\alpha) = \chi_{A_\alpha} \in F_0$, $\rho'(\nu) \leq 1$, and (iii) for any A, B, C in Σ_0 and $f_0 = \sum_{i=1}^{\infty} a_i \chi_{A_\alpha} \in L^\rho(\Sigma) \cap L^{\rho'}(\Sigma), \chi_{A_\alpha} \in F_0$,

$$\int_A f_0 \nu(B)\nu(C)d\mu = \int_B f_0\nu(A)\nu(C)d\mu = \int_C f_0\nu(A)\nu(B)d\mu, \qquad (37)$$

and when this holds $\mathcal{B}$ is the σ-algebra generated by $\{A : \nu(A) = \chi_A\}$ and completed for μ.

We shall not present a proof of this result. It can be established by showing that $T : f \mapsto \int_\Omega f d\nu$ is a contractive averaging because of

(37) and then apply Proposition 9. The details are as in the last result, although not entirely simple, but can also be found in Rao (1975). The special case that $L^\rho(\Sigma) = L^1(\Sigma)$ on a probability space, was established by Olson (1965) using a different argument. We shall instead turn to another method of conditioning.

7.4 Rényi's formulation as a specialization of the abstract version

In this section we show that the regular conditional measures formulated by Rényi can be regarded as a particularization of the disintegration problem treated in Section 5.4. We shall also discuss briefly its relation with another abstract integration process.

Let us recall from Section 4.2 that the distinguishing feature of the Rényi method is his third axiom, namely $P(\cdot|\cdot) : \Sigma \times \mathcal{B} \to \mathbb{R}^+$ satisfies (i) $P(A|B) = P(A \cap B|B)$ and (ii) $A \subset B \subset C \Rightarrow P(A|B)P(B|C) = P(A|C)$, where $\mathcal{B} \subset \Sigma$ is a class *not* containing $\emptyset$. It is also assumed that $\{P(\cdot|C), C \in \mathcal{B}\}$ is a family of probability measures such that $P(C|C) = 1$. Note that (i) above implies that $P(\cdot|B)$ has its support contained in B, and (ii) implies a certain "transition" property; namely, the family of measures $\{P(\cdot|B), B \in \mathcal{B}\}$ indicates the probability of a particle moving from A to C is the same as the probability of independently going from A to B and then from B to C (or given C) where $A \subset B \subset C$. This can also be regarded as a simple case of the disintegration of a measure into regular conditionals $P(\cdot|\cdot)$ discussed in Section 5.4. Let us present a little more detail.

In fact, if $\{B_n, n \geq 1\} \subset \mathcal{B}$ and $C \in \mathcal{B}$ such that the B_ns are disjoint and $C \subset \bigcup_n B_n$, and $C \cap B_n \in \mathcal{B}$, $n \geq 1$, then Proposition 4.2.3 implies the following:

$$P(A|C) = \sum_{n=1}^{\infty} P(A|B_n \cap C)P(B_n|C), A \in \Sigma. \qquad (1)$$

When $C = \bigcup_n B_n$, this is sometimes called the "total probabilities" formula in the elementary theory since $P(A|C) = P(A)$ when $P(\cdot|\cdot)$ is given the (original) ratio definition with $C = \Omega$. Suppose for simplicity that $\{B_n, n \geq 1\}$ is a collection whose union contains C. Consider a mapping $h : C \to \mathbb{N}$, the natural numbers, by setting $h(B_n \cap C) = n$. With $\mathcal{P}(\mathbb{N})$, the power set, as the σ-algebra of $\mathbb{N}$, let $\mathcal{B}_h(C) =$

$h^{-1}(\mathcal{P}(\mathbb{N}))$ and writing $P_C(\cdot) = P(\cdot|C)$, let $Q(\cdot) = P_C \circ h^{-1}$ so that $Q(\{n\}) = P_C(B_n \cap C) = P(B_n \cap C|C) = P(B_n|C)$. If $X : C \to \mathbb{R}$ is any integrable random variable, then we can apply the Radon-Nikodým theorem (since $Q(\cdot)$ is at most σ-finite) in the following form:

$$\int_{h^{-1}(\{n\})} X dP_C = \int_{B_n \cap C} X \, dP_C$$

$$= \int_{\{n\}} g_{X,C}(k) dQ(k)$$

$$= g_{X,C}(n) P_C(B_n). \tag{2}$$

Hence

$$g_{X,C}(n) = \frac{1}{P_C(B_n)} \int_{B_n \cap C} X \, dP_C. \tag{3}$$

Letting $X = \chi_A$, $A \in \Sigma$, and then writing $g_{X,C}$ as $g_{A,C}$ we get

$$g_{A,C}(n) = \frac{P_C(B_n \cap C \cap A)}{P_C(B_n)} = \frac{P_C(B_n \cap A)}{P_C(B_n)}. \tag{4}$$

But $g_{X,C}(n) = E(X|h^{-1}(\{n\}))$ in Kolmogorov's formulation, and taking $X = \chi_A$, this becomes $P_C(A|h^{-1}(\{n\}))$ since

$$P_C(A|B_n \cap C) = \frac{P_C(B_n \cap A)}{P_C(B_n)} \quad , \quad \text{substituting in (4).} \tag{5}$$

Since, by definition of $P_C(\cdot)$, the left side of (5) is $P_C(A|B_n)$, and (5) is the same as Rényi's Axiom III, namely

$$P(B_n \cap A|C) = P(A|B_n \cap C) P(B_n|C), \tag{6}$$

it is a direct consequence of Kolmogorov's conditioning.

We shall also derive (1) from the disintegration formula of the general theory, given in Theorem 5.4.4. Let (Ω, Σ) be a locally compact Borelian space, and $\Omega_2 = \mathbb{N}, \Sigma_2 = \mathcal{P}(\mathbb{N})$ be as in the last case. [Non topological case will be discussed later.] If $C \in \Sigma$ as before, $h : C \to \Omega_2$ such that if $\{B_n, n \geq 1\}$ is a partition covering C, then $h(B_n \cap C) = n$. We now identify various elements here with those of Theorem 5.4.4. Let $P_C(\cdot)(= P(\cdot|C))$ be given and $\nu_h(\cdot) = P_C \circ h^{-1}$. Taking $\alpha(\cdot) = \nu_h$ in that theorem, we conclude the existence of a measure $Q^h : \Sigma \times \Omega_2 \to \mathbb{R}^+$

such that (i) $\rho(Q^h(\cdot,\cdot)) = Q^h(\cdot,\cdot)$, where ρ is a lifting operator, (ii) $Q^h(\Omega, n) = \frac{d\nu_h}{d\alpha} = 1$, and (iii)

$$\int_\Omega g(t) P_C(dt) = \int_{\Omega_2} \left(\int_{\Omega_1} g(t) Q^h(dt, n) \right) d\nu_h(n), \tag{7}$$

for each continuous g vanishing off compact sets, and supp $(Q^h(\cdot, n))$ is contained in $h^{-1}(\{n\}) = B_n \cap C$. (We do not explain about lifting again than what is in Chapter 5; see, Rao, 1987, Chap. 8.) Now (7) can be extended for all bounded Borel functions vanishing outside sets of finite $P_C(\cdot)$ measure. Taking $g = \chi_A, A \in \Sigma$, the relation (7) becomes

$$P_C(A) = \sum_{n=1}^{\infty} Q^h(A, n) P_C(h^{-1}(\{n\})).$$

This is simply

$$P(A|C) = \sum_{n=1}^{\infty} Q^h(A, n) P(B_n|C). \tag{8}$$

But $Q^h(A, n) = g_{A,C}(n) = P(A|B_n \cap C)$ by (5). Putting this in (8) one sees that it again reduces to (1).

If there is no topology, then we introduce one in (Ω, Σ) by the Fréchet metric $d(\cdot,\cdot) : \Sigma \times \Sigma \to \mathbb{R}^+$ given by $d(A, B) = P_C(A \Delta B)$. Then (Σ, d) becomes a complete semimetric space in which $P_C(\cdot)$ is continuous, and one may extend (8) to this formulation. [All these, of course, depend on the fixed C.] On the other hand the property expressed by (7) [and (8)] is taken as a definition (or Axiom III) by Rényi who considered the resulting properties without invoking any topological hypotheses. This shows the generality of Kolmogorov's formulation when looked at abstractly.

We finally note an interesting relation of $P(\cdot|\cdot)$ when considered as a mapping $\tilde{P}(\cdot) : \mathcal{B} \to \mathcal{M}(\Sigma)$ where $\tilde{P}(C) = P(\cdot|C)$ is a probability measure, and $\mathcal{M}(\Sigma)$ is the Banach space of countably additive scalar measures on Σ, with total variation norm. Here $\tilde{P}(\cdot)$ is not σ-additive, but has certain monotonicity properties. We now show that (1) (and (8)) can be interpreted as a (vector) form of the abstract Burkill-Kolmogorov integral of cell functions as generalized by Romanovski (1941) from intervals to metric spaces. Both of these authors only treated scalar functions, and the last author's work was extended to vector valued functions by Soloman (1969). Still a further extension is needed in our context. We shall indicate the desirable ideas for the present work which, hopefully, will induce some future research in this direction.

As noted above we consider $\tilde{P}(\cdot) : \mathcal{B} \to \mathcal{M}(\Sigma)$, and for any finite system $\mathcal{I}$ of non overlapping (i.e., interiors are disjoint in a topological context when the given measure gives zero mass for the boundaries) sets from $\mathcal{B}$, define, if $\mathcal{I} = \{B_1, \ldots, B_n\} \subset \mathcal{B}$,

$$\mathcal{P}(\mathcal{I}) = \sum_{k=1}^{n} \tilde{P}(B_k), \tag{9}$$

which need not be additive. Given a set $R_0 \in \mathcal{B}$, and $\mathcal{I}$ as above let $x^* \in (\mathcal{M}(\Sigma))^*$, and consider

$$x^*(\mathcal{P}(\mathcal{I})) = \sum_{k=1}^{n} x^*(\tilde{P}(B_k)). \tag{10}$$

We set, with $\mu : \Sigma \to \bar{\mathbb{R}}^+$ a given localizable measure,

$$x^*(\mathcal{P}(R_0)) = \lim_{\delta \downarrow 0} \sup\{x^* \circ \mathcal{P}(\mathcal{I}) : \quad \text{all possible } \mathcal{I} \subset \mathcal{B} \text{ with } B_k \subset R_0,$$

and $\max_{k \leq n} \mu(B_k) \leq \delta\}$.

This may be termed the (weak) upper Burkill-Kolmogorov integral, denoted $\bar{B}(x^* \circ \tilde{P}, R_0, \mu)$. Similarly, using 'inf' in lieu of 'sup' one gets a corresponding (weak) lower integral, denoted $\underline{B}(x^* \circ \tilde{P}, R_0, \mu)$. If $-\infty < \underline{B}(x^* \circ \tilde{P}, R_0, \mu) = \bar{B}(x^* \circ \tilde{P}, R_0, \mu) < \infty$ for each $x^* \in (\mathcal{M}(\Sigma))^*$, and if there is a $\bar{P}(R_0) \in \mathcal{M}(\Sigma)$ such that $(x^* \circ \bar{P})(R_0) = B(x^* \circ \tilde{P}, R_0, \mu)$, the common value above, then $\bar{P}$ will be called the (weak) Burkill-Kolmogorov (vector) integral of $\tilde{P}$ on R_0 of $\mathcal{B}$. The existence of the limit and the properties of the integral, assuming it exists, will shed considerable light on $\tilde{P}(\cdot)$. These conditions are not known at present. The work has to be obtained by considering various modifications of the available (very few) vector treatments such as Soloman's (1969). The defined integral here is motivated by the formula (1) [or (8)] of Rényi's method. Traditionally μ was taken to be a nonatomic regular measure on $\mathcal{B}$ when Ω is a locally compact second countable space. The general case that is motivated by Rényi's specialized approach has to use μ as a localizable measure and the set $\mathcal{B}$ should be "rich enough" to admit some Vitali conditions of differentiation theory (see Hayes and Pauc,

1970). We thus raise the problem, based on the abstract conditioning ideas, for a possible future research topic, and terminate the discussion at present.

7.5 Conditional measures and differentiation

The preceding analysis and the work of Chapters 3 and 5 already showed that abstract differentiation is unavoidable for the conditioning theory. In this section we therefore present a few special methods and results for calculation of conditional expectations of the type $E(X|Y)$ when Y takes values in a group, complementing the last two sections of Chapter 5. Especially, we shall present here an extension of Corollary 5.6.5 to the case of certain locally compact groups including the general linear group $GL(n, \mathbb{R})$.

Definition 1. Let G be a locally compact group with a (left) Haar measure λ on it. A decreasing sequence $\{U_k, k \geq 1\}$ of λ-measurable subsets of G is called a *D-sequence* ('D' stands for 'differentiation') if there is an absolute constant $0 < C < \infty$ such that

$$0 < \lambda(U_k \cdot U_k^{-1}) < C\lambda(U_k), \quad k \geq 1, \tag{1}$$

where $(a, b) \to a.b$ and $a \to a^{-1}$ are the group operations. If each U_k is Borel (open, closed, compact or relatively compact) then the set $\{U_k, k \geq 1\}$ is termed a *Borel* (respectively open, etc.) *D-sequence*. If, moreover, each neighborhood of the identity of G contains some U_k, then the $\{U_k, k \geq 1\}$ is called a *D'-sequence*. If further there is a λ-measurable V_k for each k, such that

$$V_k \cup (V_k V_k^{-1}) \subset U_k, \quad \text{and} \quad \lambda(U_k) < C'\lambda(V_k) \tag{2}$$

for an absolute $0 < C' < \infty$, then the $\{U_k, k \geq 1\}$ is called a *D''-sequence*. (Thus $D'' \Rightarrow D' \Rightarrow D$ for a sequence.)

The point of these D-sequences is that they enable us to obtain an analog of the classical Vitali covering lemma which is fundamental to the (Lebesgue) differentiation theory. We now present this, adapting the Edwards and Hewitt (1965) procedure, and then apply these results to calculate exactly conditional expectations of the type $E(X|Y)$ when Y is a G-valued random variable (or vector), G with a D-sequence.

The following result gives an idea of the class of groups admitting D or rather D''-sequences.

Theorem 2. *Every Lie group G admits a D''-(hence D'-and D-) sequence.*

Proof. We recall, for the present purpose, that a topological group G is a Lie group if the mapping $(x, y) \to xy^{-1}$ of $G \times G$ onto G, is infinitely differentiable (or a C^∞-map). Such a group has a sequence of compact neighborhoods [can be taken decending] $\{W_k, k \geq 1\}$ of the identity e such that $\underset{k}{\cap} W_k = \{e\}$, and can assume to satisfy $\lambda(W_k W_k^{-1} W_k W_k^{-1}) \leq C\lambda(W_k)$ for some fixed $C > 0$. This is a consequence of the basic structure theory of Lie groups, but a brief construction is sketched below. (A good source on this subject is Montgomery and Zippin, 1955.) If we take $U_k = W_k W_k^{-1}$, $V_k = W_k$ in Definition 1, then $\{U_k, k \geq 1\}$ satisfies the conditions of a D''-sequence. We shall show how the W_k may be found in G.

A Lie group has all its basic neighborhoods homeomorphic to those of a Euclidean space $\mathbb{R}^m$ for some $m(< \infty)$, called its dimension. Thus let m be the dimension of G, and let $(t_1, \ldots, t_m)$ be a local coordinate system of a neighborhood N ($\bar{N}$ compact) of 'e' such that $t_j(e) = 0$ so that identity goes to identity. Then the mapping $T : g \mapsto (t_1(g), \ldots, t_m(g))$ of N onto a neighborhood of $\mathbb{R}^m$, can be extended to be a homeomorphism of G onto $\mathbb{R}^m$. If $f : G \to \mathbb{R}$ is a continuous compactly based function whose support is contained in N, then the change of variables formula of real analysis shows

$$\int_G f \, d\lambda = \int_N f \, d\lambda = \int_{\mathbb{R}^m} (f \circ T^{-1})(\mathbf{x})F(\mathbf{x})d\mathbf{x} \qquad (3)$$

where $F(\mathbf{x})$ is the (strictly) positive Jacobian of the transformation which is continuous on $\mathbb{R}^m$, $d\mathbf{x}$ being the Lebesgue measure on $\mathbb{R}^m$. If now we consider an ε-cube, $Q_\varepsilon = \{\mathbf{x} \in \mathbb{R}^m : |x_i| \leq \varepsilon, 1 \leq i \leq m\}$ then, using the differentiability of the coordinate functions, we have $W_k = T^{-1}(Q_{\varepsilon_k})$ where $\varepsilon_k \downarrow 0$ is any sequence, as desired. $\square$

This result implies that $GL(n, \mathbb{R}^n), \mathbb{R}^n, \mathbb{T}^n$, and several other groups have D''-sequences. It can also be shown that every *finite* dimensional compact group G admits a D''- sequence; and also locally compact 0-dimensional groups satisfying the first countability axiom for the neighborhood system have D''-sequences (Edwards and Hewitt, 1965). [Here

0-dimensional means the neighborhood system consists of clopen ($=$ closed-open) sets.]

We have the following result, adapted from Edwards and Hewitt:

Theorem 3. *Let $f \in L^2(G, \lambda)$, and $\{U_k, k \geq 1\}$ be a D'-sequence in G. Then*

$$\lim_{k \to \infty} \frac{1}{\lambda(U_k)} \int_{xU_k} f \, d\lambda = f(x), a.e.[\lambda], \ x \in G. \tag{4}$$

Proof. First we assume, as we may, that the U_k are relatively compact. Since f is compactly based and $\lambda(\bar{U}_k) < \infty$, we can consider a sequence $\{A_n, n \geq 1\}$ of nonempty open sets with $\lambda(A_n) < \infty$, and restrict to $L(\tilde{A}, \lambda) = \{f \in L^1(G, \lambda) : \text{supp}\,(f) \subset \underset{n}{\cup} A_n = \tilde{A}, \text{ a } \sigma\text{-finite set }\}$. Consider the semi-norm, $\|\cdot\| : f \mapsto \int_{A_k} |f| d\lambda, k \geq 1$. Then $L(\tilde{A}, \lambda)$ will be a Fréchet space relative to this countable collection of norms, and $C_c(\tilde{A})$, the set of compactly based continuous funtions on $\tilde{A}$, is dense in $L(\tilde{A}, \lambda)$. Since $\lambda(U_k) \to 0$, it follows that for $f \in C_c(\tilde{A})$, (4) holds for each $x \in G$. In order to extend it to $L(\tilde{A}, \lambda)$, we need to appeal to (an analog of) a theorem of Banach (see Dunford and Schwartz, 1958, Thm. IV 11.3, and its current formulation by Edwards and Hewitt (1965), Thm. 1.6) is as follows. Let

$$(T_k f)(x) = \Big| \frac{1}{\lambda(U_k)} \int_{xU_k} f(x) d\lambda(t) - f(x) \Big|. \tag{5}$$

The key point here is to show that $\sup_k (T_k f)(x) < \infty$, a.e., for each $f \in L(\tilde{A}, \lambda)$, i.e., the sequence $\{T_k, k \geq 1\}$ of sublinear operators is pointwise bounded, uniformly in $k \geq 1$. This depends upon an extension of the Vitali covering lemma to the D'-sequences, and is not entirely simple. (The necessary computation are given by Edwards and Hewitt in the paper cited above.) Since for each $x \in G$, $\lim_{k \to \infty} (T_k f)(x) = 0$, the Banach Theorem noted above implies that $(T_k f)(x) \to 0$, a.e. for all $f \in L(\tilde{A}, \lambda)$ as $k \to \infty$. This is the assertion of (4). $\quad\square$

We have omitted the computations leading to the Vitali covering lemma. With that result, Theorem 3 is really an extension of Lebesgue's classical differentiation theorem with $G = \mathbb{R}$ (or $\mathbb{R}^k$), (see Rao, 1987, p. 237). Using similar ideas, Edwards and Hewitt (1965) also proved the following statement.

Proposition 4. *Let G be a locally compact group admitting a D'-sequence, $\{U_k, k \geq 1\}$. If μ is a signed measure on G such that $\mu \perp \lambda$ (i.e., μ is singular relative to λ), then one has*

$$\lim_{k \to \infty} \frac{\mu(xU_k)}{\lambda(U_k)} = 0, \quad a.e. \ [\lambda], \quad on \quad G. \tag{6}$$

With these differentiation results, we can establish the following general assertion on the evaluation of the conditional expectation (hence measure), of interest in many applications.

Theorem 5. *Let (Ω, Σ, P) be a probability space, G be a locally compact group admitting a D'-sequence $\{U_k, k \geq 1\}$, $X : \Omega \to \mathbb{R}$ be an integrable random variable and $Y : \Omega \to G$ be any measurable mapping (i.e., a G-valued random variable). If $P_Y (= P \circ Y^{-1})$ is the image measure of P on G, let $f_Y = \frac{dP_Y^c}{d\lambda}$, the density of the absolutely continuous part of P_Y relative to λ, which exists. Then the conditional expectation $E(X|Y)$ can be evaluated as:*

$$E(X|Y)(r)f_Y(r) = \lim_{k \to \infty} \frac{1}{\lambda(U_k)} E(\chi_{rU_k}(Y) \cdot X), \quad a.e.[\lambda], \tag{7}$$

so that $E(X|Y)$ is uniquely calculable on the set $\{r : f_Y(r) > 0\}$.

Proof. The main idea here is to use the identity $E(Z) = E(E^{\mathcal{B}}(Z))$, for any integrable random variable Z and any σ-algebra $\mathcal{B} \subset \Sigma$. We now choose Z and $\mathcal{B}$ suitably (see Prop. 2.2.1, regarding this property), which was also the basis for Theorem 5.6.1. Let h be a bounded Borel function, $h : G \to \mathbb{R}$, and consider $Z = Xh(Y)$ in the above with $\mathcal{B} = \mathcal{B}_Y = Y^{-1}(\mathcal{G})$, where $\mathcal{G}$ is the Borel σ-algebra of G. Then

$$
\begin{aligned}
E(Z) &= E(E(Xh(Y)|\mathcal{B}_Y)) \\
&= E(h(Y)E(X|\mathcal{B}_Y)), \text{ since } h(Y) \text{ is } \mathcal{B}_Y\text{-adapted and} \\
&\qquad\qquad \text{apply Prop. 2.2.1,} \\
&= \int_\Omega h(Y)E(X|Y)dP_{\mathcal{B}_Y}, \text{ by definition,} \\
&= \int_G h(r)E(X|Y)(r)dP_Y(n), \text{ by the fundamental law}
\end{aligned}
$$

of Probability (Rao, 1987, p.155). $\hspace{4cm}$ (8)

Now P_Y is a (finite) measure on G and λ is localizable being a (left) Haar measure. Hence we can apply the Lebesgue decomposition of P_Y to get $P_Y = P_Y^c + P_Y^s$ where $P_Y^c \ll \lambda$ and $P_Y^s \perp \lambda$. (See Dunford-Schwartz, 1958, Thm. III.4.14.) Then by the general form of the Radon-Nikodým theorem (Rao, 1987, p.275), $\frac{dP_Y^c}{d\lambda} = f_Y \geq 0$ exists a.e. $[\lambda]$, and (8) can be expressed as:

$$E(Xh(Y)) = \int_G h(r)E(X|Y)(r)f_Y(r)d\lambda + \int_G h(r)E(X|Y)(r)dP_Y^s(r).$$
(9)

Let $h(r) = \frac{1}{\lambda(U_n)}\chi_{xU_n}(r)$, $r \in G$ in the above for an arbitrarily fixed n. Then (9) becomes (using the left translation invariance of λ)

$$\frac{1}{\lambda(U_n)}E(X\chi_{xU_n}(Y)) = \frac{1}{\lambda(U_n)}\int_{xU_n} E(X|Y)(r)d\lambda(r)$$
$$+ \frac{1}{\lambda(U_n)}\int_{xU_n} E(X|Y)(r)dP_Y^s(r).$$
(10)

If the last integral in its indefinite form is denoted by $\mu(\cdot)$, then μ is a signed measures, $\mu \perp \lambda$ and by Theorem 3 and Proposition 4, the limits on the right side of (10) exist as $n \to \infty$ since $\{U_n, n \geq 1\}$ is a D'-sequence. So the left side limit must also exist, and we get

$$E(X|Y)(x)f_Y(x) = \lim_{n\to\infty} \frac{1}{\lambda(U_n)}E(X\chi_{xU_n}(Y)), \quad a.a.(x).$$

This gives (7), as desired. $\square$

In particular, if $P_Y \ll \lambda$ in the above, one has the following:

Corollary 6. *Under the conditions of the theorem, if $P_Y \ll \lambda$, then one gets*

$$E(X|Y)(x) = \lim_{n\to\infty} \frac{1}{\lambda(U_n)f_Y(x)}E(X\chi_{xU_n}(Y)), \quad a.a.(x), \qquad (11)$$

for all x for which $f_Y(x) > 0$. In the special case that $P_Y \sim \lambda$ (is equivalent to λ), then (11) holds for a.a. (x) on G.

A simple illustration of formula (11) is Example 5.6.6 wherein $G = \mathbb{R}$ and $U_n = (-a_n, a_n)$ where $a_n \downarrow 0$ as $n \to \infty$, and $a_n \leq c\, a_{n+1}, c > 0$. The examples of Section 3.3 have sequences that are *not* of D'-or of D-type.

Although these results cover a large class of groups G in applications, they do not exhaust all groups. For instance, the infinite cartesian product of the torus group does not admit a D-sequence. But in the case of locally compact abelian groups we can apply the method of Fourier transforms as shown in Section 5.6. Thus one has to employ various (often nontrivial) ideas to evaluate the conditional expectations. We therefore conclude this abstract study at this point.

7.6 Bibliographical notes

The work presented in this chapter is an abstraction of the Kolmogorov model and forms an ellucidation of its functional analytic character. The conditional Jensen inequality for general (non-finite) measures and convex functions, as given in Proposition 2.2, is adapted from Y.S. Chow (1960), and the first complete characterization of conditional expectations is due to S.C. Moy (1954), followed by G.C. Rota (1960) in the context of L^p-spaces on a probability space. Because of their applicational potential we treated the material mostly on general measure spaces and on function spaces $L^p(\Sigma)$. The presentation in the text in both Sections 2 and 3, follows the author's papers (Rao (1975, 1976)), with some simplications and elaborations.

The comparison of Rényi's work in Section 4 is briefly noted earlier (Rao, 1981). It leads to a study of abstract Burkill-Kolmogorov-Denjoy-Peron vector integration, the ramifications of which are largely unknown at this time. A good account of the latter for *scalar* measures is found in the monograph by V. Čelidze and A. Džvaršcišvile (1989).

The evaluation of conditional expectations of the type when conditioning can be given for a group valued random variable is adapted from the recent work by the author (Rao, 1992). Here the basic role played by the results of Edwards and Hewitt (1965) was apparently not recognized in the probabilistic literature until now. It is clear that one has to devise other methods; or simplifications of the general results of differentiation theory in probabilistic settings have to be found.

It is hoped that the last part of this chapter highlights the nontrivial problems in evaluation of conditional measures or expectations that have been neglected so far in the literature. We shall show later in Chapter 9 how these problems interfere in some parts of Probability Theory for which these considerations are important. To complete the picture, we discuss the product conditional measures in the next chapter, as they play a key role in the existence theory of Markov processes and in many other applications.

CHAPTER 8: PRODUCTS OF CONDITIONAL MEASURES

The basic aim of this chapter is to formulate and prove the existence of a probability space determined by an infinite product of conditional probability measures, relating it to a projective system of measures admiting a limit. This plays a fundamental role in establishing the existence results for various classes of Markov processes (cf. the following chapter) and other stochastic functions. Some consequences and its essential equivalence with the problem of disintegration of topological measures together with the appearance of lifting operators in this subject are discussed in adequate details.

8.1 Introduction

It is already seen in the preceding chapter how an abstract formulation of conditioning can be viewed as a functional operation. On the other hand, iterated conditioning is an essential part of the theory and it is important for many applications. Indeed there is a finite product conditional measure result generalizing the classical Fubini theorem on products of marginal measures. Then there exist infinite product versions jointly extending the corresponding Fubini-Jessen and Kolmogorov-Bochner theorems. After proving the finite case, we establish a useful mechanism to formulate the infinite product case with detailed analysis in the next section. These utilize the order of the product formation crucially. In the third section a general infinite product result, based on the regularity of conditional (component) measures, when they exist, is established. Several consequences of this result are given in Section 4 and various interrelations between the regularity of conditioning on topological measurable spaces, and the disintegration of measures on these spaces along with the properties of a strong lifting that intervenes, are treated there and in the fifth section.

To motivate the problem, recall that if X is an integrable random variable on a probability space (Ω, Σ, P) and $\mathcal{B}_1 \subset \mathcal{B}_2 \subset \cdots \subset \mathcal{B}_n \subset \Sigma$ are σ-subalgebras, then we have

$$E(X) = E(E^{\mathcal{B}_1}(\cdots(E^{\mathcal{B}_n}(X))\cdots)). \tag{1}$$

Taking $X = \chi_A$, $A \in \Sigma$, (1) gives the equivalent formulation as:

$$P(A) = \int_\Omega \left(\int_\Omega \cdots \int_\Omega P^{\mathcal{B}_n}(A) dP^{\mathcal{B}_{n-1}} \cdots dP^{\mathcal{B}_1} \right) dP, \tag{2}$$

where $P^{\mathcal{B}_i}$ are the corresponding conditional probability functions and the integrals in (2), inside parentheses, are in Dunford-Schwartz sense. Here the measure $\mu : \Sigma \to \bar{\mathbb{R}}^+$ can replace P if $\mu_{\mathcal{B}_i}$ is σ-finite (or even localizable) for each i (cf. also Corollary 7.2.5). This will be extended.

The desired generalization of Fubini's theorem can be presented as follows:

Theorem 1. *Let* (Ω_i, Σ_i), $i = 1, 2$, *be measurable spaces,* $\mu_1 : \Sigma_1 \to \bar{\mathbb{R}}^+$ *be σ-additive, and* $\nu : \Sigma_2 \to L^p(\mu_1)$ *be such that (i)* $0 \le \nu(A) \le 1$ *and* $\nu(\Omega) = 1$ *a.e.* $[\mu_1]$, *and (ii)* ν *is σ-additive in the norm* $(1 \le p < \infty)$ *or order* $(p = +\infty)$ *topologies. Then there exists a σ-additive* μ *on the product measurable space* $(\Omega_1 \times \Omega_2, \Sigma_1 \otimes \Sigma_2)$ *satisfying*

$$\mu(A \times B) = \int_A \nu(B)(\omega) d\mu_1(\omega), \quad A \in \Sigma_1^0 = \{A \in \Sigma_1 : \mu_1(A) < \infty\}, \tag{3}$$

for each $B \in \Sigma_2$, *and that* $\mu(\cdot \times \Omega) = \mu_1(\cdot)$ *on* Σ_1^0. *Moreover, if* $f : \Omega_1 \times \Omega_2 \to \mathbb{R}$ *is measurable and μ-integrable then*

$$\int_{\Omega_1 \times \Omega_2} f \, d\mu = \int_{\Omega_1} \left[\int_{\Omega_2} f(\omega_1, \omega_2) \nu(d\omega_2)(\omega_1) \right] d\mu_1(\omega_1). \tag{4}$$

The inner integral on the right of (4) *is in the sense of Dunford-Schwartz [and Wright's for order topology] and all the others are Lebesgue integrals.*

Remark. If $\nu : \Sigma_2 \times \Omega_1 \to \mathbb{R}^+$ has the stronger property that $\nu(\cdot, \omega_1)$ is σ-additive for each $\omega_1 \in \Omega_1$ in addition to being $\nu(B, \cdot)$ measurable for Σ_1, so that it behaves as a regular conditional measure, then all the integrals in (4) can be regarded as Lebesgue integrals. When ν is only a vector measure, as in the statement, one has to invoke the vector integration concepts, given in the last statement above.

Proof. The result of (3) is an easy generalization of the classical case, (see Rao (1987), p. 315). We include the detail for completeness. Consider the measurable rectangle $A \times B$ of $\Sigma_1^0 \times \Sigma_1$ and define μ as in (3). Note that $\Sigma_1^0 \times \Sigma$ is a semi-ring on which μ is additive. To see that it is in fact σ-additive, consider a disjoint sequence $\{A_i \times B_i, \ i \geq 1\}$ from $\Sigma_1^0 \times \Sigma_2$ with union, say $A \times B$, in $\Sigma_1^0 \times \Sigma_2$ again. Using the fact that $\chi_{A \times B} = \chi_A \chi_B$, one has

$$\chi_{A \times B} = \sum_{i=1}^{\infty} \chi_{A_i} \chi_{B_i} = \lim_{n \to \infty} \sum_{i=1}^{n} \chi_{A_i} \chi_{B_i}. \tag{5}$$

This is a convergent series since for each point $(\omega_1, \omega_2) \in A \times B$, all but one term on the right side vanish, and the last sums are nondecreasing. In the $L^p(\mu_1)$-context, norm convergence implies order convergence for the vector measure ν, and the monotone convergence theorem holds for the integral [cf. Wright (1969)]. So integrating (5) relative to ν, and simplifying one gets

$$\chi_A \nu(B) = \lim_{n \to \infty} \sum_{i=1}^{n} \chi_{A_i} \nu(B_i). \tag{6}$$

Integrating this series of nonnegative terms relative to μ_1 gives

$$\mu(A \times B) = \sum_{i=1}^{\infty} \mu(A_i \times B_i).$$

Now applying the standard Carathéodory procedure for the pair $(\Sigma_1^0 \times \Sigma_2, \mu)$ one concludes that μ has a σ-additive extension to $\Sigma_1 \otimes \Sigma_2$ and this extension, denoted by the same symbol, is the desired measure. [Regarding this standard extension procedure, see, for instance, Rao (1987), p. 41.]

If $f = \sum\limits_{i=1}^{n} a_i \chi_{A_i \times B_i}$ is a simple function, $A_i \times B_i \in \Sigma_1^0 \times \Sigma_2$, then (4) holds. Using the monotone convergence property, it can be extended for all integrable $f \geq 0$, measurable for $\Sigma_1 \otimes \Sigma_2$. Finally if f is any μ-integrable function, and $\varepsilon > 0$ is given, then there exists a μ-integrable simple function f_ε such that

$$\int_{\Omega_1 \times \Omega_2} |f - f_\varepsilon| d\mu < \varepsilon.$$

Since (4) holds for f_ε, this approximation implies [as in the classical case, see Rao (1987), p. 325], that (4) holds as stated. $\square$

The following specialization for a conditional measure, of the above result, will be useful in applications.

Corollary 2. *Let* $(\Omega_i \Sigma_i)$, $i = 1, 2$, *be measurable spaces* $h : \Omega_1 \to \Omega_2$ *be a measurable mapping and* $\mu : \Sigma_1 \to \mathbb{R}^+$ *be a probability measure. If* $\mu_2 = \mu_1 \circ h^{-1}$ *is the image probability, then there is a conditional measure* $P_{12} : \Sigma_1 \times \Omega_2 \to \mathbb{R}^+$ *such that*

$$\mu_1(A \cap h^{-1}(B)) = \int_B P_{12}(A, \omega_2) d\mu_2(\omega_2), \quad A \in \Sigma_1, \ B \in \Sigma_2. \qquad (7)$$

Since $\mu_1(A \cap h^{-1}(\cdot))$ and $\mu_2(\cdot)$ are measures on Σ_2 for each $A \in \Sigma_1$, and the former vanishes on μ_2 null sets, $P_{12}(A, \cdot)$ is simply the Radon-Nikodým derivative of it relative to μ_2. Also the σ-additivity of $\mu_1(\cdot \cap h^{-1}(B))$ immediately gives the same property of $P_{12}(\cdot, \omega_2)$ for almost all $\omega_2 \in B$. This implies all assertions.

The above result extends to finitely many factors and the following form of it will be useful for later considerations. Note that if there are n-probability spaces, $n \geq 2$, then there are $\binom{n}{2}$ mappings h_{ij} in contrast to a single h when $n = 2$, and hence we need to connect the measures to take all these relations into account. This is a new problem here.

Proposition 3. *Let* $(\Omega_i, \Sigma_i, P_i)$, $i = 1, 2, \cdots, n$, *be a family of probability spaces and* $h_{ij} : \Omega_j \to \Omega_i$ *be* (Σ_j, Σ_i)-*measurable mappings such that for* $1 \leq i < j < k \leq n$ *(1)* $h_{ij} \circ h_{jk} = h_{ik}$, *(2)* $h_{ii} =$ *identity, and (3)* $P_i = P_j \circ h_{ij}^{-1}$. *If for each* $i < j$, $q_{ij} : \Sigma_j \times \Omega_i \to \mathbb{R}^+$ *is the function given by (cf. (7)):*

$$P_j(B \cap h_{ij}^{-1}(A)) = \int_A q_{ij}(B, \omega) \, dP_i(\omega), \quad A \in \Sigma_i, \ B \in \Sigma_j, \qquad (8)$$

then for any $1 \leq i_1 < i_2 < \cdots < i_k \leq n$, we have

$$q_{i_1 i_k}(A, \omega) =$$
$$\int_{\Omega_{i_2}} q_{i_1 i_2}(d\omega_2, \omega) \int_{\Omega_{i_3}} q_{i_2 i_3}(d\omega_3, \omega_2) \cdots \int_{\Omega_{i_k}} \chi_A q_{i_{k-1} i_k}(d\omega_k, \omega_{k-1}),$$

$$(9)$$

for almost all $\omega[P_{i_1}]$ where all are Dunford-Schwartz integrals and they are evaluated from right to left. Moreover, each vector measure q_{ij} : $\Sigma_j \to L^2(P_i)$ has total variation measure P_i and the integrals in (9) can be taken in the Lebesgue sense when each q_{ij} is regular in that $q_{ij}(\cdot, \omega)$ is σ-additive on Σ_j for each $\omega \in \Omega_i$ and $q_{ij}(A, \cdot)$ is Σ_i-measurable for each $A \in \Sigma_j$.

Proof. Observe that each $q_{ij}(A, \cdot)$ is a bounded measurable function so that all the Dunford-Schwartz integrals in (9) are well-defined. Using the relations (1)–(3) on h_{ij} and the consistency conditions on P_i, we have the inclusions:

$$\mathcal{B}_1 = h_{i_1 i_k}^{-1}(\Sigma_{i_1}) = (h_{i_1 i_2} \circ h_{i_2 i_k})^{-1}(\Sigma_{i_1}) \subset h_{i_2 i_k}^{-1}(\Sigma_{i_2}) = \mathcal{B}_2 ,$$

and similarly $h_{i_2 i_k} = h_{i_2 i_3} \circ h_{i_3 i_k}$ implies $(\mathcal{B}_3 = h_{i_3 i_k}^{-1}(\Sigma_3))$

$$\mathcal{B}_1 \subset \mathcal{B}_2 \subset \mathcal{B}_3 \subset \cdots \subset \mathcal{B}_{k-1} \subset \Sigma_{i_k}. \qquad (10)$$

Hence (9) can be verified to be another form of (2). However, we shall establish it by induction, without using (2), for variety, as in Choksi (1958).

In fact for $k = 1$, (9) is just a trivial identity for $q_{i_1 i_1}$. Suppose it is true for $k > 1$. Now by (8) with $A \in \Sigma_{i_1}$, and $B \in \Sigma_{i_k}$, one has:

$$P_{i_k}(B \cap h_{i_1 i_k}^{-1}(A)) = \int_A q_{i_1 i_k}(B, \omega_1) dP_{i_1}(\omega_1)$$
$$= \int_A dP_{i_1}(\omega_1) \int_{\Omega_{i_2}} q_{i_1 i_2}(d\omega_2, \omega_1) \cdots \times$$
$$\int_{\Omega_{i_{k-1}}} \chi_B q_{i_{k-2} i_{k-1}}(d\omega_{k_1}, \omega_{k-2}) ,$$

by the inductive hypothesis on $q_{i_1 i_k}$. $\qquad (11)$

Taking $A = \Omega_{i_1}$ here and extending it for simple functions and then for all bounded measurable $f : \Omega_{i_k} \to \mathbb{R}$, which is valid for our vector integrals, one gets the following identity from (11):

$$\int_{\Omega_{i_k}} f(\omega) dP_{i_k}(\omega) = \int_{\Omega_{i_1}} dP_{i_1}(\omega_1) \int_{\Omega_{i_2}} q_{i_1 i_2}(d\omega_2, \omega_1) \cdots \times$$
$$\int_{\Omega_{i_{k-1}}} f(\omega) q_{i_{k-1} i_k}(d\omega, \omega_{k-1}). \tag{12}$$

Apply (8) for $P_{i_{k+1}}$ with $f(\omega) = q_{i_k i_{k+1}}(B, \omega), B \in \Sigma_{i_{k+1}}$, to get:

$$P_{i_{k+1}}(B) = \int_{\Omega_{i_k}} q_{i_k i_{k+1}}(B, \omega) dP_{i_k}(\omega) \tag{13}$$

$$= \int_{\Omega_{i_1}} dP_{i_k}(\omega_1) \int_{\Omega_{i_2}} q_{i_1 i_2}(d\omega_2, \omega_1) \cdots \times$$
$$\int_{\Omega_{i_k}} q_{i_k i_{k+1}}(B, \omega_k) q_{i_{k-1} i_k}(d\omega_k, \omega_{k-1}), \tag{14}$$

where we used the change of variables formula for the Lebesgue integral in going from (13) to (14), since (13) involves only a Lebesgue inegral. But $B \in \Sigma_{i_{k+1}}$ is arbitrary and hence the integrands of (13) and (14) can be identified. Thus (9) holds for $k + 1$, and hence the result is true as stated. $\square$

   The interest in (9) is seen from the following type of application. Let $\{(\Omega_i, \Sigma_i, P_i), 0 \le i \le T\}$ be a family of measure spaces, $\mathcal{F}$ the directed (by inclusion) set of all finite subsets of $[0, T]$ and if $\alpha \in \mathcal{F}$, $\Omega_\alpha = \underset{i \in \alpha}{\times} \Omega_i$, $\Sigma_\alpha = \underset{i \in \alpha}{\otimes} \Sigma_i$ and $P_\alpha : \Sigma_\alpha \to \mathbb{R}^+$ be given such that $P_\alpha = P_\beta \circ \pi_\beta^{-1}$ where $\pi_{\alpha\beta} : \Omega_\beta \to \Omega_\alpha$ are coordinate projections ($\alpha < \beta$ so that $\alpha \subset \beta$). Let $\alpha_1 < \alpha_2 < \cdots < \alpha_n$. In this application (9) can be expressed with $0 < t_0 < t_1 < \cdots < t_n, \alpha_i = \{t_0, t_1, \ldots, t_i\}$, $A = A_0 \times \cdots \times A_n, (A_i \in \Sigma_{t_i})$ and $\omega = (\omega_{t_0}, \ldots, \omega_{t_n})$, as:

$$q_{\alpha_0 \alpha_n}(A, \omega_{t_0}) = q_{t_0, t_1, \ldots, t_n}(A_{t_1} \times A_{t_2} \times \cdots \times A_{t_n}, \omega_{t_0})$$

$$= \int_{A_{t_1}} q_{t_0, t_1}(d\omega_{t_1}, \omega_{t_0}) \int_{A_{t_2}} q_{t_0, t_1, t_2}(d\omega_{t_2}, \omega_{t_0}, \omega_{t_1}) \cdots \times$$

$$\int_{A_{t_n}} q_{t_0, \ldots, t_{n-1}, t_n}(d\omega_{t_n}, \ldots, \omega_{t_{n-1}}). \tag{15}$$

   This form of (9) has the following natural interpretation, when the conditional measures $q_{\alpha_i \alpha_{i+1}}$ are all regular. The left side of (15) is the

conditional probability of the set A at time t_n given that the system
started at ω_{t_0} at time t_0. This is obtained as the conditional probability that the system starting at state ω_{t_0} and time t_0, visited the
intermediate states $d\omega_{t_1}$ of $A_{t_1}, \ldots, d\omega_{t_{n-1}}$ of $A_{t_{n-1}}$ at intermediate
times $t_1, t_2, \ldots, t_{n-1}$ before landing at time t_n in $d\omega_{t_n}$ of A_{t_n}. Thus
$q_{\alpha_0 \alpha_n}$ is the "transition probability" of the system starting at t_0 in state
ω_{t_0} and arriving in A at time t_n. We shall exploit these ideas in the
next chapter where this language will be made precise.

8.2 A general formulation of products

The preceding discussion shows that when we consider several (possibly different) measure spaces, it is desirable to have connecting map-
pings between the spaces to formulate a useful theory. If $\Omega = \overset{n}{\underset{i=1}{\times}} \Omega_i, \Sigma = \overset{n}{\underset{i=1}{\otimes}} \Sigma_i$ and $P = \overset{n}{\underset{i=1}{\otimes}} P_i$, one can consider coordinate projections $\pi_i : \Omega \to \Omega_i$ with $P \circ \pi_i^{-1} = P_i$, the marginal measure. This situation is generalized to (not necessarily cartesian) products with a connecting set of
mappings satisfying the consistency relations as indicated in Proposition 1.3. We abstract these ideas and set down a precise description of
the concept as originally formulated by Bochner (1947) and (1955).

Definition 1. Let $\{(\Omega_i, \Sigma_i, P_i), i \in I\}$ be a family of probability spaces
where I is a (perhaps uncountably infinite) directed index set (i.e., there
is a partial ordering '$<$' in I and that for each pair i, j in I there is a
$k \in I$ such that $i < k, j < k$). Suppose for the pair i, j in I, $i < j$,
there is a mapping $h_{ij} : \Omega_j \to \Omega_i$ satisfying: (1) $h_{ii} =$ identity, (2)
$i < j < k \Rightarrow h_{ij} \circ h_{jk} = h_{ik}$, and (3) $h_{ij}^{-1}(\Sigma_i) \subset \Sigma_j$ so that each h_{ij} is
(Σ_j, Σ_i)-measurable. Then the collection $\{(\Omega_i, \Sigma_i, P_i, h_{ij})_{i<j}, i, j$ in $I\}$
is called a *projective system* of (probability) measure spaces whenever
$P_i = P_j \circ h_{ij}^{-1}$ for $i < j$. (Here P_i can even be σ-finite or localizable
measures, although for us they are mostly probability measures.) In
case each Ω_i is a (Hausdorff) topological space with Σ_i containing all
Baire sets, the h_{ij} are moreover required to be continuous and the P_i
to be Radon measures.

Let us explain the above concept by specializations which will also
show what natural restrictions need be imposed for a successful theory
useful in applications. We first observe that the h_{ij} are an abstraction

of coordinate projections. This was briefly noted for equation (15) of the last section, but we now discuss it in more detail.

Let A and B be a pair of arbitrarily given point sets and $\mathcal{F}$ be the family of all finite subsets of B. If $\alpha, \beta \in \mathcal{F}$ define $\alpha < \beta$ whenever $\alpha \subset \beta$. Then '$<$' is a partial order on $\mathcal{F}$ and for any α_1, α_2 in $\mathcal{F}$ $\alpha_3 = \alpha_1 \cup \alpha_2 \in \mathcal{F}$ and $\alpha_1 < \alpha_3, \alpha_2 < \alpha_3$ so that $(\mathcal{F}, <)$ is a directed set. Now consider A^α, the space of all A-valued mappings from a finite set α of B. If the cardinality of α is $d \geq 0$, then A^α is isomorphic to A^d, and if $\beta \in \mathcal{F}$, $\text{card}(\beta) = d'$, then $\alpha < \beta$, (so $d \leq d'$) and the coordinate mappings $h_{\alpha\beta} : A^\beta \to A^\alpha$ defined by $h_{\alpha\beta}(a_1, \ldots, a_{d'}) = (a_1, \ldots, a_d)$ satisfy $h_{\alpha\beta}(A^\beta) = A^\alpha$, $h_{\alpha\alpha} = $ identity, and for $\alpha < \beta < \gamma$ in $\mathcal{F}$, $h_{\alpha\beta} \circ h_{\beta\gamma} = h_{\alpha\gamma}$. Thus the $h_{\alpha\beta}$s are the familiar coordinate projections. On the other hand, they can be inclusion mappings for a system of subsets of a large set so that they are not in this case coordinate projections any more.

We need to analyze the generality of the concept. Thus let $\{(\Omega_\alpha, h_{\alpha\beta}), \alpha < \beta$ in $I\}$ be a system of spaces with $h_{\alpha\beta} : \Omega_\beta \to \Omega_\alpha$ as the connecting consistent family of mappings with I as a directed index set. If $\Omega_I = \underset{\alpha \in I}{\times} \Omega_\alpha$ is the cartesian product space, then the set $\Omega = \{\omega = \{\omega_\alpha, \alpha \in I\} \in \Omega_I : \omega_\alpha = h_{\alpha\beta}(\omega_\beta),$ for all $\alpha < \beta$ in $I\}$ is called the *projective limit* of the system, and is denoted $\Omega = \varprojlim(\Omega_\alpha, h_{\alpha\beta})$. If $\Omega \neq \emptyset$, for each $\omega \in \Omega$, $\omega = \{\omega_\alpha, \alpha \in I\}$, called *threads*, let us set $h_\alpha : \omega \mapsto \omega_\alpha$. Then $h_\alpha : \Omega \to \Omega_\alpha$ is a well-defined mapping and for $\alpha < \beta$ one has

$$h_\alpha(\omega) = \omega_\alpha = h_{\alpha\beta}(\omega_\beta) = (h_{\alpha\beta} \circ h_\beta)(\omega), \quad \omega \in \Omega, \tag{1}$$

so that $h_\alpha = h_{\alpha\beta} \circ h_\beta$ holds on Ω. We now show that the limit set Ω defined in this way can be identified as the cartesian product set Ω_T, when I is replaced by the directed set $\mathcal{F}$ of all finite subsets of an index T under inclusion ordering for $\{\Omega_t, t \in T\}$, discussed in the preceding paragraph. This identification is not entirely obvious and we give the details.

Let $\Omega_T = \underset{t \in T}{\times} \Omega_t, \Omega_{\mathcal{F}} = \underset{\alpha \in \mathcal{F}}{\times} \Omega_\alpha$ with $\Omega_\alpha = \underset{t \in \alpha}{\times} \Omega_t$, $\Omega = \varprojlim(\Omega_\alpha, h_{\alpha\beta})$. Note that $h_{\alpha\beta} = \pi_{\alpha\beta} : \Omega_\beta \to \Omega_\alpha$ are coordinate projections. Although $\text{card}(\mathcal{F}) \geq \text{card}(T)$, we show that Ω_T and Ω are isomorphic. Indeed, if $\pi_\alpha : \Omega_T \to \Omega_\alpha$ is the usual coordinate projection, we have $\pi_\alpha = \pi_{\alpha\beta} \circ \pi_\beta = h_{\alpha\beta} \circ \pi_\beta$. We claim that there is a bijective (= one-one and onto) mapping $u : \Omega_T \to \Omega$ such that $\pi_\alpha = h_\alpha \circ u$ where $h_\alpha : \Omega \to \Omega_\alpha$ is the canonical mapping as usual. To see this, let $\omega' = \{\omega_t, t \in T\} \in \Omega_T$ and $\tilde{\omega} = \{\omega_\alpha, \alpha \in \mathcal{F}\}$ be a thread obtained from ω'. Define $u : \omega' \mapsto \tilde{\omega}$. Then u is a mapping of Ω_T into Ω. To see that u is one-to-one, if ω' and ω'' are two elements of Ω_T such that $u(\omega') = \tilde{\omega} = u(\omega'')$ then for each $\alpha \in \mathcal{F}$, $h_\alpha(u(\omega')) = \omega_\alpha = h_\alpha(u(\omega''))$ so that the threads are identical for both ω' and ω''. In particular, considering the one point sets in $\mathcal{F}$, we conclude that ω' and ω'' have the same coordinates so that $\omega' = \omega''$. To see next that u is onto, consider $\tilde{\omega} = \{\omega_\alpha, \alpha \in \mathcal{F}\} \in \Omega$. For each $t \in T$, let $\alpha \in \mathcal{F}$ satisfy $\{t\} < \alpha$. Set $\omega_t = h_{\{t\}\alpha}(\omega_\alpha)$. If $\beta \in \mathcal{F}$ and $\{t\} < \beta$, then there is a $\gamma \in \mathcal{F}$ such that $\alpha < \gamma, \beta < \gamma$, so

$$h_{\{t\}\alpha}(\omega_\alpha) = \omega_t = h_{\{t\}\alpha} \circ h_{\alpha\gamma}(\omega_\gamma)$$
$$= h_{\{t\}\gamma}(\omega_\gamma) = h_{\{t\}\beta} \circ h_{\beta\gamma}(\omega_\gamma) = h_{\{t\}\gamma}(\omega_\gamma). \qquad (2)$$

Thus ω_t is independent of α and if $\omega' = \{\omega_t, t \in T\}$, then $\omega' \in \Omega_T$ and $u(\omega') = \tilde{\omega}$. Hence $u : \Omega_T \to \Omega$ is a bijective mapping as asserted.

The generalization involved in the projective systems is such that the limit set Ω can be empty even when all $h_{\alpha\beta}$ are onto mappings, as seen from the following example. We follow Bourbaki (1968, p. 251) here.

Let $\mathcal{F}$ again be the collection of all finite subsets of $T = [0, 1]$. To define a projective system of the "pathological kind" sought for, first note that $\mathcal{F}$ has no countable cofinal subset. Indeed, if there is such a set let it be $\{\alpha_n, n \geq 1\}$, and $J = \bigcup_{n=1}^{\infty} \alpha_n \subset T, \alpha_n \in \mathcal{F}$, so that J is countable, and if $\alpha \subset T - J, \alpha \in \mathcal{F}$, there can be no $\beta \in J$ satisfying $\alpha < \beta$ since $\alpha \cap \beta = \emptyset$. Thus J cannot be cofinal. Now define sets $\mathcal{X}$ and Ω_α as:

$$\mathcal{X} = \{x = (\alpha_1, \ldots, \alpha_{2n}) : \alpha_i \in \mathcal{F}, \alpha_{2i-1} < \alpha_{2i}, 1 \leq i \leq n \text{ and}$$
$$\alpha_{2i-1} \not< \alpha_{2j-1} \text{ for } 1 \leq j < i \leq n\}. \qquad (3)$$

Let $r(x) = \alpha_{2n-1}$, $s(x) = \alpha_{2n}$. We then say x in $\mathcal{X}$ has length n. For $\alpha \in \mathcal{F}$ if we set

$$\Omega_\alpha = \{x \in \mathcal{X} : r(x) = \alpha\} \qquad (4)$$

then $\Omega_\alpha \neq \emptyset$, since $x = (\alpha, \alpha) \in \Omega_\alpha$. If $x_\beta = (\alpha_1, \ldots, \alpha_{2n}) \in \Omega_\beta, \alpha < \beta$, let j be the first index i such that $\alpha \subset \alpha_{2i-1}$, and $= n$ if there is no such i. With each such x_β we define $x_\alpha = (\alpha_1, \ldots, \alpha_{2j-2}, \alpha, \alpha_{2j})$. Since $r(x_\alpha) = \alpha$ one has $x_\alpha \in \Omega_\alpha$; now set $g_{\alpha\beta}(x_\beta) = x_\alpha$. Then $g_{\alpha\beta} : \Omega_\beta \to \Omega_\alpha$ is well-defined. To see that it is onto, let $x \in \Omega_\alpha$. So $r(x) = \alpha$ and $x = (\alpha_1, \ldots, \alpha_{2n-2}, \alpha, \alpha_{2n})$. If the lengths of α, β are equal then $\alpha < \beta \Rightarrow x \in \Omega_\beta$, and if the lengths are unequal then they differ by a pair of elements so $x_\beta = (\alpha_1, \ldots, \alpha_{2n}, \beta, \alpha_{2n+2})$ for an $\alpha_{2n+2} \in \mathcal{F}$, $\beta < \alpha_{2n+2}$. Hence again $x_\beta \in \Omega_\beta$. Since $\alpha < \beta(j = 2n)$, $g_{\alpha\beta}(x_\beta) = x$ and $g_{\alpha\beta}(\Omega_\beta) = \Omega_\alpha$. A similar reasoning shows that for $\alpha < \beta < \gamma$ in $\mathcal{F}, g_{\alpha\beta} \circ g_{\beta\gamma} = g_{\alpha\gamma}$, and $g_{\alpha\alpha} =$ identity. Thus $\{(\Omega_\alpha, g_{\alpha\beta}), \alpha < \beta \text{ in } \mathcal{F}\}$ is a projective system. If $\Omega = \varprojlim(\Omega_\alpha, g_{\alpha\beta})$, then we verify that $\Omega = \emptyset$, as follows.

Suppose $\Omega \neq \emptyset$, and let $x = \{x_\alpha, \alpha \in \mathcal{F}\} \in \Omega$, $x_\alpha \in \Omega_\alpha, x_\beta \in \Omega_\beta$. Then there is a $\gamma \in \mathcal{F}, \alpha < \gamma, \beta < \gamma, x_\gamma \in \Omega_\gamma$, so that $x_\alpha = g_{\alpha\gamma}(x_\gamma), x_\beta = g_{\beta\gamma}(x_\gamma)$. If both x_α, x_β have the same length, then by definition (cf. (3)), $s(x_\alpha) = s(x_\beta)$ which means that they have the same last elements. If $\beta = s(x_\alpha)$ where $x_\alpha = g_{\alpha\gamma}(x_\gamma)$ consider the set of β's so obtained. It can be verified (after some computation) that this is a countable cofinal set of $\mathcal{F}$. Since there is no such cofinal set, there cannot be an x in Ω. Thus $\Omega = \emptyset$ for this projective system, even though each $\Omega_\alpha \neq \emptyset$ and the $g_{\alpha\beta}$ are onto mappings.

In order to avoid the pathology of the above example, we introduce a useful sufficient condition, following Bochner (1955), for the general case.

Definition 2. A projective system $\{(\Omega_\alpha, g_{\alpha\beta}) : \alpha < \beta \text{ in } I\}$ of nonempty spaces is said to be *sequentially maximal* (or satisfies the s.m. condition) if for every sequence $\alpha_1 < \alpha_2 < \ldots$ and any points $\omega_{\alpha_j} \in \Omega_{\alpha_j}$ such that $\omega_{\alpha_j} = g_{\alpha_j \alpha_k}(\omega_{\alpha_k})$, $\alpha_j < \alpha_k$, then there is an element $\omega = \{\omega_\alpha, \alpha \in I\}$ in Ω, the projective limit set, satisfying $g_{\alpha_j}(\omega) = \omega_{\alpha_j} (g_\alpha = g_{\alpha\infty} : \Omega \to \Omega_\alpha)$.

It is clear that g_α is the restriction of $\pi_\alpha : \Omega_I = \underset{\alpha \in I}{\times} \Omega_\alpha \to \Omega_\alpha$ and that for $\alpha < \beta$, $g_\alpha = g_{\alpha\beta} \circ g_\beta$. Also if I is regarded as a (point) set of indexes, and Ω_αs are cartesian product spaces, then the s.m. condition always holds. Further under the s.m. condition, all g_α and $g_{\alpha\beta}$ are easily seen to be onto mappings. To note the significance of the s.m. hypothesis, let us consider the case of topological spaces. The following result is an abstracton,due to Bourbaki, of a theorem of Mittag-Leffler's in complex function theory, and is stated for reference and later use. (See Bourbaki (1966, p. 187) for a proof.)

Theorem 3. (Mittag-Leffler and Bourbaki) *Let $\{(\Omega_\alpha, g_{\alpha\beta}), \alpha < \beta$ in $I\}$ be a projective system of spaces such that each Ω_α is a complete metric space and $g_{\alpha\beta}(\Omega_\beta) = \Omega_\alpha$, for each $\alpha < \beta$. If the directed index set I has a countable cofinal subset and if for each α in I, there is a $\beta_\alpha > \alpha$ such that for each $\gamma > \beta_\alpha, g_{\alpha\gamma}(\Omega_\gamma)$ is dense in $g_{\alpha\beta}(\Omega_\beta)$, then the projective limit set $\Omega = \varprojlim(\Omega_\alpha, g_{\alpha\beta})$ is nonempty and for each α and $\beta > \alpha$ satisfying the latter condition, $g_\alpha(\Omega)$ is dense in $g_{\alpha\beta}(\Omega_\beta)$.*

In light of this theorem, the following result on the subject in relation to the s.m. condition illuminates it in some respects and has special interest in our study.

Proposition 4. *Let $\{(\Omega_\alpha, g_{\alpha\beta}) : \alpha < \beta$ in $I\}$ be a projective system of spaces with I as a directed index set. Then $\Omega = \varprojlim(\Omega_\alpha, g_{\alpha\beta})$ is nonempty in each of the following cases, and moreover the system has the stated properties.*

(i) If I has a countable cofinal set, $g_{\alpha\beta}$ are onto and $\Omega_\alpha \neq \emptyset$ for each α, then the canonical mappings $g_\alpha : \Omega \to \Omega_\alpha$ are onto for each $\alpha \in I$.

(ii) If the system satisfies the s.m. condition, but I need not have a countable cofinal subset, and $\Omega_\alpha \neq \emptyset$, then $g_\alpha(\Omega) = \Omega_\alpha, \alpha \in I$.

(iii) If each Ω_α is a nonempty compact Hausdorff space, then Ω is also a nonempty compact set, and if moreover $g_{\alpha\beta}(\Omega_\beta) = \Omega_\alpha$ for each $\alpha < \beta$ in I, then $g_\alpha(\Omega) = \Omega_\alpha$ so that the s.m. condition holds automatically.

Proof. (i) This can be reduced to Theorem 3 by considering a discrete uniformity on each Ω_α, i.e., if one considered the class of all subsets of $\Omega_\alpha \times \Omega_\alpha$ that contain the diagonal, then it defines a structure in terms of which Ω_α becomes a discrete uniform space (a metric like topology) and then we can apply Theorem 3 to deduce the conclusion.

(ii) This follows from definition of the s.m. condition at once.

(iii) Let each Ω_α be a nonempty compact Hausdorff space, $\alpha \in I$. The result will be established following an argument of N.E. Steenrod's (cf. S. Lefschetz (1942), p. 32). Thus if $G_{\alpha\beta} = \{(\omega_\alpha, \omega_\beta) : \omega_\alpha = g_{\alpha\beta}(\omega_\beta)\}$, the graph of $g_{\alpha\beta}$, and $C_{\alpha\beta} = G_{\alpha\beta} \times \Omega_{I-\{\alpha,\beta\}}$, the cylinder with base $G_{\alpha\beta}$, then each $C_{\alpha\beta}$ is a closed set in the product topology of Ω_I since $g_{\alpha\beta}$ is continuous. But $\Omega = \varprojlim(\Omega_\alpha, g_{\alpha\beta}) = \cap\{C_{\alpha,\beta} : \alpha < \beta$ in $I\}$ so that Ω is also closed. If each Ω_α is compact and nonempty than so is Ω_I (by Tychonov's theorem), and $\Omega \subset \Omega_I$ is compact. To see that Ω is nonempty, we show the stronger property that the $\{C_{\alpha\beta}, \alpha < \beta$ in $I\}$ has the finite intersection property which implies, since each $C_{\alpha\beta} \neq \emptyset$, that Ω is nonempty as well as compact at the same time.

Let $\{(\alpha_i, \beta_i), \alpha_i < \beta_i, i = 1, \ldots, n\}$ be an arbitrary set of n-pairs of indices from I. By directedness, there exists a $\gamma \in I$ such that $\gamma > \beta_i > \alpha_i, i = 1, \ldots, n$. For any $\omega_\gamma \in \Omega_\gamma$, let $\omega_{\alpha_i} = g_{\alpha_i\gamma}(\omega_\gamma)$ and $\omega_{\beta_i} = g_{\beta_i\gamma}(\omega_\gamma)$, and let $\omega \in \Omega_I$ be an element with $2n$ of its components as $(\omega_{\alpha_i}, \omega_{\beta_i}), i = 1, \ldots, n$. Since $g_{\alpha_i\gamma} = g_{\alpha_i\beta_i} \circ g_{\beta_i\gamma}$ we have $\omega_{\alpha_i} = f_{\alpha_i\beta_i}(\omega_{\beta_i}), i = 1, \ldots, n$, so that (by definition of $C_{\alpha\beta}$) $\omega \in C_{\alpha_i\beta_i}$ and hence $\omega \in \cap_{i=1}^n C_{\alpha_i\beta_i}$. Thus the intersection is nonempty and then $\Omega = \cap C_{\alpha\beta} \neq \emptyset$ because each $C_{\alpha\beta}$ is also compact. Finally, if $\Omega_\alpha = g_{\alpha\beta}(\Omega_\beta)$ for each $\alpha < \beta$, we assert that the s.m. condition holds.

Consider the mapping $g_\alpha : \omega = \{\omega_\alpha, \alpha \in I\} \mapsto \omega_\alpha, \omega \in \Omega$. Then $g_\alpha(\Omega) \subset \Omega_\alpha$. To see that there is equality here, let $\omega_\alpha \in \Omega_\alpha$ and $A = g_\alpha^{-1}(\{\omega_\alpha\}) \subset \Omega_I$. Then A is a closed nonempty set since g_α is continuous and points are closed in Ω_α (being Hausdorff). We show that $A \cap \Omega \neq \emptyset$. In fact we assert that $A \cap C_{\alpha\beta}$ has the finite intersection property for all $\beta > \alpha$. For, let $\alpha < \alpha_i$, $i = 1, \cdots, n$ be any set of indices. There is a $\beta \in I$, $\beta > \alpha_i$. Since $g_{\alpha\beta}(\Omega_\beta) = \Omega_\alpha$, by hypothesis, let $\omega_\beta \in g_{\alpha\beta}^{-1}(\{\omega_\alpha\}) \subset \Omega_\beta$. Let $\omega \in \Omega_I$ be a point with ω_α as a component, so $\omega \in A$, and $\omega_{\alpha_i} = g_{\alpha_i\beta}(\omega_\beta)$. Since $\alpha < \alpha_i < \beta \Rightarrow \omega_{\alpha_i} = f_{\alpha_i\beta}(\omega_\beta)$ we get $(\omega_{\alpha_i}, \omega_\beta) \in G_{\alpha_i\beta} \Rightarrow \omega \in C_{\alpha_i\beta}$ all $i = 1, \cdots, n$. Hence $\omega \in A \cap C_{\alpha\alpha_i} \Rightarrow \omega \in A \cap \bigcap_{\beta > \alpha} C_{\alpha\beta} = A \cap \Omega$. Thus $g_\alpha(\omega) = \omega_\alpha$ and so $g_\alpha(\Omega) = \Omega_\alpha$. Finally, let $\alpha_1 < \alpha_2 < \cdots$ be any sequence from I and $\Omega = g_{\alpha_i}^{-1}(\Omega_{\alpha_i}) \neq \emptyset$. If $\omega_{\alpha_i} = g_{\alpha_i\alpha_{i+1}}(\omega_{\alpha_{i+1}})$, let $\omega \in \Omega$ be taken with ω_{α_i} as its coordinates for $i \geq 1$ which is possible by the preceding sentence. Thus the s.m. condition is always satisfied in this case. $\square$

Remark. In the general case without reference to topology, the s.m. condition is a useful sufficient requirement for nonemptyness of Ω and the onto property of the g_α. However, Millington and Sion (1973) have pointed out that the s.m. condition can be weakend and can still have the limit space $\Omega \neq \emptyset$. We shall however impose the s.m. condition in our study because of its simplicity.

8.3 A general projective limit theorem

Using the concept and analysis of projective systems of probability spaces we now establish a general theorem on (nontrivial) projective limits of such families. This will then be specialized for some existence problems of stochastic processes arising in applications. If $\{(\Omega_\alpha, \Sigma_\alpha, g_{\alpha\beta}, P_\alpha)_{\alpha<\beta} : \alpha, \beta \text{ in } I\}$ is a projective system of measure spaces and $\Omega = \varprojlim(\Omega_\alpha, g_{\alpha\beta}) \neq \emptyset$, $\Sigma = \sigma(\cup_\alpha g_\alpha^{-1}(\Sigma_\alpha))$ and if $P : \Sigma \to \mathbb{R}^+$ defined by $P \circ g_\alpha^{-1} = P_\alpha, \alpha \in I$, exists as a σ-additive function, then (Ω, Σ, P) is termed the *projective limit* of the given projective system. Such a limit need not exist in general, and the main problem is to find "good" conditions on the system for the existence of the limit.

The following result, due essentially to Choksi (1958), enables us to present a solution to the above question, and is adequate for many problems that are usually encountered in applications.

Theorem 1. *Let $\{(\Omega_\alpha, \Sigma_\alpha, \mu_\alpha, g_{\alpha\beta}) : \alpha < \beta \text{ in } I\}$ be a projective system of probability (or σ-finite) spaces satisfying the s.m. condition. For each pair α, β in $I, \alpha < \beta$, suppose the following two properties are true.*

(i) If $g_{\alpha\beta} : \Sigma_\beta \times \Omega_\alpha \to \mathbb{R}^+$ is a mapping satisfying (for $\alpha < \beta$),

$$\mu_\beta(B \cap g_{\alpha\beta}^{-1}(A)) = \int_A q_{\alpha\beta}(B, \omega) \, d\mu_\alpha(\omega), \quad A \in \Sigma_\alpha, B \in \Sigma_\beta, \quad (1)$$

then $q_{\alpha\beta}(B, \cdot)$ is Σ_α-measurable and $q_{\alpha\beta}(\cdot, \omega)$ is a probability measure for each $\omega \in \Omega_\alpha$, i.e., $q_{\alpha\beta}$ is a regular conditional measure.

(ii) For each $\beta > \alpha$, there is an $N_{\alpha\beta} \in \Sigma_\alpha, \mu_\alpha(N_{\alpha\beta}) = 0$, and if $\omega \in N_{\alpha\beta}^c$ and $B \in \Sigma_\beta$ such that $q_{\alpha\beta}(B, \omega) > 0$ then $B \cap g_{\alpha\beta}^{-1}(\{\omega\}) \neq \emptyset$.

Then the system admits a unique projective limit (Ω, Σ, μ), denoted as $\varprojlim(\Omega_\alpha, \Sigma_\alpha, \mu_\alpha, g_{\alpha\beta})$.

Proof. By the s.m. condition $\Omega = \varprojlim(\Omega_\alpha, g_{\alpha\beta})$ is nonempty, the mappings $g_\alpha : \Omega \to \Omega_\alpha$ are onto and $g_\alpha = g_{\alpha\beta} \circ g_\beta$. Let $\mathcal{A} = \cup_\alpha g_\alpha^{-1}(\Sigma_\alpha)$, the cylinder algebra, and $\Sigma = \sigma(\mathcal{A})$. We need to show, under the given conditions that $\mu = \varprojlim(\mu_\alpha, g_{\alpha\beta})$, the projective limit, exists as a measure. Let us first establish that there is a finitely additive set function on $\mathcal{A}$ and then show that it must also be σ-additive.

Since for $\alpha < \beta$, $g_\alpha^{-1}(\Sigma_\alpha) = (g_{\alpha\beta} \circ g_\beta)^{-1}(\Sigma_\alpha) = g_\beta^{-1}(g_{\alpha\beta}^{-1}(\Sigma_\alpha)) \subset g_\beta^{-1}(\Sigma_\beta)$, and each $g_\alpha^{-1}(\Sigma_\alpha)$ is a σ-algebra, it follows that $\mathcal{A}$ is an algebra. Let us define $\mu : \mathcal{A} \to \mathbb{R}^+$. If $A \in \mathcal{A}$, then $A \in g_\alpha^{-1}(\Sigma_\alpha)$ for some $\alpha \in I$. Suppose $A \in g_\beta^{-1}(\Sigma_\beta)$ also, so that there exists $B_\alpha \in \Sigma_\alpha$ and $C_\beta \in \Sigma_\beta$ satisfying $A = g_\alpha^{-1}(B_\alpha) = g_\beta^{-1}(C_\beta)$. Since I is directed, there is a $\gamma \in I$ such that $\alpha < \gamma, \beta < \gamma$ and hence $g_\alpha = g_{\alpha\gamma} \circ g_\gamma, g_\beta = g_{\beta\gamma} \circ g_\gamma$ implying

$$g_\gamma^{-1}(g_{\alpha\gamma}^{-1}(B_\alpha)) = g_\alpha^{-1}(B_\alpha) = A = g_\beta^{-1}(C_\beta) = g_\gamma^{-1}(g_{\beta\gamma}^{-1}(C_\beta)). \qquad (2)$$

Since the g_γ is onto so that g_γ^{-1} is one-to-one, (3) implies $g_{\alpha\gamma}^{-1}(B_\alpha) = g_{\beta\gamma}^{-1}(C_\beta)$. Consequently we have

$$\mu_\alpha(B_\alpha) = \mu_\gamma(g_{\alpha\gamma}^{-1}(B_\alpha)) = \mu_\gamma(g_{\beta\gamma}^{-1}(C_\beta)) = \mu_\beta(C_\beta). \qquad (3)$$

If $\mu(A) = \mu_\alpha(B_\alpha) = \mu_\beta(C_\beta)$ as the common value, then μ is uniquely defined with values in $\mathbb{R}^+$. To see that μ is additive, let $A_i \in \mathcal{A}, i = 1, 2$ disjoint. Then (by directedness of I again) there is a $\gamma \in I$ such that $A_i = g_\gamma^{-1}(C_i), C_i \in \Sigma_\gamma$ and disjoint, implying

$$\mu(A_1 \cup A_2) = \mu_\gamma(C_1 \cup C_2) = \mu_\gamma(C_1) + \mu_\gamma(C_2) = \mu(A_1) + \mu(A_2), \qquad (4)$$

since μ_γ is a measure on Σ_γ. Hence $\mu : \mathcal{A} \to \mathbb{R}^+$ is additive.

The main problem (and difficulty) is to show that μ is σ-additive, and hence has a unique extension to Σ, under the conditions (i) and (ii). Suppose, for an indirect argument, that μ is not σ-additive on $\mathcal{A}$. Then there exist $A_n \in \mathcal{A}$, $A_n \downarrow \emptyset$ and a $\delta > 0$ such that $\mu(A_n) \geq \mu(A_{n+1}) \geq \delta$. We present the argument in steps for convenience and clarity. First observe that, by directedness of I, we can find successively $\alpha_n \in I, \alpha_n < \alpha_{n+1}$ such that $A_n = g_{\alpha_n}^{-1}(C_n)$, $C_n \in \Sigma_{\alpha_n}$. Hereafter the regularity of the conditional measure plays an important role.

Step 1. By hypothesis $q_{\alpha\beta}$ of (1) is a regular conditional measure outside of a μ_α-null set $N_{\alpha\beta}$. So for the sequence $\{\alpha_n, n \geq 1\}$ for each n, and $m > n$, consider $N_{\alpha_n \alpha_m} \in \Sigma_{\alpha_n}$ and set $q_{\alpha_n \alpha_m}(\cdot, \omega) = 0$ for $\omega \in \bigcup_{m>n} N_{\alpha_n \alpha_m} \in \Sigma_{\alpha_n}$. We denote the thus modified conditional measure by the same symbol and observe that it again satisfies (1). By Part (ii) of the hypothesis, for any $\omega \in \Omega_{\alpha_n}$, $A_{\alpha_m} \in \Sigma_{\alpha_m}$, $q_{\alpha_n \alpha_m}(A_{\alpha_m}, \omega) >$

$0 \Rightarrow A_{\alpha_m} \cap g_{\alpha_n \alpha_m}^{-1} (\{\omega\}) \neq \emptyset$. But for $n < m < r$ we also have the following relation:

$$q_{\alpha_n \alpha_r} (A, \omega) = \int_{\Omega_{\alpha_{n+1}}} q_{\alpha_n \alpha_{n+1}} (d\omega_{\alpha_{n+1}}, \omega) \int_{\Omega_{\alpha_{n+2}}} q_{\alpha_{n+1} \alpha_{n+2}} (d\omega_{\alpha_{n+2}}, \omega_{\alpha_{n+1}}) \cdots$$

$$\cdots \int_{\Omega_{\alpha_r}} \chi_A q_{\alpha_{r-1} \alpha_r} (d\omega_{\alpha_r}, \omega_{\alpha_{r-1}}), \ A \in \Sigma_{\alpha_r}, \quad (5)$$

for almost all $\omega \in \Omega_{\alpha_n}$, by Proposition 1.3. Hence if $A = A_r$ of our exceptional sequence, where $A_n = g_{\alpha_n}^{-1} (C_n)$, then there is a μ_{α_n}-null set $N_{\alpha_n} (C_n) \in \Sigma_{\alpha_n}$ so that the null set N_{α_n} depends on C_n, outside of which (5) holds for all ω. Thus if $\omega \in \bigcup_{k \geq n} N_{\alpha_n} (C_k)$, we set $q_{\alpha_n \alpha_k} (\cdot, \omega) = 0, k \geq n$, and the modified $q_{\alpha_n \alpha_m}$ of this sequence again satisfies (5). This version is also denoted by the same symbol so that the new one satisfies (5) on Ω_{k_m} for $m > n$ without exceptions and $A_{\alpha_m} \cap g_{\alpha_n \alpha_m}^{-1} (\{\omega\}) \neq \emptyset$ for $m > n$ whenever $q_{\alpha_n \alpha_m} (A_{\alpha_m}, \omega) > 0$.

Step 2. Consider this (new) sequence $\{q_{\alpha, \alpha_k} (C_k, \cdot), k \geq 1\}$. We assert that this is an a.e. $[\mu_{\alpha_1}]$ decreasing sequence of nonnegative functions. Indeed let $B_n = \{\omega \in \Omega_{\alpha_1} : q_{\alpha_1 \alpha_n} (C_n, \omega) < q_{\alpha_1 \alpha_{n+1}} (C_{n+1}, \omega)\}$. Then

$$0 \geq \int_{B_n} [q_{\alpha_1 \alpha_n} (C_n, \omega) - q_{\alpha_1 \alpha_{n+1}} (C_{n+1}, \omega)] d\mu_{\alpha_1} (\omega)$$

$$= \mu_{\alpha_{n+1}} (C_n \cap g_{\alpha_n \alpha_{n+1}}^{-1} (B_n)) - \mu_{\alpha_{n+1}} (C_{n+1} \cap g_{\alpha_1 \alpha_{n+1}}^{-1} (B_n)), \text{ by (1)},$$

$$= \mu(A_n \cap g_{\alpha_{n+1}}^{-1} (B_n)) - \mu(A_{n+1} \cap g_{\alpha+1}^{-1} (B_n))$$

$$\geq 0, \quad (6)$$

since $A_n \supset A_{n+1}$ and $g_{\alpha_1} = g_{\alpha_1 \alpha_{n+1}} \circ g_{\alpha_{n+1}}$. Hence (6) implies that $\mu_{\alpha_1} (B_n) = 0$. Replacing α_1 by α_k in the above, the result is that $q_{\alpha_k \alpha_{n+1}} (C_{n+1} \cdot) \downarrow$ a.e. $[\mu_{\alpha_k}]$. Thus for each k if $N'_{\alpha_k \alpha_n}$ is the set on which the sequence does not decrease, then $\bigcup_{n > k} N'_{\alpha_k \alpha_n}$ is μ_{α_k}-null. On this set define $q_{\alpha_k \alpha_n} (C_n, \cdot) = 0$ and the new sequence is then monotone decreasing and it again satisfies (5) as well as (ii) of the hypothesis. Let $f_k(\omega) = \lim_n q_{\alpha_k \alpha_n} (C_n, \omega)$, which exists for each $k \geq 1$ and each $\omega \in \Omega_{\alpha_k}$. Also by choice

$$0 < \delta \leq \lim_n \mu_{\alpha_n} (A_n) = \lim_n \mu_{\alpha_n} (C_n)$$

$$= \lim_n \int_{\Omega_{\alpha_1}} q_{\alpha_1 \alpha_n} (C_n, \omega) d\mu_{\alpha_1} (\omega)$$

$$= \int_{\Omega_{\alpha_1}} f_1(\omega) d\mu_{\alpha_1} (\omega). \quad (7)$$

Step 3. Let $G_k = \{\omega : f_k(\omega) > 0\}$. Then $\mu_{\alpha_1}(G_1) > 0$ by (7). Choose $\omega_1 \in G_1$ such that $f_1(\omega_1) \geq \delta_1 > 0$, for some δ_1. Since the modified qs satisfy the identity of Proposition 1.3, we have

$$q_{\alpha_1 \alpha_n}(C_n, \omega_1) = \int_{\Omega_{\alpha_2}} q_{\alpha_1 \alpha_2}(d\omega_{\alpha_2}, \omega_1) q_{\alpha_2 \alpha_n}(C_n, \omega_{\alpha_2}). \qquad (8)$$

From (7) and (8) and the monotonicity of qs established in Step 2, we get $q_{\alpha_1 \alpha_n}(C_n, \omega_1) \geq \delta_1 > 0$. Hence $\omega_1 \in g_{\alpha_1 \alpha_n}(C_n), n \geq 1$. Since $q_{\alpha_1 \alpha_n}(\cdot, \omega_1)$ is now (after modifications) a measure and $\lim_n q_{\alpha_1 \alpha_n}(C_n, \omega_1) \geq \delta_1 > 0$, equation (8) implies that

$$0 < \delta_1 \leq \lim_n q_{\alpha_1 \alpha_n}(C_n, \omega_1)$$

$$= \int_{\Omega_{\alpha_2}} f_2(\omega_{\alpha_2}) q_{\alpha_1 \alpha_2}(d\omega_{\alpha_2}, \omega_1). \qquad (9)$$

Hence the set G_2 has positive $q_{\alpha_1 \alpha_2}(\cdot, \omega_1)$-measure so that there is an $\omega_2 \in g_{\alpha_2 \alpha_n}(B_2)$ such that $f_2(\omega_2) > 0$ and

$$q_{\alpha_2 \alpha_n}(C_n, \omega_2) = \int_{\Omega_{\alpha_3}} q_{\alpha_2 \alpha_3}(d\omega_{\alpha_3}, \omega_2) q_{\alpha_3 \alpha_n}(C_n, \omega_2), \qquad (10)$$

and $\omega_2 \in g_{\alpha_2 \alpha_n}(C_n)$ for $n \geq 2$. Here we can (and do) demand that $\omega_2 \in g_{\alpha_1 \alpha_2}^{-1}(\{\omega_1\})$.

Step 4. We now iterate the procedure of the above step and find $\omega_n \in \Omega_{\alpha_n}$ such that $g_{\alpha_n \alpha_{n+1}}(\omega_{n+1}) = \omega_n$ and $\omega_n \in C_n, n \geq 1$. Hence by the s.m. condition, there exists an $\omega \in \Omega = \lim_{\leftarrow}(\Omega_\alpha, g_{\alpha\beta})$ such that $g_{\alpha_n}(\omega) = \omega_n$. But this means $\omega \in A_n$ for all n so that $\omega \in \bigcap_n A_n$. This is the desired contradiction since $A_n \searrow \emptyset$ by the initial choice.

Thus μ must be σ-additive on $\mathcal{A}$ and being a finite measure has a unique extension to $\sigma(\mathcal{A}) = \Sigma$, by the classical measure theory, as asserted in the statement. The uniqueness of the projective limit measure is clear since $\mu \circ g_\alpha^{-1} = \mu_\alpha, \alpha \in I$. $\square$

The above proof is a (nontrivial) modiciation of the classical argument used by Kolmogorov (1933). [See also Doob (1953), pp. 613–615.] It is possible to consider a sort of converse to the above assertions but we shall not persue this line here, as it is not needed below.

Some specializations of the above theorem, when $N_{\alpha\beta} = \emptyset$ for all α, β and $\Omega = \Omega_I$ ($g_{\alpha,\beta}$ are then coordinate prejections) so that some arguments get simplified, will be given. The precise result stated next was proved early by C. Ionescu Tulcea (1949):

Theorem 2. *Let* $\{(S_n, \mathcal{S}_n), n \geq 1\}$ *be a family of measurable spaces,* $\Omega_n = \underset{i=1}{\overset{n}{\times}} S_i, \Sigma_n = \underset{i=1}{\overset{n}{\otimes}} \mathcal{S}_i,$ *and* $q_{n+1} : \Sigma_{n+1} \times \Omega_n \to \mathbb{R}^+$ *be given. Suppose that (i)* $q_{n+1}(A, \cdot)$ *is measurable and (ii)* $q_{n+1}(\cdot, \omega)$ *be a probability measure for each* $\omega \in \Omega_n$ *so that* q_{n+1} *is a regular conditional measure. Then there exists a probability measure* μ *such that*

$$\mu(\pi_n^{-1}(A^{(n)})) = \int_{A_1} d\mu_1(\omega_1) \int_{A_2} q_2(d\omega_2, \omega_1) \cdots \int_{A_n} q_n(d\omega_n, \omega_{n-1}),$$
(11)

where $A^{(n)} = A_1 \times \cdots \times A_n \in \Sigma_n$ *and* $\pi_n : \Omega(= \Omega_\infty) \to \Omega_n$ *is the coordinate projection. Thus* $(\Omega, \Sigma, \mu) = \lim_{\leftarrow}(\Omega_n, \Sigma_n, \mu^{(n)}, \pi_{mn})_{m<n}$ *exists where* $\mu^{(n)}(A^{(n)}) = \int_{\Omega_1} q_n(A^{(n)}, \omega_1) d\mu_1(\omega_1),$ μ_1 *is the initial probability on* $\Sigma_1 = \mathcal{S}_1,$ *and* $\pi_{mn} : \Omega_n \to \Omega_m$ *is the coordinate projection mapping* $(m \leq n).$

Proof. The hypothesis and the integration of the identity in Proposition 1.3 give in the present case $\mu^{(n)}$ as a probability, and $q_{mn} : \Sigma_m \times \Omega_n \to \mathbb{R}^+$ as a regular conditional measure for each $m \leq n$. We now deduce this result by verifying Part (ii) of the hypothesis of Theorem 1 directly since Part (i) is assumed here. Thus if $\omega_m \in \Omega_m$ and $A^{(n)} \in \Sigma_n$ is a rectangle, $m \leq n$, then

$$(A^{(n)}, \omega_m) = \chi_{A^{(m)}}(\omega_m) \int_{A_{m+1}} dq_{m+1}(d\omega_{m+1}, \omega_m) \cdots \int_{A_n} q_n(d\omega_n, \omega_{n-1}).$$
(12)

Suppose now that $q_{mn}(A^{(n)}, \omega_m) > 0$ for $\omega_m \in \Omega_m$, and $A^{(n)} \in \Sigma_n$, a rectangle. We define $\tilde{q}_{mn}$ as:

$$\tilde{q}_{mn}(A, \omega_m) = q_{mn}(\Omega_m \times A^{(n)}_{\omega_m})$$
(13)

where $A^n_{\omega_m}$ is the ω_m-section of $A^{(n)}$. Thus

$$A^{(n)}_{\omega_m} = \begin{cases} A_{m+1} \times \cdots \times A_n, & \text{if } \omega_m \in A^{(n)} \\ \emptyset, & \text{if } \omega_m \notin A^{(n)}. \end{cases}$$

If $A^{(n)}_m = A_{m+1} \times \cdots \times A_n$, then $A^{(n)}_{\omega_m} = \chi_{A^{(n)}}(\omega_m) A^{(n)}_m$. Hence for

$\omega_m \notin A^{(m)}$ we have $q_{mn}(\Omega_m \times A_{\omega_m}^{(n)}, \omega_m) = 0$. Consequently

$$\int_{\Omega_m} \tilde{q}_{mn}(A^{(m)} \times A_m^{(n)}, \omega_m)d\mu_m = \int_{A^{(m)}} q_{mn}(\Omega_m \times A_m^{(n)}, \omega_m)d\mu_m$$
$$= \mu_n((\Omega_m \times A_m^n) \cap \pi_{mn}^{-1}(A^{(m)}))$$
$$= \mu_n((\Omega_m \times A_m^{(n)}) \cap (A^{(m)} \times \Omega_m^{(n)})),$$
$$\text{where } \Omega_m^{(n)} = \Omega_{m+1} \times \cdots \times \Omega_n,$$
$$= \mu_n(A^{(n)} \times A_m^{(n)})$$
$$= \int_{\Omega_m} q_{mn}(A^{(m)} \times A_m^{(n)}, \omega_m)d\mu_m.$$

Thus $\tilde{q}_{mn}$ and q_{mn} agree on all measurable rectangles and hence on Σ_n itself. Thus $\tilde{q}_{mn}$ is a version of the regular conditional measure q_{mn} for each $m \leq n$. If $\tilde{q}_{mn}(A, \omega_m) > 0$, then the section A_{ω_m} of A cannot be empty. If $A = \Omega_m \times B$, then $B \neq \emptyset$ and

$$A \cap \pi_{mn}^{-1}(\{\omega_m\}) = (\Omega_m \times B) \cap (\{\omega_m\} \times \Omega_m^n)$$
$$= \{\omega_m\} \times B \neq \emptyset. \tag{14}$$

Thus hypothesis (ii) of Theorem 1 is satisfied, and the result then follows from that theorem. $\square$

Remark. One can give a direct proof by specializing the argument used in the demonstration of Theorem 1. In fact Tulcea's original proof is so formulated and the corresponding one given for Theorem 1 is an extension. It is possible to drop Part (ii) of the hypothesis of Theorem 1 if Part (i) is suitably strengthened. Such a result has recently been obtained by Rao and Sazonov (1993).

The following even more specialized version further illuminates the issues involved in the above results, and has independent interest.

Proposition 3. *Let $\{(\Omega_i, \Sigma_i), i \in I\}$ be a collection of measurable spaces, $\Omega = \underset{i \in I}{\times} \Omega_i$, $\Sigma = \underset{i \in I}{\otimes} \Sigma_i = \sigma(\underset{n}{\cup} \pi_n^{-1}(\Sigma_{(n)}))$ where $\Sigma_{(n)} = \overset{n}{\underset{i=1}{\otimes}} \Sigma_i$, and $\pi_{(n)} : \Omega \to \Omega_{(n)} = \overset{n}{\underset{i=1}{\times}} \Omega_i$, the coordinate projection. Let $\mathcal{F}$ be the class of all finite subsets of I ordered by inclusion, making it a directed set. Suppose that for each $F \in \mathcal{F}$, there is given a probability measure μ_F on $\Sigma_F = \underset{i \in F}{\otimes} \Sigma_i$ such that μ_F and μ_G agree on $F \cap G$ for any F, G in $\mathcal{F}$. (This is the compatibility condition with $\pi_{FG} : \Omega_G \to \Omega_F$ as* *coordinate projections for $F \subset G, \mu_F = \mu_G \circ \pi_{FG}^{-1}$.) Then the finitely additive μ on Σ defined by the class $\{(\Omega_F, \Sigma_F, \pi_{FG}, \mu_F) : F \subset G \text{ in } \mathcal{F}\}$ is σ-additive if for each $t \in I - F$, $F \in \mathcal{F}$, the conditional measure $q_F(\pi_t^{-1}(\cdot), \omega_F) : \Sigma_F \to \mathbb{R}^+$ is regular for each $\omega_F \in \Omega_F$ (cf. (1) for $q_F(\cdot, \cdot)$).*

This result follows from the preceding theorem by expressing the μ_F as $(F = (t_0, t_1, \cdots, t_n) \in \mathcal{F}$ for each $A = A_0 \times \cdots \times A_n \in \Sigma_F)$,

$$\mu_F(A) = \int_{A_0} \mu_{t_0}(d\omega) \int_{A_1} q_{t_0 t_1}(d\omega_1, \omega) \cdots \int_{A_n} q_{t_0 \cdots t_n}(d\omega_n, \omega_0, \cdots, \omega_{n-1}),$$
(15)

in which all qs are regular conditional measures. Without regularity such a representation is not always possible and μ determined by the μ_F need not be σ-additive.

8.4 Some consequences

Since regularity of conditional measures plays such a key role in the work of the preceding section, we give here some easily recognizable spaces for which the results apply. These include the Kolmogorov existence theorem when the Ω_i are topological spaces. We begin with the following case.

Proposition 1. *Let $\{(S_t, \mathcal{S}_t), t \in I\}$ be a family of Borelian spaces where S_t is a Polish space, $t \in I$. If $\mathcal{F}$ is the directed (by inclusion) set of all finite subsets of I, $S_\alpha = \underset{t \in I}{\times} S_t$ and $\mathcal{S}_\alpha = \underset{t \in \alpha}{\otimes} \mathcal{S}_t$, suppose there are given probability measures $P_\alpha : \mathcal{S}_\alpha \to \mathbb{R}^+$ such that $P_\alpha = P_\beta \circ \pi_{\alpha\beta}^{-1}$ for each $\alpha < \beta$ in $\mathcal{F}$, $\pi_{\alpha\beta} : S_\beta \to S_\alpha$ being the coordinate projections. Then there exists a unique probability $P : \mathcal{S} = \underset{t \in I}{\otimes} \mathcal{S}_t \to \mathbb{R}^+$ such that $P \circ \pi_\alpha^{-1} = P_\alpha$, $\alpha \in \mathcal{F}$, where $\pi_\alpha : S = \underset{t \in I}{\times} S_t \to S_\alpha$ is the corrdinate projection.*

Proof. By Theorem 5.3.7 the conditional measure $q_{\alpha\beta} : \mathcal{S}_\beta \times S_\alpha \to \mathbb{R}^+$ defined by the equation

$$P_\beta(B \cap \pi_{\alpha\beta}^{-1}(A)) = \int_A q_{\alpha\beta}(B, \omega) dP_\alpha(\omega), \quad A \in \mathcal{S}_\alpha, B \in \mathcal{S}_\beta, \quad (1)$$

has a regular version, since S_α and S_β are Polish spaces, $\alpha < \beta$ in $\mathcal{F}$, and $\mathcal{S}_\alpha, \mathcal{S}_\beta$ are Borel σ-algebras. Consequently Theorem 3.2 implies that $P = \lim_{\leftarrow}(P_\alpha, \pi_{\alpha\beta})$ exists and is unique as desired. $\square$

Remark. If I is uncountable, then S is not necessarily Polish and although each P_α is a Radon probability, the same need not be true of the limit measure P, except under additional conditions.

If each $S_i = \mathbb{R}$ so that $S = \mathbb{R}^I$ and P_α for $\alpha = (i_1, \cdots, i_n)$ defined by

$$P_\alpha(A) = \int \cdots \int_A dF_{i_1, \cdots, i_n}(x_1, \cdots, x_n), \quad A \in \mathcal{S}_\alpha , \tag{2}$$

where $\mathcal{S}_\alpha$ is the Borel σ-algebra of $\mathbb{R}^n$, and $P_\alpha = P_\beta \circ \pi_{\alpha\beta}^{-1}$ (so the P_α are "compatible") this result was originally established by Kolmogorov (1933). Here $\{F_{i_1, \cdots, i_n} = F_\alpha, \alpha \in \mathcal{F}\}$ is a given family of distribution functions satisfying the compatible conditions. It has become a basis for all the results on the projective limit theory. We state it for reference in a different but equivalent form.

Theorem 2. (Kolmogorov) *Let $T \subset \mathbb{R}$ be an index set and suppose that a family $\{F_{t_1, \cdots, t_n}, t_i \in T, n \geq 1\}$ of distribution functions is given satisfying the following compatibility conditions for any $t_1 < t_2 < \cdots < t_n$,*

(i) (a) $\lim_{x_n \to +\infty} F_{t_1, \cdots, t_n}(x_1, \cdots, x_n) = F_{t_1, \cdots, t_{n-1}}(x_1, \cdots, x_{n-1})$,

(b) $\lim_{x_n \to -\infty} F_{t_1, \cdots, t_n}(x_1, \cdots, x_{n-1}, x_n) = 0$,

(ii) $F_{t_{i_1}, \cdots, t_{i_n}}(x_{i_1}, \cdots, x_{i_n}) = F_{t_1, \cdots, t_n}(x_1, \cdots, x_n)$,

for each permutation $(i_1, \cdots, i_n)$ for $(1, \cdots, n)$. Let $\Omega = \mathbb{R}^T$ and $\Sigma =$ the σ-algebra generated by all sets of the form $\{\omega \in \Omega : \omega(t_i) < x_i, i = 1, \cdots, n\}$, $t_i \in T, x_i \in \mathbb{R}$ and $n \geq 1$. Then there exists a unique probability function $P : \Sigma \to \mathbb{R}^+$ such that, if $X_t(\omega) = \omega(t), t \in T$, denotes the coordinate functon, we have

$$P[\omega : X_{t_1}(\omega) < x_1, \cdots, X_{t_n}(\omega) < x_n] = F_{t_1, \cdots, t_n}(x_1, \cdots, x_n), x_i \in \mathbb{R}. \tag{3}$$

Thus $\{X_t, t \in T\}$ is a stochastic process on (Ω, Σ, P) having the given family as its finite dimensional distribution functions.

When can we conclude that the limit space (Ω, Σ, P) has some regularity properties, since $\Omega = \mathbb{R}^T$ need not be locally compact, or a Polish space? First we introduce a key concept and a related result from abstract analysis, and then connect it with our problem. At the beginning this seems to have little relevance, but will turn out later to be intimately related to our present study.

Definition 3. Let Ω be a locally compact space and $\mu(\not\equiv 0)$ be a Radon measure on it, and ρ a lifting on $L^\infty(\Omega, \mu)$. [ρ always exists since such a μ is strictly localizable, cf. Rao (1987), p. 443 and p. 429.] Then ρ is called an *almost strong lifting* if there is a μ-null set $N \subset \Omega$ such that $\rho(f)|N^c = f|N^c$ for each bounded continuous function $f : \Omega \to \mathbb{R}$.

The weakening in this definition is when supp $(\mu) \subsetneqq \Omega$. This is because when supp $(\mu) = \Omega$, then an almost strong lifting can be extended to a strong lifting. In fact, consider the Banach algebra $L^\infty(\Omega, \mu)$ and for each $\omega \in N$, consider a character ξ_ω satisfying $\xi_\omega(\tilde{f}) = f(\omega)$ for each bounded continuous $f : \Omega \to \mathbb{R}$, $\tilde{f}$ being the equivalence class containing f. If we define $\bar{\rho}$ by the equation

$$\bar{\rho}(\tilde{f})(\omega) = \rho(\tilde{f})\chi_{N^c}(\omega) + \xi_\omega(\tilde{f})\chi_N(\omega),$$

then $\bar{\rho}(f) = \rho(f)$ on N^c and $\bar{\rho}$ is a strong lifting.

A result of direct interest for us is the following:

Theorem 4. *Let Ω be a locally compact metrizable space and μ be a Radon measure on Ω such that $\emptyset \neq$ supp $(\mu) \subset \Omega$. Then each lifting on $L^\infty(\Omega, \mu)$ is almost strong and if supp $(\mu) = \Omega$, then there is actually a strong lifting on it.*

This abstract result will not be proved here. (For details see Tulcea and Tulcea (1969), pp. 128–131.) Using this we establish

Theorem 5. *Let $\{(\Omega_t, \Sigma_t), t \in I\}$ be a family of locally compact metric Borelian spaces, and $\mathcal{F}$ the directed (by inclusion) set of all finite subsets of I. If $\{(\Omega_\alpha, \Sigma_\alpha, P_\alpha, \pi_{\alpha\beta}) : \alpha < \beta$ in $\mathcal{F}\}$ is a Radon probability projective system, as in Proposition 1, formed of the given (metric) family with $\pi_{\alpha\beta} : \Omega_\beta \to \Omega_\alpha(= \underset{t \in \alpha}{\times} \Omega_t)$ as coordinate projections, then $(\Omega, \Sigma, P) = \varprojlim(\Omega_\alpha, \Sigma_\alpha, P_\alpha, \pi_{\alpha\beta})$ exists $[P = \varprojlim(P_\alpha, \pi_{\alpha\beta})]$. If each Ω_t is also compact $(\neq \emptyset)$, then μ is a Baire measure on Σ (since then Σ is a Baire σ-algebra).*

Proof. By hypothesis each $\mu_\alpha : \Sigma_\alpha \to \mathbb{R}^+$ is a Radon measure and for $\alpha < \beta$ in $\mathcal{F}$, $\mu_\alpha = \mu_\beta \circ \pi_{\alpha\beta}^{-1}$. We also have, from this,

$$\mu_\beta(B \cap \pi_{\alpha\beta}^{-1}(A)) = \int_A q_{\alpha\beta}(B, \omega)d\mu_\alpha(\omega), A \in \Sigma_{\alpha_1} B \in \Sigma_\beta. \qquad (4)$$

Here $q_{\alpha\beta} : \Sigma_\beta \times \Omega_\alpha \to \mathbb{R}^+$ has the following properties:

(i) $q_{\alpha\beta}(B,\cdot)$ is Σ_α-measurable for each $B \in \Sigma_\beta$,

(ii) $q_{\alpha\beta}(\cdot,\omega)$ is a probability measure for each $\omega \in N^c_{\alpha\beta}$, $N_{\alpha\beta} \in \Sigma_\alpha$ and $\mu(N_{\alpha\beta}) = 0$.

This important conclusion is a consequence of the fact that $(\Omega_\alpha, \mu_\alpha)$ admits an almost strong lifting (cf. Theorem 4), and when this happens $q_{\alpha\beta}$ has the stated property by the corresponding disintegration theorem (cf., Tulcea and Tulcea (1969), Theorem $5'$ on p. 154). Also from Theorem 3.2 above, for each $\omega \in N^c_{\alpha\beta}$, $q_{\alpha\beta}(B,\omega) > 0 \Rightarrow B \cap \pi_{\alpha\beta}^{-1}(\{\omega\}) \neq \emptyset$. Therefore the hypothesis of Theorem 3.1 is satisfied and we deduce from it that $\varprojlim(\Omega_\alpha, \Sigma_\alpha, \mu_\alpha, \pi_{\alpha\beta}) = (\Omega, \Sigma, \mu)$ exists, giving the first part.

Suppose now each Ω_t is also compact so that Ω_α and $\Omega(= \underset{t \in I}{\times} \Omega_t)$ are compact (by Tychonov's theorem). However, Ω is not a metric space in general. In any case, each μ_α is a Baire probability and $\mu : \Sigma \to \mathbb{R}^+$ is σ-additive. These are the only two properties that are essential for the remainder of this proof. If we show that Σ is a Baire σ-algebra, then the result that μ is a Baire measure follows from classical measure theory. [This was proved by N. Dinculeanu and I. Kluvanek (1967), even for vector measures, and independently somewhat earlier in an unpublished paper by R.P. Langlands (for scalar measures), as noted in Dunford-Schwartz (1963), Part II, p. 5 in Errata.] Then the classical results in measure theory show that μ has a unique extension to be a Radon measure on the Borel σ-algebra of Ω.

To establish the Baire property of Σ, note that the topology of Ω is determined by the neighborhood base $\{U_{i_1} \times \cdots \times U_{i_k} \times \underset{t \in I - \{i_1, \cdots, i_k\}}{\Pi} \Omega_k\}$ where U_is are open Baire sets of Ω_is. These are cylinder sets with open Baire bases and $U_i \in \Sigma_i$, the latter containing all Baire sets. Such cylinder sets form a subbase of the topology of Ω. If $C \subset \Omega$ is a compact G_δ-set, then there exist open sets $O_n \subset \Omega$ such that $C = \overset{\infty}{\underset{n=1}{\cap}} O_n$, and $O_n \supset O_{n+1}$ may be assumed. By the "sandwich theorem," for each $\omega \in C$ and n, there exists an open Baire set $B_{n,\omega} \subset O_n$ and such that $O_n \supset \underset{\omega \in C}{\cup} B_{n,\omega} \supset C$. So extracting a finite subcover $B_{n,\omega_1}, \cdots, B_{n,\omega_{k_n}}$ from this collection so that B_{n,ω_i} is an element of the subbase, we have $V_n = \overset{k_n}{\underset{i=1}{\cup}} B_{n,\omega_i} \in \Sigma$ and $C \subset V_n \subset O_n$. Hence $C \subseteq \underset{n}{\cap} V_n \subset \underset{n}{\cap} O_n = C$. This shows that $C \in \Sigma$. Since each compact G_δ of Ω is thus in Σ, it follows that Σ contains (and hence is equal to) the Baire σ-algebra of

Ω. By the earlier observation this shows that $\mu : \Sigma \to \mathbb{R}^+$ is a Baire measure. $\quad\square$

Remarks. 1. If the Ω_i are only locally compact (but $\Omega = \underset{i \in I}{\times}\, \Omega_i$ need not be) then the above argument merely shows that each compact G_δ subset of Ω is in the σ-ring generated by the cylinders with Baire bases, and one will not be able to extend it to the σ-algebra Σ without additional conditions.

2. There is a version of the projective limit theorem in which $\mu = \lim_{\leftarrow}(\mu_\alpha, \pi_{\alpha\beta})$ exists even if each Ω_i is (compact but) not metrizable. This result cannot be obtained from Theorem 3.1. (See the next section on this point.)

3. The regularity hypothesis on conditional measures in the above results allows us to treat all the integrals in the Lebesgue sense. Other reasons will be discussed at the end of the chapter.

8.5 Remarks on conditioning, disintegration, and lifting

We analyze interrelations here between the items mentioned in the section title. Thus consider the abstract measure spaces as in Theorem 3.2, so that $\{(S_n, \mathcal{S}_n), n \geq 1\}$ are measurable spaces $\Omega_n = \overset{n}{\underset{i=1}{\times}}\, S_i, \Sigma_n = \overset{n}{\underset{i=1}{\otimes}}\, \mathcal{S}_i, \pi_{mn} : \Omega_n \to \Omega_m\,(m \leq n)$ are coordinate projections. Suppose we are given an "initial" measure $\mu_1 : \Sigma_1 \to \mathbb{R}^+$ and a regular conditional (also termed a *transition*) measure $q_{nn+1} : \Sigma_{n+1} \times \Omega_n \to \mathbb{R}^+$ where $q_{nn+1}(\cdot, \omega_n)$ is a measure ($\omega_n \in \Omega_n$) and $q_{nn+1}(A, \cdot)$ is Σ_n-measurable for each $A \in \Sigma_{n+1}$. For each measurable rectangle $A_1 \times \cdots \times A_n \in \Sigma_n$, define as before a function $\mu_n : \Sigma_n \to \mathbb{R}^+$ by the equation:

$$\mu_n(A_1 \times \cdots \times A_n) = \int_{A_1} d\mu_1(\omega_1) \int_{A_2} q_{12}(d\omega_2, \omega_1) \cdots \times$$
$$\int_{A_n} q_{n-1n}(d\omega_n; \omega_1 \cdots \omega_{n-1}). \quad (1)$$

Then μ_n is a measure and for $m < n, \mu_m = \mu_n \circ \pi_{mn}^{-1}$. If on the other hand each $q_{mn} : \Sigma_n \times \Omega_m \to \mathbb{R}^+$ is independent of the second variable so that it is a constant on $\Omega_m\,(m \leq n)$, say $\tilde{q}_n(\cdot)$, then (1) gives for $n \geq 2$.

$$\mu_n(A_1 \times \cdots \times A_n) = \mu_1(A_1)\tilde{q}_2(A_2) \cdots \tilde{q}_n(A_n). \quad (2)$$

Taking $A_1 = \Omega_1 (= S_1), A_i = S_i, i \geq 2$, one gets from (2) that $\mu_n (S_1 \times A_2 \times S_3 \times \cdots \times S_n) = \tilde{q}_2 (A_2)$. So by iteration we have

$$\mu_n (A_1 \times \cdots \times A_n) = \mu_1 (A_1) \prod_{i=2}^{n} \tilde{q}_2 (A_i). \qquad (3)$$

Since these $\tilde{q}_i$ trivially satisfy the hypothesis of Theorem 3.2, their projective limit exists. This is the classical Jessen's extension of Fubini's theorem and can be stated for comparison as:

Theorem 1. (Fubini-Jessen) *Suppose* $\{(\Omega_n, \Sigma_n, \mu_n, \pi_{mn}) 1 \leq m \leq n < \infty\}$ *be a system of probability spaces satisfying (2) or (3). Then there is a unique measure* $\mu : \Sigma \rightarrow \mathbb{R}^+$, *($\Sigma = $ cylinder σ-algebra) such that* $\mu \circ \pi_n^{-1} = \mu_n, n \geq 1$, *(which is (3) for every measurable rectangle).*

Thus when the q_{mn} are simply product (probability) measures, then no topological conditions are needed on the measure spaces for the existence of a projective limit. The same is true if the existence of *regular* conditional measures satisfying (1) is assured, as seen in Tulcea's theorem (cf. Theorem 3.2 or Proposition 3.3). This latter assurance is available if the Ω_i's are locally compact metric spaces and the μ_i are Radon measures. The work of Chapter 5 shows that, in the abstract case, we need to demand that the μ_i should be perfect, compact or pure for the regularity of conditional measures. Since (1) can also be considered as a (generalized) disintegration of $\mu_{n-1}, \cdots, \mu_1$ successively, some relation between these two theories can be suspected.

One may ask as to why should this measure problem be connected with the (strong) lifting question which (on the surface at least) is a certain multiplicative linear operator on the Banach algebra $L^\infty (\mu)$? We analyze this relation here to gain further insight into the structure of conditional measures.

To put the question in perspective, we need to restate the disintegration problem. Let $(\Omega_i, \Sigma_i), i = 1, 2$, be a pair of measurable spaces and $f : \Omega_1 \rightarrow \Omega_2$ be a measurable mapping. If $\mu : \Sigma_1 \rightarrow \mathbb{R}^+$ is a measure and $\nu_f = \mu \circ f^{-1} : \Sigma_2 \rightarrow \mathbb{R}^+$ is the image measure of μ, then for each B in Σ_2, $\nu_f (\cdot \cap f^{-1} (B))$ is σ-additive on Σ_1, and $\nu_f (A \cap f^{-1} (\cdot))$ is a measure on Σ_2 for each $A \in \Sigma_1$. Also $\nu_f (\cdot \cap f^{-1} (B))$ is μ-continuous so that

$$\nu_f (A \cap f^{-1} (B)) = \int_A q(B, \omega_1) d\mu(\omega_1), A \in \Sigma_1, B \in \Sigma_2. \qquad (4)$$

This equation may be written symbolically as:

$$\nu_f(A \cap \cdot) = \int_A q(\cdot, \omega_1) d\mu(\omega_1), \ A \in \Sigma_1, \tag{5}$$

as an "identity" on the σ-algebra $\tilde{\Sigma}_2 = f^{-1}(\Sigma_1) \subset \Sigma_2$. But what does this signify? Since $\omega \mapsto q(\cdot, \omega)$ is a vector function, the symbol in (5) is not a Lebesgue integral, and one needs to attach a meaning in some "generalized" way. Thus if $g : \Omega_2 \to \mathbb{R}$ is a $(\tilde{\Sigma}_2\text{-})$ measurable bounded function, we consider

$$\int_{\Omega_2} g(\omega_2) \nu_f(A, d\omega_2) = \int_{\Omega_2} g(\omega_2)[\int_A q(d\omega_2, \omega_1) d\mu(\omega_1)]$$
$$= \int_A \int_{\Omega_2} g(\omega_2) q(d\omega_2, \omega_1) \ d\mu(\omega_1), \tag{6}$$

if the change of order can be justified. When this is true, then the vector integral in (5) has a meaning in a "weak" sense which can be made precise. It therefore becomes necessary to relate the properties of $q(\cdot, \omega_1)$ with those of the vector integral. If Ω_2 is a topological space and g is a bounded continuous function on Ω_2 one can hope to describe the properties of $q(\cdot, \omega_2)$ and of $(\Omega_2, \Sigma_2, \nu_f)$. For this it will be useful to be able to interpret all integrals in (6) in Lebesgue's sense, and select (continuous) functions from equivalence classes of bounded measurable functions in $L^\infty(\Omega_2, \nu_f)$. This brings us to look at the selection problem which is just the lifting operator on the latter space. It is not enough, however, to have a simple lifting that does not relate to the topologies of Ω_1 and Ω_2. This question leads to a deep study taking one into the existence and analysis of (almost) strong lifting maps. It is a happy conclusion that, in this topological case, one finds the actual equivalence of the existence of regular conditional measures, supported by slices such as $f^{-1}(\{\omega_2\})$, with the disintegration of μ (relative to ν_f) and that in turn with the existence of an almost strong lifting operator, making the material in the monograph of Tulcea and Tulcea (1969) indispensable in this work.

Thus although on the surface, there appears to be no connection between these concepts in reality the technical aparatus needed to prove the existence of one of them leads to the existence of the other and conversely. In this way one sees that the regularity property of conditioning is a deep part of the subject and must be studied as such.

We now point out that the existence of regular conditional measures, used in Theorem 3.1, is a good sufficient condition for the existence of projective limits of probability spaces, but it is not a necessary condition. From the equivalence of this with strong lifting problem noted above, it follows that the nonexistence of regular conditioning is the same as the nonexistence of a strong lifting. It has been shown recently by Losert (1979) that there exist Radon measure spaces (Ω, μ), with Ω a compact Hausdorff space, that do not admit any strong lifting. In view of this, the following result is of interest and it is essentially taken from Bochner (1955):

Theorem 2. *Let $\{(\Omega_\alpha, \Sigma_\alpha, P_\alpha, g_{\alpha\beta}) : \alpha < \beta \text{ in } I\}$ be a projective system of Radon probability spaces, Ω_α's being compact Hausdorff spaces and the $g_{\alpha\beta}$ are also onto. Then the projective limit of the system exists as a Baire probability space.*

Proof. Since $P_\alpha(\Omega_\alpha) = 1$, each Ω_α is a nonempty compact Hausdorff space and $g_{\alpha\beta}(\Omega_\beta) = \Omega_\alpha, \alpha < \beta$. By Proposition 2.4, $\Omega = \lim_{\leftarrow}(\Omega_\alpha, g_{\alpha\beta})$ exists as a nonempty compact Hausdorff space and $g_\alpha : \Omega \to \Omega_\alpha$ are also onto (and continuous). By the first part of the proof of Theorem 3.1 there exists an additive $P : \mathcal{A} = \bigcup_\alpha g_\alpha^{-1}(\Sigma_\alpha) \to \mathbb{R}^+$ such that $P_\alpha = P \circ g_\alpha^{-1}, \alpha \in I$. We thus need to show that P is σ-additive on $\mathcal{A}$ so that it has a unique extension to be a probability measure on $\sigma(\mathcal{A}) = \Sigma$. When this is shown the fact that (Ω, Σ, P) is a Baire probability space follows from the proof of the last part of Theorem 4.5, since we are dealing with compact spaces.

For the σ-additivity of the finitely additive P, it suffices to show for each $A_n \in \mathcal{A}$, $A_n \searrow \emptyset$, that $\lim_n P(A_n) = 0$. Suppose this is false. Then there exists a $\delta > 0$ and a sequence $A_n \searrow \emptyset$ in $\mathcal{A}$ such that $P(A_n) \geq \delta$. By the directedness of I, we can choose indices $\alpha_1 < \alpha_2 < \cdots$ such that $A_n \in g_{\alpha_n}^{-1}(\Sigma_{\alpha_n}), n \geq 1$. Hence there are $B_n \in \Sigma_{\alpha_n}$ such that $A_n = g_{\alpha_n}^{-1}(B_n)$, and $P(A_n) = P_{\alpha_n}(B_n)$. Since each P_α is a Radon measure, it is inner regular. So there exists a compact set $C_n \subset B_n$ such that $C_n \in \Sigma_{\alpha_n}$ (being a Borel set) and

$$P_{\alpha_n}(C_n) > P_{\alpha_n}(B_n) - \frac{\delta}{2^{n+1}}, n \geq 1. \tag{7}$$

Let $\tilde{C}_n = g_{\alpha_n}^{-1}(C_n)(\subset g_{\alpha_n}^{-1}(B_n) = A_n)$. Since C_n, being compact, is closed and g_{α_n} is continuous, it follows that $\tilde{C}_n$ is a closed subset of

Ω which is compact. Hence the $\tilde{C}_n$ are compact, but not necessarily monotone. Define $C'_n = \overset{n}{\underset{k=1}{\cap}} \tilde{C}_k$. Then $C'_n (\subset A_n)$ is compact and decreasing. So $\underset{n}{\cap} C'_n \subset \cap A_n = \emptyset$. However, we also have

$$C'_n = \tilde{C}_n - \overset{n-1}{\underset{k=1}{\cup}} (A_k - \tilde{C}_k) \subset A_n, \tag{8}$$

so that by the finite additivity property of P (and $C'_n \in \mathcal{A}$),

$$P(C'_n) \geq P(\tilde{C}_n) - \overset{n-1}{\underset{k=1}{\Sigma}} P(A_k - \tilde{C}_k)$$

$$= P_{\alpha_n}(C_n) - \overset{n-1}{\underset{k=1}{\Sigma}} [P_{\alpha_k}(B_k) - P_{\alpha_k}(C_k)]$$

$$> P_{\alpha_n}(B_n) - \frac{\delta}{2^{n+1}} - \overset{n-1}{\underset{k=1}{\Sigma}} \frac{\delta}{2^{k+1}}$$

$$= P(A_n) - \overset{n}{\underset{k=1}{\Sigma}} \frac{\delta}{2^{k+1}} > \delta - \frac{\delta}{2} = \frac{\delta}{2} > 0. \tag{9}$$

Hence C'_n cannot be empty for any $n \geq 1$ and being compact and decreasing will have a nonempty intersection, contradicting the earlier fact. Thus P must be σ-additive. $\quad\square$

In the above result, Ω_α need not be metrizable, and, as noted already, regular conditional measures need not exist in this case. It shows that the abstract projective limit theory is more general than that covered by Theorem 3.1. On the problem of constructing (regular) conditional measures (discussed in earlier chapters), it can be better appreciated when its equivalence with (strong) lifting is recalled. It is known that the existence of a lifting operator depends on a use of the axiom of choice (at least one needs to use the existence of ultrafilters), and hence a constructive definition cannot be expected in general. Only under various separability conditions this can be done. The construction problem is thus a nontrivial aspect of the theory of conditioning and this fact has not (yet) been well appreciated in its manifold applications. We shall draw the distinction vividly in the work of the next chapter.

8.6 Bibliographical notes

The study of product conditional measures is a natural extension of the iterated conditional expectations, as noted in the text, and plays a fundamental role in several parts of Probability Theory and its applications. The connection with projective systems of measures is also intimate. We illustrated these aspects in this chapter. The abstract formulation of the problem is due to Bochner (1955) who introduced the useful sequential maximality condition, and proved a more general form of Theorem 5.2. (He did not require the compactness hypothesis.) However, his fundamental formulation of projective systems, as an abstraction of in Kolmogorov's original basic existence theorem, is at the root of all the extensions of this chapter and hence of stochastic analysis.

The basic result, Theorem 3.1 here, is due to Choksi (1958), which is an extension of an important prior result by C. Ionescu Tulcea (1949), given here as Theorem 3.2. Also in the case of topological projective systems of measures, our treatment follows the influencial monograph of Tulcea and Tulcea (1969). Here we emphasized the relations between regular conditioning and (almost) strong lifting to focus the difficult parts of conditioning theory and to illuminate and contrast it with Theorems 3.1 and 3.2. The nonexistence result, due to Losert (1979), on strong lifting clarifies the role of regular conditioning in applications as well as the theory, and helps to distinguish the projective limit study with that of disintegration and hence of regular conditioning.

It is also useful as well as important to note the reasons with our preoccupation with regularity in the conditioning study. We have already seen that conditional measures in general are vector valued (σ-additive) set functions and one has to use the Dunford-Schwartz (or Wright-) integration in their analysis. Except for simple functions, there are no usable methods of evaluation of these integrals for applications such as Markov processes. With regularity of conditional measures, one is able to utilize the Lebesgue integration and its familiar techniques. Although the Dunford-Schwartz type integration helps to simplify the exposition of the general theory and streamlines its treatment, it is necessary for most applications requiring explicit evaluations to demand regularity of conditional measures. That is why we spent considerable space in these pages for this property. However, the actual computation of conditional expectations (or measures) is still a problem of nontrivial content and awaits further research.

CHAPTER 9: APPLICATIONS TO MARTINGALES AND MARKOV PROCESSES

This chapter is devoted to applications of conditioning to two of the most important areas of Probability Theory, namely martingales and Markov processes. The former uses the general properties of conditional expectations while the latter depends essentially on the *regularity* of conditional measures. Here basic results on the mean and pointwise convergence of (directed indexed) martingale limit theorems, as well as structural properties of Markov processes are presented. These include the existence and continuity properties of Markov processes under different conditions. However, the problems on the evaluation of conditional expectations of functionals on these processes is still not settled satisfactorily.

9.1 Introduction

The notion of conditioning is fundamental to both martingales and Markov processes even to introduce the concepts. The first one uses conditional expectations and their properties without demanding special restrictions for the most part, whereas the second one (Markov processes) needs the regularity of conditional probabilities from the very beginning. Therefore, to understand the structural analysis of both these classes, we have to utilize the results on the subject developed in the preceding chapters. Only the general principles of these processes can be discussed here as each has grown to warrant a monograph length treatment.

To motivate the idea of a martingale, let $X_1, X_2, \cdots$ be a sequence of integrable random variables on (Ω, Σ, P). The X_n may be thought of as the fortune of a gambler at time n. It is considered to be a fair (favorable) game if the expected fortune on the $(n+1)^{\text{st}}$ play, having known the previous n outcomes is exactly (at least) the amount on the nth. Here one interprets the "previous knowledge" of the outcomes as the σ-algebra generated by $(X_1, \cdots, X_n)$, and the expected fortune then as the conditional expectation, calling the fair (favorable) game a (sub) martingale. A precise description is stated as:

Definition 1. Let $X_1, X_2, \cdots$ be a sequence of integrable random variables on (Ω, Σ, P) and $\mathcal{B}_n$ be the σ-algebra generated by $(X_1, \cdots, X_n)$. Then the sequence is a *(sub) martingale* if

$$E^{\mathcal{B}_n}(X_{n+1}) = (\geq)X_n, \text{ a.e.}[P], \ n \geq 1. \tag{1}$$

This is also written as $E(X_{n+1}|X_1, \cdots, X_n) = (\geq)X_n$, a.e. or

$$\int_A X_{n+1}\, dP = (\geq) \int_A X_n\, dP_{\mathcal{B}_n}, A \in \mathcal{B}_n, n \geq 1. \tag{2}$$

Here the equivalence of (1) and (2) follows from the definition of conditional expection. But (2) shows that the concept also makes sense if P is an infinite measure such that $P_{\mathcal{B}_n}$ is σ-finite (or localizable), since the existence of such a sequence depends directly on the Radon-Nikodým theorem. If one considers a continuous parameter case, i.e. $\{X_t, t \in T \subset \mathbb{R}\}$ from $L^1(P)$, then (1) may be expressed as follows: if $\{\mathcal{B}_t, t \in T\}$ is an increasing family of σ-subalgebras of Σ such that X_t is $\mathcal{B}_t$-measurable (also termed $\mathcal{B}_t$-adapted), then it is a *(sub) martingale* provided for $s < t$ and $A \in \mathcal{B}_s \subset \mathcal{B}_t$,

$$E^{\mathcal{B}_s}(X_t) = (\geq)X_s, \text{a.e.} \left[\int_A X_t dP = (\geq) \int_A X_s dP_{\mathcal{B}_s} \right]. \tag{3}$$

Here $\{\mathcal{B}_t, t \in T\}$ is called a *stochastic base* of the (sub) martingale and $\sigma(X_s, s \leq t) \subset \mathcal{B}_t$. If the inequality in (1) or (3) is reversed, then one has the concept of a *supermartingale*. Thus a martingale is simultaneously a sub- and a supermartingale. We also write these sequences as $\{X_t, \mathcal{B}_t, t \in T\}$.

Let us present some basic inequalities for these processes to use them later in the convergence theorems. Since a sequence $\{X_t, t \in T\}$ is a submartingale iff $\{-X_t, t \in T\}$ is a supermartingale, for the following work, only one of them need be considered.

Lemma 2. *Let $\{X_t, \mathcal{B}_t, t \in T \subset \mathbb{R}\}$ be a (sub) martingale and φ be a convex (increasing) function on $\mathbb{R}$ such that $\varphi(X_t) \in L^1(P), t \in T$. Then $\{\varphi(X_t), \mathcal{B}_t, \ t \in T\}$ is a submartingale.*

Proof. Let $s < t$ in T. Since $\varphi(X_t)$ is clearly $\mathcal{B}_t$-adapted, by the conditional Jensen inequality we have

$$E^{\mathcal{B}_s}(\varphi(X_t)) \geq \varphi(E^{\mathcal{B}_s}(X_t)), \quad \text{a.e.,}$$
$$= (\geq)\varphi(X_s),$$

under the given conditions. Hence $\{\varphi(X_t), \mathcal{B}_t, \ t \in T\}$ is a submartingale. $\square$

Interesting special cases are if $\varphi(x) = |x|^p, p \geq 1$ or $\varphi(x) = \max(x, 0)$, giving the following:

Corollary 3. *If $\{X_t, \mathcal{B}_t, \ t \in T\}$ is a martingale, $X_t \in L^p(P)$, then (i) $\{X_t^{\pm}, \mathcal{B}_t, t \in T\}$, and (ii) $\{|X_t|^p, \mathcal{B}_t, \ t \in T\}$ are submartingales.*

Another specialization of interest is:

Corollary 4. *If $\{X_t^{(i)}, \mathcal{B}_t, \ t \in T\}$, $i = 1, 2$, are submartingales, then (i) $\{X_t^{(1)} \vee X_t^{(2)}, \ \mathcal{B}_t, t \in T\}$ and (ii) $\{\varphi(X_t^{(i)}), \mathcal{B}_t, \ t \in T\}$ are submartingales, where '$\vee$' = max, and $\varphi(u) = u \ \log^+(u)(\log^+ u = 0$ for all $u \leq 1)$.*

Remark. So far we have only used the partial ordering in T. In the case of partially ordered index sets, in the definitions of (sub or super) real martingales (1) and (2) will be replaced by (3).

The following concept of 'stopping time' will be useful in streamlining some of the work below. It is also a key tool in the subject.

Definition 5. *If $\{\mathcal{F}_t, t \in I\}$ is an increasing net of σ-subalgebras of (Ω, Σ, P) where I is partially ordered (and $\mathcal{F}_t$ may be completed for P), then a mapping $\tau : \Omega \to I$ is a stopping time (or an optional) of the net if for each t in I, both $\{\omega : \tau(\omega) \leq t\}$ and $\{\omega : \tau(\omega) \geq t\}$ are* in $\mathcal{F}_t$ for each t. [If I is linearly ordered, the second condition may be omitted and if moreover I is countable, it is enough to assume that $\{\omega : \tau(\omega) = t\} \in \mathcal{F}_t.$] An increasing sequence or net $\{\tau_j, j \in J\}$, with J another partially ordered set, is called a *stopping time process* (or *net*) of $\{\mathcal{F}_t, t \in I\}$.

The qualification "of the net" is usually omitted if it is clear from context. The next result when $J = \mathbb{N}$, the natural numbers, is often called the *optional sampling theorem*, and is due to Doob in that case. The present generalization is due to Chow (1960) and our proof is adapted from Hunt (1966).

Theorem 6. *Let $\{X_t, \mathcal{F}_t, t \in I\}$ be a submartingale and $\{\tau_j, j \geq 1\}$ be a stopping time process where each τ_j is finitely valued and I is linearly ordered. Then $\{Y_j = X \circ \tau_j, \mathcal{G}_j, j \geq 1\}$ is a submartingale where*

$$\mathcal{G}_j = \sigma\{A \cap [\tau_j \geq i] : A \in \mathcal{F}_i, i \in I\}, j \geq 1 . \tag{4}$$

[Here $\mathcal{G}_j$ is called the σ-algebra of events "prior to τ_j" and is also denoted by $\mathcal{F}(\tau_j).$]

Proof. Observe that $\mathcal{G}_j \subset \mathcal{G}_{j+1}$. Indeed for any $A \in \mathcal{F}_i$, $A \cap [\tau_j \geq i]$ being a generator of $\mathcal{G}_i$ and $A \cap [\tau_j \geq i] = A \cap [\tau_j \geq i] \cap [\tau_{j+1} \geq i]$, since $\tau_j \leq \tau_{j+1}$, this last set is in $\mathcal{F}_i$. But $A \cap [\tau_{j+1} \geq i]$ is a generator of $\mathcal{G}_{j+1}$, so that each generator of $\mathcal{G}_j$ is in $\mathcal{G}_{j+1}$, and the stated inclusion holds. Next note that Y_j is $\mathcal{G}_j$-adapted. To see this, for each $x \in \mathbb{R}$, and j,

$$[Y_j < x] \cap [\tau_j = i] = [X_i < x] \cap [\tau_j = i] \in \mathcal{F}_i, i \in I,$$

and then

$$[Y_j < x] \cap [\tau_j \geq i] = [X_i < x] \cap [\tau_j = i] \cap [\tau_j \geq i] \in \mathcal{G}_j, i \in I.$$

Hence one has (since the following union is at most countable)

$$[Y_j < x] = \bigcup_{i \in I}[Y_j < x] \cap [\tau_j \geq i] \in \mathcal{G}_j, \text{ by } (4).$$

Let us now prove the main assertion that the Y_j-process is a (sub) martingale. Since I is linearly ordered and τ_j is finitely valued, let $i_1 < i_2 < \cdots < i_{n_j}$ be the range of $\tau_j (< \tau_{j+1})$, and consider

$$Y_j = \sum_{k=1}^{n_n} X_{i_k} \chi_{[\tau_j = i_k]} , (\in L^1(P)).$$

If $A \in \mathcal{G}_j$, let $A_k = A \cap [\tau_j = i_k]$. Then $A_k \in \mathcal{G}_j$, $\overset{n_j}{\underset{k=1}{\cup}} A_k = A$ so that if we show

$$\int_{A_k} Y_{j+1} \, dP \geq \int_{A_k} Y_j dP, \ 1 \leq k \leq n_j, \tag{5}$$

and adding over k the resulting inequality implies the desired conclusion. Thus let $B_i = A_k \cap [\tau_{j+1} = i]$ and $B^i = A_k \cap [\tau_{j+1} > i]$, $i \in I$. Then $B_i \cap B^i = \emptyset$ and $B_{i_k} \cup B^{i_k} = B^{i_{k-1}}$, $k > 1$. We have

$$\int_{A_k} Y_j dP = \int_{B_k} X_k \, dP, \text{ since } \tau_j = k \text{ on } B_k,$$

$$= \int_{B_{i_k}} Y_{j+1} \, dP + \int_{B^{i_k}} X_{i_k} \, dP, \text{ by definition of } Y_{j+1},$$

$$\leq \int_{B_{i_k}} Y_{j+1} \, dP + \int_{B^{i_k}} X_{i_{k+1}} \, dP, \text{ since } X_k \text{ is a submartingale}$$

$$\text{and } B^{i_k} \in \mathcal{F}_{i_k},$$

$$= \int_{B_{i_k}} Y_{j+1} + \int_{B_{i_{k+1}}} Y_{j+1} \, dP + \int_{B^{i_{k+1}}} X_{i_{k+1}} \, dP$$

$$\leq \cdots \leq$$

$$= \int_{\underset{\ell \geq k}{\cup} B_{i_\ell}} Y_{j+1} \, dP = \int_{A_k} Y_{j+1} \, dP, \tag{6}$$

with equality throughout in the martingale case. This establishes (5), and thus the result follows. $\square$

Remark. If $I = \mathbb{N}$ and $\tau_n : \Omega \to \mathbb{N}$ is a stopping time process of $\{\mathcal{F}_n, n \geq 1\}$ then the above result is immediately extended by truncation, i.e., letting $T_m = \tau_n \chi_{[\tau_n \in I_m]} + m \chi_{[\tau_n \notin I_m]}$, where $I_m = \{1, \cdots, m\} \nearrow \mathbb{N}$. We leave the details to the reader, but utilize this extension later.

Using the above theorem, we obtain maximal inequalities for submartingales, extending the classical Kolmogorov inequality, for independent random variables with finite variances. This will be useful for the pointwise convergence study of these processes.

Theorem 7. *Let* $\{X_k, \mathcal{B}_k, k \geq 1\}$ *be a submartingale and consider the sets*

$$A_\lambda = \{\omega : \sup_{k \geq 1} X_k(\omega) \geq \lambda\} \text{ and } B_\lambda = \{\omega : \inf X_k(\omega) < \lambda\}, \lambda \in \mathbb{R}.$$

Then

$$(a)\lambda P(A_\lambda) \leq \liminf_n \int_{A_\lambda} X_n^+ \, dP \leq \lim_n \int_{A_\lambda} X_n^+ \, dP, \qquad (7a)$$

$$(b)\lambda P(B_\lambda) \geq \lim_n E(X_1 - X_n^+) + \limsup_n \int_{B_k} X_n^+ \, dP. \qquad (7b)$$

Proof. For each $n > 1$, $\{X_k, \mathcal{B}_k, 1 \leq k \leq n\}$ is a submartingale and let $\tau_1 = \inf\{k \geq 1 : X_k > \lambda\}$ and $= n$ if $\{\ \} = \emptyset$. Then τ_1 is a stopping time of $\{\mathcal{B}_k, 1 \leq k \leq n\}$ since

$$[\tau_1 = r] = \bigcup_{k=1}^{r} [X_j \leq \lambda, 1 \leq j \leq r-1, X_k > \lambda] = \bigcup_{k=1}^{r} A_\lambda^k = B_\lambda^r$$

$$\text{(say)}, (\in \mathcal{B}_r),$$

where for $r = 1$ one takes $A_\lambda^1 = [X_1 > \lambda]$. Let $\tau_2 = n$, the constant time on Ω so that $\tau_1 \leq \tau_2$. If $Y_i = X \circ \tau_i$ and $\mathcal{G}_1 = \mathcal{B}(\tau_1)$ as in (4) and $\mathcal{G}_2 = \mathcal{B}_n$, then by Theorem 6, $\{Y_i, \mathcal{G}_i\}_{i=1}^{2}$ is a submartingale. Hence $E^{\mathcal{G}_1}(Y_2) \geq Y_1$, a.e. This gives

$$\int_{B_\lambda^r} X_n \, dP = \int_{B_\lambda^r} Y_2 \, dP$$

$$\geq \int_{B_\lambda^n} Y_1 \, dP, \text{ by the submartingale property,}$$

$$= \sum_{j=1}^{n} \int_{A_\lambda^j} Y_1 \, dP = \sum_{j=1}^{n} \int_{A_\lambda^j} X_j \, dP$$

$$\geq \lambda P(\bigcup_{i=1}^{n} A_\lambda^j) = \lambda P(B_\lambda^n). \qquad (8)$$

Letting $n \to \infty$ in (8) and noting that $B_\lambda^n \uparrow A_\lambda$, one gets

$$\lambda P(A_\lambda) \leq \liminf_n \int_{B_\lambda^n} X_n^+ \, dP \leq \liminf_n \int_{A_\lambda} X_n^+ \, dP$$

$$\leq \lim_n E(X_n^+), \qquad (9)$$

since $\{X_n^+, \mathcal{B}_n, n \geq 1\}$ is a submartingale and hence its expectations are nondecreasing. This gives (a). We can again use a similar argument for (b).

Let $\bar{\tau}_1 = \inf\{k : X_k < \lambda\}$, and $= n$ if $\{\ \} = \emptyset$. Set $\bar{\tau}_0 = 1$ on Ω so that $\bar{\tau}_0 \leq \bar{\tau}_1$ and both are stopping times of $\{\mathcal{B}_n, n \geq 1\}$ as in (a).

Then $\{Y_i, \mathcal{G}_i\}_{i=0}^1$ is a submartingale where $Y_0 = X_1, Y_1 = X \circ \bar{\tau}_1$ and $\mathcal{G}_0 = \mathcal{B}_0, \mathcal{G}_1 = \mathcal{B}(\bar{\tau}_1)$. Thus $E(Y_0) \leq E(Y_1)$. Let $B_\lambda^k = \{X_j \geq \lambda, 1 \leq j \leq k-1, X_k < \lambda\}$ so that $\overset{r}{\underset{k=1}{\cup}} B_\lambda^k \in \mathcal{B}_r$ and

$$E(X_1) = E(Y_0) \leq E(Y_1) = \int_{\underset{k=1}{\overset{n}{\cup}} B_\lambda^k} Y_1 \, dP + \int_{(\underset{k=1}{\overset{n}{\cup}} B_\lambda^k)^c} Y_1 \, dP$$

$$\leq \sum_{j=1}^n \int_{B_\lambda^j} \lambda \, dP + \int_{(\underset{k=1}{\overset{n}{\cup}} B_\lambda^k)^c} X_n \, dP$$

$$= \lambda P(\underset{j=1}{\overset{n}{\cup}} B_\lambda^j) + E(X_n) - \int_{(\underset{k=1}{\overset{n}{\cup}} B_\lambda^k)} X_n \, dP. \tag{10}$$

Letting $n \to \infty$ in (10) one gets

$$\lambda P(B_\lambda) \geq E(X_1) + \limsup_n \int_{B_\lambda} X_n^+ - \lim_n E(X_n^+).$$

This gives (b). Inequalities (8) and (10) themselves are also useful. $\square$

If $\{X_n, n \geq 1\}$ are independent random variables on (Ω, Σ, P) with zero means and finite variances $\{\sigma_n^2, n \geq 1\}$, then $\{X_n^2, \mathcal{B}_n, n \geq 1\}$ is a submartingale with $\mathcal{B}_n = \sigma(X_1, \cdots, X_n)$. Applying (8) to this sequence one obtains the classical *Kolmogorov inequality* as follows:

Corollary 8. *Let $X_n, n \geq 1\}$ be real independent random variables on (Ω, Σ, P) with zero means and variances $\{\sigma_n^2, n \geq 1\}$. Then for each $\varepsilon > 0$, one has*

$$P[\max_{1 \leq k \leq n} |S_k| \geq \varepsilon] = P[\max_{1 \leq k \leq n} S_k^2 \geq \varepsilon^2] \leq \frac{1}{\varepsilon^2} \sum_{k=1}^n \sigma_k^2,$$

where $S_n = \sum_{k=1}^n X_k$.

Further properties of martingales will be needed for their convergence theory. Although we only considered point martingales so far, it is of interest to treat set martingales in the development. This will be done now and we use the projective limit theory of the preceding chapter for the purpose.

9.2 Set martingales

The notion of a set martingale helps to relate the ordinary (point) martingale theory with that of projective systems and product integration. We thus introduce the concept as follows:

Definition 1. If (Ω, Σ, μ) is a measure space, and $\{\mathcal{B}_\alpha, \alpha \in I\}$ is a net of (increasing) σ-subalgebras of Σ, let $P_\alpha : \mathcal{B}_\alpha \to \mathbb{R}$ be a σ-additive function for each $\alpha \in I$. Then $\{P_\alpha, \mathcal{B}_\alpha, \alpha \in I\}$ is termed a *set (sub) martingale* if for each $\alpha < \beta$ in I one has $P_\beta|\mathcal{B}_\alpha = (\geq)P_\alpha$. Here the net $\{\mathcal{B}_\alpha, \alpha \in I\}$ is referred to as the *base* of the set (sub) martingale.

Note that if $\{X_\alpha, \mathcal{B}_\alpha, \alpha \in I\}$ is a (sub) martingale on (Ω, Σ, μ) and $P_\alpha : A \mapsto \int_A X_\alpha d\mu$, $A \in \mathcal{B}_\alpha$, then $\{P_\alpha, \mathcal{B}_\alpha, \alpha \in I\}$ is a set (sub) martingale. There is a simple relation between a set martingale and a projective system as seen from

Proposition 2. *Let $\{(\Omega_\alpha, \Sigma_\alpha, P_\alpha, g_{\alpha\beta}) : \alpha < \beta \text{ in } I\}$ be a projective system of scalar measure spaces admitting the projective limit, where the earlier concepts of these systems for positive measures apply verbatim. Let $\Omega = \varprojlim(\Omega_\alpha, g_{\alpha\beta}), g_\alpha : \Omega \to \Omega_\alpha$ be the canonical mapping such that for $\alpha < \beta, g_\alpha = g_{\alpha\beta} \circ g_\beta, g_\alpha(\Omega) = \Omega_\alpha$, and $\Sigma_0 = \sigma(\bigcup_\alpha g_\alpha^{-1}(\Sigma_\alpha))$. If $\tilde{P}_\alpha : \tilde{\Sigma}_\alpha = g_\alpha^{-1}(\Sigma_\alpha) \to \mathbb{R}$ is defined by the equation $\tilde{P}_\alpha \circ g_\alpha = P_\alpha, \alpha \in I$, then $\{\tilde{P}_\alpha, \tilde{\Sigma}_\alpha, \alpha \in I\}$ is the associated set martingale on (Ω, Σ_0). Conversely, if $\{(\Omega_\alpha, \tilde{\Sigma}_\alpha, P_\alpha), \alpha \in I\}$ is a set martingale, then it determines a projective system $\{(\Omega_\alpha, \Sigma_\alpha, P_\alpha, g_{\alpha\beta} : \alpha < \beta \text{ in } I\}$ by putting $\Omega_\alpha = \Omega, \Sigma_\alpha = \tilde{\Sigma}_\alpha, g_{\alpha\beta} = $ identity for all α, β in I.*

Proof. If $\alpha < \beta$ then $g_\alpha = g_{\alpha\beta} \circ g_\beta$ implies

$$\tilde{\Sigma}_\alpha = g_\alpha^{-1}(\Sigma_\alpha) = g_\beta^{-1}(g_{\alpha\beta}^{-1}(\Sigma_\alpha)) \subset g_\beta^{-1}(\Sigma_\beta) = \tilde{\Sigma}_\beta \subset \Sigma_0, \qquad (1)$$

and $\tilde{P}_\alpha \circ g_\alpha^{-1} = P_\alpha$ uniquely defines $\tilde{P}_\alpha$. Indeed, there is always an additive P on $\bigcup_\alpha \tilde{\Sigma}_\alpha$ such that $P_\alpha = P \circ g_\alpha^{-1}$ so we set $\tilde{P}_\alpha = P|\tilde{\Sigma}_\alpha$. Also $\tilde{P}_\beta \circ g_\beta^{-1} = P \circ g_\beta^{-1} = P_\beta$ gives $P \circ g_\beta^{-1} \circ g_{\alpha\beta}^{-1} = P_\beta \circ g_{\alpha\beta}^{-1} = P_\alpha$, and hence

$$P|\tilde{\Sigma}_\beta = \tilde{P}_\beta, \quad \tilde{P}_\beta|\tilde{\Sigma}_\alpha = (P|\tilde{\Sigma}_\beta)|\tilde{\Sigma}_\alpha = P|\tilde{\Sigma}_\alpha = \tilde{P}_\beta, \text{ by (1)}. \qquad (2)$$

So $\{\tilde{P}_\alpha, \tilde{\Sigma}_\alpha, \alpha \in I\}$ is a set martingale. The converse is simple. $\square$

Remark. Although every martingale defines a set martingale, the converse is not necessarily true. In fact for a set martingale $\{\mu_\alpha, \mathcal{B}_\alpha, \alpha \in I\}$ there need not exist a dominating (localizable) measure μ even if all μ_α are finite measures, if I is uncountable. Analogous results can be given for submartingales and subprojective systems. We omit its discussion here.

The above observation furnishes interest to the following result which is an extension of the classical Jordan decomposition of a single scalar measure into positive components, efficiently, to the case of set martingales. This result is a slight modification of that of the classical Vitali-Hahn-Saks theorem. [The proof is suggested by an argument in Dunford-Schwartz (1958), p. 294.] We use this in showing that a projective system of scalar measures can be decomposed (efficiently) into similar systems of positive measues, and the martingale case is deduced as an immediate consequence.

Proposition 3. *Let* $\{(\Omega, \Sigma_\alpha, \mu_\alpha), \alpha \in I\}$ *be a bounded set martingale so that* $c_0 = \sup_\alpha |\mu_\alpha|(\Omega) < \infty$. *Then there is a unique decomposition of the set martingale such that*

$$\mu_\alpha = \xi_\alpha^{(1)} - \xi_\alpha^{(2)}, \ \xi_\alpha^{(1)}(\Omega) + \xi_\alpha^{(2)}(\Omega) = c_0, \ \alpha \in I , \qquad (3)$$

where $\{(\Omega, \Sigma_\alpha, \xi_\alpha^{(i)}), \alpha \in I\}, i = 1, 2$ *are positive set martingales.*

Proof. Fix an $\alpha \in I$, arbitrarily, and define $\xi_\alpha^{(1)} : \Sigma_\alpha \to \mathbb{R}^+$ as:

$$\xi_\alpha^{(1)}(A) = \sup\{\mu_\beta^+(A) : \ \beta > \alpha\} = \lim_\beta \mu_\beta^+(A), \qquad (4)$$

where μ_β^+ is the positive variation measure of the real μ_β and 'sup' can be replaced by 'lim' from the set martingale property. Indeed,

$$\mu_\alpha^+(A) = \sup\{\mu_\alpha(B) : B \subset A, \ B \in \Sigma_\alpha\}$$
$$\leq \sup\{\mu_\beta(B) : B \subset A, B \in \Sigma_\beta \supset \Sigma_\alpha\}, \text{ for } \alpha < \beta,$$
$$\text{since } \mu_\alpha(B) = \mu_\beta(B), \ B \in \Sigma_\alpha,$$
$$= \mu_\beta^+(A), \qquad (5)$$

so that $\mu_{(\cdot)}^+(A)$ is monotone on I. Also $\mu_\alpha^+(\cdot)$ is σ-additive, because μ_α is. Hence (4) implies $\xi_\alpha^{(1)} \geq 0$, and additive on Σ_α. If $I = \mathbb{N}$, the σ-additivity of $\xi_\alpha^{(1)}$ is a simple consequence of the Vitali-Hahn-Saks theorem. In the present (general) case I is only directed. So we need another argument as follows.

Suppose $\xi_\alpha^{(1)}$ is not σ-additive. Then there is an $\varepsilon > 0$ and a sequence $A_n \in \Sigma_\alpha, A_n \searrow \emptyset$, such that $\lim_n \xi_\alpha^{(1)}(A_n) \geq \varepsilon$. Since I is directed, we can select indices $\alpha < \beta_1 < \beta_2 < \cdots$ such that

$$\mu_{\beta_n}^+(A_n) > \varepsilon, \ n \geq 1, \qquad (6)$$

[see (4)]. By the boundedness of μ_αs we get

$$\xi_\alpha^{(1)}(A) = \lim_{\beta > \alpha} \mu_\beta^+(A) \le \sup_\alpha |\mu_\alpha|(\Omega) = c_0 < \infty. \tag{7}$$

Define a finite measure $\zeta : \Sigma_\alpha \to \mathbb{R}^+$ by the equation

$$\zeta(A) = \sum_{n=1}^\infty \frac{1}{2^n} \frac{\mu_{\beta_n}^+(A)}{1 + \mu_{\beta_n}^+(\Omega)}, \quad A \in \Sigma_\alpha. \tag{8}$$

Then $\mu_{\beta_n}^+$ is ζ-continuous and by the Radon-Nikodým theorem (and the monotonicity of $\mu_{\beta_n}^+$) there is a measurable (for Σ_α) $f_n : \Omega \to \mathbb{R}^+$ such that $f_n = \frac{d\mu_{\beta_n}^+}{d\zeta} \le f_{n+1}$. If $f = \lim_n f_n$, then by the monotone convergence theorem,

$$\int_\Omega f \, d\zeta = \lim_n \int_\Omega f_n \, d\zeta = \lim_n \mu_{\beta_n}^+(\Omega) \le c_0 < \infty.$$

Hence $\{f_n, n \ge 1\}$ which is dominated by the integrable function f, is uniformly integrable on $(\Omega, \Sigma_\alpha, \zeta)$ so that

$$\lim_{n\to\infty} \mu_{\beta_n}^+(A_n) = \lim_{n\to\infty} \int_{A_n} f_n \, d\zeta \le \lim_{n\to\infty} \int_{A_n} f \, d\zeta = 0, \tag{9}$$

since $0 \le f_n \le f$. But this contradicts (6), and hence $\xi_\alpha^{(1)}$ must be σ-additive. Also we have

$$\xi_\alpha^{(1)}(A) = \lim_{\beta > \alpha} \mu_\beta^+(A) \ge \mu_\alpha^+(A) \ge \mu_\alpha(A), \tag{10}$$

and since μ_β^+ is increasing,

$$\lim_\alpha \xi_\alpha^{(1)}(\Omega) = \lim_\alpha \lim_{\beta > \alpha} \mu_\beta^+(\Omega) = \lim_\beta \mu_\beta^+(\Omega). \tag{11}$$

Let $\xi_\alpha^{(2)} = \xi_\alpha^{(1)} - \mu_\alpha$. Then by (10) $\xi_\alpha^{(2)} \ge 0$ and is σ-additive so that $\mu_\alpha = \xi_\alpha^{(1)} - \xi_\alpha^{(2)}$. We claim that this is the desired decomposition.

Since $\{\mu_\alpha, \Sigma_\alpha, \alpha \in I\}$ is a set martingale, so is its negative. Also $c_0 = \lim_\alpha |\mu_\alpha|(\Omega) = \lim_\alpha(\mu_\alpha^+(\Omega) + \mu_\alpha^-(\Omega))$, and $-\mu_\alpha = \mu_\alpha^- - \mu_\alpha^+$ so that $\lim_\alpha \xi_\alpha^{(2)}(\Omega) = \lim_\alpha(\xi_\alpha^{(1)} - \mu_\alpha^+ + \mu_\alpha^-)(\Omega) = 0 + \lim_\alpha \mu_\alpha^-(\Omega)$, using (11). But $\xi_\alpha^{(1)}(\Omega)$ and $\xi_\alpha^{(2)}(\Omega)$ are independent of α, using the set martingale property. Hence $c_0 = \sup_\alpha |\mu_\alpha|(\Omega) = \sup_\alpha(\mu_\alpha^+ + \mu_\alpha^-)(\Omega) = \xi_\alpha^{(1)}(\Omega) +$

$\xi_\alpha^{(2)}(\Omega)$. This yields (3) and only the uniqueness of decomposition remains to be verified.

Let $\mu_\alpha = \tilde{\xi}_\alpha^{(1)} - \tilde{\xi}_\alpha^{(2)}$ be another such decomposition. Then $\tilde{\xi}_\alpha^{(1)} \geq \mu_\alpha^+$, and $\tilde{\xi}_\alpha^{(1)}(\Omega) \geq \lim_\alpha \mu_\alpha^+(\Omega) = \xi_\alpha^{(1)}(\Omega)$, by (11). The preceding paragraph implies $\tilde{\xi}_\alpha^{(2)}(\Omega) \geq \xi_2^{(2)}(\Omega)$ and hence by addition,

$$c_0 = \xi_\alpha^{(1)}(\Omega) + \xi_\alpha^{(2)}(\Omega) = \tilde{\xi}_\alpha^{(1)}(\Omega) + \tilde{\xi}_\alpha^{(2)}(\Omega) . \tag{12}$$

These relations give that $\xi_\alpha^{(i)}(\Omega) = \tilde{\xi}_\alpha^{(i)}(\Omega)$, $i = 1, 2$. Now the martingale property implies $\xi_\alpha^{(i)}(A) = \xi_\beta^{(i)}(A)$, all $\beta > \alpha$, $A \in \Sigma_\alpha$, and similarly for $\tilde{\xi}_\alpha^{(i)}$. But $\xi_\alpha^{(i)}(A) \leq \tilde{\xi}_\alpha^{(i)}(A)$ by (4). Thus (12) and additivity of these functions give $\xi_\alpha^{(i)} = \tilde{\xi}_\alpha^{(i)}$, $\alpha \in I$, as asserted.   □

The above decomposition can be used to obtain the corresponding result for projective systems of scalar measures. Thus if $\{(\Omega_\alpha, g_{\alpha\beta}), \alpha < \beta$ in $D\}$ is a projective system of spaces satisfying the s.m. condition so that $\Omega = \varprojlim(\Omega_\alpha, g_{\alpha\beta})$ is nonempty and $g_\alpha : \Omega \to \Omega_\alpha$ is onto, consider the scalar projective measure system $\{(\Omega_\alpha, \Sigma_\alpha, \mu_\alpha, g_{\alpha\beta}) : \alpha < \beta$ in $I\}$. Moreover on $\Sigma = \sigma(\cup_\alpha g_\alpha^{-1}(\Sigma_\alpha))$, there is an additive function $\mu : \Sigma \to \mathbb{R}$ such that $\mu \circ g_\alpha^{-1} = \mu_\alpha$, $\alpha \in I$. Let $\Sigma_\alpha^* = g_\alpha^{-1}(\Sigma_\alpha) \subset \Sigma$, $(g_\alpha = g_{\alpha\beta} \circ g_\beta)$, and $\nu_\alpha = \mu|g_\alpha^{-1}(\Sigma_\alpha)$. Then $\{\nu_\alpha, \Sigma_\alpha^*, \alpha \in I\}$ is a set martingale on (Ω, Σ) and it is bounded if μ is. Assuming this boundedness, we have $\nu_\alpha = \xi_\alpha^{(1)} - \xi_\alpha^{(2)}$ by the preceding proposition with $(\xi_\alpha^{(i)}, \Sigma_\alpha), i = 1, 2$, being positive set martingales. Since $\mu_\alpha = \mu \circ g_\alpha^{-1} = \xi_\alpha^{(1)} \circ g_\alpha^{-1} - \xi_\alpha^{(2)} \circ g_\alpha^{-1} = \mu_\alpha^{(1)} - \mu_\alpha^{(2)}$ (say), we get $\{(\Omega_\alpha, \Sigma_\alpha, \mu_\alpha^{(i)}, g_{\alpha\beta}), \alpha < \beta$ in $I\}$, $i = 1, 2$, to be positive projective systems whose difference is the given scalar system. This result may be stated as:

Proposition 4. *Let* $\{(\Omega_\alpha, \Sigma_\alpha, \mu_\alpha, g_{\alpha\beta}) : \alpha < \beta$ in $I\}$ *be a projective system of signed measures such that* $\sup_\alpha |\mu_\alpha|(\Omega_\alpha) \leq k_0 < \infty$, *and the s.m. hold. Then there exist* $\mu_\alpha^i : \Sigma_\alpha \to \mathbb{R}^+$ *such that* $\mu_\alpha = \mu_\alpha^1 - \mu_\alpha^2$ *and* $\{(\Omega_\alpha, \Sigma_\alpha, \mu_\alpha^{(i)}, g_{\alpha\beta}) : \alpha, \beta$ in $I\}$, $i = 1, 2$, *are projective systems of positive bounded measures.*

This result shows that it suffices to study positive projective systems and then the signed measure case can be deduced from it at once. We thus state the corresponding result to Theorem 8.4.5 for reference:

Theorem 5. *Let $\{(\Omega_t, \Sigma_t), \; t \in T\}$ be a family of locally compact metrizable Borelian spaces, $\mathcal{F}$ the class of all finite subsets of T directed by inclusion and for $\alpha \in \mathcal{F}$ set $\Omega_\alpha = \underset{t \in \alpha}{\times} \Omega_t$, $\Sigma_\alpha = \underset{t \in \alpha}{\otimes} \Sigma_t$. Suppose that $\mu_\alpha : \Sigma_\alpha \to \mathbb{R}$ are signed (necessarily regular Baire) measures such that (i) $\underset{\alpha}{\sup} |\mu_\alpha|(\Omega_\alpha) \leq k_0 < \infty$, and (ii) $\alpha < \beta$ in $\mathcal{F}$, $\mu_\alpha = \mu_\beta \circ \pi_{\alpha\beta}^{-1}$ where $\pi_{\alpha\beta} : \Omega_\beta \to \Omega_\alpha$ are coordinate projections. If $\pi_\alpha : \Omega = \underset{t \in T}{\times} \Omega_t \to \Omega_\alpha$ is the canonical projection, then the system $\{(\Omega_\alpha, \Sigma_\alpha, \mu_\alpha, \pi_{\alpha\beta}), \; \alpha < \beta$ in $\mathcal{F}\}$ admits the projective limit (Ω, Σ, μ) where $\Sigma = \underset{t \in T}{\otimes} \Sigma_t$ and $\mu = \lim_{\leftarrow}(\mu_\alpha, \mu_{\alpha\beta})$. If each Ω_t is compact (but not necessarily metrizable), then the same conclusion holds with μ also as a Baire measure.*

By the preceding proposition we can express each $\mu_\alpha = \mu_\alpha^{(1)} - \mu_\alpha^{(2)}$, with $\mu_\alpha^{(i)}$ as positive measures on the Borel σ-algebras Σ_α, and then they are automatically Baire regular, as a consequence of the classical measure theory [this is a very special case of Dinculeanu and Kluvanek (1967)]. The conclusion now follows from Theorem 8.4.5 itself.

To proceed now with further applications, suppose we have a pair of probability measures P, Q on a measurable space (Ω, Σ) and it is desired to compare them for their mutual singularity or equivalence. This problem is of importance in the statistical inference and also in the convergence theory of (set) martingales. The following concept of "Hellinger distance" is useful in solving it. Let $\mu = P + Q$ so that P, Q are μ-continuous, and let $f = \frac{dP}{d\mu}$, and $g = \frac{dQ}{d\mu}$, the Radon-Nikodým derivatives. Then the above "distance" between P and Q is defined as

$$H(P, Q) = \int_\Omega \sqrt{fg} \, d\mu \left(= \int_\Omega \sqrt{dPdQ} \right). \tag{13}$$

The functional $H(\cdot, \cdot)$ does not depend on μ. In fact if $\tilde{\mu}$ is another dominating measure for P and Q, then it also dominates μ. Now if $\tilde{f} = \frac{dP}{d\tilde{\mu}}, \tilde{g} = \frac{dQ}{d\tilde{\mu}}$, we have

$$H(P, Q) = \int_\Omega \sqrt{\frac{dP}{d\mu} \cdot \frac{dQ}{d\mu} \cdot \frac{d\mu}{d\tilde{\mu}}} \cdot d\tilde{\mu} = \int_\Omega \sqrt{\tilde{f}\tilde{g}} d\tilde{\mu}. \tag{14}$$

Thus $H(P, Q)$ remains unaltered by a change of μ to $\tilde{\mu}$. By the CBS-inequality, one has $0 \leq H(P, Q) = H(Q, P) \leq 1$, and $H(P, Q) = 0$ iff $P \perp Q$ and $H(P, Q) = 1$ iff $P = Q$ (by the equality conditions in the CBS-inequality).

For computational facility, we give an approximation formula for $H(P,Q)$ of (13):

Lemma 6. *For $H(\cdot,\cdot)$ of* (13) *one has*

$$H(P,Q) = \inf\{\sum_k \sqrt{(P(A_k)Q(A_k)} : \{A_k\}_{-\infty}^{\infty} \subset \Sigma \text{ is a partition of } \Omega\}.$$
(15)

Proof. Let $\{A_k, -\infty < k < \infty\} \subset \Sigma$, be a partition of Ω. Then

$$H(P,Q) = \int_{\underset{k}{\cup} A_k} \sqrt{fg} \, d\mu$$

$$= \sum_k \int_{A_k} \sqrt{fg} d\mu \le \sum_k \left(\int_{A_k} f d\mu \right)^{\frac{1}{2}} \left(\int_k g d\mu \right)^k,$$

by the disjointness of A_k and the CBS-inequality,

$$= \sum_k \sqrt{P(A_k)Q(A_k)} \ .$$

Hence

$$H(P,Q) \le \inf\{\sum_k \sqrt{P(A_k)Q(A_k)} : \{A_k\}_{-\infty}^{\infty} \subset \Sigma, \text{ a partition of } \Omega\} \ .$$
(16)

For the opposite inequality, consider for $1 < t < \infty$ and integers m,n,

$$A_{mn} = \{\omega : t^{2(m-1)} \le f(\omega) < t^{2m}, \ t^{2(n-1)} \le g(\omega) < t^{2n}\} \ .$$

Then

$$\int_{A_{mn}} \sqrt{fg} d\mu \ge t^{m+n-2} \, \mu(A_{mn}) \ .$$
(17)

But $P(A_{mn}) \le t^{2m} \mu(A_{mn})$ and $Q(A_{mn}) \le t^{2n} \mu(A_{mn})$. If $B = \underset{m,n}{\cup} A_{mn}$, a disjoint union, we have

$$H(P,Q) = \sum_{m,n} \int_{A_{mn}} \sqrt{fg} d\mu + 0.\mu(B^c)$$

$$\ge \frac{1}{t^2} \sum_{m,n} t^{m+n} \, \mu(A_{mn})$$

$$\ge \frac{1}{t^2} \sum_{m,n} \sqrt{P(A_{mn} Q(A_{mn})}, \text{ by (17)}.$$
(18)

Then taking infimum on all such $\{A_{mn}\}$ and letting $t \downarrow 1$ in (18) we get the opposite inequality to (16). These imply (15) as asserted. $\quad\square$

Let $(\Omega_i, \Sigma_i, {}^{P_i}_{Q_i})$, $i = 1, 2$ be a pair of probability spaces and $T : \Omega_1 \to \Omega_2$ a measurable mapping such that $P_1 = P_2 \circ T^{-1}$, $Q_1 = Q_2 \circ T^{-1}$. If $\Sigma_1^* = T^{-1}(\Sigma_2) \subset \Sigma_1$, a σ-subalgebra, then $\tilde{P}_1 = P_1 | \Sigma_1^*, \tilde{Q}_1 = Q_1 | \Sigma_1^*$ are measures for which (15) becomes

$$H(P_1, Q_1) \leq H(\tilde{P}_1, \tilde{Q}_1) = H(P_2, Q_2) \tag{19}$$

since there are fewer partitions in Σ_1^* than in Σ_1 of Ω. In particular, if $\Omega_1 = \Omega_2, \Sigma_2 \subset \Sigma_1$ and $Q_2 = Q_1 | \Sigma_2, P_2 = P_1 | \Sigma_2$, then (19) is satisfied. We remark that (19) can also be obtained directly using Jensen's inequality for concave functions φ (here $\varphi(x) = \sqrt{x}$) and the fact that $\varphi(fg) = \varphi(f)\varphi(g)$. This alternative proof is left to the reader.

With (15), we can present a comparison result for singularity or equivalence of a pair of set martingales having the same base. In view of Proposition 2, we present the assertions for projective systems so as to use it in some other applications also.

Theorem 7. Let $\{(\Omega_\alpha, \Sigma_\alpha, {}^{P_\alpha}_{Q_\alpha}, g_{\alpha\beta}) : \alpha \leq \beta \text{ in } I\}$ be a pair of projective systems of probability measures, on the same base, admitting limits $(\Omega, \Sigma, {}^P_Q)$. Then $H(P, Q) = \lim_\alpha H(P_\alpha, Q_\alpha)$, so that $P \perp Q$ iff $H(P, Q) = 0$, and the latter holds automatically if $P_\alpha \perp Q_\alpha$ for some α in I.

Proof. For $\alpha < \beta$ in I, we have $P_\alpha = P_\beta \circ g_{\alpha\beta}^{-1}$, $Q_\alpha = Q_\beta \circ g_{\alpha\beta}^{-1}$ and since the projective limits exist, $P_\alpha = P \circ g_\alpha^{-1}$, $Q_\alpha = Q \circ g_\alpha^{-1}$ so that P_α, Q_α are image measures of (P_β, Q_β) as well as of (P, Q). Then (19) implies

$$H(P, Q) \leq H(P_\beta, Q_\beta) \leq H(P_\alpha, Q_\alpha). \tag{20}$$

From this it follows immediately that when $P_\alpha \perp Q_\alpha$ for some α in I, then $H(P_\alpha, Q_\alpha) = 0$ and hence there is equality in (20) so that $P \perp Q$. We now exclude this case and assume that $H(P_\alpha, Q_\alpha) > 0$ for all α. Then by the monotonicity of $H(\cdot, \cdot)$ (20) gives

$$H(P, Q) \leq \lim_\alpha H(P_\alpha, Q_\alpha). \tag{21}$$

We assert that there is equality in (21) to complete the argument.

Let $\varepsilon > 0$ be given, and find a partition $\{A_k\}_{-\infty}^\infty$, by Lemma 6, such that

$$\sum_k \sqrt{P(A_k)Q(A_k)} \leq H(P, Q) + \frac{\varepsilon}{4}. \tag{22}$$

Since $P(\underset{k}{\cup} A_k) = 1 = Q(\underset{k}{\cup} A_k)$, there is an $n_0(= n_0(\varepsilon))$ such that

$$\underset{|n| \geq n_0}{\Sigma} P(A_k) < \frac{\varepsilon}{4} \quad , \quad \underset{|k| \geq n_0}{\Sigma} Q(A_k) < \frac{\varepsilon}{4} . \qquad (23)$$

But by the continuity of the square root function on $\mathbb{R}^+$, we can find $0 < \eta_k < \frac{\varepsilon}{3} \cdot 2^{-(|k|+2)}$, such that for all $|k| < n_0$

$$([P(A_k) + \eta_k][Q(A_k) + \eta_k])^{\frac{1}{2}} \leq (P(A_k)Q(A_k))^{\frac{1}{2}} + \frac{\varepsilon}{3} \cdot 2^{-(|k|+2)} \quad . \quad (24)$$

Also since P, Q are finite measures on $\Sigma = \sigma(\underset{\alpha}{\cup} g_\alpha^{-1}(\Sigma_\alpha))$ and $A_k \in \Sigma$, we approximate A_k by a cylinder with base in Σ_{α_k}. In fact by the initial assumption $H(P_\alpha, Q_\alpha) > 0$ for all α, we may assume that $P(A_k) > 0$ and $Q(A_k) > 0$, for all k, by relabelling if necessary. Then letting $\mu = P + Q$ we get $P \leq \mu, Q \leq \mu$. Since $\Sigma = \sigma(\mathcal{A}), \mathcal{A} = \underset{\alpha}{\cup} g_\alpha^{-1}(\Sigma_\alpha)$, and μ is a finite measure on Σ, we can find for any $A_k \in \Sigma$, a $B_k \in \mathcal{A}$ such that $\mu(A_k \Delta B_k) < \eta_k$. Hence we have

$$P(A_k \Delta B_k) < \eta_k \quad , \quad Q(A_k \Delta B_k) < \eta_k . \qquad (25)$$

But $B_k \in \mathcal{A}$ implies $B_k \in g_{\alpha_k}^{-1}(\Sigma_{\alpha_k})$, for some $\alpha_k \in I$. Since I is directed and $|k| < n_0$ is finite, there is a $\gamma \in I$ such that $\alpha_k < \gamma$ for all $|k| < n_0$. Now $g_{\alpha_k}^{-1}(\Sigma_{\alpha_k}) \subset g_\gamma^{-1}(\Sigma_\gamma)$, $|k| < n_0$. So there exist sets $C_k(\in \Sigma_\gamma)$ as bases of B_k, $B_k = g_\gamma^{-1}(C_k)$. Let $\{E_k, |k| < n_0\}$ be a disjunctification of the C_k. (This is used only to produce a partition of Ω_γ.) Since $P \circ g_\gamma^{-1} = P_\gamma$, we have

$$
\begin{aligned}
P_\gamma \Big(\underset{|k| < n_0}{\cup} E_k \Big) &= P_\gamma \Big(\underset{|k| < n_0}{\cup} C_k \Big) \\
&= P \Big(\underset{|k| < n_0}{\cup} g_\gamma^{-1}(C_k) \Big) = P \Big(\underset{|k| < n_0}{\cup} B_k \Big), \\
&\geq P \Big(\underset{|k| < n_0}{\cup} (B_k \cap A_k) \Big) \\
&= \underset{|k| < n_0}{\Sigma} P(B_k \cap A_k), \text{ since } A_k\text{'s are disjoint,} \\
&= \underset{|k| < n_0}{\Sigma} (P(A_k) - P(A_k \Delta B_k)) \\
&> \underset{|k| < n_0}{\Sigma} (P(A_k) - \eta_k) \quad , \quad \text{by (25),} \\
&\geq (1 - \frac{1}{4}\varepsilon) - \Sigma_k \frac{\varepsilon}{3} \cdot 2^{-(|k|+2)} = 1 - \frac{\varepsilon}{2}. \qquad (26)
\end{aligned}
$$

A similar computation for the Q-measures with (25) gives

$$Q_\gamma\Big(\bigcup_{|k|<n_0} E_k \Big) > 1 - \frac{\varepsilon}{2} . \tag{27}$$

Letting $E_{n_0} = \Omega_\gamma - \bigcup_{|k|<n_0} E_k$, one gets from (26) and (27)

$$P_\gamma(E_{n_0}) < \frac{\varepsilon}{2}, \quad Q_\gamma(E_{n_0}) < \frac{\varepsilon}{2} . \tag{28}$$

Since $\{E_k, \ -n_0 \le k \le n_0\}$ is a partition of Ω_γ, we can estimate $H(P_\gamma, Q_\gamma)$ as

$$
\begin{aligned}
H(P_\gamma, Q_\gamma) &\le \sum_{k=-n_0}^{n_0} \sqrt{P(E_k)Q(E_k)} \\
&\le \sum_{|k|<n_0} \sqrt{P_\gamma(C_k)Q_\gamma(C_k)} + \frac{\varepsilon}{2}, \ \text{by (28)}, \\
&= \sum_{|k|<n_0} \sqrt{P(B_k)Q(B_k)} + \frac{\varepsilon}{2} \\
&\le \sum_{|k|<n_0} \sqrt{(P(A_k)+\eta_k)(Q(A_k)+\eta_k)} + \frac{\varepsilon}{2}, \ \text{by (25)}, \\
&\le \sum_{|k|<n_0} \Big(\sqrt{P(A_k)Q(A_k)} + \frac{\varepsilon}{3}\cdot 2^{-(|k|+2)} \Big) + \frac{\varepsilon}{2}, \ \text{by (24)}, \\
&\le \sum_k \sqrt{P(A_k)Q(A_k)} + \frac{\varepsilon}{4} + \frac{\varepsilon}{2} \\
&\le H(P,Q) + \frac{\varepsilon}{4} + \frac{\varepsilon}{4} + \frac{\varepsilon}{2} = H(P,Q) + \varepsilon, \ \text{by (22)}.
\end{aligned}
$$

Since $\varepsilon > 0$ is arbitrary this shows that there is equality in (21), as asserted. □

The above result which is a specialization from Brody (1971), gives conditions for singularity of a pair of probability measures determined by a couple of projective systems of such measures. However the non-singularity does not generally imply mutual absolute continuity. We analyze this and the related (point) martingale convergence results, as applicaitons, in the next section.

9.3 Martingale convergence

One of the basic aspects of (sub) martingale analysis is the pointwise and norm convergence assertions. The latter holds under general conditions even when the index set is partially ordered but the pointwise results hold only under further additional restrictions. Here we present some of the key theorems using the preceding work.

Theorem 1. (Doob) *Let $\{X_n, \mathcal{F}_n, n \geq 1\}$ be a martingale such that* $\sup_n E(|X_n|) < \infty$. *Then $X_n \to X_\infty$ a.e., and $E(|X_\infty|) \leq \lim_n E(|X_n|)$.*

Proof. If $\mu_n : A \mapsto \int_A X_n \, dP, A \in \mathcal{F}_n$, then $\{\mu_n, \mathcal{F}_n, n \geq 1\}$ is a bounded set martingale. Hence $\mu_n = \mu_n^{(1)} - \mu_n^{(2)}$, with $\{\mu_n^{(i)}, \mathcal{F}_n, n \geq 1\}$ as positive set martingales, $i = 1, 2$, and $\mu_n^{(i)}$ is P-continuous, as a consequence of Proposition 2.3. By the Radon-Nikodým theorem, if $X_n^{(i)} = \frac{d\mu_n^{(i)}}{dP}$, then $\{X_n^{(i)}, \mathcal{F}_n, n \geq 1\}$ is a positive martingale, $i = 1, 2$. Thus the result follows if we establish that a positive martingale converges a.e. So let $X_n \geq 0$ hereafter for notational convenience.

Suppose the convergence statement is false. Then there exist $-\infty < a < b < \infty$ such that $P(A) > 0$, where

$$A = \{\omega : \liminf_n X_n(\omega) < a < b < \limsup_n X_n(\omega)\} . \tag{1}$$

We now show that this leads to a contradiction of the already proved optional sampling theorem (cf. Theorem 1.6 and the remark after its proof).

Thus let $\tau_0 = 1$ and $\tau_1 = \inf\{n > 1, X_n > b\}$. For $k \geq 1$ define

$$\tau_{2k} = \inf\{n > \tau_{2k-1} : X_n < a\}, \quad \tau_{2k+1} = \inf\{n > \tau_{2k} : X_n > b\} . \tag{2}$$

As usual, we take $\inf\{\emptyset\} = +\infty$ here. Since a τ_k is determined by the random variables $X_1, \cdots, X_k$, it is $\mathcal{F}_k$-measurable so that $\{\tau_k, k \geq 1\}$ is a stopping time process of $\{\mathcal{F}_k, k \geq 1\}$. But by hypothesis the X_n-process is in a ball of $L^1(P)$ so that $\liminf_n X_n$ and $\limsup_n X_n$ are finite a.e. and since the sequence does not converge by supposition, we can assume that $\tau_n < \infty$ a.e. for $n \geq 1$. Now $\tau_{2k} < \infty$ implies $X(\tau_{2k}) < a$, and $\tau_{2k+1} < \infty$ implies $X(\tau_{2k+1}) > b$. Consequently

$$\begin{aligned} E(X(\tau_{2k})\chi_{[\tau_{2k} < \infty]}) &< a \, P[\tau_{2k} < \infty] \\ &\leq a \, P[\tau_{2k-1} < \infty] \\ &< b \, P[\tau_{2k-1} < \infty] \\ &\leq E(X(\tau_{2k-1})\chi_{[\tau_{2k-1} < \infty]}) . \end{aligned} \tag{3}$$

But by Theorem 1.6, with $I = \mathbb{N}$ (and the remark after its proof), we get $\{X(\tau_n), \mathcal{F}(\tau_n), n \geq 1\}$ to be a (positive) martingale. Hence

$E(X(\tau_n)) = \alpha$ a positive constant, independent of n. Thus we have

$$0 = E(X(\tau_{2k-1})) - E(X(\tau_{2k}))$$

$$= \int_{[\tau_{2k-1} < \infty]} X(\tau_{2k-1}) dP - \int_{[\tau_{2k} < \infty]} X(\tau_{2k}) dP$$

$$\geq b\, P[\tau_{2k-1} < \infty] - a\, P[\tau_{2k-1} < \infty], \text{ by (3)}.$$

$$= (b-a)P[\tau_{2k-1} < \infty] \geq (b-a)P(A) > 0 . \tag{4}$$

This contradiction shows that $P(A) = 0$ must hold. Hence $X_n \to X_\infty$, a.e. The final statement now follows from Fatou's lemma. $\square$

The above result can be extended to submartingales by a simple device given by:

Proposition 2. (Doob's decomposition) *Let* $\{X_n, \mathcal{F}_n,\ n \geq 1\}$ *be a submartingale. Then it can be uniquely decomposed as:*

$$X_n = X'_n + A_n,\ A_1 = 0,\ n \geq 1 \tag{5}$$

where $\{X'_n, \mathcal{F}_n,\ n \geq 1\}$ *is a martingale and* $0 \leq A_n \leq A_{n+1}$ *with the* A_{n+1} *as* $\mathcal{F}_n$*-adapted.*

Proof. With $A_1 = 0$, define A_n for $n \geq 2$, inductively by setting

$$A_n = E^{\mathcal{F}_{n-1}}(X_n - X_{n-1}) + A_{n-1}\ (\in L^1(P)). \tag{6}$$

Set $X'_n = X_n - A_n$, and claim that this satisfies (5). In fact, the submartingale hypothesis implies that $0 \leq A_n \leq A_{n+1}$, and A_n is $\mathcal{F}_{n-1}$-adapted. Moreover,

$$E^{\mathcal{F}_{n-1}}(X'_n) = E^{\mathcal{F}_{n-1}}(X_n - [E^{\mathcal{F}_{n-1}}(X_n - X_{n-1}) + A_{n-1}]),\ \text{by (6)},$$

$$= E^{\mathcal{F}_{n-1}}(X_{n-1} - A_n) = X_{n-1} - A_{n-1} = X'_{n-1},\, \text{a.e.}$$

Thus $\{X'_n, \mathcal{F}_n, n \geq 1\}$ is a martingale. For uniqueness let $X_n = Y_n + B_n$ be another such decomposition. Then $X'_n - Y_n = B_n - A_n$, and the left side is a martingale while the right side is $\mathcal{F}_{n-1}$-adapted. Hence

$$E^{\mathcal{F}_{n-1}}(X'_n - Y_n) = X'_{n-1} - Y_{n-1} = X'_n - Y_n\ \ \text{a.e.}$$

This implies $B_n - A_n = B_{n-1} - A_{n-1} = \cdots = B_1 - A_1 = 0$, a.e. Consequently, $X'_n = Y_n$ a.e. and then $A_n = B_n$ a.e. That every representation given by (5) is a submartingale relative to $\{\mathcal{F}_n, n \geq 1\}$ is immediate. $\square$

With this decomposition we can deduce the submartingale convergence statement easily as in:

Theorem 3. *If* $\{X_n, \mathcal{F}_n, n \geq 1\}$ *is a submartingale satisfying* $K_0 = \sup_n E(|X_n|) < \infty$, *then* $X_n \to X_\infty$ *a.e., and* $E(|X_\infty|) \leq \liminf_n E(|X_n|)$.

Proof. By the preceding proposition $X_n = X_n' + A_n, n \geq 1$ and

$$0 \leq E(A_n) = E(X_n) - E(X_n')$$
$$\leq \sup_n E(|X_n|) - E(X_1') = K_1 < \infty.$$

Hence

$$0 \leq \lim_n E(A_n) = E(\lim_n A_n) \leq K_1 < \infty,$$

by the monotone convergence theorem. So $\lim_n A_n = A_\infty$ exists a.e. and is integrable. Also

$$E(|X_n'|) \leq E(|X_n|) + E(A_\infty) \leq K_0 + K_1 < \infty .$$

So by Theorem 1, $X_n' \to X_\infty'$ a.e., and then $X_n = X_n' + A_n \to X_\infty = X_\infty' + A_\infty$, a.e. and by Fatou's lemma, $E(|X_\infty|) \leq \liminf_n E(|X_n|)$, as desired. $\square$

The corresponding statements, when the indexing is a partially ordered set, are more involved and true only under further restrictions. We present the following two results with suitable supplementary conditions. The first one using Theorem 2.7 is on mean convergence.

Theorem 4. *Let* $\{X_i, \mathcal{F}_i, i \in I\}$ *be a directed index martingale such that*

(i) $E(|X_i|) \leq K < \infty$, $i \in I$, *and*

(ii) for any $\varepsilon > 0$, *there is an* $i_0(= i_0(\varepsilon))$ *and an* $\alpha(= \alpha_\varepsilon)$ *such that*

$$\int_{[|X_i|>\alpha]} |X_i| dP < \varepsilon , \quad \text{uniformly in } i > i_0. \tag{7}$$

Then there is an $X \in L^1(P)$ *for which* $E(|X_i - X|) \to 0$ *as* $i \to \infty$ *and* $X_i = E^{\mathcal{F}_i}(X)$ *a.e., all* $i \in I$.

Proof. If $\mu_i : A \mapsto \int_A X_i dP$, $A \in \mathcal{F}_i$, then $\{\mu_i, \mathcal{F}_i, i \in I\}$ is a bounded set martingale by (i). By Proposition 2.3, we may and do assume for this proof that μ_i, or equivalently X_i, is nonnegative.

Since $\mathcal{F}_i \subset \mathcal{F}_{i'}$, for $i < i'$ in I, $\mathcal{F}_0 = \bigcup_{i \in I} \mathcal{F}_i$ is an algebra. Moreover, $\mu : \mathcal{F}_0 \to \mathbb{R}^+$ given by $\mu(A) = \mu_i(A)$ for $A \in \mathcal{F}_i$ is uniquely defined as shown in the proof of Theorem 8.3.1, and is finitely additive. By (ii) of the hypothesis, if $A_n \searrow \emptyset$, $A_n \in \mathcal{F}_0$ so that $A_n \in \mathcal{F}_{i_n}$ for some i_n (and $i_n < i_{n+1}$ can be assumed by the directedness of I), we get

$$\lim_{P(A_n) \to 0} \mu(A_n) = \lim_{P(A_n) \to 0} \int_{A_n} X_{i_n} dP = 0.$$

Hence μ is σ-additive on $\mathcal{F}_0$, has a unique σ-additive extension (denoted by the same symbol) to $\sigma(\mathcal{F}_0)$, and is clearly P-continuous. By the Radon-Nikodým Theorem, $X = \frac{d\mu}{dP}$ exists and $X \in L^1(P)$. We assert that this X satisfies the requirements of the theorem.

Indeed let $i_0 \in I$ be fixed and consider $0 \leq Y \in L^1(\Omega, \mathcal{F}_{i_0}, P)$. If $\nu : A \mapsto \int_A Y \, dP$, $A \in \Sigma$, and $\nu_i = \nu|\mathcal{F}_i$, then $\frac{d\nu_i}{dP} = Y$ for $i \geq i_0$. Since $\mu_i = \mu|\mathcal{F}_i$, one has the "Hellinger distance" as:

$$H(\mu, \nu) = \int_\Omega \sqrt{XY} \, dP, \quad \text{and} \quad H(\mu_i, \nu_i) = \int_\Omega \sqrt{X_i Y} dP, \quad i \geq i_0 . \quad (8)$$

But by Theorem 2.7, $H(\mu_i, \nu_i) \to H(\mu, \nu)$ as $i \nearrow \infty$. This means, by writing the inner product notation for (8), $H(\mu, \nu) = (\sqrt{X}, \sqrt{Y})$ etc.,

$$(\sqrt{X_i}, \sqrt{Y}) \to (\sqrt{X}, \sqrt{Y}), \quad \text{as} \quad i \nearrow \infty . \quad (9)$$

Since $i_0 \in I$ is arbitrary, (9) holds for all $0 \leq Y \in \bigcup_i L^1(\Omega, \mathcal{F}_i, P)$, and then by density of the latter in $L^1(\Omega, \sigma(\mathcal{F}_0), P)$, (8) is true for all $0 \leq Y \in L^1(\Omega, \sigma(\mathcal{F}_0), P)$. This holds, by considering the positive and negative parts, for all Y in the space, i.e., $\sqrt{X_i} \to \sqrt{X}$ weakly in $L^2(\Omega, \sigma(\mathcal{F}_0), P)$. In particular, (take $Y = X$) $\sqrt{X_i} + \sqrt{X} \to 2\sqrt{X}$ weakly. But,

$$\|\sqrt{X_i}\|_2^2 = \int_\Omega X_i dP = \mu_i(\Omega) = \mu(\Omega) = \int_\Omega X \, dP = \|\sqrt{X}\|_2^2 . \quad (10)$$

On the other hand for $Z \in L^2(\Omega, \sigma(\mathcal{F}_0), P)$ we have

$$|(2\sqrt{X}, Z)| = \lim_i |(\sqrt{X_i} + \sqrt{X}, Z)|$$

$$\leq \limsup_i \|Z\|_2 \|\sqrt{X_i} + \sqrt{X}\|_2 , \quad \text{by the CBS inequality},$$

$$\leq \|Z\|_2 \lim_i (\|\sqrt{X_i}\|_2 + \|\sqrt{X}\|_2)$$

$$= 2\|Z\|_2 \sqrt{\mu(\Omega)}, \quad \text{by (10)} . \quad (11)$$

Taking the suprema of (11) for $\|Z\|_2 \leq 1$, the left side becomes $\|2\sqrt{X}\|_2 = 2\|\sqrt{X}\|_2 = 2\sqrt{\mu(\Omega)}$. Hence there is equality in (11) implying that $\|\sqrt{X_i} + \sqrt{X}\|_2 \rightarrow 2\sqrt{\mu(\Omega)}$. Now using the parallelogram identity in Hilbert space, we get

$$\|\sqrt{X_i} - \sqrt{X}\|_2^2 + \|\sqrt{X_i} + \sqrt{X}\|_2^2 = 2(\|\sqrt{X_i}\|_2^2 + \|\sqrt{X}\|_2^2) = 4\mu(\Omega) . \quad (12)$$

Letting $i \nearrow \infty$ as in (12), one gets $\|\sqrt{X_i} - \sqrt{X}\|_2 \rightarrow 0$. This implies

$$\int_\Omega |X_i - X| \, dP = \int_\Omega |\sqrt{X_i} - \sqrt{X}| \, |\sqrt{X_i} + \sqrt{X}| dP$$
$$\leq \|\sqrt{X_i} - \sqrt{X}\|_2 \, \|\sqrt{X_i} + \sqrt{X}\|_2, \quad \text{(CBS inequality)},$$
$$\rightarrow 0, \quad \text{as } i \nearrow \infty .$$

Hence $X_i \rightarrow X$ in $L^1(\Omega, \sigma(\mathcal{F}_0), P)$. Since $\mu_i = \mu|\mathcal{F}_i$, this gives $X_i = E^{\mathcal{F}_i}(X)$, a.e., and all the assertions are established. $\square$

If I is linearly ordered and countable, then the mean convergence of a martingale implies, by Theorem 1, the pointwise convergence. But for the general directed index martingales this is not true, and new conditions must be imposed. The desired restrictions are recalled in:

Definition 5. Let (Ω, Σ, P) be a probability space and $\mathcal{F}_i \subset \Sigma, i \in I$, be an increasing (or right filtering) net of σ-subalgebras. Then a family $\{K_i, i \in I\}$, $K_i \in \mathcal{F}_i$ is an *essential fine covering* of $A(\in \Sigma)$ if for each $i_0 \in I$, $A \subset \bigcup_{i \geq i_0} K_i$ outside of a P-null set. (If the null set is empty, then "essential" is dropped here.) The net $\{\mathcal{F}_i, i \in I\}$ satisfies the *Vitali condition* V_0 if for each $A \in \Sigma$, and each essential fine covering $\{K_i, i \in I\}$ of A and $\varepsilon > 0$, there is a finite subcollection $\{K_{i_j}, 1 \leq j \leq n_\varepsilon\}$ and a.e. disjoint $L_j \in \mathcal{F}_{i_j}$, $L_j \subset K_{i_j}$ a.e., such that $P(A - \bigcup_{j=1}^{n_\varepsilon} L_j) < \varepsilon$.

The reader should compare this with Definition 3.4.2. In fact this is a restatement of the earlier one, in a form useful for martingale convergence. Also it can be shown that the V_0-condition is automatically verified if I is linearly ordered and countable for probability spaces. We have the following result on pointwise convergence, due to Chow (1960):

Theorem 6. *Let (Ω, Σ, P) be a complete probability space and $\{\mathcal{F}_i, i \in I\}$ be a completed right filtering net of σ-subalgebras of Σ satisfying the*

Vitali condition V_0, as in Definition 5. If $\{X_i, \mathcal{F}_i, \ i \in I\}$ is an L^1-bounded martingale (so $\sup\limits_i E(|X_i|) < \infty$), then there is an $X \in L^1(P)$ such that $X_i \to X$ a.e. as $i \nearrow \infty$.

We omit a proof of this result. The method is to consider an indirect argument, as in Theorem 1 above. For most applications Theorem 4 suffices.

In Theorem 2.7 we showed that $H(P, Q) = 0$ iff $P \perp Q$ and remarked that $H(P, Q) > 0$ does not generally imply that P, Q are equivalent. However, if $I = \mathbb{N}$ and P_α, Q_α are product measures then there is a positive solution, due to Kakutani (1948). This is of interest in statistical hypothesis testing problems and related work. We present the result, after a preliminary observation to eliminate a trivial aspect of the desired statement.

For a satisfactory solution it is necessary to assume that each P_α is equivalent to $Q_\alpha, \alpha \in I$. In fact, if for some $\alpha_0 \in I$, $P_{\alpha_0}(A) = 0$ but $Q_{\alpha_0}(A) > 0$, $A \in \Sigma_{\alpha_0}$, then $P(\pi_{\alpha_0}^{-1}(A)) = P_{\alpha_0}(A) = 0$ whereas $Q(\pi_{\alpha_0}^{-1}(A)) = Q_{\alpha_0}(A) > 0$ so that P and Q are not equivalent. If we take I to be the directed set of all finite subsets of T, $\Omega_\alpha = \underset{t \in \alpha}{\times} \Omega_t, \Sigma_\alpha = \underset{t \in \alpha}{\otimes} \Sigma_t$ where $\{(\Omega_t, \Sigma_t), \ t \in T\}$ is a family of measurable spaces, and $P_\alpha = \underset{t \in \alpha}{\otimes} P_t, \ Q_\alpha = \underset{t \in \alpha}{\otimes} Q_t$, with $\frac{P_t}{Q_t} : \Sigma_t \to \mathbb{R}^+$ as probabilities, then $P_\alpha \equiv Q_\alpha$ (equivalence) $\alpha \in I$ iff $Q_t \equiv P_t, t \in T$. The following simple example explains the situation clearly. Let $(\Omega_t, \Sigma_t, P_t)$ be the Lebesgue unit intervals for $t = 2, 3 \cdots$ but for $t = 1$, let $P_1(A) = \int_A \frac{3}{2}\chi_{[0, \frac{2}{3}]} \, d\mu, \ Q_1(A) = \int_A \frac{3}{2}\chi_{[\frac{1}{3}, 1]} \, d\mu$, where μ is the Lebesgue measure on $\Omega_1 = [0, 1]$. Let $P = \overset{\infty}{\underset{t=1}{\otimes}} P_t$, $Q = \overset{\infty}{\underset{t=1}{\otimes}} Q_t$, the infinite product measures (by the Fubini-Jessen theorem). Then $P \not\equiv Q$ but $H(P, Q) = \frac{1}{2}$, and neither $P \perp Q$ nor $P \equiv Q$. Thus to exclude this pathology we assume hereafter that $P_t \equiv Q_t, t \in T = \mathbb{N}$. This does not imply that $P \equiv Q$. The necessary strengthening is provided by:

Theorem 7. (Kakutani) *Let $\{(\Omega_t, \Sigma_t, \frac{P_t}{Q_t}), t \in \mathbb{N}\}$ be a sequence of probability spaces with $P_t \equiv Q_t$ for each $t \in \mathbb{N}$. If $\alpha_n = (1, 2, \cdots, n), \Omega_{\alpha_n}, \Sigma_{\alpha_n}$, and P_{α_n} are the products defined above, and $(\Omega, \Sigma, \frac{P}{Q})$ are the corresponding infinite product spaces (given by the Fubini-Jessen theorem), then $P \equiv Q$ or $P \perp Q$, accordingly as $H(P, Q) > 0$ or $H(P, Q) = 0$ respectively.*

Proof. Letting $\pi_{mn} : \Omega_{\alpha_n} \to \Omega_{\alpha_m}$, $m < n$, and $\pi_n : \Omega \to \Omega_{\alpha_n}$ be the canonical (or coordinate) projections with $\alpha_n = (1, 2, \cdots, n)$, we have $\Sigma = \sigma(\bigcup_{n=1}^{\infty} \pi_n^{-1}(\Sigma_{\alpha_n}))$ and $P = \lim_{\leftarrow}(P_{\alpha_n}, \pi_{mn})$, so that $(\Omega, \Sigma, \frac{P}{Q})$ is the projective limit of the spaces concerned. Then $H(P_{\alpha_n}, Q_{\alpha_n}) \to H(P, Q)$ by Theorem 2.7 and $P \perp Q$ iff $H(P, Q) = 0$. So consider the case $H(P, Q) > 0$. We may now assume that $P_{\alpha_n} \equiv Q_{\alpha_n}$ since $P_t \equiv Q_t$ for each $t \in \mathbb{N}$, and P_α, Q_α are just (finite) product measures.

Now by the Radon-Nikodým theorem applied to these products, one gets

$$u_n^2(\tilde{\omega}_n) = \frac{dQ_{\alpha_n}}{dP_{\alpha_n}}(\tilde{\omega}_n) = \prod_{t=1}^n \frac{dQ_t}{dP_t}(\omega_t), \quad \tilde{\omega}_n = (\omega_1, \cdots, \omega_n) \in \Omega_{\alpha_n}. \quad (13)$$

Hence we have by Fubini's theorem,

$$\begin{aligned} H(P_{\alpha_n}, Q_{\alpha_n}) &= \int_{\Omega_{\alpha_n}} u_n(\tilde{\omega}_n) dP_{\alpha_n} \\ &= \prod_{t=1}^n \int_{\Omega_t} \left(\frac{dQ_t}{dP_t}(\omega_t) \right)^{1/2} dP_t \\ &= \prod_{t=1}^n H(P_t, Q_t). \end{aligned} \quad (14)$$

Since $H(P_{\alpha_n}, Q_{\alpha_n}) \searrow H(P, Q) > 0$ and $0 < H(P_{\alpha_n}, Q_{\alpha_n}) \leq 1$ for all n, the (infinite) product in (14) converges as $n \to \infty$, so that $\prod_{t=m}^n H(P_t, Q_t) \to 1$ as $m, n \to \infty$. Under these conditions it is asserted that $P \equiv Q$.

Since $\pi_n : \Omega \to \Omega_{\alpha_n}$, let $u_n^* : \Omega \to \mathbb{R}^+$ be defined as $u_n^* = u_n \circ \pi_n$. Then (14) implies (i) u_n^* is measurable for Σ, and (ii) $u_n^* \in L^2(P)$. Moreover, $u_n^*(\omega) = \prod_{k=1}^n u_k(\pi_k(\omega))$, and since $H(P_{\alpha_n}, Q_{\alpha_n}) = (\sqrt{u_n}, \sqrt{u_n})_{L^2(P_{\alpha_n})}$, one has for $m < n$,

$$\begin{aligned} \|u_n^* - u_m^*\|_{2,P} &= 2 - 2(u_n^*, u_m^*)_{2,P} \\ &= 2(1 - (u_n, u_m)_{2, P_{\alpha_n}}) \\ &= 2(1 - \int_{\Omega_{\alpha_m}} u_m^2 dP_{\alpha_m} \int_{\Omega_{m+1} \times \cdots \times \Omega_{\alpha_n}} \left(\prod_{t=m+1}^n \frac{dQ_t}{dP_t} \right)^{1/2} \times \\ & \qquad dP_{m+1} \cdots dP_n) \\ &= 2(1 - \prod_{t=m+1}^n H(P_t, Q_t)), \end{aligned}$$

$$\text{since} \int_{\Omega_{\alpha_n}} u_m^2 dP_{\alpha_m} = 1 \text{ and then use (14).} \quad (15)$$

Thus $H(P,Q) = \lim_{n\to\infty} H(P_{\alpha_n}, Q_{\alpha_n}) = \Pi_{t=1}^\infty H(P_t, Q_t) > 0$ so that the right side of (15) tends to zero and $\{u_n^*, n \geq 1\}$ is a Cauchy sequence in $L^2(P)$. Let $u^*(\cdot)$ be the limit of this sequence, which belongs to $L^2(P)$. If $\mathcal{F}_n = \pi_n^{-1}(\Sigma_{\alpha_n}) \subset \mathcal{F}_{n+1}$, we have for $A = \pi_{\alpha_n}^{-1}(A_n) \in \mathcal{F}_m$,

$$\int_A (u_n^*)^2 dP = \int_{\pi_{mn}^{-1}(A_m)} u_n^2 dP_{\alpha_n} \,, \quad \text{(image law and } \pi_m = \pi_{mn} \circ \pi_n),$$

$$= \int_{A_m} u_m^2 dP_{\alpha_n}, \quad \text{by (13) and Fubini's theorem,}$$

$$= \int_{\pi_m^{-1}(A_m)} (u_m^*)^2 dP, \quad \text{by the image law.} \tag{16}$$

Since u_m^* is $\mathcal{F}_m$-adapted, (16) implies $(u_m^*)^2 = E^{\mathcal{F}_m}((u_n^*)^2)$ which forms a uniformly integrable sequence by (15). Now we may apply Theorem 4 to conclude that $(u_n^*)^2 = E^{\mathcal{F}_n}((u^*)^2)$, and hence

$$\int_A (u^*)^2 dP = \int_A E^{\mathcal{F}_n}((u^*)^2)dP = \int_A (u_n^*)^2 dP, \quad A \in \mathcal{F}_n,$$

$$= \int_{A_n} u_n^2 dP_{\alpha_n} = Q_n(A_n), \quad A = \pi_n^{-1}(A_n),$$

$$= Q(\pi_n^{-1}(A_n)) = Q(A) \,. \tag{12}$$

Since Q is σ-additive (17) holds for all $A \in \cup_{n=1}^\infty \mathcal{F}_n$ and then for all A in Σ. This shows that $\frac{dQ}{dP} = (u^*)^2$ and Q is P-continuous. But $Q_{\alpha_n} \equiv P_{\alpha_n}$ and hence by symmetry we can use the above computation to conclude that P is Q-continuous also so that $P \equiv Q$ holds. $\square$

Remark. It should be noted that by (15) and (16), $\{(u_n^*)^2, \mathcal{F}_n, n \geq 1\}$ is a martingale, but $\{u_n^2, \Sigma_{\alpha_n}, n \geq 1\}$ is *not*, since the latter set is defined on different spaces as n increases and the Σ_n are *not* nested.

We present a specialization and an application of the above result to illustrate its potential uses. Let $X_1, X_2, \cdots$ be a sequence of independent identically distributed $(i \cdot i \cdot d \cdot)$ random variables (or k-vectors) so that $(x_1, \cdots, x_n)$ can be regarded as a random sample for a given n. Suppose that their common distribution is either P or Q on a space (Ω_0, Σ_0). Let μ be a (σ-finite) measure on Σ_0 that dominates both P and Q, (e.g., $\mu = P + Q$ can be taken). Let $p = \frac{dP}{d\mu}, q = \frac{dQ}{d\mu}$, be their densities. Then $H(P,Q) = \int_{\Omega_0} \sqrt{pq}\,d\mu$ and if $P^{(n)}, Q^{(n)}$ are the (product) probability measures of $(X_1, \cdots, X_n)$, one has $H(P^{(n)}, Q^{(n)}) = \prod_{i=1}^{n} H(P,Q) = H^n(P,Q)$ since the X_i are $i \cdot i \cdot d$. From the fact that $0 < H(P,Q) \leq 1$ when P and Q are not mutually singular, it follows that $\lim_{\overleftarrow{n}} H(P^{(n)}, Q^{(n)}) = 0$ if $P \neq Q$ [even when $P \equiv Q$ we can have $H(P,Q) < 1$, and $= 1$ iff $P = Q$]. Thus if $\tilde{P} = \lim_{\overleftarrow{}}(P^{(n)}, \pi_n)$, $\tilde{Q} = \lim_{\overleftarrow{}}(Q^{(n)}, \pi_n)$ on $(\Omega = \Omega_0^{\mathbb{N}}, \Sigma = \otimes_{\mathbb{N}} \Sigma_0)$, one always has $\tilde{P} \perp \tilde{Q}$. Let us illustrate this further by the following concrete example.

Let P, Q in the above be k-variate normal distributions with means $\alpha, \beta (\alpha \neq \beta)$ and the same nonsingular convariance matrix Λ. Thus with μ as the Lebesgue measure in $\mathbb{R}^k$ $(= \Omega_0)$, we have $p(\cdot)$ and $q(\cdot)$ to be $N(\alpha, \Lambda)$ and $N(\beta, \Lambda)$ respectively, the multinormal densities:

$$p(\mathbf{x}) = (2\pi)^{-\frac{k}{2}} (\det \Lambda)^{-\frac{1}{2}} \exp\{-\frac{1}{2}(\mathbf{x} - \alpha)'\Lambda^{-1}(\mathbf{x} - \alpha)\}, \qquad (18)$$

and a similar expression for $q(\cdot)$ with β in place of α and $(\mathbf{x} - \alpha)'$ denotes the transpose of the k-vector $(\mathbf{x} - \alpha)$. Then using a standard formula for the integral of this function [cf., e.g. (11.12.1a) in Cramér (1946), p. 118] we can simplify $H(P,Q)$ to get

$$H(P,Q) = \exp\{-\frac{1}{8}(\alpha - \beta)'\Lambda^{-1}(\alpha - \beta)\}. \qquad (19)$$

Hence $H(P^{(n)}, Q^{(n)}) = H^n(P,Q) \to 0$ as $n \to \infty$ if $\alpha \neq \beta$, since Λ is positive definite. Thus $\tilde{P} \perp \tilde{Q}$. (Note that $P \equiv Q$ here.) Accordingly in a statistical application one takes a sufficiently large sample of observations $(x_1, \cdots, x_n)$, computes their "likelihood ratio" $\frac{q_n}{p_n}$, and finds a constant τ such that if the observed sample point falls into the set $A_\tau = \{(x_1, \cdots, x_n) : \frac{q_n(x_1, \cdots, x_n)}{p_n(x_1, \cdots, x_n)} \geq \tau\}$, then P is declared as the correct probability measure governing the experiment or model and in the opposite case Q is decided on. Several applications of this type are considered by Kraft (1955).

It should be pointed out that the dichotomy given by Theorem 7 is also valid in general for all Gaussian probability measures which are not necessarily of product type. But this situation demands additional nontrivial analysis. [Details and references are given by the author, cf. Rao (1981), pp. 212-217.] We shall not consider it further here.

9.4 Markov processes: some basic results

In contrast to the preceding application of abstract conditioning to the formulation and convergence of martingales, the very concept of Markov processes depends on the regularity of conditional measures. Although we discussed conditional independence in Section 2.5 and some of its applications there, the Markov dependence needs a finer structure of conditioning as seen below. Also an immediate concern will be the ability to exactly evaluate these measures in a given situation. We intend to illuminate these points of the subject here.

Definition 1. A stochastic process $\{X_t, t \in I \subset \mathbb{R}\}$ on (Ω, Σ, P) is *Markovian* if for any set of points $t_1 < t_2 < \cdots < t_n < t_{n+1}$, $t_i \in I$, one has

$$P(X_{t_{n+1}} < x | X_{t_1}, \cdots, X_{t_n}) = P(X_{t_{n+1}} < x | X_{t_n}), \quad a.e., \quad x \in \mathbb{R}. \quad (1)$$

Recall that $P(A|X)$ is a shortened notation for $P^{\sigma(X)}(A)$ where $\sigma(X)$ is the σ-algebra generated by the random variable X, and $A \in \Sigma$. In the case that the conditional probabilities in (1) are regular, then the concept can be given a verbal description. Thus a process $\{X_t, t \in I\}$ is Markovian if the probability of the event $[X_{t_{n+1}} < x]$ given the past $X_{t_1}, \cdots, X_{t_n}$ (for any n) depends only on the most recent past X_{t_n}. Clearly in any application, one should be able to evaluate exactly these conditional probabilities and our work in Chapters 5, 7, and 8 will be useful in this situation, and thus the very definition demands a resolution of the computational problem. In the literature one finds that such regular conditional measures are assumed as part of the data, and then the subject is moved forward.

The Markovian concept has the following alternative formulation:

Proposition 2. *For a process $\{X_t, t \in I \subset \mathbb{R}\}$ on (Ω, Σ, P) the following are equivalent.*

(i) The process is Markovian in the sense of Definition 1.

(ii) For each $u < v$ in I one has

$$P[X_v < x | X_s, s \leq u] = P[X_v < x | X_u], \quad a.e., \quad x \in \mathbb{R}.$$

(iii) For any $s_1 < s_2 < \cdots < s_m < t < t_1 < < t_2 < \cdots < t_n$ from I,

one has

$$P[X_{s_i} < x_i, \; X_{t_j} < y_j, 1 \le i \le m, \; 1 \le j \le n | X_t] =$$
$$P[X_{s_i} < x_i, \; 1 \le i \le m | X_t] P[X_{t_j} < y_j, 1 \le j \le n | X_t], \quad a.e., \quad (2)$$

for all $x_i \in \mathbb{R}$, $y_j \in \mathbb{R}$, so that $X_{s_i}, 1 \le i \le m$, and $Y_{t_j}, 1 \le j \le n$ are conditionally independent given X_t.

Proof. We show that (ii) $\Leftrightarrow$ (i) $\Leftrightarrow$ (iii). Since (ii) $\Rightarrow$ (i) is clear, suppose (i) holds so that (1) is valid for all finite subsets of I. Note that $\sigma(X_t, t \le u) = \sigma(\cup[\sigma(X_{t_1}, \cdots, X_{t_n})|t_1 < \cdots < t_n \le u, n \ge 1])$ where $\sigma(\cdot)$ is the σ-algebra generated by the set of random variables shown. Then

$$\int_A P[X_v < x | X_u] dP = \int_A P[X_v < x | X_{t_1}, \cdots, X_{t_n}, X_u] dP, \quad \text{by (1)},$$
$$= P(A \cap [X_v < x]), \quad A \in \sigma(X_{t_1}, \cdots, X_{t_n}, X_u), \quad (3)$$

by definition of conditional probability. But (3) also holds for all A in the union of σ-algebras shown, and then, since both sides are σ-additive measures on this union, the same holds on the σ-algebra generated by this class. Thus it holds for all $A \in \sigma(X_t, t \le u)$, which is (ii).

To see that (i) $\Rightarrow$ (iii), let $A = \{X_{s_i} < x_i, \; 1 \le i \le m\}$, $B = \{X_{t_j} < y_j, 1 \le j \le n\}$, $\mathcal{B}_n = \sigma(X_{t_1}, \cdots, X_{t_n})$, $\tilde{\mathcal{B}}_n = \sigma(X_{s_1}, \cdots, X_{s_m})$, and $\mathcal{B} = \sigma(X_t)$. Then with the properties of conditional expectations (cf. Proposition 2.2.1), we have

$$P(A|\mathcal{B})P(B|\mathcal{B}) = E^{\mathcal{B}}(\chi_A) E^{\mathcal{B}}(\chi_B)$$
$$= E^{\mathcal{B}}(\chi_A E^{\mathcal{B}}(\chi_B)) = E^{\mathcal{B}}(\chi_A P(B|\mathcal{B}))$$
$$= E^{\mathcal{B}}(\chi_A P(B|\mathcal{B}, \tilde{\mathcal{B}}_m)), \quad \text{by (1)},$$
$$= E^{\mathcal{B}}(\chi_A E^{\sigma(\mathcal{B}, \tilde{\mathcal{B}}_n)}(\chi_B))$$
$$= E^{\mathcal{B}}(\chi_A \cdot \chi_B), \quad \text{since } \mathcal{B} \subset \sigma(\mathcal{B}, \tilde{\mathcal{B}}_n) \text{ and } A \in \sigma(\mathcal{B}, \tilde{\mathcal{B}}_n),$$
$$= P(A \cap B|\mathcal{B}).$$

Thus (iii) is valid. For the converse implication, let (2) hold and consider $A \in \tilde{\mathcal{B}}_n$, $C \in \mathcal{B}$ so that $A \cap C \in \sigma(\mathcal{B}, \tilde{\mathcal{B}}_m)$ and is a generator of the latter. We get with (2) and B as above,

$$\int_{C \cap B} E^{\sigma(B,\tilde{B}_m)}(\chi_B)dP = \int_{C \cap A} \chi_B dP = \int_C \chi_{A \cap B}\, dP$$

$$= \int_C E^B(\chi_{A \cap B})dP = \int_C P(A \cap B|B)dP,$$

$$= \int_C P(A|B)P(B|B)dP \ , \quad \text{by (2)},$$

$$= \int_C E^B(\chi_A)E^B(\chi_B)dP$$

$$= \int_C E^B(\chi_A E^B(\chi_B))dP$$

$$= \int_{C \cap A} E^B(\chi_B)dP \ .$$

Since the extreme integrands are $\sigma(B, \tilde{B}_m)$-measurable and $C \cap A$ is a generator, the integrands can be identified to get

$$E^B(\chi_B) = P(B|B) = E^{\sigma(B,\tilde{B}_m)}(\chi_B) = P(B|B, \tilde{B}_m), \quad a.e.$$

This is (i) and hence (i) $\Leftrightarrow$ (iii) as desired. $\square$

Discussion 3. The Markovian concept expressed in terms of (iii) of the above proposition has an interesting interpretation and a useful consequence. If the conditional measures are regular, then the statement is equivalent to saying that the process $\{X_t, t \in I\}$ is Markovian iff for each $t \in I$ the collections $\{X_s, s < t\}$ and $\{X_r, r > t\}$ are conditionally independent given X_t. In other words the "past" and "future" are independent given the present. Since "past" and "future" depend only on the ordering of I, it shows that if $\{X_t, t \in I\}$ is a Markov process, so is $\{X_t, t \in \tilde{I}\}$ where $I = \tilde{I}$ but have opposite orderings. In particular, $\{X_t, -a \le t \le a\}$ is Markovian implies the same of $\{X_{-t}, -a < t < a\}$. Note that the regularity of all the conditional measures requires restrictions on the underlying measure space (cf. Chapter 5), or on the ranges of all X_t. If all the X_t take values in a fixed countable set (then the process is called a *chain*). Then the condition of regularity is automatic. In general one needs restrictions such as $X_t(\Omega) \subset \mathbb{R}$ is a Borel set for each t, or (Ω, Σ, P) is a Polish space and the like.

Assuming regularity as part of the definition of a Markov process, one can proceed to a detailed analysis of this class. In fact, the subject has grown and enormous constructs have been erected on that foundation. However, the *existence* of such processes generally need a (nontrivial) proof. We first give an example, and then establish the desired existence result.

Example 4. Let $X_1, X_2, \cdots$ be a sequence of independent random variables on (Ω, Σ, P). Then the partial sums $\{S_n = \sum_{i=1}^{n} X_i, \ n \geq 1\}$ form a Markov process.

Proof. Let $\mathcal{B}_n = \sigma(S_1, \cdots, S_n)$. This is also the same as $\sigma(X_1, \cdots, X_n)$. We now use the fact that all bounded measurable (for Σ) functions are integrable relative to the conditional measure $P^{\mathcal{B}}(\cdot)$, even if regularity is not assumed provided we use the Dunford-Schwartz integral as discussed in Section 7.2. Thus the following holds a.e. (P):

$$E^{\mathcal{B}_n}(e^{itS_{n+1}}) = E^{\mathcal{B}_n}(e^{itS_n} e^{itX_{n+1}})$$

$$= e^{itS_n} E^{\mathcal{B}_n}(e^{itX_{n+1}}), \text{ since } S_n \text{ is } \mathcal{B}_n\text{-adapted,}$$

$$\tag{4}$$

$$= e^{itS_n} E(e^{itX_{n+1}}),$$

$$\text{since } X_{n+1} \text{ is independent of } \mathcal{B}_n,$$

$$= E^{\sigma(S_n)}(e^{itS_n}) E^{\sigma(S_n)}(e^{itX_{n+1}}),$$

$$\text{since } X_{n+1} \text{ is independent of } S_n,$$

$$= E^{\sigma(S_n)}(e^{itS_n} E^{\sigma(S_n)}(e^{itX_{n+1}}))$$

$$= E^{\sigma(S_n)}(e^{it(S_n + X_{n+1})}), \text{ by Proposition 2.2.1,}$$

$$= E^{\sigma(S_n)}(e^{itS_{n+1}}). \tag{5}$$

Since (5) holds for all t, by dominated convergence for $P^{\mathcal{B}}$-measures, we can deduce from (5) that it holds for all bounded Borel functions g on $\mathbb{R}$ (instead of only the trigonometric functions), so that we have for (5),

$$E^{\mathcal{B}_n}(g(S_{n+1})) = E^{\sigma(S_n)}(g(S_{n+1})), \quad a.e. \tag{6}$$

In particular, if $g = \chi_A$, $A = (-\infty, x)$, then (6) implies

$$P(S_{n+1} < x | \mathcal{B}_n) = P(S_{n+1} < x | S_n), \quad a.e. \tag{7}$$

so that $\{S_n, \ n \geq 1\}$ is a Markov process as asserted. $\quad \square$

We note that (4) can be written as:

$$E^{\mathcal{B}_n}(e^{itS_{n+1}})(\omega) = e^{itS_n(\omega)} E(e^{itX_{n+1}}), \quad a.a.(\omega),$$

$$= E(e^{it(X_{n+1} + S_n(\omega))}), \quad a.a.(\omega). \tag{8}$$

When all the conditional measures are regular, and if F_n is the distribution function of X_n, then by the image probability law (8) can be expressed (as in (6) with $g = \chi_A$) as:

$$P(S_{n+1} < x | S_1, \cdots, S_n)(\omega) = F_{n+1}(x - S_n(\omega)), \quad a.a.(\omega) . \qquad (9)$$

Remark. If the given process $\{X_n, n \geq 1\}$ on (Ω, Σ, P) is replaced by its function space representation triple $(\tilde{\Omega}, \tilde{\Sigma}, \tilde{P})$ provided by the Kolmogorov existence theorem (cf. Theorem 8.4.2) then the new process $\{\tilde{X}_n, n \geq 1\}$ has the same finite dimensional distributions as the given one and hence has the same probabilitic structure. But in the new representation $\tilde{X}_n(\tilde{\Omega}) = \mathbb{R}$, a Borel set for each $n \geq 1$, and hence the resulting conditional probabilities, for (8) and (9), are automatically regular. So the formula (9) holds for the new variables rather than the original ones. A different proof of (9), using the latter representation was given in [Doob (1953), p. 85].

Note that if $E(X_n)$ exists and $= 0$, then $\{S_n, \mathcal{B}_n, n \geq 1\}$ is also a martingale in Example 4. Thus there are families which are both martingales and Markov processes at the same time, and in fact Brownian motilion is a prime example of this fact. But the Markov concept does not demand the existence of any moments as in the latter case, and the martingale notion does not need the complete distributional structure as in the former case. Thus neither concept implies the other in general.

Before turning to the existence problem, some basic properties of these processes are recorded to get a better feeling. The fundamental relation, known as the *Chapman-Kolmogorov* equation is given by:

Lemma 5. *Let* $\{X_t, \ t \in T \subset \mathbb{R}\}$ *be a Markov process on* (Ω, Σ, P) *and* $u < t < v$ *be points from* T. *Then for any* $x \in \mathbb{R}$, *one has*

$$P(X_v < x | X_u) = E(P(X_v < x | X_t) | X_u) a.e. \qquad (10)$$

Moreover, if $p(\xi, u; A, v) = P(X_v \in A | X_u = \xi)(= P(X_v \in A | \sigma(X_u))(\xi))$, *is regular, then for all Borel sets* A, *and* $\xi \in \mathbb{R}$, *one has*

$$p(\xi, u; A, v) = \int_{\mathbb{R}} p(\zeta, y; A, v) p(\xi, u; d\zeta, t), \quad a.a.(\xi). \qquad (11)$$

Proof. Taking $n = 2$ in (1), $t_1 = u, t_2 = t$ and $t_3 = v$, $A = (-\infty, x)$ we get

$$P[X_v < x | X_u, X_t] = P[X_v < x | X_t], \quad a.e. \tag{12}$$

Applying the operator $E^{\sigma(X_u)}(\cdot)$, to both sides of (12) and remembering $E^{\sigma(X_u)} E^{\sigma(X_u, X_t)} = E^{\sigma(X_u)}$, one gets

$$E^{\sigma(X_u)} \left(P[X_v < x | X_t) = P(X_v < x | X_u), \quad a.e.[P \circ X_u^{-1}].$$

Since all such intervals generate the Borel σ-algebra of $\mathbb{R}$, this is simply (10). Writing the integral representation of $E^B(\cdot)$ in (10), it reduces to (11) with Lebesgue integrals when P^B is regular. $\square$

Let us introduce some familiar terminology often used in the subject:

Definition 6. A function $p(\cdot, \cdot; \cdot, \cdot) : B \times T \times \mathcal{B} \times T \to \mathbb{R}^+, (B, \mathcal{B})$ being a measurable space, is called a *transition probability function* if:

(i) $p(\xi, s; \cdot, t) : \mathcal{B} \to \mathbb{R}^+$ is a probability measure for $s, t \in T, \xi \in B$,

(ii) $p(\cdot, s; A, t) : B \to \mathbb{R}^+$ is $\mathcal{B}$-measurable for $s, t \in T, A \in \mathcal{B}$, with $p(x, s; A, s) = 1$ if $x \in A$, and

(iii) for each $s < t$, this p satisfies the Chapman-Kolmogorov equation (11) for *all* $\xi \in B$, there.

Moreover, if $p(\xi, s; A, t) = \tilde{p}(\xi, t - s; A)$, $s \le t(t - s \in T)$ $A \in \mathcal{B}$, then $\tilde{p}$ is termed a *stationary transition probability function*, $(T \subset \mathbb{R}^+)$.

Remark. In case there is a $(\sigma-)$ finite measure μ on $\mathcal{B}$ relative to which (i) and (iii) hold for a.a. (ξ), then $p(\cdot, \cdot; \cdot, \cdot)$ may be called a *generalized transition function*. Thus for a Markov process of Definition 1, there is only a generalized transition function with the range space $(B, \mathcal{B})$, termed a *state space*. Note that (i) and (iii) imply the relation $p(x, s; A - \{x\}, s) = 0$.

The following pair of simple examples indicate how one can manufacture various transition functions, without reference to (Ω, Σ, P).

Example 7. Let $B = \mathbb{N}$, $\mathcal{B} = \mathcal{P}(\mathbb{N})$-the power set and $T = \mathbb{R}^+$. Define p for $s \le t$ in T as:

$$p(i, s; A, t) = \sum_{j \in A} a_{st}(i, j) \quad, \quad A \in \mathcal{P}(\mathbb{N}) \,, \quad i \in \mathbb{N},$$

where $a_{st} \ge 0$, $\sum_{j=1}^{\infty} a_{st}(i, j) = 1$ and $a_{ss}(i, j) = \delta_{ij}$, the Kronecker data. These are probabilities of a "chain." Suppose we also have for

$0 \leq s \leq u \leq t$ in T,

$$\sum_{j=1}^{\infty} a_{su}(i,j)\, a_{ut}(j,k) = a_{su}(i,k).$$

Then it is easily verified that p is a (nonstationary) transition probability function.

Example 8. Let $B = \mathbb{R}^n$, $\mathcal{B} =$ Borel σ-algebra of B, $T = \mathbb{R}^+$ and define p for $0 \leq s \leq t$ in T by the following expression:

$$p(\xi, s; A, t) = \begin{cases} \left(\frac{1}{2\pi(t-s)}\right)^{1/2} \int_A \exp\{-\frac{|x-\xi|^2}{2(t-s)}\} dx, & \text{if } s < t, \\ \chi_A(\xi), & \text{if } s = t, \end{cases}$$

where $|\cdot|$ denotes the Euclidean distance and $A \in \mathcal{B}$. One can verify that $p(\xi, s; A, t) = \tilde{p}(\xi, t-s; A)$ and is a stationary transition function. It corresponds to the standard Brownian motion. In both examples these functions satisfy Chapman-Kolmogorov equation identically (i.e., without exceptional sets).

Having thus seen that transition probability functions can be independently prescribed, we now address the converse problem, namely the existence of a Markov process on some (Ω, Σ, P) with the given functions as its transition probabilities. We shall show how a positive solution is obtained from Kolmogorov's theorem itself (cf. Thm. 8.4.2).

Observe that a Markov process $\{X_t, t \in T\}$ on (Ω, Σ, P) determines not only its (generalized) transition functions by Lemma 5, but it also gives the (finite dimensional) marginal measures $\mu_{t_1, \cdots, t_n} : A_1 \times \cdots \times A_n \mapsto P[X_{t_i} \in A_i, 1 \leq i \leq n]$. We can express these measures as follows:

$$\mu_{t_1, \cdots, t_n}(A_1 \times \cdots \times A_n) = \int_{A_1} \cdots \int_{A_n} \mu_{t_1, \cdots, t_n}(dx_1, \cdots, dx_n)$$

$$= \int_{A_1} \mu_{t_1}(dx_1) \int_{A_2} q_{t_1, t_2}(x_1, dx_2) \cdots \times$$

$$\int_{A_n} q_{t_1, \cdots, t_n}(x_1, x_1, \cdots, x_{n-1}, dx_n),$$

$$= \int_{A_1} \mu_1(dx_1) \int_{A_2} q_{t_1 t_2}(x_1, dx_2)$$

$$\int_{A_3} q_{t_2 t_3}(x_2, dx_3)$$

$$\cdots \times \int_{A_n} q_{t_{n-1} t_n}(x_{n-1}, dx_n), \tag{13}$$

by the Markovian property and the change of measures in the Radon-Nikodým theory. This gives

$$\mu_{t_i}(A_i) = \mu_{t_1, \cdots, t_i, \cdots, t_n}(\Omega_1 \times \cdots \times A_i \times \cdots \times \Omega_n)$$

and similarly other marginals are obtained. On the other hand if the transition probabilities (i.e. $q_{t_i t_{i+1}}$) are given, one needs only the one-dimensional marginals as (13) shows. The higher-dimensional ones follow from the formula (13). Because of this, the one-dimensional measures are termed the *initial distributions* (or *measures*).

We first present a special, but useful, existence theorem and later extend it for the general case. Thus let $T_0 \subset \mathbb{R}$ having an initial point t_0. Let $\mathcal{F}(t_0)$ be the collection of all finite subsets of T_0 each having t_0 as its first element. If $\alpha, \beta \in \mathcal{F}(t_0)$, we say that $\alpha < \beta$ iff $\alpha \subset \beta$ and α is an initial segment of β. This makes $\mathcal{F}(t_0)$ a directed set. Moreover, with $\alpha = (t_0, t_1, \cdots, t_m), \beta = (t_0, t_1, \cdots, t_m, t_{m+1}, \cdots, t_n)(m < n)$ so that $\alpha < \beta$, we have

$$\mu_\beta(A_1 \times \cdots \times A_n) = \int_{A_1} \mu_{t_0}(d\omega_0) \int_{A_2} p_{t_0, t_1}(\omega_0, d\omega_1) \cdots \times$$

$$\int_{A_{m+1}} p_{t_m t_{m+1}}(\omega_m, d\omega_{m+1}) \cdots \int_{A_n} p_{t_{n-1} t_n}(\omega_{n-1}, d\omega_n)$$

$$= \int_{A_1} \mu_{t_0}(d\omega_0) \cdots \int_{A_m} p_{t_{m-1} t_m}(\omega_{m-1}, d\omega_m) \times$$

$$\mu_{t_n, \cdots, t_n}(\omega_m, A_{m+1}, \times \cdots \times A_n),$$

by Proposition 8.1.3 [cf. (15) in Section 8.1)].

$$(14)$$

Letting $A_k = \Omega_k, k = m + 1, \cdots, n$, the right side of (14) becomes $\mu_\alpha(A_1 \times \cdots \times A_m)$. Thus the compatibility conditions are satisfied. Since all the conditional measures are regular, being transition probability measures, we can invoke Theorem 8.3.2 to conclude that the system $\{(\Omega_\alpha, \Sigma_\alpha, \mu_\alpha, \pi_{\alpha\beta}) : \alpha < \beta \text{ in } \mathcal{F}(t_0)\}$ admits a unique projective limit (Ω, Σ, P) where $\Omega = \underset{i \in T_0}{\times} \Omega_i$, $\Sigma = \underset{i \in T_0}{\otimes} \Sigma_i$, $P \circ \pi_\alpha^{-1} = \mu_\alpha$: $\Omega_\alpha = \underset{i \in \alpha}{\times} \Omega_i \to \mathbb{R}^+$ is given by (13). Thus we have the following result.

Theorem 9. Let $\{(\Omega_i, \Sigma_i), i \in T_0 \subset \mathbb{R}\}$ be a family of measurable spaces with $p(\cdot, s; \cdot, t) = p_{st}(\cdot, \cdot) : \Sigma_j \times \Omega_i \to \mathbb{R}^+, i < j$ and $s < t$, as transition probability functions and μ_{t_0} as an initial probability at t_0.

Let $\mathcal{F}(t_0)$ be the collection of all finite ordered tuples of T_0, ordered by inclusion starting with t_0, as defined above. For $\alpha = (t_0, \cdots, t_n) \in \mathcal{F}(t_0)$ we set

$$\mu_\alpha(A_0 \times \cdots \times A_n) = \int_{A_0} \mu_{t_0}(\omega_0) \int_{A_1} p_{t_0 t}(\omega_0, d\omega_1) \cdots \int_{A_n} p_{t_{n-1} t_n}(\omega_{n-1}, d\omega_n)$$

$$(15)$$

Then $\{(\mu_\alpha, \pi_{\alpha\beta}) : \alpha < \beta$ in $\mathcal{F}(t_0)\}$ is a compatible family of probability measures on $\{(\Omega_\alpha, \Sigma_\alpha, \mu_\alpha, \pi_{\alpha\beta}) : \alpha < \beta$ in $\mathcal{F}(t_0)\}$ admitting a limit (Ω, Σ, P) where the cartesian product measurable space (Ω, Σ) is defined, as above, and $P : \Sigma \to \mathbb{R}^+$ is a probability such that $P \circ \pi_\alpha^{-1} = \mu_\alpha, \alpha \in \mathcal{F}(t_0)$. Moreover, if $X_t : \Omega \to \Omega_t, t \in T_0$, is defined as $X_t(\omega) = \omega(t) \in \Omega_t$, then $\{X_t, t \in T_0\}$ is a Markov process on (Ω, Σ, P) with values in $\{\Omega_t, t \in T_0\}$ and the given $\{p_{st}(\cdot, \cdot) : s < t$ in $T_0\}$ as its family of transition probabilities, having μ_{t_0} as the initial distribution.

Proof. From what precedes the statement, only the Markovian property remains to be verified. Since X_t is the coordinate variable, $X_t^{-1}(\Sigma_t) \subset \Sigma$ so that X_t is an r.v. Let $A \in \Sigma_t$ and $B = [X_t \in A]$. By Proposition 2, it suffices to verify that (1) is true for $t_0 < t_1 < \cdots < t_n < t_{n+1} = t$ in T_0. Let $C \in \sigma(X_{t_0}, \cdots, X_{t_n}) = \sigma(\bigcup_{i=0}^{n} X_{t_i}^{-1}(\Sigma_{t_i})) = \mathcal{S}_n$ (say), and take C as a generator, i.e., $C = \bigcap_{i=0}^{n} X_{t_i}^{-1}(A_i)$, $A_i \in \Sigma_{t_i}$. Then the left side of (1) is $P^{\mathcal{S}_n}(B)$ with $B = [X_t \in A]$, $A \in \Sigma_t$. Hence

$$\int_C P^{\mathcal{S}_n}(B)dP = \int_C E^{\mathcal{S}_n}(\chi_B)dP, \ (= P(C \cap B)),$$

$$= \int_{\bigcap_{i=0}^{n} X_{t_i}^{-1}(A_i)} \chi_{X_t^{-1}(A)} \, dP$$

$$= P(X_t \in A, \ X_{t_i} \in A_i, \ 0 \le i \le n)$$

$$= \int_{A_0} \mu_{t_0}(d\omega_0) \int_{A_1} p_{t_0 t_1}(\omega_0, d\omega_1) \cdots \times$$

$$\int_{A_n} p_{t_{n-1} t_n}(\omega_{n-1}, d\omega_n) p_{t_n t}(\omega_n, A),$$

since $P \circ \pi_\alpha^{-1} = \mu_\alpha, \alpha = (t_0, \cdots, t_n),$

$$= \int_{A_0 \times \cdots \times A_n} \mu_\alpha(d\omega_0, d\omega_1, \cdots, d\omega_n) p_{t_n t}(\omega_n, A)$$

$$= \int_C P(B|X_{t_{n-1}})dP, \quad (16)$$

by the image law on using $P \circ \pi_\alpha^{-1} = \mu_\alpha$. Identifying the extreme integrands which are S_n-measurable, we get (1). $\quad\square$

A natural and related question here is to ask whether there is a Markov process $\{X_t, t \in T\}$ on a probability space (Ω, Σ, P) where $T \subset \mathbb{R}$ does not necessarily have an "initial" or minimal point. This is proper since in the general existence theory of processes (e.g. Kolmogorov's or Choksi's extension considered in Chapter 8), no such minimal point of the index set T was demanded. A positive solution for this problem can be given, based directly on our study of Chapter 8 related to such existence questions, a point which is often ignored in the literature on Markov processes.

A precise statement is given by the following:

Theorem 10. *Let* $\{(\Omega_t, \Sigma_t, \mu_t), \ t \in T \subset \mathbb{R}\}$ *be a family of probability spaces and for* $s, t \in T, s < t$, *let* $p_{st} : \Sigma_t \times \Omega_s \to \mathbb{R}^+$ *be a set of regular conditional measures such that the following properties obtain:*

" *(a)*" *if* $\Omega = \underset{t \in T}{\times} \Omega_t, \Sigma = \underset{t \in T}{\otimes} \Sigma_t, \pi_t : \Omega \to \Omega_t$ *and* $\pi_{st} : \Omega_s \times \Omega_t \to \Omega_s$ *for* $s < t$ *are canonical projections, then*

$$\mu_{st}(A \cap \pi_{st}^{-1}(B)) = \int_B p_{st}(\omega, A)\mu_s(d\omega), \ A \in \Sigma_t, \ B \in \Sigma_s, \quad (17)$$

(b) if $r < s < t$, *then we have the Chapman-Kolmogorov equation:*

$$p_{rt}(\omega, A) = \int_{\Omega_s} p_{rs}(\omega, d\omega')p_{st}(\omega', A), \ a.e.[\mu_r], \ A \in \Sigma_t. \quad (18)$$

Then, for each $t_1 < t_2 < \cdots < t_k$ *of* $T, \mu_{t_1, \cdots, t_k} : \Sigma_{t_1} \times \cdots \times \Sigma_{t_k} \to \mathbb{R}^+$ *defined by*

$$\mu_{t_1, \cdots, t_k}(A_1 \times \cdots \times A_k) = \int_{A_1} \mu_{t_1}(\omega_1) \int_{A_2} p_{t_1 t_2}(\omega_1, d\omega_2) \cdots \times$$

$$\int_{A_k} p_{t_{k-1} t_k}(\omega_{k-1}, d\omega_k), \quad (19)$$

we have $\mu_{t_1, \cdots, t_{k+1}} : \Sigma_{t_1} \times \cdots \times \Sigma_{t_{k+1}} \to \mathbb{R}^+, \ t_k < t_{k+1}$, *satisfying*

$$\mu_{t_1, \cdots, t_{k+1}}(A_1 \times \cdots \times A_{k+1}) =$$

$$\int_{A_1 \times \cdots \times A_k} p_{t_k t_{k+1}}(\omega_k, A_{k+1})\mu_{t_1, \cdots, t_k}(d\omega_1, \cdots, d\omega_k), \quad (20)$$

and there is a unique probability $\mu : \Sigma \to \mathbb{R}^+$ such that $\mu \circ \pi_t^{-1} = \mu_t, \mu \circ \pi_\alpha^{-1} = \mu_\alpha$ for $\alpha = (t_1, \cdots, t_k)$ with $t_1 < \cdots < t_k$ from T, $\pi_\alpha : \Omega \to \Omega_\alpha$. Moreover there is a Markov process $\{X_t, t \in T\}$ defined by $X_t(\omega) = \omega(t) \in \Omega_t, t \in T$, satisfying $\mu[X_{t_1} \in A_1, \cdots, X_{t_n} \in A_n] = \mu_{t_1, \cdots, t_n} (A_1 \times \cdots \times A_n)$ of (19) for which the transition probability functions are precisely the given family.

Here we only need the Chapman-Kolmogorov equation to be satisfied a.e.$[\mu_t]$ which is (10), and is a weakening of the transition probability condition. Since we are considering Ω_α, Ω as cartesian products, condition (ii) of Theorem 8.3.1 is automatically satisfied and (i) is a consequence of (20) which follows from (19), the definition of μ_α and the classical Radon-Nikodým theorem. Hence this result is a consequence of Theorem 8.3.1 *whose general form is needed here* since no single initial measure can be given. This generality is useful for applications.

9.5 Further properties of Markov processes

Having presented general existence results of Markov processes in the preceding section, we shall turn to a few deeper structural properties of them and this involves additional (nontrivial) analysis. We illustrate this by giving conditions for the existence of Markov processes with *continuous* sample paths.

Let us start with the following:

Definition 1. If $\{X_t, t \in T\}$ is a stochastic process on (Ω, Σ, P) with values in a topological space $(S, \mathcal{S})$, then almost all of its sample paths are continuous at $t_0 \in T$, where T is also a topological space, if the sets $\{\omega : \liminf_{t \to t_0} X_t(\omega) \in A\}$ and $\{\omega : \limsup_{t \to t_0} X_t(\omega) \in A\}$ are measurable for (Σ), for each $A \in \mathcal{S}$, and their symmetric difference is a P-null set. If this property holds for all t_0 in T, then the process has almost all continuous sample paths on T.

Without further restrictions on $(S, \mathcal{S})$ and T the above sets may not be measurable, and so a solution is not possible. Here we give reasonable sufficient conditions, useful for various applications. Let us establish the following general result in this regard: (from now on the order of appearance of variables in $p(\cdot, \cdot; \cdot, \cdot)$ is slightly changed from Definite 4.6 for convenience)

Theorem 2. *Let Ω_0 be a Polish space and (Ω_0, Σ_0) be its Borelian space with $I \subset \mathbb{R}$ as a set. If $\Omega = \Omega_0^I$, $\Sigma = \underset{t \in I}{\otimes} \Sigma_t, \Sigma_t = \Sigma_0$, let $p : I \times \Omega_0 \times I \times \Sigma_0 \to \mathbb{R}^+$ be a transition probability function (cf. Definition 4.6) such that for some $\alpha > 0, \beta > 0, K > 0$, and for each pair of points t_1, t_2 in I we have*

$$\int_{\Omega_0} |x - y|^\beta p(t_1, x; t_2, dy) \leq K |t_1 - t_2|^{1+\beta}, \qquad (1)$$

uniformly in x. Then, for any family $\mu_t : \Sigma_0 \to \mathbb{R}^+$, $t \in I$, of marginal probability measures [i.e., (17) of Section 4 holds] there exists a unique $P : \Sigma \to \mathbb{R}^+$ such that $\mu_t = P \circ \pi_t^{-1}$, $[\pi_t : \Omega \to \Omega_t (\equiv \Omega_0)$ is the canonical projection] and a Markov process $X_t : \Omega \to \mathbb{R}$, with continuous sample paths $(X_t(\omega) = \omega(t) \in \Omega_0)$, having the given p and $\mu_t, t \in I$, as its transition and marginal probability functions respectively.

Proof. First it should be noted that if the transition function p and the marginal family $\{\mu_t, \ t \in I\}$ are given to satisfy (17) of Section 4 above, then by Theorem 4.10 there is a Markov process $\{X_t, t \in I\}$ on (Ω, Σ, P) with the $p(\cdot)$ and μ_ts as its associated functions, as required in the statement. If in addition the transition functions satisfy (1), then using the two-dimensional measures μ_{t_1, t_2} determined with μ_{t_1} and p defined by (19) of Section 4, we get (by the image law theorem) that the left side of (1) to be precisely $E^{\mathcal{B}}(|X_{t_1} - X_{t_2}|^\beta)$ where $\mathcal{B} = \pi_{t_1 t_2}^{-1}(\Sigma_{t_1} \otimes \Sigma_{t_2})$ and $\pi_{t_1 t_2} : \Omega \to \Omega_0 \times \Omega_0$ is the usual projection. Then on integrating the resulting expression in (1) one has:

$$E(|X_{t_1} - X_{t_2}|^\beta) = E(E^{\mathcal{B}}(|X_{t_1} - X_{t_2}|^\beta))$$
$$\leq KE(|t_1 - t_2|^{1+\alpha}) = K|t_1 - t_2|^{1+\alpha}. \qquad (2)$$

This inequality implies the desired continuity of sample paths by the following general theorem due to Kolmogorov.

Theorem 3. *Let $\{X_t, t \in I \subset \mathbb{R}\}$ be a stochastic process on (Ω, Σ, P) with values in $(S, \mathcal{S})$ where S is a Polish space with $d(\cdot, \cdot)$ as its distance function. If there exists constants $\alpha > 0, \beta > 0$, and $K > 0$ such that for any pair of points t_1, t_2 in I and $\varepsilon > 0$ we have*

$$P[\omega : d(X_{t_1}(\omega), dX_{t_2}(\omega)) \geq \varepsilon] \leq K|t_1 - t_2|^{1+\alpha} \varepsilon^{-\beta}, \qquad (3)$$

then the process has almost all continuous sample paths.

A proof of this result will not be included here. It is available in many places [cf., e.g., Rao (1979), pp. 189-191]. Thus in our case $d(x, y) = |x - y|$, and since by Markov's inequality

$$P[\omega : |X_{t_1} - X_2|^{\beta}(\omega) > \varepsilon^{\beta}] \leq \varepsilon^{-\beta} E(|X_{t_1} - X_{t_2}|^{\beta})$$
$$\leq K\varepsilon^{-\beta} |t_1 - t_2|^{1+\alpha}, \quad \text{by (2)},$$

it follows that (3) holds and the result of Theorem 2 follows. $\square$

Remarks. 1. Instead of using Theorem 4.10, one could invoke Theorem 4.9 if only one initial measure μ_0 is given. In this case the transition probability functions can be modified to have $X_{t_0} = x_0$:

$$\tilde{p}(t_1, x; t_2, \cdot) = \begin{cases} \delta_{x_0}(\cdot), & \text{if } t_1 < t_2 \leq t, \\ p(t_1, x_0; t_2, \cdot), & \text{if } t_1 < t \leq t_2, \\ p(t_1, x; t_2, \cdot), & \text{if } t \leq t_1 < t_2, \end{cases} \quad (4)$$

and then define P by expression (15) of the last section. In this case P is replaced by $P_{x_0} : \Sigma \to \mathbb{R}^+$ as our measure on (Ω, Σ).

2. The condition given for (1) is satisfied for the transition function of Example 9 of Section 4, with $\alpha = 1, \beta = 4$ and $K = 3$ (or $= 3n$, if the range is $\mathbb{R}^n$). Thus the Brownian motion process is a Markov process with almost all continuous sample paths. The transition function depends only on $t_2 - t_1$ (the time homogeneity), and the initial distribution can be taken as δ_0 to say that the process starts at the origin.

3. If the transition probability $p(s, x; t, dy) = \tilde{p}(x, t - s, dy)$ for $s < t$, then p was termed a stationary transition function (cf. Definition 4.6). However, this does not imply that the Markov process itself is *stationary* in the sense that its finite dimensional distributions are unchanged if the time is shifted by a fixed unit. However, as seen from (19) of Section 4, this will be true if all marginal distributions are identical. Thus a Brownian motion has only stationary transition probability functions, but itself is not a stationary process.

Because of the analytical tractability of the transition functions of a Brownian motion process, it is possible to consider a very detailed analysis of this process which is also a martingale. There is a large literature on it. For a recent account one may refer to the monograph by D. Revuz and M. Yor (1991).

It should now be noted that the transition probability function also plays a crucial role in relating the Markov process theory with abstract analytical study of evolution operators. Thus an important aspect of probability theory, with regular conditional measures, is transferable into an important part of modern analysis (the so-called abstract evolution operators and eventually into partial differential equations)—all of this is made possible because of the Chapman-Kolmogorov equation in the following form:

Proposition 4. *Let $p : \mathbb{R} \times I \times \mathcal{B} \times I \to \mathbb{R}^+$ be a transition probability function and $B(\mathbb{R})$ be the Banach space, under uniform norm, of all real bounded Borel functions on $\mathbb{R}$. Then for each $s < t$ of $I(\subset \mathbb{R})$, the operator $U(s,t)$ defined by*

$$(U(s,t)\,f)(x) = \int_{\mathbb{R}} f(y)\, p(s,x;t,dy) \ , \quad f \in B(\mathbb{R}), \qquad (5)$$

is a positive linear contraction on $B(\mathbb{R})$ into itself preserving constants, and satisfying the **evolution identity** *for $s < r < t$:*

$$U(s,r)\, U(r,t) = U(s,t) \ . \qquad (6)$$

Conversely, every such $U(\cdot,\cdot)$ of (6) determines uniquely a transition probability function and hence a Markov process which is unique if an initial distribution is also given. In particular, if p is stationary, then $U(s,t) = V(t-s)$ and $\{V(t), t \geq 0\}$ forms a strongly continuous contractive semigroup of operators on $B(\mathbb{R})$.

Proof. If $U(s,t)$ is given by (5) then it clearly satisfies all the stated conditions. Only (6) need be verified, using the Chapman-Kolmogorov equation. Indeed for $s < r < t$ we have

$$(U(s,t)f)(x) = \int_{\mathbb{R}} f(v)[\int_{\mathbb{R}} p(s,x;r,dy)\, p(r,y;t,dv)]$$

$$= \int_{\mathbb{R}} [\int_{\mathbb{R}} f(v)p(r,y;t,dv)\, p(s,x;r,dy], \text{by Fubini's theorem,}$$

$$= \int_{\mathbb{R}} (U(r,t)f)(y)\, p(s,x;r,dy)$$

$$= U(s,r)(U(r,t)f)(x) \ , \quad f \in B(\mathbb{R}). \qquad (7)$$

From this (6) follows.

Conversely, if $\{U(s,t), s < t$ in $I\}$ is such a family of operators on $B(\mathbb{R})$, then $(U(s,t)f)(x)$ is a positive linear functional in $f(\in B(\mathbb{R}))$ and by the Riesz representation theorem there is a unique additive $p(s,x;t,\cdot)$ on the Borel σ-algebra of $\mathbb{R}$ such that $p(s,\cdot;t,A)$ is Borel, $p(s,x;t,\mathbb{R}) = 1$, and (5) holds provided $p(s,x;t,\cdot)$ is shown to be σ-additive. But this is a consequence of the fact that for each continuous $f_n \downarrow 0, f_n$ having compact supports, $\|U(s,t)f_n\| \leq \|f_n\| \to 0$ as $n \to \infty$, the latter following from Dini's theorem. This implies that $p(s,x;t,\cdot)$ is σ-additive, and the Chapman-Kolmogorov equation is a consequence of (6).

Finally if $V(t-s) = U(s,t)$, then (6) gives $V(h_1+h_2) = V(h_1)V(h_2)$ for all $h_1, h_2 \geq 0$. Since $U(\cdot,\cdot)$ is strongly continuous to start with, $\{V(t), t \geq 0\}$ is the desired semigroup keeping constants invariant. $\square$

To emphasize the functional analytic character of the transition probability, further, we now show how one can associate uniquely, on a *different* space, a stationary transition function with a given general transition function. Consider $p : I \times \Omega \times I \times \Sigma \to \mathbb{R}^+$, $I \subset \mathbb{R}$, a transition function. Let $\widetilde{\Omega} = I \times \Omega, \widetilde{\Sigma} = \mathcal{B} \otimes \Sigma$ where $\mathcal{B}$ is the Borel σ-algebra of I. For each $B \in \widetilde{\Sigma}$ we have $B_{t_1} = \{y : (t_1,y) \in \widetilde{\Sigma}\}$, $t_1 \in I$, a measurable section. If $t_1 < t_1 + h = t_2, (h > 0)$ then we let $\tilde{p} : \mathbb{R}^+ \times \widetilde{\Omega} \times \widetilde{\Sigma} \to \mathbb{R}^+$ be defined by

$$\tilde{p}(h,x,B) = p(((t_1,y); t_1 + h, B_{t_1+h}) , \quad x = (t_1,y) . \tag{8}$$

If it is assumed that $p(\cdot,\cdot,;A,\cdot)$ is $\mathcal{B} \otimes \Sigma \otimes \mathcal{B}$-measurable, then one can easily verify that $\tilde{p}$ is a stationary transition function. Thus with a general (jointly measurable) transition function it is possible to associate uniquely a stationary such function on a new space $(\widetilde{\Omega}, \widetilde{\Sigma})$. However, this space is more complicated then the original one to use it effectively in many problems of interest. It is nevertheless useful to have the possibility of associating such a (conceptually simpler) representation, which is somewhat analogous to representing an nth order linear (scalar) differential equation as a first order vector equation.

This type of representation can also be generalized for processes that are not Markovian to begin with. Indeed, Johnson (1970, 1979) has given algebraic representations of an arbitrary separable stochastic process $\{X_t, t \in T\}$ on (Ω, Σ, P), where $\Omega = \Omega_0^T$ with Ω_0 as a compact metric space, and Σ is the Borel σ-algebra of Ω. Here P is the governing probability measure, and the represented process is an image of a Markov process with stationary transition functions. Thus some qualitative information about the structural properties of the given general process can be obtained from a knowledge of the "accompanying" Markov process with stationary transition probabilities. For the detailed analysis of the given processes, however, one usually has to study the original family with its individual features, but the various representations can provide an overview of the subject.

Finally, we note that if in a problem it is desired to evaluate the conditional probability, or expectation, of some functional given another functional (as seen in the examples at the end of Chapter 5 where the basic process is even Brownian motion so that it is Markovian), there is no algorithm for such an evaluation even with all the voluminus known work. In fact, exact evaluations are needed in translating the properties of transition probability functions to those of sample paths of the related Markov processes. This can be seen, for instance, in Loève (1963), pp. 595-603, where formal manipulations are used, but their rigorous derivations are not simple and not attempted. The only exact evaluation methods that are known seem to be those that we have outlined in the last parts of Chapters 5 and 8. Thus as noted before, this evaluation problem has not received adequate attention even in such very active areas as Markov processes.

9.6 Bibliographical notes

The (set) martingale theory can now be considered classical. Theorem 2.7 is an extension of a well-known result due to Kakutani (1948) and is adapted from Brody (1971). The proof of Theorem 3.4, for the mean convergence of directed index martingales, as a consequence of Theorem 2.7, is due to the author. The existence theory of Markov processes is also classical. The general result (Theorem 4.10) on the existence of Markov processes without a fixed initial distribution, but has a family of "marginals," should be (but has not been) clearly presented in the literature. The other results and their sources are already indicated in the text. The fact that a general Markov process can be represented on a different space by a process with stationary transition probability functions was remarked by Dynkin (1960). This point of view, apparently without the knowledge of Dynkin's earlier observation, has been further developed and analyzed in detail in a series of papers by Johnson (1970, 1979). It would be interesting to explore the full potential of this type of representation.

Markov processes as solutions of diffusion equations has a large theory. A general and intense development is given in the book by Stroock and Varadhan (1979), where references to related work can be found. We shall indicate other types of applications in the final two chapters. In particular, we include some results on (probabilistic) potential theory and also illustrate the key role played by the conditional expectation operators in function and operator algebras of modern analysis.

CHAPTER 10: APPLICATIONS TO MODERN ANALYSIS

Conditioning as a general concept has applications not only in probability but also in many other contexts in modern analysis. In this chapter some of these ideas are discussed showing in particular that they are well-suited as (subMarkov) kernels in extending the classical potential theory, analyzing bistochastic operators as well as Reynolds operators arising in turbulence theory, and elsewhere. Also it is shown how conditional expectations can be used as models for describing general (contractive) projections in many function spaces. Almost all these applications are based on regular conditional measures, and the theory is then executed.

10.1 Introduction and motivation

Let $(\Omega_i, \Sigma_i, P_i), i = 1, 2$, be a pair of probability spaces and $h : \Omega_1 \to \Omega_2$ be a measurable mapping, so $h^{-1}(\Sigma_2) \subset \Sigma_1$. Consider the function $q_{12} : \Sigma_1 \times \Omega_2 \to \mathbb{R}^+$ given by

$$P_1(A \cap h^{-1}(B)) = \int_B q_{12}(A, \omega) dP_2(\omega), \quad A \in \Sigma_1, B \in \Sigma_2 \quad . \quad (1)$$

Then $q_{12} : \Sigma_1 \to L^1(\Sigma_2)$ is a conditional measure. Suppose it is regular, so that it has a version satisfying, (i) $0 \leq q_{12}(A, \omega) \leq 1$, (ii) $q_{12}(\cdot, \omega)$ is a measure on Σ_1 for each $\omega \in \Omega_2$. Such a positive q_{12} is called a *kernel* when $\Omega_1 = \Omega_2 = \Omega$ and $\Sigma_2 \subset \Sigma_1, h = $ identity. Motivated by this property, one can consider $q_{12}(\Omega_1, \omega) \leq 1$, called the subMarkovian kernel, in place of equality when it is termed Markovian. This will lead to studying nonfinite kernels satisfying: if $A_n \in \Sigma$, $\overset{\infty}{\underset{n=1}{\cup}} A_n = \Omega$, and $q_{12} : \Sigma(A_n) \times A_n \to \mathbb{R}^+$ is a subMarkov kernel for each n. Such a q_{12} is termed a "proper" kernel. [$\Sigma(A)$ is the trace of Σ on A, as usual.]

Following the work in the preceding chapter, one can define

$$\nu(A) = \int_\Omega q(A, \omega)d\mu(\omega) , \qquad (2)$$

a new measure for any $\mu : \Sigma \to \mathbb{R}^+$, as an "initial" measure. If q_1, q_2 are a pair of (proper) kernels on $\Sigma \times \Omega \to \mathbb{R}^+$, then their composition

$$q_0(A, \omega) = (q_1 * q_2)(A, \omega) = \int_\Omega q_1(A, \omega')q_2(d\omega', \omega) \qquad (3)$$

is another kernel of the same kind. If $q_1 = q_2 = q$, then q_0 of (3) may be written as $q^2(= q * q)$, and by iteration one has $q^n = q * q^{n-1}$ with $q^0 = $ identity. In this way we can introduce:

Definition 1. If (Ω, Σ) is a measurable space, $f : \Omega \to \mathbb{R}^+$ is a positive measurable function [$\mathbb{R}^+$ is endowed with its Borel σ-algebra] then f is termed *excessive* (*invariant*) relative to a kernel $q : \Sigma \times \Omega \to \mathbb{R}^+$ if for each $\omega \in \Omega$,

$$\infty \geq f(\omega) \geq q(f)(\omega) = \int_\Omega f(\omega')q(d\omega', \omega), [f(\omega) = q(f)(\omega)]. \qquad (4)$$

Thus $q(\cdot, \cdot)$ is a (sub)Markovian kernel iff the constant function 1 is excessive (invariant) for q. This function is a generalization of the classical Newtonian kernel of potential theory. The connection is discussed in more detail in the next section. By iteration, therefore, one has for each excessive f relative to q,

$$q^{n+1}(f) \leq q^n(f) \leq \cdots \leq f \qquad (5)$$

so that $\lim_{n \to \infty} q^n f = q^\infty(f)$ always exists although q^∞ need not be a kernel itself. Letting

$$p = \overset{\infty}{\underset{n=0}{\Sigma}} q^n , \qquad (6)$$

p is called a *potential kernel* associated with q. For a positive measurable function f on $\Omega, (pf)(\cdot)(= (I + pq)f = (I + qp)f)$ is termed the *potential* of f. These concepts already lead to some nontrivial conclusions on the behavior of functions. Here is an illustration.

Proposition 2. *Let q be a kernel on (Ω, Σ) with p as its associated potential kernel. Then for any positive measurable f, $g = pf$ is an excessive function for q and one has*

$$(q^\infty g)(\omega) = 0 \ or \ +\infty, \ for \ all \ \omega \in \Omega . \tag{7}$$

In the opposite direction, every excessive function $g : \Omega \to \bar{\mathbb{R}}^+$, satisfying (7) is the potential of some positive measurable f. Further, if g is an excessive fuction (relative to q) such that $q^\infty g < \infty$, it admits a unique decomposition

$$g = f_1 + f_2, \tag{8}$$

*called the **Riesz decomposition**, where f_1 is invariant and f_2 is a potential.*

Proof. Since $p = I + qp$ we have for any measurable $f : \Omega \to \bar{\mathbb{R}}^+$

$$g = pf = f + qpf \geq q(pf) = q(g)$$

so that g is excessive. If $(q^\infty g)(\omega) < \infty$ for some $\omega \in \Omega$, then

$$(q^k g)(\omega) = q^k \Big(\sum_{j=0}^{\infty} (q^j f) \Big)(\omega)$$

$$= \sum_{j=k}^{\infty} (q^j f)(\omega) . \tag{9}$$

Since $(q^\infty g)(\omega) = \lim_{k \to \infty} (q^k g)(\omega) < \infty$, there is a $k > 1$ such that $(q^k g)(\omega) < \infty$ so that by (9), the tail part of the series must tend to zero. Hence $(q^\infty g)(\omega) = 0$ for all $\omega \in \Omega$ for which it is finite. This gives (7).

On the other hand if g is excessive and $(q^\infty g)(\omega)$ takes only 0 and ∞, let $f : \Omega \to \bar{\mathbb{R}}^+$ be given by

$$f = (g - qg)\chi_{[g<\infty]} + \infty\chi_{[g=+\infty]} . \tag{10}$$

We assert that $g = pf$. Indeed, since $g \geq pg$, $f \geq 0$ and $pf = f + p(qf) \geq f$, we have $(pf)(\omega) = f(\omega)$ for $\omega \in [g = +\infty]$, while for $\omega \in [g < \infty]$, $(qg)(\omega) \leq g(\omega) < \infty$ so that $(q^k g)(\omega) \leq (q^{k-1} g)(\omega) \leq g(\omega) < \infty$ for all $k \geq 1$. Hence

$$(qf)(\omega) = (qg)(\omega) - (q^2 g)(\omega) \leq g(\omega) - (q^2 g)(\omega) \leq g(\omega) < \infty .$$

By iteration one has $(q^k f)(\omega) < \infty$ for all $k \geq 1$, and then by (10),

$$\sum_{k=0}^{n} (q^k f)(\omega) = \sum_{k=0}^{n} ((q^k g)(\omega) - (q^{k+1} g)(\omega)) = g(\omega) - (q^{n+1} g)(\omega) . \quad (11)$$

Letting $n \to \infty$, the left side of (11) becomes $(pf)(\omega) \geq 0$ so that the right side gives $(q^\infty g)(\omega) < \infty$. The first part now implies $(q^\infty g)(\omega) = 0$. It follows that $(pf)(\omega) = g(\omega)$ for $\omega \in [g < \infty]$, and so in both cases $pf = g$ holds.

Regarding the decomposition (8), let $\omega \in \Omega$ be fixed for which $(q^\infty g)(\omega) < \infty$ for an excessive g. Then $(q^k g)(\omega) < \infty$ for some k. However, letting $\tilde{q}(A, \omega) = \int_\Omega q(A, \omega') d\mu_\omega(\omega')$, we have

$$(q^k g)(\omega) = \int_\Omega (q^{k-1} g)(\omega') \tilde{q}(d\omega', \omega), \quad (12)$$

where $\mu_\omega(\cdot)$ is a point measure at ω. Since $\tilde{q}(\cdot, \omega) = (q\mu_\omega)(\cdot)$ is a measure, (12) implies that $(q^{k-1} g)(\cdot)$ is integrable relative to $\tilde{q}(\cdot, \omega)$. Then by the dominated convergence theorem, since $(q^n g)$ is decreasing, we deduce that $(q^\infty g)(\omega) < \infty$ and is invariant for $\tilde{q}$. Let $f_1 = q^\infty g$ and $f_2 = g - f_1$. Since $qf_1 = f_1$ (the invariant function) $(q^\infty f_2) = 0$ by the first part, so f_2 is a potential of g. It remains to prove uniqueness.

Let $g = f_1 + f_2 = f_1' + f_2'$ be two decompositions where f_1, f_1' are invariant and g is excessive. Thus $q^\infty g = f_1' + q^\infty f_2' \geq f_1'$. If f_2' is also a potential of g, then there is equality here, and it holds iff $q^\infty f_2 = 0$. But then, $f_1 - f_1' = f_2' - f_2$ so that

$$f_1 - f_1' = q^\infty (f_1 - f_1') = q^\infty (f_2' - f_2) = 0.$$

Hence $f_1 = f_1'$ and it follows that $f_2 = f_2'$, proving uniqueness of the decomposition.   □

This result shows and motivates many other developments, for genral kernels, in portential theory. The excessive and invariant functions are also known as *super harmonic* and *harmonic functions* in the classical theory. They correspond to super martingale and martingale concepts in probability, discussed in the preceding chapter. Thus if $\mathcal{F}_n \subset \mathcal{F}_{n+1}$ is a filtering sequence of a probability space (Ω, Σ, P), let $\tilde{\Omega} = \mathbb{N} \times \Omega, \tilde{\Sigma}$ the σ-algebra generated by the collection $\{\{n\} \times A_n : A_n \in \mathcal{F}_n, n \in \mathbb{N}\}$, and $\tilde{K} : \tilde{\Sigma} \times \tilde{\Omega} \to \mathbb{R}^+$ be defined as

$$\tilde{K}(\tilde{A}, \tilde{\omega}) = K(A_n, \omega) \otimes \delta(n', n),$$

where $\tilde{A} = \{n\} \times A_n \in \tilde{\Sigma}$, $\delta(\cdot, n) : \mathcal{P}(\mathbb{N}) \to \mathbb{R}^+$, the delta function ($\delta(n', n) = 1$ if $n' = n$, and $= 0$ otherwise). Let $X : \tilde{\Omega} \to \mathbb{R}$ be given by $X(\tilde{\omega}) = X_{n+1}(\omega)$ for $\tilde{\omega} = (n, \omega) \in \tilde{\Omega}$. Then by definition,

$$
\begin{aligned}
(\tilde{K}X)(\tilde{\omega}) &= \int_{\tilde{\Omega}} X(\tilde{\omega})\tilde{K}(d\tilde{\omega}', \tilde{\omega}) \\
&= \int_{\mathbb{N}} \int_{\Omega} X_{n'+1}(\omega')K(d\omega', \omega)\delta(dn', n) \\
&= \int_{\Omega} X_{n+1}(\omega')K(d\omega', \omega) \\
&= E^{\mathcal{F}_n}(X_{n+1})(\omega), \text{ by Prop. 7.2.1,} \\
&\leq X_n(\omega) = X(\tilde{\omega}),
\end{aligned}
\tag{12}
$$

iff $\{X_n, \mathcal{F}_n, n \geq 1\}$ is a supermartingale, or X is an excessive function for the "kernel" $\tilde{K}$. Note that $\tilde{K}$ will be a true kernel only if $K(\cdot, \cdot)$ is a regular conditional measure. In any case, taking $K(A, \omega) = P^{\mathcal{F}}(A)(\omega), A \in \Sigma, \omega \in \Omega$ and $\delta : \mathcal{P}(\mathbb{N}) \times \mathbb{N} \to \{0, 1\}$, we see that the (super) martingale theory is closely related to the potential kernels. However, martingales impose less restrictions although a result in one area suggests one in the other. We now analyze these ideas in detail.

10.2 Conditional measures and kernels

As seen above, a kernel defined by a regular conditional measure leads to an interesting class, called *potential kernels*. If there are several such kernels, forming a semigroup, then the interrelations between this calss, and their role in (classical) potential theory should be analyzed for a better knowledge of this function class. The probabilistic connection with potential theory was first established by S. Kakutani (1944) and a far reaching generalization of the latter subject based on subMarkov kernels was made by G.A. Hunt in middle 1950s. For a clear understanding and appreciation of these relations, let us present the equivalence of Brownian motion with classical (i.e., Newtonian) potential theory.

This theory starts with the so-called Coulomb's law, according to which two charges at a distance r units apart having magnitudes m_1 and m_2 attract each other with a force F given by

$$
F = c\frac{m_1 m_2}{r^2}
$$

where c is some constant. The potential P at a point $\mathbf{x}$ in $\mathbb{R}^3$, due to a charge q is the work necessary to bring the unit charge from infinity to $\mathbf{x}$, and is given by the quantity $\frac{1}{2\pi}q(|\mathbf{x} - \mathbf{x}_0|^{-1})$ where $\mathbf{x}_0$ is the position of the charge, $|\cdot|$ being the norm in $\mathbb{R}^3$. If there are n charges $(q_1, \cdots, q_n)$ at $(\mathbf{x}_1, \cdots, \mathbf{x}_n)$ in $\mathbb{R}^3$ then the potential is defined as:

$$\frac{1}{2\pi} \sum_{i=1}^{n} q_i(|\mathbf{x} - \mathbf{x}_i|^{-1}) . \tag{2}$$

If the charge distribution (a signed measure) is $\mu(\cdot)$ on the Borel σ-algebra $\mathcal{B}(\mathbb{R}^3)$ of $\mathbb{R}^3$ and n is large, the potential g_μ is approximated by the integral:

$$g_\mu(\mathbf{x}) = \frac{1}{2\pi} \int_{\mathbb{R}^3} \frac{d\mu(\mathbf{y})}{|\mathbf{x} - \mathbf{y}|} . \tag{3}$$

The total charge $\mu(\mathbb{R}^3)$ is always finite. We observe that $|g_\mu(\mathbf{x})| < \infty$ for almost all $\mathbf{x}$ in $\mathbb{R}^3$, relative to the Lebesgue measure λ. To see this since $\mathbb{R}^3$ is σ-compact, it suffices to verify the assertion for a closed ball B_n of $\mathbb{R}^n$. Thus $|g_\mu(\mathbf{x})| < \infty$ for a.a.$(\mathbf{x})$ in B_n, $n \geq 1$. But this follows from the computation:

$$\int_{B_n} |g_\mu(\mathbf{x})| d\lambda(\mathbf{x}_0) = \frac{1}{2\pi} \int_{B_n} \int_{\mathbb{R}^3} \frac{d|\mu|(\mathbf{y})}{|\mathbf{x} - \mathbf{y}|} d\lambda(\mathbf{x})$$

$$\leq \frac{1}{2\pi} \int_{\mathbb{R}^3} d|\mu|(\mathbf{y}) \int_{B_n} \frac{d\lambda(\mathbf{x})}{|\mathbf{x} - \mathbf{y}|}$$

$$\leq c_n |\mu|(\mathbb{R}^3) < \infty,$$

where c_n is the value of the integral over B_n and $|\mu|(\mathbb{R}^3)$ is the variation norm of μ. The operator $\Pi : \mu \mapsto g_\mu$ mapping a charge μ into its potential g_μ is called the *potential operator*. Thus $(\Pi\mu)(\mathbf{x}) = g_\mu(\mathbf{x})$ is given by (3). We let $B(\mathbb{R}^3)$ be the set of real Baire functions on $\mathbb{R}^3$.

Consider now the transition kernel of the standard Brownian motion, described in Example 9.4.8.

Then the *transition operator* family $\{P_t, t > 0\}$ is defined by

$$(P_t\mu)(A) = \int_{\mathbb{R}^3} \left(\int_A (2\pi t)^{-\frac{3}{2}} \exp\{-\frac{|\mathbf{x} - \xi|^2}{2t}\} d\lambda(\mathbf{x}) \right) \mu(d\xi),$$

$$= \int_A \int_{\mathbb{R}^3} (2\pi t)^{-\frac{3}{2}} \exp\{-\frac{1}{2t}|\mathbf{x} - \xi|^2\} \mu(d\xi) d\lambda(\mathbf{x}). \tag{4}$$

The key relation between Brownian motion and the classical potential theory is now given by

Proposition 1. *If $g_\mu(\cdot)$ is the potential of the charge (signed) measure μ given by (3), then*

$$g_\mu(\mathbf{x}) = \lim_{T\to\infty} \int_0^T \frac{d(P_t\mu)}{dx}(x)dt, \tag{5}$$

where $(P_t\mu)(\cdot)$ is the measure defined by (4).

Proof. Replacing μ by its variation $|\mu|$, we can and do assume that $\mu \geq 0$ for this proof. Then substituting for $(P_t\mu)(\cdot)$ from (4) we get

$$\int_0^T \frac{d(P_t\mu)}{dx}(x)dt = \int_0^T (2\pi t)^{-\frac{3}{2}} \int_{\mathbb{R}^3} \exp(-\frac{1}{2t}|\mathbf{x}-\xi|^2)d\mu(\xi)dt$$

$$= \int_{\mathbb{R}^3} \int_{|\mathbf{x}-\xi|/\sqrt{T}} 2(2\pi)^{-\frac{3}{2}}|\mathbf{x}-\xi|^{-1}e^{-\frac{u^2}{2}} du\, d\mu(\xi),$$

by changing the order and setting $|\mathbf{x}-\xi|^2 = t\,u^2$,

$$\to \frac{1}{2\pi} \int_{\mathbb{R}^3} (|\mathbf{x}-\xi|^{-1})\, d\mu(\xi), \quad \text{as } T \to \infty,$$

$$= g_\mu(\mathbf{x}), \quad a.a.(\mathbf{x}).$$

This is the desired result. $\square$

It is thus interesting that the classical potential operators are obtainable from a study of Brownian motion and probabilistic analysis. But, surprisingly, the converse implication is also true. Namely the Brownian transistion functions and hence the operators P_t can be obtained from the classical potential theory. This implication demands several detailed computations and a somewhat intricate analysis. We sketch the main ideas and refer the reader to Knapp (1965) whose presentation is being adapted in this equivalence demonstration.

Let S be the (Schwartz) space of real infinitely differentiable functions on $\mathbb{R}^3$ which together with all their (partial) derivatives remain bounded when multiplied by any polynonial. Thus $f \in S$ iff f and all derivatives tend to zero faster than any power of $|\mathbf{x}|$, which means $\lim_{|\mathbf{x}|\to\infty} ||\mathbf{x}|^k f^{(m)}(\mathbf{x})| = 0$ for each $k = 1, 2, \cdots$, and all $m \geq 1$. It is then known that the Fourier transform $\mathcal{F} : S \to S$ is a bijective map. Let $\mathcal{M}(\mathbb{R}^3)$ be the Banach space of signed measures on $\mathcal{B}(\mathbb{R}^3)$ with variation norm. Our converse implication is then given by

Theorem 2. *Let $\Pi : \mathcal{M}(\mathbb{R}^3) \to B(\mathbb{R}^3)$ be a potential operator, as defined by (3). Then the transition operator family $\{P_t, t > 0\}$ of the*

Brownian motion, defined by (4), is determined by Π *as follows: Let* $\Pi_S = \Pi|S$, *where* $S(\subset \mathcal{M}(\mathbb{R}^3))$ *consists of absolutely continuous signed measures* μ *such that* $\frac{d\mu}{d\lambda} \in S$, λ *being Lebesque measure. Then (a)* $\Pi_S : S \to S$ *is one-to-one and if* $\mu \in$ range $(\Pi_S), g_\mu$ [cf. (3)] *is the potential of* μ; *and (b) if* $-L : g_\mu \to \frac{d\mu}{d\lambda}$, *there exists a family* $\{Q_t, t > 0\}$ *of strongly continuous semigroup of positive linear operators on* $C_0(\mathbb{R}^3) \to C_0(\mathbb{R}^3)$, *the Banach space under uniform norm of continuous real functions vanishing at infinity, having the following properties:*

(i) $\|Q_t\| = 1$,   (ii) $\lim_{t\downarrow 0} \|Q_t f - f\| = 0$, $f \in C_0(\mathbb{R}^3)$,

(iii) *L is the infinitesimal generator of this family, and*

(iv) *the adjoints* $P_t = Q_t^* : \mathcal{M}(\mathbb{R}^3) \to \mathcal{M}(\mathbb{R}^3)$ *form the desired semigroup* $\{P_t, t > 0\}$ *of transition operators.*

The proof involves a construction of $\{Q_t, t > 0\}$ from the desired $\{P_t, t > 0\}$ having the asserted properties. Since the adjoint space of $C_0(\mathbb{R}^3)$ is $\mathcal{M}(\mathbb{R}^3)$ and P_t^* acts on $(\mathcal{M}(\mathbb{R}^3))^*$ one consideres $Q_t = P_t^*|C_0(\mathbb{R}^3)$ where $C_0(\mathbb{R}^3)$ is identified as a subspace of its second adjoint $(\mathcal{M}(\mathbb{R}^3))^*$. Then one can verify the following:

Proposition 3. Q_t *is a mapping of* $C_0(\mathbb{R}^3)$ *into itself, and*

$$(Q_t f)(\xi) = \int_{\mathbb{R}^3} f(\mathbf{x}) p_t(d\mathbf{x}, \xi), \xi \in \mathbb{R}^3, \qquad (6)$$

where $p_t(\cdot, \cdot) : \mathcal{B}(\mathbb{R}^3) \times \mathbb{R}^3 \to \mathbb{R}^+$ *is the Brownian transition function which satisfies (b) (i) and (b) (ii) of the above theorem.*

We present the main ideas of the proof without details as popositions, refering the omitted work to Knapp's exposition noted above. Thus if L is the infinitesimal generator of the (positive) semigroup so that $\lim_{t\downarrow 0} \|\frac{1}{t}(Q_t f - f) - Lf\| = 0, f \in C_0(\mathbb{R}^3)$, then the classical Hille-Yosida theory of such a family implies domain of L is a dense subset of $C_0(\mathbb{R}^3)$ and on it $Q_t L = L Q_t$. Moreover $(Q_{(\cdot)} f) : \mathbb{R}^+ \to C_0(\mathbb{R}^3)$ and L is a closed (unbounded) operator. The semigroup $\{Q_t, t > 0\}$ and the generator L uniquely determine each other in the following sense.

Proposition 4. *For each* $t > 0$, $f \in C_0(\mathbb{R}^3)$ *and* $\alpha > 0$, *one has*

$$\lim_{\alpha \to \infty} \| \exp(t\alpha L(\alpha - L)^{-1})f - Q_t f\| = 0$$

where $\exp B = \sum_{n=0}^{\infty} \frac{B^n}{n!}$, *the series converging in the norm topology.*

The details of this operator theory can be found in Dunford and Schwartz (1958, Chapter VIII, Section 1, cf. especially VIII 3.15).

The connection with potential theory is obtained through an operator G defined on $C_0(\mathbb{R}^3)$ by

$$\lim_{T\to\infty} \left\| Gf - \int_0^T (Q_t f)dt \right\| = 0, \quad f \in C_0(\mathbb{R}^3) ,$$

and then showing that G is the inverse of $-L$, i.e., $f \in dom(L) \Rightarrow Lf \in dom(G)$ and $G(-Lf) = f$, and $f \in dom(G) \Rightarrow Gf \in dom(L)$ and $(-L)(Gf) = f$. Next these operators are extended to $L^1(\mathbb{R}^3)$ by verifying: (I) the Schwartz space $S \subset C_0(\mathbb{R}^3) \cap L^1(\mathbb{R}^3)$ and $Q_t(C_0(\mathbb{R}^3) \cap L^1(\mathbb{R}^3)) \subset dom(L)$, $f \in C_0(\mathbb{R}^3) \cap L^1(\mathbb{R}^3) \Rightarrow (LQ_t)f(\xi) = \int_{\mathbb{R}^3} \frac{\partial}{\partial t}((2\pi t)^{-\frac{3}{2}} \exp(-\frac{1}{2t}|\mathbf{x} - \xi|^2))f(\mathbf{x})d\lambda(\mathbf{x})$.

(II) $f \in C_0(\mathbb{R}^3) \cap L^1(\mathbb{R}^3) \subset dom(G)$ and

$$(Gf)(\xi) = \frac{1}{2\pi} \int_{\mathbb{R}^3} \frac{f(\mathbf{x})}{|\mathbf{x} - \xi|} d\lambda(\mathbf{x}).$$

(III) $f \in S$ (the Schwartz space) $\Rightarrow Lf = 2\nabla^2 f$ where $\nabla^2 = \frac{\partial^2}{\partial x_1^2} + \frac{\partial^2}{\partial x_2^2} + \frac{\partial^2}{\partial x_3^2}$, the Laplacian. Thus L is proportional to ∇^2, and that $(I - \nabla^2)y = f, f \in S$ has a solution in S.

With all this work, one shows that the L defined above satisfies (a) of Theorem 2, and in fact Π_S is the composition of $(-L)$ and G. The other parts of the theorem then follow from this analysis easily.

It should be noted that the theorem is valid for $\mathbb{R}^n, n \geq 3$, if g_μ of (3) is replaced by

$$g_\mu(\mathbf{x}) = \frac{1}{2\pi^{\frac{n}{2}}\Gamma(\frac{n-2}{2})} \int_{\mathbb{R}^n} \frac{d\mu(y)}{|\mathbf{x} - \mathbf{y}|^{n-2}}.$$

For $n = 1,2$ one has to consider different kernels, and we omit further discussion of this classical case and turn to an outline of the modern aspects with general (sub) Markov kernels motivated by the above work, of which the Brownian family forms an important particular case.

Definition 5. Let $\{p_t, t > 0\}$ be a set of subMarkov kernels on a measurable space (Ω, Σ). Then it is called a (subMarkov) semigroup if (i) the mapping $(s, x) \to p_s(x, A)$ is $\mathcal{B} \otimes \Sigma$-measurable for each $A \in \Sigma$, $\mathcal{B}$ being the Borel σ-algebra of $\mathbb{R}^+$, and (ii) for each $s, t > 0$ and $x \in \Omega$,

$$(p_s(p_t f))(x) = \int_\Omega p_s(x, dy) \int_\Omega f(z)p_t(y, dz) = (p_{s+t} f)(x), \qquad (7)$$

for all bounded measurable $f : \Omega \to \mathbb{R}$.

Taking $f = \chi_A$ in (7), it reduces to

$$p_{s+t}(x, A) = \int_{\Omega} p_t(y, A) p_s(x, dy), \qquad (8)$$

which is just the Chapman-Kolmogorov equation for stationary Markov transition functions. Note that $p_{s+t}(x, A) = p_{t+s}(x, A)$, by Fubini's theorem (compare with Proposition 9.5.4 and the discussion after its proof). The relation (7) or (8) can be put in perspective when we specialize Theorem 9.4.9 (or 9.4.10), and adapt the latter to the present situation on the existence of a Markov process.

Theorem 6. *Let Ω be a Polish space, Σ its Borel σ-algebra and $T \subset \mathbb{R}^+$ be a set. Let $p_t : \Omega \times \Sigma \to \mathbb{R}^+$ be a Markov kernel. If $\Omega_0 = \Omega^T$ and $\Sigma_0 (= \bigotimes_T \Sigma)$ is the cylinder σ-algebra of Ω_0, then for each $x \in \Omega$, there is a unique probability $p(x, \cdot) : \Sigma_0 \to \mathbb{R}^+$ such that for each $A = A_{t_1} \times \cdots \times A_{t_n} \times \Omega^{T - \{t_1, \cdots, t_n\}}$,*

$$p(x, A) = \int_{A_{t_1}} p_{t_1}(x, dy_1) \int_{A_{t_2}} p_{t_2}(y_1, dy_2) \cdots \int_{A_{t_n}} p_{t_n}(y_{n-1}, dy_n), \quad (9)$$

holds for each finite subset $t_1, \cdots, t_n$ of T. If there is an initial probability μ on Σ, then there is a Markov process $\{X_t, t \in T\}$ on $(\Omega_0, \Sigma_0, \nu)$, with stationary transition probabilities $\{p_t(\cdot, \cdot), t \in T\}$, where $X_t(\omega) = \omega(t) \in \Omega_0$, and

$$\nu(A) = \int_{\Omega} p(x, A) d\mu(x), \quad A \in \Sigma_0 . \qquad (10)$$

It may be noted that the operators Q_t defined by

$$(Q_t f)(x) = \int_{\Omega} f(y) p_t(x, dy), \quad t \in T, \qquad (11)$$

for $f \in B(\Omega_0)$, the Banach space (under uniform norm) of bounded Baire functions, forms a semigroup of positive contractions, because of (8), (cf., Prop. 9.5.4). In (9) if $\tau_0 < \tau_1 < \cdots < \tau_n$ from T are considered, then $t_1 = \tau_1 - \tau_0, \cdots, t_n = \tau_n - \tau_{n-1}$ will be taken in order to compare this with the work of Chapter 9.

The analysis of these families is largely accomplished by studying their Laplace transforms, called the resolvents in this context. These are defined for each $\alpha > 0$ and $T = \mathbb{R}^+$ as:

$$(R_\alpha f)(x) = \int_0^\infty e^{-\alpha t} (Q_t f)(x) d\mu(t), \quad f \in B(\Omega_0). \qquad (12)$$

The corresponding concept for kernels is given by

Definition 7. Let $\{p_t(\cdot, \cdot), t > 0)\}$ be a semigroup of subMarkov kernels (cf. Definition 5) satisfying $\lim_{t \downarrow 0} p_t(x, A) = \chi_A(x), \quad A \in \Sigma$. Then the *resolvent kernel* of this semigroup is given for $\alpha > 0$ by:

$$R_\alpha(x, A) = \int_0^\infty e^{-\alpha t} p_t(x, A) d\mu(t), \quad A \in \Sigma, \quad \alpha \in \Omega, \qquad (13)$$

where μ is a σ-finite Borel measure on $\mathbb{R}^+$. If μ is the Lebesgue measure, then $R_\alpha(\cdot, \cdot)$ is termed the α-*potential kernel* and $R_0(\cdot, \cdot)$ is simply the *potential kernel*.

When μ is the Lebesgue measure, $R_\alpha(x, \Omega) = 1$ iff $p_t(x, \Omega) = 1$ for all t. If $e^{-\alpha t} d\mu(t)$ is replaced by a bounded Borel measure $d\mu_1$ then the resulting function R_{μ_1} of (13) may be written as:

$$(R_{\mu_1} f)(x) = \int_0^\infty \int_\Omega f(y) p_t(x, dy) d\mu_1(t), \quad f \in B(\Omega). \qquad (14)$$

If μ_1, μ_2 are two such measures, then we have by Fubini's theorem

$$(R_{\mu_1}(R_{\mu_2} f))(x) = \int_0^\infty \int_\Omega (R_{\mu_2} f)(y) p_t(x, dy) d\mu_1(t)$$

$$= \int_0^\infty \int_0^\infty \int_\Omega \int_\Omega f(z) p_s(y, dz) p_t(x, dy) d\mu_2(s) d\mu_1(t)$$

$$= \int_0^\infty \int_0^\infty \int_\Omega f(z) p_{s+t}(x, dz) d\mu_2(s) d\mu_1(t),$$

by (8), (9) and (14),

$$= \int_0^\infty \int_\Omega f(y) p_\tau(x, dy) d(\mu_1 * \mu_2)(\tau),$$

$\mu_1 * \mu_2$ being the convolution,

$$= (R_{\mu_1 * \mu_2} f)(x) = (R_{\mu_2 * \mu_1} f)(x). \qquad (15)$$

Thus $R_{\mu_1} R_{\mu_2} = R_{\mu_1 * \mu_2} = R_{\mu_2 * \mu_1}$ is a valid operator identity on $B(\Omega)$.

We observe that if $\mu_1 = \delta_t$, the point mass at t, and $f = \chi_A$, then (14) reduces to $R_{\delta_t}(x, A) = p_t(x, A)$ and (13) becomes

$$R_t^{(\alpha)}(x, A) = e^{-\alpha t} p_t(x, A), \tag{16}$$

so that $\{R_t^{(\alpha)}, t > 0\}$ also forms a semigroup of kernels for each $\alpha > 0$. This family is richer than the original subMarkovian class. One defines α-supermedian functions $f : \Omega \to \mathbb{R}^+$, measurable, for $\{p_t, t > 0\}$ (and its resolvent) if f satisfies

$$R_t^{(\alpha)} f \leq f, \ t \geq 0 \quad (t R_{\alpha+t} f \leq f, \ t > 0) \tag{17}$$

and f is α-excessive if there is equality in (17) as $t \downarrow 0$.

Since $p(x, \cdot)$ is a subMarkovian kernel, the α-potential kernel of (13) satisfies $|\alpha R_\alpha(x, A)| \leq 1$ for all $x \in \Omega, A \in \Sigma$ and $\alpha > 0$. Moreover, one has the *resolvent identity*: for all $\alpha, \beta > 0$,

$$R_\alpha(x, A) - R_\beta(x, A) = (\alpha - \beta)(R_\alpha R_\beta)(x, A), \ x \in \Omega, A \in \Sigma . \tag{18}$$

This is verified by the following computation. Let μ_1, μ_2 be the measures defined by $d\mu_1(t) = e^{-\alpha t} dt, \ d\mu_2(t) = e^{-\beta t} dt$. Then

$$(\alpha - \beta)(R_\alpha R_\beta)(x, A) = (\alpha - \beta) R_{\mu_1 * \mu_2}(x, A), \quad \text{by (15)},$$

$$= (\alpha - \beta) \int_0^\infty p_t(x, A) d_t \left(\int_0^t \mu_1(t - \tau) d\mu_2(\tau) \right)$$

$$= (\alpha - \beta) \int_0^\infty p_t(x, A) \left(\frac{e^{-\alpha t} - e^{-\beta t}}{\alpha - \beta} \right) dt,$$

$$\text{by a simple evaluation,}$$

$$= R_\alpha(x, A) - R_\beta(x, A).$$

Abstracting this relation one can define subMarkovian resolvent kernels directly (to gain some generality) as a class $\{R_\alpha(x, A) : \alpha > 0, x \in \Omega, A \in \Sigma\}$ of nonnegative functions such that (i) $(\alpha, x) \mapsto R_\alpha(x, A)$ is $\mathcal{B} \otimes \Sigma$-measurable for each $A \in \Sigma$, (ii) $R_\alpha(x, \cdot)$ is σ-additive for $\alpha > 0, x \in \Omega$, and (iii) the identity (18) holds; and $\alpha R_\alpha(x, \Omega) \leq 1$ for all $\alpha > 0, x \in \Omega$. It is *Markovian* if there is equality in (iii) here.

Since each α-potential kernel satisfies the identity (18), it is natural to find conditions on a resolvent kernel to be an α-potential. This is a nontrivial question. The "best" conditions under which a solution of this problem is available for a wide-class, is called *Ray resolvent kernels* and were isolated by D.B. Ray (1959). His work is extended and perfected by others. We state a general result here to give a clear picture of the subject and then relate it to Markov processes.

Definition 8. Let $C(D)$ be the space of real continuous functions on a compact metrizable sapce D, and $R_\alpha(\cdot, \cdot)$ be a subMarkovian resolvent kernel on the Borelian space $(D, \mathcal{D})$. Then R_α is called a *Ray resolvent* if the associated integral operator $\tilde{R}_\alpha : f \mapsto \int_D f(y) R_\alpha(\cdot, dy)$, $f \in C(D)$, satisfies the conditions:

(i) $\tilde{R}_\alpha(C(D)) \subset C(D)$, and

(ii) $\tilde{R}_{\alpha+1} f \leq f$ for all $\alpha > 0$ and $0 \leq f \in C_1(D)(\subset C(D))$, a countable collection separating points of D.

Here $C_1(D)$ is the subset of $C(D)$, determined by the 1-super median functions of R_α. We now present a characterization of this class.

Theorem 9. *Let $\{R_\alpha, \alpha > 0\}$ be a Ray resolvent family of kernels on $(D, \mathcal{D})$. Then there is a subMarkovian semigroup of kernels $\{p_t, t \geq 0\}$ such that*

(i) *for each $f \in C(D)$, the mapping $(x, t) \mapsto \int_D f(y) p_t(x, dy)$ is $\mathcal{D}$-measurable in x for each $t > 0$, and is right continuous on $[0, \infty)$ with left limits on $(0, \infty)$ in t for each $x \in D$, and*

(ii) *R_α is representable as*

$$R_\alpha(x, A) = \int_0^\infty e^{-\alpha t} p_t(x, A) dt, \quad \alpha > 0, x \in D, \ A \in \mathcal{D} \ . \qquad (19)$$

A proof of this result is based on several auxiliary facts and the reader is referred to Dellacherie and Meyer [(1988), p. 293 ff].

The importance of this representation and its consequences are appreciated if its probabilistic content is exihibited. The latter is related to the existence of strong Markov processes. We first recall this concept.

If $\{X_t, \mathcal{F}_t, t \geq 0\}$ is an adapted process on (Ω, Σ, P), we recall that it is termed a Markov process (cf. Definition 9.4.1) if for any $u > t$

$$P[X_u < x | \mathcal{F}_t] = P[X_t < x | \sigma(X_t)] \quad a.e. \ [P] \ . \qquad (20)$$

Now suppose T is a stopping time of the filtration $\{\mathcal{F}_t, t \geq 0\}$ and let $\mathcal{F}(T)$ be the σ-algebra of events prior to T so that $\mathcal{F}(T) = \sigma(A \cap [T \leq$

$t] \in \mathcal{F}_t : t \geq 0, \ A \in \Sigma$). If $X(t, \omega)(= X_t(\omega))$ is jointly measurable for $\mathcal{B}([0, t]) \otimes \mathcal{F}_t$ and $\mathcal{B}(\mathbb{R})$, then $X_T = X \circ T$ on $[T < \infty]$ is well-defined and measurable for $\mathcal{F}(T)$ (and $\mathcal{B}(\mathbb{R})$) as a composition of measurable functions. The same is true of X_{t+T} for each $t \geq 0$ on the set $[T < \infty]$. Then one can ask if the new process $\{X_{t+T}, t \geq 0\}$ is again a Markov process in the sense that the following analog of (20) holds for all $u > t$,

$$P[X_{u+T} < x | X_{r+T}, r \leq t] = P[X_{u+T} < x | X_{t+T}] \quad a.e., \qquad (21)$$

on $[T < \infty]$. This reduces to (20) if T is a constant, and will also hold if T takes only a finite number of values. But it is not true in general. The process is said to have the *strong Markov property* if (21) holds for all finite stopping times of $\{\mathcal{F}_t, t \geq 0\}$, i.e. $P[T < \infty] = 1$. This says that the process stopped at a random instant T, starts afresh as a Markov process having the same transition probability functions. It can be shown that Brownian notion has the strong Markov property. A natural question is to find a condition, applicable to a wide class of Markov processes, to have this property. The following result furnishes a solution.

Theorem 10. *Let $\{X_t, \mathcal{F}_t, t \geq 0\}$ be a Markov process with stationary transition probability functions $\{p_t(\cdot, \cdot), t > 0\}$ and the resolvent kernels $\{R_\alpha(\cdot, \cdot), \alpha > 0\}$. Suppose the latter satisfy the Ray resolvent condition (Definition 8), and that the function*

$$x \mapsto (P_t f)(x) = \int_{\mathbb{R}} f(y) p_t(x, dy), \quad f \in C_0(\mathbb{R}) ,$$

is right continuous. Then the given process has a modification (with values in $\bar{\mathbb{R}}$) having the strong Markov property.

This useful theorem is an extension of Ray's (1959) original result. We omit its long proof and refer to H. Kunita and T. Watanabe (1967) where it is established and extended if the range (or state) space of the process is a locally compact separable Hausdorff space.

The above two theorems show how a study of resolvent kernels associated with Markov processes with stationary transitions plays an important role. It is an interplay of abstract analysis, potential theory and (regular) conditional measures. We indicate, in the following section, some other applications in which a resolvent equation arises in a different context but has analogous properties involving conditional measures.

10.3 Reynolds operators and conditional expectations

In working with dynamical systems of incompressible fluids, O. Reynolds in 1895 discovered an algebraic operator identity which plays an important role there. Later J. Kampé de Feriet has studied the properties of this operator and G.-C. Rota (1964) obtained its structure in L^2, giving several earlier references on the subject. The operator in question can be described in the present-day terminology as follows.

Definition 1. A linear operator $R : L^p(P) \to L^p(P), p \geq 1$, which maps the set of bounded elements into itself, is called a *Reynolds operator* if it satisfies the identity:

$$R(fg) = (Rf)(Rg) + R[(f - Rf)(g - Rg)], \quad f, g \in L^\infty \cap L^p(P) . \quad (1)$$

Since (1) may be simplified to the form

$$R(fRg) + R(gRf) = (Rf)(Rg) + R[(Rf)(Rg)], \quad (2)$$

it follows that this class of Reynolds operators contains the averaging operators that we considered in Chapter 7 as a subset, since the latter are also projections. More important is the relation of R (or rather $R_\alpha, \alpha > 0$) with resolvent operators discussed in the preceding section. To make this comparison clear, let us present their integral representation to contrast it with Theorem 2.9 above, which in turn will suggest further analysis of this class.

Theorem 2. *Let $R_\alpha : L^p(P) \to L^p(P), 1 \leq p < \infty, \alpha > 0$, be a set of contractive Reynolds operators such that $R_\alpha 1 = 1$ and $R_\alpha^* 1 = 1$, a.e., where R_α^* is the adjoint of R_α and 1 is the constant function one in the $L^p(P)$ space on (Ω, Σ, P). Then there exist (a) a σ-algebra $\mathcal{B}_\alpha \subset \Sigma$, (b) a strongly continuous semigroup $\{V(t), t \geq 0\}$ of linear operators on $L^p(\mathcal{B}_\alpha)$ induced by a measure preserving mapping $\tau : \Omega \to \Omega$ [so $V(t) : f \to f \circ \tau^t, f \in L^p(\mathcal{B}_\alpha)$ and $P \circ \tau^{-1} = P$] in terms of which the following [strong or Bochner] vector integral representation holds:*

$$R_\alpha f = \int_0^\infty e^{-\alpha t} V(t) E^{\mathcal{B}_\alpha}(f)dt , \quad f \in L^p(\Omega, \Sigma, P), \quad (3)$$

where the range of R_α is dense in $L^p(\mathcal{B}_\alpha)$ and moreover

$$R_\alpha f = (R_\alpha E^{\mathcal{B}_\alpha})(f) = E^{\mathcal{B}_\alpha}(R_\alpha f), \; \alpha > 0, \; f \in L^p(\Sigma) \;, \tag{4}$$

so that R_α and $E^{\mathcal{B}_\alpha}$ commute.

This result was established on $L^2(P)$ by Rota (1964) and was extended in Rao [(1970b) and (1975)] to the L^p-spaces which are discussed in Section 7.3. The long proof, which also uses the now-classical Hille-Yosida theory of semigroups and other results, will not be given here. It is available in a book [cf. Rao (1981), pp. 98-103 and p. 262] and the reader can consult it.

It may be noted that R_α is a positive operator, although this is not assumed in Definition 1, but is seen *after* (3) is derived. Letting $f = \chi_A$ and writing $P(\omega, A)$ for $E^{\mathcal{B}_\alpha}(\chi_A)(\omega)$ in (3), we set

$$R_\alpha(\omega, A) = \int_0^\infty e^{-\alpha t} V(t)(P(\omega, A))dt \;, \quad A \in \Sigma \;. \tag{5}$$

Since $P(\omega, A)$ need not be regular in general, let (Ω, Σ, P) be, for instance, a Polish measure space. Then by Theorem 5.3.7, $P(\cdot, \cdot)$ will be a regular conditional measure and qualifies to be a kernel. If we let $p_t(\omega, A) = (V(t)P(\cdot, A))(\omega)$, then $\{p_t, t > 0\}$ becomes a family of transition kernels since $V(\cdot)$ is a nice semigroup. Also $|\alpha R_\alpha(\omega, A)| \leq 1$ [cf. Rao (1981), p. 102, eq. 916] so that it becomes a (subMarkov) resolvent kernel. It is not clear, however, as to its relation with Ray resolvents, and will be of interest to establish this connection to study its possible role in potential theory.

Another property of Reynolds operators is important in this context, which is discussed after introducing the relevant concept.

Definition 3. Let $\{R_\alpha, \alpha > 0\}$ be a commuting family of Reynolds operators on $L^p(\Sigma), 1 \leq p < \infty$. Then $\{R_\alpha f, \alpha > 0\}$ is termed a *generalized martingale* if the following identity holds for all $0 < \alpha < \beta$:

$$(\beta R_\alpha - \alpha R_\beta)R_\beta f = (\beta - \alpha)R_\alpha R_\beta^2 f, \quad f \in L^p \cap L^\infty(P) \;. \tag{6}$$

Since $R_\alpha = E^{\mathcal{B}_\alpha}, \mathcal{B}_\alpha \supset \mathcal{B}_\beta$ (for $\alpha < \beta$) satisfies (6) and $\{E^{\mathcal{B}_\alpha}(f), \alpha > 0\}$ is a (decreasing) martingale in the sense of Section 9.3, the general case $\{R_\alpha f, \alpha > 0\}$ can be (justifiably) called by the above name as in Definition 3. In this connection the following result has some interest.

Theorem 4. *Let* $\{R_\alpha f, \alpha > 0\}$ *be a generalized martingale for* $f \in$ $L^p(\Sigma), 1 < p < \infty$, *where each* R_α *satisfies the hypothesis of Theorem 2. Then* $\lim_{\alpha \to 0} R_\alpha f$ *exists a.e. and in norm. The same holds for* $p = 1$, *if moreover, each* R_α *is weakly compact and* f *is in* $L \log L$ *space on* (Ω, Σ, P).

A proof of this result is again found in the last referenced book and will not be included here. The Reynolds operators are also useful in ergodic theory as well as unifying the latter with martingale theory. They could be of interest in potential theory. We shall not discuss these ideas further, since they are not yet fully developed.

10.4 Bistochastic operators and conditioning

The properties of (regular) conditional measures and kernels find an interesting application to bistochastic operators, leading to certain structural (and later limit) analysis. Let us introduce the relevant terminology.

Definition 1. Let $T : L^p(P) \to L^p(P), p \geq 1$, be a linear operator which is a contraction on both L^1 and L^∞-norms, so that (by the Riesz-Thorin theorem) T is defined and is contractive on all $L^p(P), 1 \leq p \leq \infty$. The class of such T is called the Dunford-Schwartz operator class, denoted $\mathcal{D}$. If moreover T is positive and $T1 = 1$ a.e., then it is called a *bistochastic* (or *doubly stochastic*) operator.

Here we shall present an interesting connection between potential kernels and bistochastic operators, after obtaining the structure of the latter. If P is supported by a finite number of points of Ω so that $L^p(P)$ is isomorphic to $\mathbb{R}^n$, then a bistochastic operator is simply an nth order matrix of nonnegative numbers with row and column sums each adding to unity. Also if $T^2 = T$ then $T \in \mathcal{D}$ and $T1 = 1$ already imply that $T = E^{\mathcal{B}}$ for some σ-algebra $\mathcal{B} \subset \Sigma$, by Proposition 7.3.9. However, T need not be a projection in Definition 1.

We have the following structure and representation:

Theorem 2. *Let* (Ω, Σ, P) *be a complete separable probability space. Then the following are mutually equivalent statements:*

(i) T *is a bistochastic operator on* $L^p(P), p \geq 1$.

(ii) T *admits an integral representation with kernel* $K : \Omega \times \Sigma \to \mathbb{R}^+$

such that

(a) $K(\omega, \cdot)$ *is σ-additive, vanishes on P-null sets, $K(\omega, \Omega) = 1$ for almost all $\omega \in \Omega, [P]$,*

(b) *the mapping $\nu_f : A \mapsto \int_\Omega K(\omega, A)f(\omega)dP(\omega), A \in \Sigma$, is σ-additive, P-continuous, $\nu_f(A) \geq 0$ for $f \geq 0$, and $\frac{d\nu_f}{dP} = 1$ a.e. for $f = 1$ a.e.,*

(c) $(Tf) = \frac{d}{dP}(\int_\Omega K(\omega', \cdot)f(\omega')dP(\omega')), \ f \in L^1(P), \|T\| \leq 1,$ (1)

$$(T^*g)(\omega) = \int_\Omega K(\omega, d\omega')g(\omega'), g \in L^\infty(P), a.a.(\omega), \qquad (2)$$

(iii) T *is an $L^1(P)$-contraction $T1 = 1$ and $T^*1 = 1$ a.e.*

Finally in (1) the integral and derivative can be interchanged iff T is weakly compact (i.e., T maps bounded sets into relatively weakly compact sets in its range).

Proof. (i) $\Rightarrow$ (ii). By definition, a bistochastic operator T is positive and $T1 = 1$ a.e. Also $\|Tf\|_1 \leq \|f\|_1$ and $\|Tf\|_\infty \leq \|f\|_\infty$. This implies, as already noted, $\|Tf\|_p \leq \|f\|_p, 1 \leq p \leq \infty$. For each $A \in \Sigma$ define

$$|\|T_A|\|_1 = \sup\{\sum_{i=1}^n \|T\chi_{A_i}\|_1 : A_i \in \Sigma(A), \text{ disjoint}\},$$

and

$$\|T_A\|_1 = \sup\{\|\sum_{i=1}^n a_i T\chi_{A_i}\|_1 : A_i \in \Sigma(A), \text{ disjoint}, |a_i| \leq 1, a_i \in \mathbb{R}\} .$$

Clearly $\|T_A\|_1 \leq |\|T_A|\|_1$. The opposite inequality is also true, since

$$\sum_{i=1}^n |a_i| \|T\chi_{A_i}\|_1 \leq \|\sum_{i=1}^n |a_i| \|T_{A_i}\| |\chi_{A_i}\|_1$$

$$\leq \|T_A\| \|\sum_{i=1}^n |a_i| \chi_{A_i}\|_1,$$

because the L^1-norm is additive on positive elements

and $\|T_{A_i}\| \leq \|T_A\|$ for $A_i \subset A$,

$$= \|T_A\| \|f\|_1 , \quad \text{with } f = \sum_{i=1}^n a_i \chi_{A_i} .$$

Taking supremum on both sides as $\|f\|_1 \leq 1$ vary one gets $|\|T_A|\|_1 \leq \|T_A\|_1$. Hence $|\|T_A|\|_1 = \|T_A\|_1 \leq 1$. Then by a well-known representation theorem [cf. Dunford-Schwartz (1958), VI.8.6, and especially Dinculeanu (1967), p. 279, Theorem 8], there exists a function

$G : \Omega \to (L^\infty(P))^* \cong ba(\Omega, \Sigma, P)$, the space of bounded additive set functions vanishing on P-null sets, such that

$$\langle Tf, g \rangle = \int_\Omega \langle (G(\omega)f, g \rangle dP(\omega), \quad f \in L^1(P), g \in L^\infty(P), \qquad (3)$$

where $\langle \cdot, \cdot \rangle$ is the duality pairing with $\|G(\omega)\|\chi_A = \||T_A|\|_1$. The norm symbol on G is the total variation of $G(\omega) \in ba(\Omega, \Sigma, P)$. However, (3) can be written explicitly as:

$$\int_\Omega (Tf)g \, dP = \int_\Omega f(\omega) \int_\Omega G(\omega, d\omega')g(\omega')dP(\omega) . \qquad (4)$$

In particular, taking $g = \chi_A, A \in \Sigma$, and the inside integral on the right of (4) is defined relative to $G(\omega, \cdot)$ in the sense of Dunford-Schwartz [(1958), III.2], we get on using $T1 = 1$ with $f = 1$,

$$P(A) = \int_\Omega G(\omega, A)dP(\omega) . \qquad (5)$$

From this equation, we deduce that $G : \Omega \times \Sigma \to \mathbb{R}^+$ satisfies: (i) $G(\omega, \Omega) = 1$, a.a.(ω) and (ii) $G(\omega, \cdot)$ is σ-additive for a.a.(ω). But now using the fact that (Ω, Σ, P) is a complete separable space, we can find a regular version $\tilde{G}(\omega, \cdot)$ to be a measure for all $\omega \in \Omega - N$ with $P(N) = 0$. So we can assume $G(\omega, \cdot) \in ca(\Omega, \Sigma, P) \subset ba(\Omega, \Sigma, P)$ and (a) and (b) of (ii) hold. Since $\frac{d}{dP}(\int_{(\cdot)} Tf \, dP) = Tf$ a.e., (1) is true with $K = G$. Thus (ii) is established.

We next observe, with the above work, that (3) may be written as,

$$\int_\Omega f(T^*g)dP = \int_\Omega f(\omega)(\int_\Omega g(\omega')G(\omega, d\omega')) \, dP, \qquad (6)$$

where T^* is the adjoint of T. Since f is arbitrary in $L^1(P)$, (6) implies

$$(T^*g)(\omega) = \int_\Omega g(\omega')G(\omega, d\omega'), \ a.a.(\omega), \qquad (7)$$

so that T^* is represented by a Markov kernel. Thus (2) is true.

   (ii) $\Rightarrow$ (i) is immediate from the properties of the kernel $G(\cdot, \cdot)$.
   (i) $\Rightarrow$ (iii). This is clear by Definition 1.
   (iii) $\Rightarrow$ (i). Let $T \in \mathcal{D}$ and $T1 = 1$. Then T is positive. Indeed,

$$1 - \int_\Omega T\chi_A dP = \int_\Omega T(1 - \chi_A)dP$$

$$\leq \|T\| \, \|1 - \chi_A\|_1 \leq 1 - \int_\Omega \chi_A dP .$$

Hence

$$P(A) = \int_\Omega \chi_A dP \leq \int_\Omega (T\chi_A) dP \leq \|T\| \int_\Omega \chi_A dP \leq P(A), \quad A \in \Sigma \ .$$

Thus $T\chi_A \geq 0$, and so by linearity $Tf \geq 0$ for all simple f and then for all $0 \leq f \in L^1(P)$. Hence T^* is also positive since by definition

$$\int_\Omega (Tf)g \, dP = \int_\Omega f(T^*g) dP, \quad f \in L^1(P), g \in L^\infty(P).$$

To see that $T^*1 = 1$, let $A_0 = \{\omega : (T^*1)(\omega) < 1\}$. If $P(A_0) > 0$, then

$$1 = \int_\Omega T1 dP = \int_\Omega (T^*1)\chi_{A_0} \, dP + \int_\Omega T\chi_{A_0^c} dP$$
$$< P(A_0) + \|T\chi_{A_0^c}\|_1 \leq P(A_0) + P(A_0^c) = 1.$$

This contradiction shows that $P(A_0) = 0$. Hence $T^*1 = 1$ a.e. so that T is a bistochastic operator.

Finally the last statement is a consequence of the standard results in abstract analysis [cf., e.g. Dunford-Schwartz (1958), VI.8.10], since $L^1(P)$ is separable and $L^\infty(P)$ is its adjoint space. $\square$

Remark. This result is taken from [Rao (1979), Theorem 2] where another characterization of bistochastic operators, with references to related work may be found. A different argument using a Stone space representation of general (Ω, Σ, P) is found in Doob (1963).

We can now present the following result due originally to Rota(1961), and our demonstration is based on the paper by Starr (1966).

Theorem 3. *Let (Ω, Σ, P) be as in the above theorem and $\{T_n, n \geq 1\}$ be a sequence of bistochastic operators on $L^p(P), 1 < p < \infty$. Then*

$$\lim_{n \to \infty} T_{1n}^* T_{1n} f \quad \text{exists a.e. and in} \quad L^1(P), \quad f \in L^p(P), \quad (8)$$

where $T_{1n} = T_n T_{n-1} \cdots T_1$ and $T_{1n}^* = T_1^* T_2^* \cdots T_n^*$.

Proof. Since by (2) there is a kernel $q_k : \Omega \times \Sigma \to \mathbb{R}^+$ such that

$$(T_k^* f)(\omega) = \int_\Omega f(\omega') q_k(\omega, d\omega'), \quad f \in L^p(P) \ ,$$

where q_k can be regarded as a regular conditional measure, we have for the product $T_{1n}^* f$, for any bounded measurable

$$(T_{1n}^* f)(\omega_0) = \int_{\Omega_1} q_1(\omega_0, d\omega_1) \cdots \int_{\Omega_n} q_n(\omega_{n-1}, d\omega_n) f(\omega_n) \ . \quad (9)$$

The right side products define a compatible class since $q_k(\omega, \Omega) = 1$, and hence by Theorem 8.3.2 there is a unique probability measure Q on $(\tilde{\Omega}, \tilde{\Sigma})$ where $\tilde{\Omega} = \overset{\infty}{\underset{i=1}{\times}} \Omega_i = \Omega^{\mathbb{N}}$ (since $\Omega_i \equiv \Omega$) and $\tilde{\Sigma}$ is the cylinder σ-algebra generated by $(\Omega_i, \Sigma_i)(= (\Omega, \Sigma))$. Here P is also the initial measure. Hence for each cylinder $\pi_n^{-1}(A^{(n)})$ where $A^{(n)} = A_1 \times \cdots \times A_n$, $\pi_n : \tilde{\Omega} \to \overset{n}{\underset{i=1}{\times}} \Omega_i$ is the coordinate projection,

$$Q(\pi_n^{-1}(A^{(n)})) = \int_{A_1} dP(\omega_1) \int_{A_2} q_2(\omega_1, d\omega_2) \cdots \int_{A_n} q_n(\omega_{n-1}, d\omega_n).$$
(10)

If $h : \Omega \to \mathbb{R}$ is a bounded measurable (for Σ) function, and X_n is the nth coordinate function so $X_n(\tilde{\omega}) = \omega_n \in \Omega_n, \tilde{\omega} \in \tilde{\Omega}$, then (10) implies

$$\int_{\tilde{\Omega}} h \circ X_n \, dQ = \int_{\Omega} P(d\omega_0) \int_{\Omega_1} dq_1(\omega_0, d\omega_1) \cdots \int_{\Omega_n} h(\omega_n) q_n(\omega_{n-1}, d\omega_n).$$
(11)

Let $\mathcal{B}_0 = X_0^{-1}(\Sigma) \subset \tilde{\Sigma}$. For any bounded Borel function $g : \mathbb{R} \to \mathbb{R}$ and $A \in \Sigma$, so that $X_0^{-1}(A) \in \mathcal{B}_0$, consider with (10) and (11),

$$\int_{\tilde{\Omega}} \chi_{X_0^{-1}(A)} (g \circ X_n) dQ = \int_{X_0^{-1}(A)} g(X_n) dQ$$

$$= \int_{X_0^{-1}(A)} E^{\mathcal{B}_0}(g(X_n)) dQ.$$
(12)

The left side integral can also be written using (9) and (11) as:

$$\int_{X_0^{-1}(A)} g(X_n) dQ = \int_{A \times \Omega^{(1)}} g(X_n) dQ, \quad \Omega^{(n)} = \underset{i \geq n}{\times} \Omega_i,$$

$$= \int_A dP(\omega_0) \int_{\Omega_1} q_1(\omega_0, d\omega_1) \cdots \int_{\Omega_n} g(\omega_n) q_n(\omega_{n-1}, d\omega_1)$$

$$= \int_A (T_{1n}^* g)(\omega_0) dP(\omega_0)$$

$$= \int_{X_0^{-1}(A)} (T_{1n}^* g)(X_0(\omega)) dQ(\omega),$$
(13)

by the image law of probability. Since the right side of (12) and (13) have $\mathcal{B}_0$-measurable integrands and $X_0^{-1}(A) \in \mathcal{B}_0$ is arbitrary, they can be identified a.e. $(Q|\mathcal{B}_0)$. Hence we have the key equation:

$$(T_{1n}^* g)(X_0) = E^{\mathcal{B}_0}(g(X_n)) = E(g(X_n)|X_0), \quad a.e.$$
(14)

We now simplify $T_{1n} f$ in a similar way to relate it to another conditional expectation. We observe now that $Q \circ X_n^{-1} = P$. To see this consider for $A \in \Sigma(= \Sigma_n)$:

$$
\begin{aligned}
Q \circ X_n^{-1}(A) &= \int_{\tilde{\Omega}} \chi_{X_n^{-1}(A)} \, dQ \\
&= \int_{\tilde{\Omega}} E^{B_0}(\chi_{X_n^{-1}(A)}) dQ \\
&= \int_{\tilde{\Omega}} E^{B_0}(\chi_A \circ X_n) dQ \\
&= \int_{\tilde{\Omega}} (T_{1n}^* \chi_A)(X_0) dQ, \quad \text{by (14) with} \quad g = \chi_A, \\
&= \int_{\Omega_0} (T_{1n}^* \chi_A)(\omega_0) dP(\omega_0), \quad \text{by the image law,} \\
&= \int_{\Omega} \chi_A(T_{1n} 1) dP, \quad \text{since} \quad \Omega_n = \Omega, \ n \geq 0, \\
&= \int_{\Omega} \chi_A dP = P(A), \quad \text{since} \quad T_k 1 = 1 \quad \text{a.e.} \quad (15)
\end{aligned}
$$

Next we assert the *all* important property of T_{1n}, namely

$$(T_{1n} f)(X_n) = E(f(X_0)|X_n) = E(f(X_0)|X_k, \ k \geq n) \quad \text{a.e. ,} \quad (16)$$

for each bounded Borel $f : \mathbb{R} \to \mathbb{R}$.

Let us assume (16). Then the proof can be completed quickly as follows. Thus if $\mathcal{F}_n = \sigma(X_k, k \geq n)$, then (16) shows that for each bounded f, since $\mathcal{F}_n \supset \mathcal{F}_{n+1}$ and $E^{\mathcal{F}_{n+1}}(E^{\mathcal{F}_n}(f(X_0))) = E^{\mathcal{F}_{n+1}}(f(X_0))$ for all $n \geq 1$, we get $\{E^{\mathcal{F}_n}(f(X_0)), n \geq 1\}$ to be a decreasing martingale sequence and $f(X_0)$ is bounded. Hence by the corresponding (decreasing) martingale convergence theorem this sequence converges a.e. and in L^p-norm, [cf., e.g. Rao (1981), p. 194]. The martingale is also uniformly integrable. From (14) and (16) together with this (uniform) property we get:

$$
\begin{aligned}
\lim_{n \to \infty} (T_{1n}^* T_{1n} f)(X_0) &= \lim_{n \to \infty} E((T_{1n} f)(X_n)|X_0) \\
&= \lim_{n \to \infty} E^{B_0}(E^{\mathcal{F}_n}(f(X_0))) \\
&= E^{B_0}(\lim_{n \to \infty} E^{\mathcal{F}_n}(f(X_0))), \quad \text{exists a.e.,} \\
&\hspace{8cm} (17)
\end{aligned}
$$

provided $E^{\mathcal{F}_n}(f(X_0))$ is dominated by an integrable function. [This interchange will not hold in general, cf., e.g., Rao (1984), p. 100.] Note that we can let f be any Borel function such that $f(X_0) \in L^\Phi(P)$ with $\Phi(x) = |x| \log^+ |x|$ and in particular if $\Phi(x) = |x|^p, p > 1$. It remains to establish (16), and its proof follows.

Since $(T_{1n} f)(X_n)$ is $\mathcal{B}_n (= X_n^{-1}(\Sigma))$-measurable, we have

$$\int_{X_n^{-1}(A)} (T_{1n} f)(X_n) dQ = \int_A (T_{1n} f) d(Q \circ X_n^{-1}), \quad A \in \Sigma,$$

$$= \int_{\Omega_n} (T_{1n} f) \cdot \chi_A dP, \quad \text{by (15)},$$

$$= \int_{\Omega_n} (T_{1n}^* \chi_A) \cdot f \, dP, \quad (\Omega_n = \Omega),$$

$$= \int_{\tilde{\Omega}} T_{1n}^* (\chi_A \circ X_n) f(X_0) \, dQ,$$

$$\text{by the image law,}$$

$$= \int_{\tilde{\Omega}} E(f(X_0) \chi_A(X_n) | \mathcal{B}_0) \, dQ, \quad \text{by (14)},$$

$$= \int_{\tilde{\Omega}} f(X_0) \chi_A(X_n) \, dQ$$

$$= \int_{X_n^{-1}(A)} f(X_0) dQ$$

$$= \int_{X_n^{-1}(A)} E^{\mathcal{B}_n} (f(X_0)) dQ. \tag{18}$$

Since the extreme integrands are $\mathcal{B}_n$-measurable and $A \in \Sigma$ is arbitrary they can be identified. This is the first half of (16).

For the second half, we observe that (9) defines a stationary transition function by Theorem 9.4.9. Hence $\{X_n, n \geq 0\}$ is a Markov process with stationary transition functions and the initial distribution P. [Since the initial distribution is independent of n, as seen in (15), the X_n-process is also (strictly) stationary.] But the Markov property is symmetric relative to the past and future (cf. Proposition 9.4.2 and Definition 9.4.3), so the right side of (16) follows. $\square$

Using known maximal inequalities for conditional operators and also for (sub) martingales (cf. Theorem 9.2.7), the following result can be derived for $1 < p < \infty$. We use the notation of the above proof.

$$\int_\Omega \sup_{n\geq 0} |T_{1n}^* T_{1n} f|^p dP = \int_{\tilde\Omega} \sup_{n\geq 0} |E^{B_0}(E^{\mathcal{F}_n}(f(X_0)))|^p dQ$$

$$\leq \int_{\tilde\Omega} \sup_{n\geq 0} E^{B_0}(|E^{\mathcal{F}_n}(f(X_0))|^p)dQ,$$

by conditional Jensen's inequality,

$$\leq \int_{\tilde\Omega} E^{B_0}\left(\sup_{n\geq 0}|E^{\mathcal{F}_n}(f(X_0))|^p\right) dQ, f(X_0) \in L^p(Q),$$

$$= \int_{\tilde\Omega} \sup_{n\geq 0}|E^{\mathcal{F}_n}(f(X_0))|^p dQ$$

$$\leq \left(\frac{p}{p-1}\right)^p \int_{\tilde\Omega} |f(X_0)|^p \, dQ,$$

(cf., e.g., Rao (1984), p. 173),

$$= \left(\frac{p}{p-1}\right)^p \int_\Omega |f(\omega)|^p dP, \text{ by the image law.}$$
(19)

Remark. The above proof of the theorem shows that (Ω, Σ, P) can be replaced by $(\Omega_n, \Sigma_n, P_n)$ and $T_k : L^p(P_{k-1}) \to L^p(P_k)$. Only simple notational modifications are needed for this change. This case was treated both in Doob (1963) and Starr (1966).

If in the above work we merely assume $T^*1 \leq 1$, instead of equality, the resulting analysis leads to subMarkovian kernels. Next one can analyze the positive and then the general operators of the Dunford-Schwartz class $\mathcal{D}$. This was indeed considered by Starr, and the following statements hold.

Theorem 4. *Let $T_n \in \mathcal{D}$, and be positive. Then for $f \in L^p(P)$, $1 < p < \infty$ (or only $\int_\Omega |f| \log^+ |f| \, dP < \infty$), we have*

$$\lim_{n\to\infty} T_{1n}^* T_{1n} f \qquad \text{exists a.e.,}$$

where T_{1n} and T_{1n}^* are as in Theorem 3.

The idea of proof here is to associate a bistochastic operator $\tilde{T}_k$ with each T_k, by changing the underlying measure P slightly, on (Ω, Σ) and then use the result of Theorem 3. But the computations need care. The methods show that (Ω, Σ, P) can be σ-finite in all these statements, and the reader can find details in the paper by Starr (1966). If the positivity of T_n is also dropped, one can use the modulus of these operators which are positive and in $\mathcal{D}$, but only has the following:

Theorem 5. *Let $T_n \in \mathcal{D}$, and $f \in L^p(P)$. Then*

$$\int_\Omega \sup_{n \geq 0} |T_{1n}^* T_{1n} f|^p dP \leq \begin{cases} \left(\frac{p}{p-1}\right)^p \int_\Omega |f|^p dP, & p > 1 \\ \frac{e}{e-1} \int_\Omega (1 + |f| \log^+ |f|)dP, & p = 1 \end{cases}.$$

The question of pointwise convergence of $\{T_{1n}^* T_{1n} f, n \geq 1\}$ in this case is still not settled. There are continuous parameter versions of these results, but they will not be discussed here.

The following (slightly weaker) form of Theorem 3 for $\mathcal{D}$ has been established by Starr.

Proposition 6. *Let $T_n \in \mathcal{D}$. Then there is an operator $S \in \mathcal{D}$ such that, $S = S^*$ on $L^2(P)$ and, for each $f \in L^p(P)$, $1 < p < \infty$, we have $\|(T_{1n}^* T_{1n} - S)f\|_p \to 0$ as $n \to \infty$. [For $p = 1$ there is a version with change of norms.]*

We omit the proof of both these results, and consider another aspect of conditioning and its relation to projections which need not be in $\mathcal{D}$, and need not preserve constants.

10.5 Contractive projections and conditional expectations

We show in this section that conditional expectation operators can serve as models for contractive projections in $L^p(P)$-spaces. To make this precise, we need to establish the structure of contractive projections in these spaces. The following result for $p = 1$ is due to R.G. Douglas (1965) and for $1 < p < \infty$ to T. Andô (1966). Later we shall comment on its extensions to more general function spaces.

Theorem 1. *Let T be a contractive projection on $L^p(P), 1 \leq p < \infty$, into itself, with range $\mathcal{M}$. Then $\mathcal{M}$ contains a maximal element f_0 with support S_0 of $\mathcal{M}$ (cf., Definition 7.3.4), and*
(i) if $p \neq 2$,

$$Th = \frac{f_0 E^{\mathcal{B}_0}(hf_0^{p-1})}{E^{\mathcal{B}_0}(|f|^p)} + Vh, \quad h \in L^p(P), \tag{1}$$

where $V = 0$ for $p > 1$, and is a contractive nilpotent operator of order 2 for $p = 1$, satisfying $V(L^1(P)) \subset \mathcal{M}$ and $V(\mathcal{M}) = \{0\}$. Here

$\mathcal{B}_0$ is the σ-subalgebra of the trace $\Sigma(S_0)$ generated by the elements of $\mathcal{M}$, $f^{p-1} = |f|^{p-1} sgn(f)$, and

(ii) if $p = 2$, T is equivalent to the conditional expectation $E^{\mathcal{B}_0}$.

In particular if $T1 = 1$, then $V = 0$ in (1), $f_0 = 1, S_0 = \Omega$ and $T = E^{\mathcal{B}}, 1 \leq p \neq 2 < \infty$. The same conclusion holds for $p = 2$ if T is positive in addition, and in any case we have

(iii) for $1 < p < \infty$, a contractive projection is isometrically equivalent to a conditional expectation operator under a change of norm on $L^p(P)$. [The 'equivalence' of (ii) and (iii) is explained in the proof.]

A detailed proof of this result is given after establishing two auxiliary propositions of some independent interest. The argument below extends to certain Orlicz spaces, and the structure will be illuminated.

Proposition 2. If (Ω, Σ, μ) is a σ-finite measure space and $\mathcal{M}$ is a closed subspace of $L^p(\mu), 1 \leq p < \infty$, then $\mathcal{M}$ has a support S and contains an element f_0 with support S. If $\mathcal{M}$ is moreover a lattice we can choose $f_0 > 0$ a.e. on S.

Proof. Let $\mathcal{S} = \{S_f, f \in \mathcal{M}\}$ where S_f is the support of f. Since (Ω, Σ, μ) is σ-finite, and every σ-finite measure is "localizable" such a collection $\mathcal{S}$ has a supremum, by a classical result [cf., e.g., Rao (1987), p. 70], which has the asserted property. We restate this for a construction of f_0. Thus let $\mathcal{F} = \{\chi_{S_f}, f \in \mathcal{M}\}$. Then the former statement is equivalent to saying that χ_S is an a.e. supremum of $\mathcal{F}$. By σ-finiteness of μ, there exists a sequence $\{f_n, n \geq 1\}$ in $\mathcal{M}$, such that $\lim_{n \to \infty} \chi_{S_{f_n}} = \chi_S$ so that $S = \bigcup_{n=1}^{\infty} S_{f_n}$ a.e., $\mu(S_{f_n}) > 0, n \geq 1$. Now define f_0 as

$$f_0 = \sum_{n=1}^{\infty} (f_n / 2^n \|f_n\|_p) . \qquad (2)$$

Clearly $f_0 \in \mathcal{M}$ since the latter is closed, and $S = S_{f_0}$ a.e.

Finally if $\mathcal{M}$ is also a lattice, then with each element g in $\mathcal{M}$, both g^+ and g^- (hence $|g|$) is in it. So in (2) we may replace f_n by $|f_n|$ and $f_0 > 0$, a.e. on S, as desired. $\square$

Remark. Since the range of a positive contractive projection T is a complete vector lattice, for each f in $\mathcal{M}$ one has

$$\|f\|_p = \| |f| \|_p = \| |Tf| \|_p \leq \|T(|f|)\|_p \leq \|f\|_p ,$$

whence $T(|f|) = |f|$ a.e., and $|f| \in \mathcal{M}$. So there is $f_0 \in \mathcal{M}, f_0 > 0$ a.e. with support S which is that of $\mathcal{M}$ itself.

Taking $\rho(f) = \|f\|_p, 1 \leq p < \infty$ in Theorem 7.3.6, we get the following result after a reduction, since now $\rho(\cdot)$ is also an absolutely continuous norm.

Theorem 3. *Let T be a contractive projection on $L^p(P), 1 \leq p < \infty, p \neq 2$, and T be also positive in case $p = 2$. Then $T1 = 1$ a.e. implies that $T = E^{\mathcal{B}}$, for a σ-subalgebra $\mathcal{B}$ of (Ω, Σ, P).*

Proof. As seen in the proof of Theorem 4.2 (iii) above, if $T1 = 1$ and $p = 1$, then the contractive T is also positive. Thus the result follows in cases $p = 1$ and $p = 2$ from Theorem 7.3.6. So let $1 < p < 2$. We assert that the contractive T on $L^p(P)$ has an extension to $L^1(P)$ to be a contractive projection there so that the result holds again.

Indeed let $\varphi(x) = \frac{1}{p}|x|^p, \psi(x) = \frac{1}{q}|x|^q, q = p/(p-1)$. Then $\varphi'(x)$ is increasing and concave, but $\psi'(x)$ is convex and increasing. Let $f \in \mathcal{M} = T(L^p(P)), \|f\|_p = 1$, so that $\varphi'(f) = |f|^{p-1} \operatorname{sgn}(f)$, and hence $\varphi'(f) \in L^q(P)$. We now have

$$1 = \int_\Omega f\varphi'(f)dP = \int_\Omega (Tf)\varphi'(f)dP, \text{ since } f \in \mathcal{M},$$
$$= \int_\Omega fT^*(\varphi'(f))dP$$
$$\leq \|f\|_p \ \|\varphi'(f)\|_q = 1, \quad \text{by Hölder's inequality.} \tag{3}$$

Here we used the contractivity of T^* also. Since there is equality in Hölder's inequality, using the equality conditions in the Young (hence Hölder) inequality, we must have either $f = \psi'(g)$ or $g = \varphi'(f)$ where $g = T^*(\varphi'(f))$. [See, e.g., Rao and Ren (1991), p. 80.] Thus $T^*(\varphi'(f)) = \varphi'(f)$ a.e. holds. The key idea now is to show that $\varphi'(f) \in \mathcal{M}$ also. For this, given $0 < \varepsilon \leq 1$, consider

$$h_\varepsilon = \frac{1}{\varepsilon}[\psi'(1 + \varepsilon\varphi'(f)) - 1]. \tag{4}$$

Since $\psi'(1) = 1$ and $\psi''(1) > 0$ we obtain from (4) that for each $\omega, \lim_{\varepsilon \to 0} h_\varepsilon(\omega) = \psi''(1)\varphi'(f)(\omega)$. But from the properties of ψ', noted above it follows that $h_\varepsilon(\omega)$ is increasing as $\varepsilon \downarrow 0$, [cf., e.g., Dunford-Schwartz (1958), V.9.1]. Thus $h_1 g \in L^1(P)$ for any $g \in L^q(P)$, and this dominates $h_\varepsilon g$. Hence by the Lebesgue dominated convergence theorem we can interchange the limit to get $\lim_{\varepsilon \downarrow 0} h_\varepsilon \in L^p(P)$ since $h_\varepsilon \in L^p(P)$ for each $\varepsilon > 0$. It follows that $\varphi'(f) \in L^p(P)$ as well as in $T^*(L^q(P))$.

We now iterate the procedure by writing $\varphi''(f) = \varphi'(\varphi'(f))$ and $\varphi^{(n)}(f) = \varphi'(\varphi^{(n-1)}(f))$ to conclude that $\varphi^{(n)}(f) \in L^p(P) \cap T^*(L^q(P))$, for $n \geq 1$. But for each ω for which $f(\omega) > 1$, we get $\varphi^{(n)}(f) \downarrow 1$ a.e., and if $0 < f(\omega) < 1$, then $\varphi^{(n)}(f) \uparrow 1$, a.e. Considering negative values of $f(\omega)$, we finally obtain that $\varphi^{(n)}(f) \to \operatorname{sgn}(f)$ a.e. (since $0 < p-1 < 1$), and dominatedly. Hence if $f \in L^p(P)$, and $h = Tf \in \mathcal{M}$, then

$$\int_\Omega |h| dP = \int_\Omega h \operatorname{sgn}(h) \, dP$$
$$= \lim_{n \to \infty} \int_\Omega (Tf) \varphi^{(n)}(h) dP$$
$$= \lim_{n \to \infty} \int_\Omega f \, T^*(\varphi^{(n)}(h)) dP$$
$$= \lim_{n \to \infty} \int_\Omega f \varphi^{(n)}(h) dP, \quad \text{by the above simplification,}$$
$$= \int_\Omega f \operatorname{sgn}(h) dP$$
$$\leq \int_\Omega |f| dP. \tag{5}$$

This implies that T is also a contraction in $L^1(P)$-norm. Since $T1 = 1$ and $1 \in L^1(P)$ we conclude that T is a positive contraction, and so $T = E^B$.

Thus we have now shown that T is a positive contraction on $L^p(P)$ for $p = 1$ and $1 < p < 2$. It follows by a result of Ackoglu and Chacón (1965) that it is in $\mathcal{D}$ of Definition 4.1, so that $T = E^B$ holds for $1 \leq p \leq \infty$. However, we give a simple independent proof that T is contractive on $L^p(P), 2 < p < \infty$ using a duality argument, since T^* now acts on $L^q(P), 1 < q < 2$ and the analysis of the last section applies. Now by (3), taking $f = 1$ (so $\varphi'(1) = 1$) we conclude that $T^*1 = 1$ a.e. Hence by (5), T^* is a positive contractive projection on $L^1(P)$ also. This implies T has the same property, since

$$\int_\Omega |Tf| dP = \int_\Omega (Tf) \operatorname{sgn}(Tf) dP$$
$$= \int_\Omega f \, T^*(\operatorname{sgn}(Tf)) dP$$
$$\leq \int_\Omega |f| \, T^*(1) dP = \int_\Omega |f| dP.$$

Thus $T = E^B$ again. Since the additional hypothesis of positivity for $p = 2$ implies the same result, the statement holds for T in all $L^p(P), 1 \leq p < \infty$. $\square$

We can now turn to the proof of the original result:

Proof of Theorem 1. (i) Let $1 \leq p \neq 2 < \infty$, and $\mathcal{M} = T(L^p(P))$. Let S be the support of $\mathcal{M}$ and f_0 in $\mathcal{M}$ be a function supported by S, as assured by Proposition 2. Consider the (finite) measure $\mu(= \mu_{f_0})$ defined by $d\mu = |f_0|^p dP$ and let $\Sigma(S)$ be the trace of Σ on S. Then $L^p(\mu)$ is the subspace of $L^p(P)$ whose elements vanish off S, and are $\Sigma(S)$-measurable. Let $\tilde{T} : L^p(\mu) \to L^p(\mu)$ be defined by

$$\tilde{T}(h) = \frac{1}{f_0} T(f_0 h), \quad h \in L^p(\mu) . \tag{6}$$

Then $\tilde{T}$ is a contractive projection, as it is the restriction of T to $L^p(\mu)$. Indeed,

$$\tilde{T}^2(h) = \frac{1}{f_0} T(f_0 \tilde{T}(h))$$

$$= \frac{1}{f_0} T(T(f_0 h)) = \frac{1}{f_0} T(f_0 h) = \tilde{T}(h), \quad (T^2 = T \text{ is used})$$

and

$$\int_S |\tilde{T}h|^p d\mu = \int_S |\frac{1}{f_0} T(f_0 h)|^p |f_0|^p dP$$

$$= \int_S |T(f_0 h)|^p dP$$

$$\leq \int_S (f_0 h)^p dP = \int_S |h|^p d\mu .$$

Also $\tilde{T}1 = 1$ a.e. since $Tf_0 = f_0 \in \mathcal{M}$. Hence there is $\mathcal{B}_0 \subset \Sigma(S)$, a σ-algebra such that $\tilde{T} = E_\mu(\cdot|\mathcal{B}_0)$, by Theorem 3. Thus for any $A \in \mathcal{B}_0$,

$$\int_A (\tilde{T}h)|f_0|^p dP = \int_A \tilde{T} h \, d\mu = \int_A h d\mu$$

$$= \int_A h|f_0|^p \, dP, \ h \in L^p(\mu). \tag{7}$$

Then by Theorem 7.2.6, applied to $(S, \Sigma(S), P)$, we have from (7),

$$\int_A E^{\mathcal{B}_0}(\tilde{T}h \cdot |f_0|^p) \, dP = \int_A E^{\mathcal{B}_0}(h|f_0|^p) \, dP, \quad A \in \mathcal{B}_0 .$$

Since the integrands (as well as $\tilde{T}h$) are $\mathcal{B}_0$-measurable, this implies

$$\tilde{T}h \cdot E^{\mathcal{B}_0}(|f_0|^p) = E^{\mathcal{B}_0}(h \cdot |f_0|^p), \quad a.e. \tag{8}$$

But f_0 vanishes on S^c, so (6) and (8) give for all $h \in L^p(P)$

$$T(h) = \frac{f_0 E^{\mathcal{B}_0}(h|f_0|^{p-1})}{E^{\mathcal{B}_0}(|f_0|^p)} + T(\chi_{S^c} h). \tag{9}$$

Let $V = T \circ \chi_{S^c}$. Then V is a contraction and $V(\mathcal{M}) \subset \{0\}$. Also $V(L^1(P)) = T(\chi_{S^c} L^1(P)) \subset \mathcal{M}$. So $V^2 h = V(T\chi_{S^c} h) \in V(\mathcal{M}) = \{0\}$, so that it is nilpotent of order 2. Thus (9) gives the representation of (1).

If $p \neq 1$, then setting $g = \chi_{S^c} h$, for any $\varepsilon > 0$, one has

$$(1+\varepsilon)^p \int_\Omega |Tg|^p dP = \int_\Omega |(1+\varepsilon)(Tg)|^p dP$$

$$= \int_\Omega |T(Tg + \varepsilon g)|^p dP$$

$$\leq \int_\Omega |Tg + \varepsilon g|^p dP = \int_\Omega |Tg|^p dP + \varepsilon^p \int_\Omega |g|^p dP,$$

since Tg and g have disjoint supports.

Hence

$$\frac{1}{\varepsilon^{p-1}} \frac{(1+\varepsilon)^p - 1}{\varepsilon} \int_\Omega |Tg|^p dP \leq \int_\Omega |g|^p dP = \int_{S^c} |h|^p dP. \tag{10}$$

Letting $\varepsilon \searrow 0$, the left side is unbounded since $p > 1$, and (10) can hold only if $Tg = (T\chi_{S^c} h) = Vh = 0$ a.e. for any h. Thus (1) holds as stated.

In the converse direction let T be given by (1). Then the stated property and the fact that $E^{\mathcal{B}_0}$ is a conditional expectation imply that $T^2 h = Th$, $h \in L^p(P)$. We only need to verify its contractivity. To see this consider for $p = 1$

$$\int_\Omega |Th| dP = \int_\Omega \left| \frac{f_0 E^{\mathcal{B}_0}(h)}{E^{\mathcal{B}_0}(|f_0|)} + Vh \right| dP$$

$$\leq \int_S E^{\mathcal{B}_0} \left(\frac{|f_0|\, |E^{\mathcal{B}_0}(h)|}{E^{\mathcal{B}_0}(|f_0|)} \right) dP + \int_{S^c} |Vh| dP$$

$$\leq \int_S E^{\mathcal{B}_0}(|h|) dP + \int_{S^c} |Vh| dP$$

$$\leq \int_S |h| dP + \int_{S^c} |h| dP = \int_\Omega |h| dP.$$

If $1 < p \neq 2 < \infty$, then by the conditional Hölder inequality (cf. Theorem 2.2.4),

$$|E^{\mathcal{B}_0}(h\, f_0^{p-1})| \leq (E^{\mathcal{B}_0}(|h|^p))^{\frac{1}{p}}(E^{\mathcal{B}_0}(|f_0|^{(p-1)q}))^{\frac{1}{q}}$$

$$= (E^{\mathcal{B}_0}(|h|^p))^{\frac{1}{p}}(E^{\mathcal{B}_0}(|f_0|^p))^{\frac{1}{q}} \ , \quad q = \frac{p}{p-1} \ . \tag{11}$$

Since $V = 0$, we get from (1) and (11)

$$\int_\Omega |Tf|^p \, dP \leq \int_\Omega \frac{|f_0|^p |E^{\mathcal{B}_0}(h f_0^{p-1})|^p}{(E^{\mathcal{B}_0}(|f_0|^p))^p} \, dP$$

$$= \int_\Omega \frac{E^{\mathcal{B}_0}(|f_0|^p) E^{\mathcal{B}_0}(|E^{\mathcal{B}_0}(h f_0^{p-1})|^p)}{(E^{\mathcal{B}_0}(|f_0|^p))^p} \, dP$$

$$\leq \int_\Omega \frac{E^{\mathcal{B}_0}(E^{\mathcal{B}_0}(|h|^p))(E^{\mathcal{B}_0}(|f_0|^p))^{p-1})}{(E^{\mathcal{B}_0}(|f_0|^p))^{p-1}} \, dP, \text{ by (11)},$$

$$= \int_\Omega E^{\mathcal{B}_0}(|h|^p) \, dP = \int_\Omega |h|^p \, dP \ .$$

Hence T is a contractive projection.

If $p = 2$, $L^2(P)$ being a Hilbert space each subspace is again a Hilbert space, and it is well-known that such a space is isomorphic to an $L^2(\tilde{\Omega}, \tilde{\Sigma}, \tilde{P})$, [cf., e.g., Rao (1979), p. 414]. Hence (ii) follows and the particular case is included in Theorem 3.

Finally for (iii) let $1 < p \neq 2 < \infty$, since $p = 2$ is already noted above. Let $\tilde{P}(\cdot) = \mu(S \cap \cdot) + P(S^c \cap \cdot)$ where $d\mu = |f_0|^p dP$. Then $\tilde{P}$ is a finite measure and if $U : h \mapsto h(f_0 + \chi_{S^c})$, the operator $U : L^p(\tilde{P}) \to L^p(P)$ is a linear isometry and $T = U E^{\mathcal{B}_0} U^{-1}$ is the desired isometric equivalence. Thus all the assertions hold. $\square$

Remark. If T is a positive contractive projection in $L^p(P)$, then f_0 in (1) can be taken positive (and V is a positive contraction for $p = 1$). Several extensions and ramifications of the above theorem are possible. Note that if $f_0 > 0$ a.e. on Ω is given then evidently $S = \Omega$, and only then. In this case $V = 0$. In the earlier work of Chapter 8, we have specified $f_0 = 1$ and $T1 = 1$. The effect of dropping the latter assumption is thus clarified by Theorem 1 here.

The following equivalence, of the preceding results, is a geometric description and its details are omitted.

Theorem 4. *A closed subspace of an $L^p(P)$-space, $1 \leq p < \infty$, is the range of a contractive projection iff it is isometrically equivalent to an L^p-space on some measure space.*

This characterization is natural for linear prediction problems in the L^p-spaces. Indeed, if $\mathcal{M} \subset L^p(\mu)$ is a closed subspace $f \in L^p(\mu)$, then an element $g_f \in \mathcal{M}$ is called a *best predictor* of f relative to the L^p-norm, if

$$\|f - g_f\|_p = \inf\{\|f - h\|_p : h \in \mathcal{M}\}.$$

A set $\mathcal{M}$ is termed a *Čebyšev subspace* if there exists a unique g_f for each f. The mapping $T : f \mapsto g_f$ is thus well-defined, but is generally not linear when $p \neq 2$. The subspaces for which T is linear are clearly of interest for (linear) prediction theory. It can be seen, as a consequence of the above theorem for $1 < p < \infty$, that T is linear iff $L^p(P) = \mathcal{M} \oplus \mathcal{N}$, the direct sum, where $\mathcal{N}$ is the range of a contractive projection or equivalently it is isometrically equivalent to an $L^p(\mu)$-space. This problem was considered if the L^p-spaces are replaced by certain (subclass of) Orlicz spaces in Rao (1974), and the function spaces L^ρ-considered in Chapter 7 are also relevant in this study but the necessary computations are more involved. Extending the methods of R.G. Douglas (1965), a characterization analogous to Theorem 1 for a class of L^ρ-type spaces was given in V.G. Kulakova (1983) which the reader may consult.

10.6 Bibliographical notes

The assumption of regularity of conditional measures allows one to use Lebesgue's theory of integration directly and hence to use it in several areas of standard analysis. As seen in the preceding chapter, these specialized (conditional) measures are related to transition probability functions of Markov processes. A slight extension of this concept leads to subMarkov kernels, and the work of the present chapter may be regarded as an introduction for several areas of modern analysis. An important application is to potential theory and we illustrated an aspect of it is Sections 1 and 2.

The equivalence of the classical (Newtonian) potential theory in $\mathbb{R}^n, n \geq 3$, with that obtainable from the Brownian motion transition function was first established by S. Kakutani (1944) and our exposition follows Knapp (1950). But the final generalization reaching its rapid growth with Markov processes (culminating in the middle 1950s) is achieved in the work of G.A. Hunt (1966, and references). We have discussed a few of these ideas in Section 2, and a thorough treatment of the subject with related references can be found in the treatise by Dellacherie and Meyer (1988). It is interesting that Ray resolvents play a key role in this study. But the analogous Reynolds operators appearing in turbulence theory much earlier and worked on by J. Kampé de Fériet, and analyzed by G.-C. Rota (1964), later by the author (1975), have not yet been studied from the potential theoretic point of view although both have very similar structure. It will be useful to investigate this aspect and we have indicated a possibility in Section 3.

If the regular conditional measure $p(\omega, \cdot)$ is absolutely continuous relative to a σ-finite measure and if $f(\omega, \omega') = \frac{dp(\omega, \cdot)}{d\mu}(\omega')$ then $f \geq 0$ and is a potential kernel, as in the Newtonian case. If f is replaced by a positive definite kernel (but not necessarily positive), then this leads to another aspect analyzed in depth by N. Aronszajn (1950), associating a reproducing kernel Hilbert space with it. This plays again an important role in the covariance analysis of second order stochastic processes, differential equations and elsewhere. If the positive definiteness requirement is dropped but instead the square integrability of $f(\cdot, \cdot)$ on $(\Omega \times \Omega, \Sigma \otimes \Sigma, \mu \otimes \mu)$ is demanded, then it leads to integral equations and the Hilbert-Schmidt theory. These latter directions have also been extensively developed, but are somewhat distant from conditional measures. We therefore do not include any of their accounts or references.

However, retaining positivity of the kernels has other aspects, applicable to the Dunford-Schwartz and bistochastic operators. This part is briefly discussed in Section 4, and the structural analysis follows the paper by Rao (1979) where other characterizations and references to the literature by other authors may be found. The limit theory of Rota and Doob, has been extended in the paper of Starr (1966). It has deep connections with martingale and ergodic theories.

The pointwise convergence results of Theorems 4.3 and 4.4 are not valid if L^p is replaced by L^1 there, even if $T_n = E^{\mathcal{B}_n}$, and $T_{1n} = (T_1 T_2)^n$. See Ornstein (1968) on this point with counterexamples. One can also consider the corresponding results when the operators are indexed by directed sets, and to asymptotic martingales (or "amarts") and a detailed exposition of the latter is available in the recent book by G.A. Edgar and L. Sucheston (1992). Finally, in the last section, we have analyzed the structure of contractive projection operators in $L^p(P)$-spaces, and showed that they are isomorphic to conditional expectations under changes of measures.

Thus conditional operators serve as "models" for projections on such function spaces. In addition to functional characterizations given in Chapter 7, numerous extensions of these ideas are available, beginning with the works of R.G. Douglas (1965), T. Andô (1966), and applications to (linear) prediction theory [cf. Andô-Amemiya (1965), Rao (1965, 1974, 1975), V.G. Kulakova (1983), and others].

A few different types of applications of conditioning will be given in the next (and final) chapter to show how the Kolmogorov definition of this concept, contrasted with other axiomatic attempts, proves to be so important. However, the computational questions of all these conditional expectations, raised in most of the earlier chapters, largely remain for future investigations.

CHAPTER 11: CONDITIONING IN GENERAL STRUCTURES

In this final chapter we indicate applications of conditioning in certain algebras of functions and operators as well as in cones of (real) function spaces. Some characterizations of averaging projections on such spaces, noncommutative conditioning and related martingale convergence together with an application to sufficiency (in the noncommutative case) are discussed. Here conditional expectations rather than measures often take a central place. The treatment is intended to show the far reaching impact of Kolmogorov's extension of conditioning in different areas of mathematics.

11.1 Introduction

Although the concept of conditioning, as defined by Kolmogorov, is based on an underlying measure space leading to positive (linear) operators, especially projections, on function spaces, one can also start with the latter aspect (of conditional expectations) and study the consequences in structures where measure does not appear *apriori*. Sometimes the full vector space property is not necessary. If there is a positive cone in the space [using only the available partial order], one can introduce conditional expectations and still extend (or utilize) the work of the preceding chapters. We discuss both these aspects here and show their use in applications based on such structures. This will illuminate the fundamental role played by conditioning in different branches of mathematical analysis.

We first consider conditioning in cones of function lattices and present a characterization essentially based on the work of Moy (1954), (cf. also Rao (1975)). After that, in the next section, we study the relations between averaging and projection operators on continuous function spaces in which measures do not appear in a formulation of the results. This leads to conditioning on algebras of operators. We study that problem in Section 4, where we indicate the ideas started by H. Umegaki (1954). We include two types of martingale convergence theorems in Section 5. The operator conditioning leads us to extend the concept of sufficiency; and a formulation of some of the results of Chapter 6 is also given there.

In considering the relations between conditional expectations and averaging operators on function spaces based on a measure space (Ω, Σ, P) in Chapter 7, the following properties have played a key role. Suppose we replace the $L^p(\Sigma)$-space as well as the subspace $\mathcal{M}$ in Theorem 7.3.5 by the sets $\mathcal{N}$ and $\mathcal{C}$ defined as: $\mathcal{N}$ is the set of all nonnegative extended real valued measurable functions on (Ω, Σ, P), and $\mathcal{C} \subset \mathcal{N}$ is a set satisfying the conditions

1. $f_1, f_2 \in \mathcal{C} \Rightarrow a_1 f_1 + a_2 f_2 \in \mathcal{C}$ for all $a_1, a_2 \geq 0$, and if $f_1 \leq f_2$ then $f_2 - f_1 \in \mathcal{C}$,

2. $1 \in \mathcal{C}$ [or there is a g.w.u. $F_0 \subset \mathcal{C}$]

3. $f_1, f_2 \in \mathcal{C} \Rightarrow f_1 f_2 \in \mathcal{C}$, $[3'. f_1 \vee f_2 \in \mathcal{C}]$, and

4. $f_n \in \mathcal{C}, f_n \leq g$ a.e., g bounded, $f_n \to f$ a.e. $\Rightarrow f \in \mathcal{C}$, $[4'. f_n \uparrow f$ a.e. $\Rightarrow f \in \mathcal{C}]$.

If $\mathcal{S} \subset \mathcal{C}$ is a set of all bounded functions satisfying 1.–4. [or 1., 2., 3.' and 4.'] above, then the argument of the proof of Theorem 7.3.5 can be adapted directly to show that the σ-algebra $\mathcal{B}$ generated by $\mathcal{S}$ is the same as that generated by all the bounded elements of $\mathcal{C}$. Moreover $\mathcal{B}$ coincides with the σ-algebra generated by $\mathcal{C}$ itself iff each element of $\mathcal{C}$ is a pointwise limit of some sequence from $\mathcal{S}$ (such as that in 4.' above). Thus one has the following result.

Proposition 1. *Let $\mathcal{C}$ and $\mathcal{N}$ be sets of nonnegative functions on (Ω, Σ, P) as defined above. Then $\mathcal{B} = \sigma(\mathcal{S}) = $ the σ-algebra generated by all the bounded elements of $\mathcal{C}$. If $\mathcal{M}$ is the class of all nonnegative $\mathcal{B}$-measurable functions on Ω, then $\mathcal{M} \subset \mathcal{C}$, and equality holds iff each* *f in $\mathcal{M}$ is a pointwise limit of some bounded sequence from $\mathcal{C}$.*

This measure theoretical description of subsets of $\mathcal{N}$ will be useful in characterizing averaging operators that are additive on the cone $\mathcal{N}$. Note that $\mathcal{N}$ is partially ordered (pointwise) but has no topology, and hence one can demand that the operators on it be order continuous.

11.2 Averagings in cones of positive functions

The following characterization is valid even if the measure space is nonfinite as in Section 7.3. For simplicity we take it as a probability space.

Theorem 1. *Let $\mathcal{N}$ be the set of nonnegative extended real valued measurable functions on (Ω, Σ, P) and $T : \mathcal{N} \to \mathcal{N}$ be a mapping such that*

(i) $T(a_1 f_1 + a_2 f_2) = a_1(Tf_1) + a_2(Tf_2)$, $a_i \geq 0$, $f_i \in \mathcal{N}$, $i = 1, 2$,

(ii) $T(\mathcal{N} \cap L^\infty(\Omega, \Sigma, P)) \subset \mathcal{N} \cap L^\infty(\Omega, \Sigma, P)$,

(iii) $T(fTg) = (Tf)(Tg)$, $f, g \in \mathcal{N}$,

(iv) $T\, 1 = 1$, and

(v) $f_n \uparrow f$, $f_n \in \mathcal{N} \Rightarrow Tf_n \uparrow Tf$.

Then $T^2 = T$, and there is a P-unique $h : \Omega \to \bar{\mathbb{R}}^+$, measurable for Σ, and a σ-algebra $\mathcal{B} \subset \Sigma$ such that

$$Tf = E^{\mathcal{B}}(fh), \quad f \in \mathcal{N}, \quad \text{with} \quad E^{\mathcal{B}}(h) = 1. \tag{1}$$

[All the above relations hold a.e. (P).]

Proof. First we produce a subset $\mathcal{C}$ of $\mathcal{N}$, to satisfy Conditions 1.–4. of the preceding section, based on the properties of T. Consider

$$\mathcal{C} = \{f \in \mathcal{N} \ : \ T(fg) = f\, T(g), \quad \text{for all} \ g \in \mathcal{N}\} \ .$$

Clearly $1 \in \mathcal{C}$, and by *(iii)* $T(\mathcal{N}) \subset \mathcal{C}$. If $f_i \in \mathcal{C}$ then

$$T(f_1 f_2 g) = f_1 T(f_2 g) = f_1 f_2 T(g) \ , \quad g \in \mathcal{N} \ ,$$

since $f_1 f_2 g \in \mathcal{N}$. So $f_1 f_2 \in \mathcal{C}$ and similarly $a_1 f_1 + a_2 f_2 \in \mathcal{C}$ for any $a_i \geq 0, i = 1, 2$. Thus conditions 2. and 3. hold for $\mathcal{C}$. For 1., let

$f_i \in \mathcal{C}$, $i = 1, 2$, $f_1 \leq f_2$, and $f_2 < \infty$ a.e. Then $f_2 - f_1 \in \mathcal{N}$ so that for $g \in \mathcal{N}$, bounded,

$$f_2 T(g) = T(g(f_2 - f_1) + g f_1) = T(g(f_2 - f_1)) + f_1 T(g) \ .$$

Since $T(g)$ is bounded by hypothesis, and $f_1 < \infty$, one has

$$(f_2 - f_1) T(g) = T(g(f_2 - f_1)) \ , \tag{2}$$

so that $f_2 - f_1 \in \mathcal{C}$. If g is not bounded, let $g_n \uparrow g$, g_n simple, and by (2) and (v) of the hypothesis, the relation holds so that 1.-3. are verified. Finally let $f_n \in \mathcal{C}$, $f_n \to f$ a.e. and $f_n \leq g$, bounded. Then $f_n g \to f g$ a.e., and let $h_n = \inf\{f_i, i \geq n\}$. We get $h_n \uparrow f$ a.e., $h_n \leq f_n \leq g$ so that

$$T(f_n) = T(f_n - h_n) + T(h_n) \geq T(h_n)$$

and by (v) this yields

$$T(f) = \lim_n T(h_n) \leq \liminf_n T(f_n) \leq \limsup_n T(f_n). \tag{3}$$

Since $g - f \geq 0$, this and property 1. already verified imply

$$0 \leq T(g) - T(f) \leq \liminf_n T(g - f_n) \ . \tag{4}$$

But g is bounded $f, f_n \leq g$ so that

$$T(g) = T(g - f) + T(f) \Rightarrow T(g) - T(f) = T(g - f),$$

and (4) simplifies to

$$T(g) - T(f) \leq \liminf_n [T(g) - T(f_n)] = -\limsup_n [T(f_n) - T(g)]$$
$$\leq T(g) - \limsup_n T(f_n) \ .$$

Hence

$$\limsup_n T(f_n) \leq T(f) \leq \limsup_n T(f_n), \quad \text{by (3).} \tag{5}$$

Thus there is equality, and for $h \in \mathcal{N}$, $f_n h \to f h \leq f g$ a.e. so that

$$T(fh) = \lim_n T(f_n h) = \lim_n f_n T(h) = f T(h), \quad \text{by (5).}$$

From this we conclude that $\mathcal{C}$ satisfies 1.–4. of the last section.

Now let $\mathcal{B}$ be the σ-algebra generated by the bounded elements of $\mathcal{C}$. Thus by Proposition 1.1, the set of all $\bar{\mathbb{R}}^+$ valued $\mathcal{B}$-measurable functions belong to $\mathcal{C}$. Hence for any bounded, and then for all, h in $\mathcal{N}$, $T(h)(\in \mathcal{C})$ is $\mathcal{B}$-measurable. Now define $\nu : A \mapsto \int_\Omega T(\chi_A)dP$, $A \in \Sigma$. Then $\nu(\cdot)$ is additive and P-continuous. If $A_n \in \Sigma$, disjoint, we get $T(\chi_{\underset{i=1}{\overset{n}{\cup}} A_i}) \uparrow T(\chi_{\underset{i=1}{\overset{\infty}{\cup}} A_i})$ by hypothesis and so

$$\nu(\bigcup_n A_n) = \int_\Omega \lim_n T(\chi_{\underset{i=1}{\overset{n}{\cup}} A_i})dP$$

$$= \lim_n \sum_{i=1}^n \nu(A_i) \ .$$

Hence ν is σ-additive, and by the Radon-Nikodým theorem there is a P-unique $h \geq 0$, such that for each $A \in \Sigma$

$$\int_\Omega T(\chi_A)dP = \nu(A) = \int_A h \ dP = \int_\Omega h\chi_A dP \ .$$

By linearity, this can be extended first for simple and then for all f in $\mathcal{N}$:

$$\int_\Omega T(f) \ dP = \int_\Omega fh \ dP, \quad f \in \mathcal{N} \ . \tag{6}$$

If $B \in \mathcal{B}$, then $\chi_B \in \mathcal{C}$ and $T(\chi_B f) = \chi_B T(f)$, $f \in \mathcal{N}$ so that (6) gives

$$\int_B (Tf)dP = \int_B fh \ dP = \int_B E^{\mathcal{B}}(fh)dP_{\mathcal{B}}, \quad f \in \mathcal{N} \ . \tag{7}$$

Since $Tf(\in \mathcal{C})$ is $\mathcal{B}$-measurable, this implies $T(f) = E^{\mathcal{B}}(fh)$, a.e. and $1 = T1 = E^{\mathcal{B}}(h)$. $\square$

Remark. The last part is also obtainable from Theorem 7.3.7 by extending T to $L^\infty(\Sigma)$, by setting $\tilde{T}(f) = T(f^+) - T(f^-)$ where $f = f^+ - f^-, f \in L^\infty(\Sigma)$. The above independent computation is simpler. The hypothesis on T is not strong enough to allow us to conclude that $h = 1$ a.e. The following is a sufficient condition (as in Proposition 7.3.9).

Corollary 2. *Let* $T : \mathcal{N} \to \mathcal{N}$ *be as in the above theorem and* $\tilde{T}$ *be its linear extension to* $L^\infty(\Omega, \Sigma, P)$ *as noted in the remark. Then* $\tilde{T}$ *is a bounded linear operator, and* $h = 1$ *iff* $\tilde{T}^* P = P$ *where* $\tilde{T}^*$ *is the adjoint of* $\tilde{T}$ *acting on* $(L^\infty(\Omega, \Sigma, P))^*$ *invariant on the subspace* $L^1(\Omega, \Sigma, P)$.

The corresponding characterization related to Šidák's identity is as follows and its proof, being similar to that of Theorem 7.3.7, is omitted.

Proposition 3. *Let $T : \mathcal{N} \to \mathcal{N}$ satisfy conditions (i), (ii), (iv), (v) of Theorem 1 and (iii'), i.e., $T((Tf) \vee (Tg)) = (Tf) \vee (Tg)$, for all f, g in $\mathcal{N}$. Then there is a measurable h (for Σ) such that*

$$Tf = E^{\mathcal{B}}(fh), \quad f \in \mathcal{N}, \quad E^{\mathcal{B}}(h) = 1, \quad a.e., \tag{8}$$

where $\mathcal{B}$ is the σ-algebra generated by the bounded elements of $T(\mathcal{N})$.

The preceding results concern the averaging operators and each is also a projection, acting on function spaces such as $L^\infty(P)$. The latter space is isometrically isomorphic to $C(Q)$, the space of continuous functions on a compact space Q. Thus it will be instructive to analyze the structure of these operators specifically on these spaces in relation to conditioning, and we now consider precisely this aspect.

11.3 Averaging operators on function algebras

In this section we present an extension of characterizations of contractive projections of Section 7.3 to the particular spaces $C(Q)$. The next result is due to Lloyd (1966), and a further extension is discussed later. It is also independently obtained by Seever (1966).

Theorem 1. *Let $T : C(Q) \to C(Q)$ be a positive contractive projection where $C(Q)$ is the space of real continuous functions on a compact Hausdorff space Q, $C(Q)$ with uniform norm. Then T is representable as*

$$T(f \cdot Tg) = T((Tf) \cdot (Tg)), \quad f, g \in C(Q) . \tag{1}$$

Moreover T is an averaging iff its range is an algebra.

Proof. We present Seever's argument which is illuminating. It uses a classical representation of abstract M-spaces, due to Kakutani (1941). We already recalled and used this idea at the end of Section 6.2. The result is proved in two steps.

Step 1. Let Q be a Stone space and suppose that, for each increasing net $\{f_i, i \in I\}$ of $C(Q)$ with $f = \sup_i f_i (\in C(Q))$, we have $Tf = \sup_i Tf_i$.

Thus if $\{f_i, i \in I\} \subset \mathcal{M} = T(C(Q))$, with $f = \sup_i f_i$ in $C(Q)$, then by the above assumption $Tf = \sup_i Tf_i = \sup_i f_i = f$. Let $u = T1$. The order interval $U = \{f \in \mathcal{M} : 0 \le f \le u\}$ has the property that its

extreme points [cf. e.g., Day (1962)] span a dense subset of $\mathcal{M}$. By the
continuity of multiplication and linearity of T, we only need to show
that $T(fg) = T(fTg)$ for any extreme point f of U, and g of $C(Q)$
satisfying $0 \leq g \leq 1$. For this purpose, consider $h = T(fg) - T(fTg)$,
and observe that $1 + g - Tg \geq 0$ since T is a positive contraction,
$0 \leq f \leq 1$, and hence $0 \leq T(f(1 + g - Tg)) \leq T(1 + g - Tg) = u$. This
implies, since $Tf = f$, $f + h \in U$; and similarly $f - h \in U$. Since f is
an extreme point, this can be true only if $h = 0$.

Step 2. We now consider the general case and reduce it to Step 1.

Since $C(Q)$ can be identified as an M-space (i.e. a Banach lattice
with norm satisfying $\|f \vee g\| = \|f\| \vee \|g\|$ for $f, g \geq 0$, and $\|f \pm g\|$ is un-
changed if $f \wedge g = 0$, cf., Kakutani (1941) or Day (1962), p. 100), its second adjoint space $C^{**}(Q)$ is also an M-space and hence is isometrically
isomorphic to $C(S)$ where S is a compact Stone space. Let $\{f_i, i \in I\}$
be an increasing net in $C^{**}(Q)$, with $f = \sup_i f \in C^{**}(Q)$. If ℓ is a positive linear functional on $C(Q)$, so $\ell \in C^*(Q)$, $f(\ell) = \lim_i f_i(\ell)$. Then the
same holds for any ℓ in $C^*(Q)$ by decomposing it as $\ell = \ell^+ - \ell^-$. But
we also note that T is weak*-continuous, i.e. it is $\sigma(C^{**}(Q), C^*(Q))$-
continuous. Hence T^{**}, the second adjoint operator, has the same
property, is a projection and $\|T^{**}\| = \|T^*\| = \|T\| \leq 1$. It has the
further property that $T^{**}f = \lim_i T^{**}f_i$ by the weak*-continuity, and
the monotonicity of $\{T^{**}f_i, i \in I\}$. So T^{**} and $\{f_i, i \in I\}$ satisfy the
hypothesis of Step 1. Note that the natural embedding $\tau : C(Q) \to$
$C^{**}(Q)$ is linear and multiplicative. Hence we have for any f, g in $C(Q)$,

$$
\begin{aligned}
\tau(T(fTg)) &= T^{**}(\tau(fTg)) = T^{**}(\tau(f) \cdot \tau(Tg)) \\
&= T^{**}(\tau(f) \cdot (T^{**}\tau(g))) \\
&= T^{**}(T^{**}(\tau(f)) \cdot T^{**}(\tau(g))), \quad \text{by Step 1,} \\
&= T^{**}(\tau(Tf) \cdot \tau(Tg)) \\
&= T^{**}(\tau(Tf \cdot Tg)) = \tau(T(Tf) \cdot T(g))) .
\end{aligned} \tag{2}
$$

But this is equivalent to (1).

Finally $T(Tf \cdot Tg) = Tf \cdot Tg$ implies the range is an algebra and the
converse is obviously true. $\square$

The positivity of T played a key role in the above proof. Let us
extend this result to a class of general contractive projections including

the positive ones. [It does not hold for arbitrary contractions as easy examples show.] Analyzing the above proof, the following condition on the range of T was formulated by Wulbert (1969). A set $A \subset C(Q)$ is said to have a *weakly separting quotient* if for each pair $p, q \in Q, \hat{p}, \hat{q}$ being the evaluation functionals on $C(Q)$ [i.e., $\hat{p}(f) = f(p)$, $f \in C(Q)$] $\hat{p}_A = t\hat{q}_A$ for some $t \neq 1$ implies $\hat{p}_A$ is *not* an extreme point of the unit ball of $A^* \subset C^*(Q)$, when $p \neq q$ and $\hat{p}_A = \hat{p}|_A$, the restriction to A. If A is nonempty then it is known that the unit ball of A^* is nonempty. If A is the range of a positive projection on $C(Q)$, then A has a weakly separating quotient. With this concept Wulbert's result is the following:

Proposition 2. *Let T be a contractive projection on $C(Q)$ whose range has a weakly separating quotient. Then T admits the representation (1).*

We omit the proof. Another formulation of the result can be given. It is in the same spirit as the work of Section 10.5 and is a simple extension of the earlier theorem. Indeed if $u : C(Q) \to C(Q)$ is an algebraic isometric isomorphism such that $\Pi = uTu^{-1}$ is a positive projection then we have the representation (1) again. For Π is clearly a contractive projection and then

$$
\begin{aligned}
T(fTg) &= u^{-1} \Pi u(f u^{-1} \Pi u g) \\
&= u^{-1} \Pi u(u^{-1} \tilde{f} u^{-1} \Pi u u^{-1} \tilde{g}), \quad \text{where} \quad \tilde{f} = uf, \; \tilde{g} = ug, \\
&= u^{-1} \Pi(\tilde{f} \Pi \tilde{g}), \quad \text{since} \quad uu^{-1} = I \text{ and } u(fg) = uf \cdot ug, \\
&= u^{-1} \Pi(\Pi \tilde{f} \cdot \Pi \tilde{g}) \quad \text{by (1) since } \Pi \text{ is a positive} \\
&\qquad \text{contractive projection,} \\
&= T(Tf \cdot Tg) \, . \tag{3}
\end{aligned}
$$

The weakness of the above two conditions is that there is no easily verifiable method to see whether the weakly separating quotient of the range of T or the existence of an algebraic isomorphism on $C(Q)$ is present. Since T^* acts on $C^*(Q)$ which is an abstract $(L-)$ space, hence is isomorphic to an $L^1(S, \mathcal{B}, \mu)$, one can give conditions on T^* to satisfy some of these relations using the ideas of Section 10.5. However, we shall not pursue this study further. Instead we consider an analysis of conditioning in certain operator algebras.

11.4 Conditioning in operator algebras

In the preceding sections conditional operators are considered on function algebras such as $L^\infty(P)$ or $C(\Omega)$. It was noted in the last part of Theorem 1.1 that an averaging projection has its range (essentially) isometrically isomorphic to an algebra of the same kind (see Proposition 2.3 also). But $L^\infty(P)$ can clearly be regarded as an algebra of operators acting on $L^2(P)$, defined by $T_k : f \mapsto kf$, $f \in L^2(P)$ and $k \in L^\infty(P)$. Then $\{T_k,\ k \in L^\infty(P)\} = \mathcal{A}$ is a subalgebra of $B(\mathcal{H})$, the algebra of all continuous linear operators on the Hilbert space $\mathcal{H} = L^2(P)$. Thus an averaging projection $S : \mathcal{A} \to \mathcal{A}$ can be thought of as a contractive projection with range $\mathcal{B} = S(\mathcal{A})$ which is again of the same type. This is a canonical example of a (commutative) operator algebra and admits a (not necessarily commutative) generalization where one may define a conditional expectation on it. We include an aspect of this set of ideas in this and the following sections. The object in question, motivated by some interesting and deep applications to quantum mechanics and related areas, is a W^* (or von Neumann)-algebra of operators on a Hilbert space $\mathcal{H}$ and it has further extensions to C^*-algebras. We wish to show how the concept of conditioning (of the Kolmogorov model) enables an analysis, analogous to that of the preceding chapters, even in such an abstract context. Here we need to use some basic properties of Hilbert space, especially the spectral theorem, and the reader should familiarize with it. We often identify $\mathcal{H}$ and its adjoint space $\mathcal{H}^*$.

We recall the concept and some properties of W^*-algebras, keeping $L^\infty(P)$ as a reference with its *-condition (to be called involution) as complex conjugation. Let $B(\mathcal{H})$ be the space of all bounded linear operators on a Hilbert space $\mathcal{H}$. For each T in $B(\mathcal{H})$ the adjoint T^* is also in $B(\mathcal{H})$, and $\|T\| = \|T^*\|$ where $\|\cdot\|$ is the operator (or uniform) norm of $B(\mathcal{H})$. The mapping $T \to T^*$, called *involution* or *-*operation* satisfies: (i) $(T^*)^* = T^{**} = T$, (ii) $(\lambda T_1 + T_2)^* = \bar{\lambda} T_1^* + T_2^*$ (λ-complex) (iii) $(T_1 T_2)^* = T_2^* T_1^*$. These are immediate from definition. Further, (iv) $\|T^* T\| = \|T\|^2$ where $\mathcal{H}$ is identified with $\mathcal{H}^*$. To see the truth of (iv) since the norm evidently satisfies $\|T_1 T_2\| \leq \|T_1\| \|T_2\|$ we get $\|T^* T\| \leq \|T\|^2$. On the other hand $\|T\|^2 = \sup\{(Tx, Tx)/\|x\|^2 : x \neq 0\} = \sup\{(T^* Tx, x)/\|x\|^2 : x \neq 0\} \leq \|T^* T\|$ by the CBS-inequality where $(\cdot, \cdot)$ is the inner product of $\mathcal{H}$. Thus (iv) is proved. Any Banach algebra with a *-operation satisfying (i) - (iv) is called a C^*-*algebra*, and $B(\mathcal{H})$ is an example. A C^*-algebra is called a W^*-*algebra or a von Neumann algebra*, if it is the dual of a Banach space. Thus $L^\infty(\mu)$ is a W^*-algebra only when it is a dual space (of $L^1(\mu)$, of course) iff μ is localizable, by a classical theorem of Segal's [cf. e.g. Rao (1987), Section 5.4 especially p. 276]. Although in general the predual of a Banach space need not be unique, for W^*-algebras it is. An alternative characterization of W^*-algebras which is basic in their analysis is due to J. von Neumann and can be stated as follows: If $\mathcal{A}$ is a set of linear operators on $\mathcal{H}$ then $\mathcal{A}'(\subset B(\mathcal{H}))$ is the set each of whose members commutes with each member of $\mathcal{A}$, called the *commutator*. The desired result is that a *-algebra $\mathcal{A}$ on $\mathcal{H}$ is a W^*-algebra iff $\mathcal{A} = \mathcal{A}''$ $(= (\mathcal{A}')')$, i.e., it is equal to its bicommutant. Hereafter $\mathcal{A}$ is assumed to have identity I, since it can otherwise be adjoined. An element $A \in \mathcal{A}$ is termed positive $(A \geq 0)$ iff $(x, Ax) = (Ax, x) \geq 0$ for all $x \in \mathcal{H}$, and consequently, the spectrum of A is contained in $\bar{R}^+$. A linear functional ℓ on $\mathcal{A}$ is a *state* if $\ell(I) = 1$, is *positive* if $\ell(A^* A) \geq 0$ for all A in $\mathcal{A}$, is *faithful* if $\ell(A^* A) = 0$ implies $A = 0$, and *normal* if $A_\alpha \uparrow A$ implies $\ell(A_\alpha) \to \ell(A)$ for any increasing net. Further $\ell(\cdot)$ is *semifinite* if

$$\ell(A) = \sup\{\ell(B) : 0 \leq B \leq A , \ \ell(B) < \infty\} \tag{1}$$

where $B \leq A$ means $B - A$ is positive. (This is the "finite subset property" for measures.) A faithful normal state $\ell(\cdot)$ on a W^*-algebra $\mathcal{A}$ is called a *trace* functional if for each A in $\mathcal{A}$ and unitary element u in $\mathcal{A}$ one has $\ell(A) = \ell(uAu^*)$.

The preceding concepts also apply for a C^*-algebra defined abstractly as a Banach *-algebra with identity. However, a classical theorem of Gel'fand and Naĭmark states that every C^*-algebra can be realized as a subalgebra of $B(\mathcal{H})$ on some Hilbert space $\mathcal{H}$. Hence we restrict ourselves to the case of $B(\mathcal{H})$ and then to von Neumann subalgebras contained in them. [For an introduction to these subjects, see Arveson (1976), and Sakai (1971).]

Let $\mathcal{A}$ be a W^*-algebra on $\mathcal{H}$, assumed separable. It is known that $\mathcal{A}$ has at least one semi-finite trace functional, and contains sufficiently many orthogonal projections. We now introduce the necessary integration on $\mathcal{A}$ and the L^p-spaces relative to a semi-finite trace $\tau(\cdot)$ which will allow us to define a conditional expectation on $\mathcal{A}$.

Since unbounded operators appear in integration theory and in applications, we need to recall a few more concepts and some related facts from Segal (1953). Let $(\mathcal{H}, \mathcal{A}, \tau)$ be a triple in which $\mathcal{H}$ is a Hilbert space, $\mathcal{A}$ a W^*-algebra over $\mathcal{H}$, and τ a faithful normal semifinite trace on $\mathcal{A}$. A (possibly unbounded) linear operator T on $\mathcal{H}$ is *affiliated* to $\mathcal{A}$ if T commutes with every unitary operator in $\mathcal{A}'$, written $T\eta\mathcal{A}$. [If T is bounded, then this implies $T \in \mathcal{A}$.] A linear set $\mathcal{D} \subset \mathcal{H}$ is *associated* with $\mathcal{A}$ ($\mathcal{D}\eta\mathcal{A}$) if each unitary $u \in \mathcal{A}$ is invariant on $\mathcal{D}$, i.e. $u(\mathcal{D}) \subset \mathcal{D}$. If $\mathcal{D}$ is also closed, then by the spectral theorem, $\mathcal{D}\eta\mathcal{A}$ means that each (orthogonal) projection on $\mathcal{H}$ onto $\mathcal{D}$ belongs to $\mathcal{A}$. A linear set $\mathcal{D}$ is *strongly dense* in $\mathcal{H}$ relative to $\mathcal{A}$ if $\mathcal{D}\eta\mathcal{A}$ and there are linear spaces $\mathcal{T}_n \subset \mathcal{D}$ such that $\mathcal{T}_n\eta\mathcal{A}, \mathcal{T}_n^\perp (= \mathcal{H} \ominus \mathcal{T}_n)$ is finite dimensional, and $\mathcal{T}_n^\perp \downarrow \{0\}$ as $n \to \infty$. A linear operator T on $\mathcal{H}$ is *measurable* relative to $\mathcal{A}$, if $T\eta\mathcal{A}$ and T has a strongly dense domain: Alternatively, if $T\eta\mathcal{A}$ then $T = u(|T|)$ where $|T| = (T^*T)^{\frac{1}{2}}$ and u is a partial isometry. The latter means u^*u and uu^* are projections onto the domain and ranges of u respectively. Thus by the spectral theorem

$$|T| = \int_0^\infty \lambda E(d\lambda), \quad E([\lambda_1, \lambda_2)) \in \mathcal{A}, \ 0 \le \lambda_1 \le \lambda_2 < \infty, \qquad (2)$$

where $\{E(\lambda), \ 0 \le \lambda < \infty\}$ is a resolution of the identity. Then T is τ-*measurable* if $\tau(E((\lambda, \infty))) < \infty$ for some $\lambda \ge 0$. The following result illuminates the concept of measurability of T:

Proposition 1. *Let (Ω, Σ, μ) be a localizable measure space and $L^2(\mu)$ be the Hilbert space on it, $\mathcal{A}$ a ring of operators on $L^2(\mu)$, and τ, a simifinite trace on it. Then an operator $T : L^2(\mu) \to L^2(\mu)$ is τ-measurable iff it is expressible as $(Tf)(\omega) = k(\omega)f(\omega), \quad \omega \in \Omega$, for some measurable $k : \Omega \to \mathbb{R}$ and domain of T as $\{f \in L^2(\mu) : kf \in L^2(\mu)\}$.*

We shall omit a proof of this result, from Segal (1953), as it is given for illustration but will not be needed below.

Let $L^0(\mathcal{A}, \tau)$ be the set of all τ-measurable operators on $(\mathcal{H}, \mathcal{A}, \tau)$ and $p \ge 1$. Define the norm $\| \cdot \|_p$ for $T \in L^0(\mathcal{A}, \tau)$ as:

$$\|T\|_p = \begin{cases} (\tau(|T|^p))^{\frac{1}{p}}, & \text{if } p < \infty \\ \|T\|, & (\text{operator norm}) \text{ if } p = +\infty. \end{cases} \qquad (3)$$

where $|T|^p = \int_0^\infty \lambda^p E(d\lambda)$, by the spectral theorem. Let $L^p(\mathcal{A}, \tau) = \{T : \|T\|_p < \infty\}$ for $1 \le p < \infty$ and $L^\infty(\mathcal{A}, \tau) = \mathcal{A}$. It can be shown (nontrivially) that $\{L^p(\mathcal{A}, \tau), \|\cdot\|_p\}$ are Banach spaces. (See Kunzi (1959), Sec. 2.)

We are now ready to introduce the concept of conditioning in a W^*-algebra $\mathcal{A}$, (containing I) with a faithful normal trace τ on it.

Definition 2. Let $\mathcal{B}(\subset \mathcal{A})$ be a W^*-subalgebra with a unit. Then a mapping $E^{\mathcal{B}} : \mathcal{A} \to \mathcal{B}$ is called a *conditional expectation* relative to τ if (i) $E^{\mathcal{B}}$ is a contractive positive linear projection, and (ii) $\tau(E^{\mathcal{B}}(x)) = \tau(x)$, $x \in \mathcal{A}$.

The mapping $E^{\mathcal{B}}$ has the following existence and structural properties, first established by H. Umegaki (1954) who based it on a Radon-Nikodým theorem due to Dye and Segal. We give another proof.

Proposition 3. *There exists a unique conditional expectation $E^{\mathcal{B}}$ on $\mathcal{A}$ onto $\mathcal{B}$ and it satisfies:*

> *(i) $(E^{\mathcal{B}}(x))^*(E^{\mathcal{B}}(x)) \le E^{\mathcal{B}}(x^*x)$, $x \in \mathcal{A}$,*
> *(ii) $E^{\mathcal{B}}(yxz) = yE^{\mathcal{B}}(x)z$, for all $y, z \in \mathcal{B}$ and $x \in \mathcal{A}$,*
> *(iii) $x_\alpha \uparrow x \Rightarrow E^{\mathcal{B}}(x_\alpha) \uparrow E^{\mathcal{B}}(x)$, (normality)*
> *(iv) $x \ge 0 \Rightarrow E^{\mathcal{B}}(x) \ge 0$, with equality iff $x = 0$,*
> *(v) $E^{\mathcal{B}} : \mathcal{A} \to \mathcal{B}$ is both strongly and weakly continuous on their unit balls, and*
> *(vi) $\|x - E^{\mathcal{B}}(x)\| = \inf\{\|x - y\| : y \in \mathcal{B}\}$, $x \in \mathcal{A}$.*

Just as in the earlier work, the existence result depends on a Radon-Nikodým theorem which is the characteristic feature of Kolmogorov's extension. We therefore present a version of this result for a class of *-algebras that include both the C^*-and W^*-algebras in our context, and it is more probabilistic in approach then the Dye-Segal method. This is due to Gudder and Hudson (1978) which we now discuss.

We start with a *-algebra $(\mathcal{A}, *)$ which is an associative, distributive complex algebra with a unit I, and an involution $*$, satisfying

$$x^{**} = x, \ (xy)^* = y^*x^*, (\lambda x + y)^* = \bar{\lambda}x^* + y^*, \ \lambda \in \mathbb{C}, \ x, y \in \mathcal{A}. \quad (4)$$

A linear functional $\varphi : \mathcal{A} \to \mathbb{C}$ is a *faithful state* if it is positive (i.e., $\varphi(x^*x) \ge 0$, $x \in \mathcal{A}$), $\varphi(I) = 1$ and $\varphi(x^*x) = 0$ iff $x = 0$. We consider

$(\mathcal{A}, {}^{*}, \varphi)$ as given and call it a probability algebra, and define an inner product $\langle \cdot, \cdot \rangle$ on $\mathcal{A}$ by $\langle x, y \rangle = \varphi(y^{*}x)$. We then have for x, y, z in $\mathcal{A}$, $\langle y, x^{*}z \rangle = \varphi((x^{*}z)^{*}y) = \varphi(z^{*}xy) = \langle xy, z \rangle$. Let $\mathcal{H}$ be the Hilbert space spanned by $(\mathcal{A}, \langle \cdot, \cdot \rangle)$. For each $x \in \mathcal{A}$, let $\pi_{x} : \mathcal{A} \to \mathcal{H}$ be the operator defined by $\pi_{x}y = xy \in \mathcal{A} \subset \mathcal{H}$, so that $\pi : \mathcal{A} \to B(\mathcal{H})$ is a representation of $\mathcal{A}$ with a domain $D_{\pi_{x}} = \mathcal{A}$ dense in $\mathcal{H}$. Hence $D_{\pi} = \cap_{x} D_{\pi_{x}} = \mathcal{A}$. The adjoint transformation of π_{x}, denoted $(\pi_{x})^{*}$, is defined and we set

$$D_{\pi^{*}} = \cap \{ D_{(\pi_{x})^{*}} : x \in \mathcal{A} \}, \quad \text{and} \quad \pi_{x}^{*} = (\pi_{x^{*}})^{*} | D_{\pi^{*}}, \ x \in \mathcal{A}. \qquad (5)$$

Since $D_{(\pi_{x})^{*}} \supset \mathcal{A}$, so that $D_{\pi} = \mathcal{A} \subset D_{\pi^{*}}$, we see that π^{*} is an extension of the representation π. Note that π satisfies, $\pi(I) = I$ [writing π_{x} as $\pi(x)$ from now on] and the following:

(i) $\pi(\lambda_{1}x + \lambda_{2}y)z = \lambda_{1}(\pi(x)z + \lambda_{2}\pi(y)z$, $x, y \in \mathcal{A}$, $z \in D_{\pi}$ and $\lambda_{1}, \lambda_{2} \in \mathbb{C}$,

(ii) $\pi(x)D_{\pi} \subset D_{\pi}$, $x \in \mathcal{A}$; $\pi(x)\pi(y)z = \pi(xy)z$, $x, y \in \mathcal{A}$, $z \in D_{\pi}$,

(iii) $\langle z_{1}, \pi(x)z_{2} \rangle = \langle \pi(x^{*})z_{1}, z_{2} \rangle$, $z_{1}, z_{2} \in D_{\pi}$ and $x \in \mathcal{A}$, [i.e., $\pi(x^{*}) \subset (\pi(x))^{*}$].

Let $L(\mathcal{A})$ be the set of all linear mappings on $\mathcal{A}$ into $\mathcal{H}$ and $B(\mathcal{H})$ be the usual bounded operators on $\mathcal{H}$. An element $A \in L(\mathcal{A})$ is *symmetric* (nonnegative) if $\langle Ax, y \rangle = \langle x, Ay \rangle$ ($\langle Ax, y \rangle = \langle x, Ay \rangle \geq 0$). The *unbounded commutant* of π is defined as:

$$\pi(\mathcal{A})^{c} = \{ A \in L(\mathcal{A}) : \langle A\pi(x)y, z \rangle = \langle Ay, \pi(x^{*})z \rangle, \ x, y, z \in \mathcal{A} \}. \qquad (6)$$

Thus A in $\pi(\mathcal{A})^{c}$ commutes with all elements of $\mathcal{A}(= D_{\pi}$ here). The usual commutant $\pi(\mathcal{A})'$ is then defined as, $(A|\mathcal{A}$ is the restriction of A to $\mathcal{A})$:

$$\pi(\mathcal{A})' = \{ A \in B(\mathcal{H}) : (A|\mathcal{A}) \in \pi(\mathcal{A})^{c} \}. \qquad (7)$$

From these concepts we see that for $A \in \pi(\mathcal{A})^{c}$ and for x, y, z in $\mathcal{A}$,

$$\begin{aligned} \langle c\pi(x)y, z \rangle &= \langle cxy, z \rangle = \langle xcy, \ z \rangle \\ &= \langle cy, x^{*}z \rangle = \langle cy, \pi(x^{*})z \rangle, \quad \text{by definition of } \pi_{x}, \\ &= \langle \pi(x^{*})^{*}cy, z \rangle. \end{aligned}$$

Hence $c\pi(x) = (\pi(x^*))^* c$ and $cx = xc1 = \pi(x)(c1) \in D_{\pi(x)}$, $x \in \mathcal{A}$. So we conclude that $c(\mathcal{A}) \subset \cap_x D_{\pi(x)} = \mathcal{A} \subset D_{\pi^*}$ and $c\pi(x) = \pi(x)^* c$ since $(\pi(x))^* = (\pi(x^*))^* | D_{\pi^*}$. Similar computations will be used later.

If $\ell : \mathcal{A} \to \mathbb{C}$ is a continuous linear functional and φ is a state of $\mathcal{A}$, as above, then ℓ is said to be absolutely continuous relative to φ, denoted $\ell << \varphi$, if $\varphi(x_i^* x_i) \to 0$ implies $\ell(yx_i) \to 0$ for all $y \in \mathcal{A}$, and ℓ is *dominated* by φ, written $\ell \prec \varphi$, if $|\ell(x^* x)| \le k_0 \varphi(x^* x)$ for some constant $k_0 > 0, x \in \mathcal{A}$. Expressing ℓ into its nonnegative components, $\ell = \ell_1 - \ell_2 + i(\ell_3 - \ell_4)$ and using the CBS inequality we see that $\ell \prec \varphi$ implies $\ell << \varphi$. (Compare with Daniell integration where $\mathcal{A}$ is the set of all simple functions and $\varphi(\chi_A) = \mu(A)$ gives (Ω, Σ, μ) a measure space.) We then have the Radon-Nikodým type result as:

Theorem 4. *Let* $\ell : \mathcal{A} \to \mathbb{C}$ *be a continuous linear functional. Then* $\ell << \varphi$ *iff there is a unique* $A \in \pi(\mathcal{A})^c$ *such that* $\ell(x) = \overline{\varphi(Ax^*)}, x \in \mathcal{A}$, *[A is denoted symbolically as* $\frac{d\ell}{d\varphi}$*] and A is positive if* ℓ *is also positive.*

Proof. Let $\ell << \varphi$ and y in $\mathcal{A}$ be arbitrarily fixed. Then the mapping $\tilde{\ell} : x \mapsto \ell(y^* x)$ is a continuous linear functional on $\mathcal{A}$ and has a unique continuous extension to $\mathcal{H}$. Then by the Riesz representation, $\ell(y^* x) = \tilde{\ell}(x) = \langle x, y_1 \rangle$ for a unique $y_1 \in \mathcal{H}$. Let $A : y \mapsto y_1$. Then A is a well-defined linear operator. To see that $A \in \pi(\mathcal{A})^c$, consider for x, y, z in $\mathcal{A}$, using the fact that $\ell(y^* x) = \langle x, Ay \rangle$, the following:

$$\langle A\pi(x)y, z \rangle = \langle z, A\pi(x)y \rangle = \langle z, Axy \rangle$$
$$= \overline{\ell((xy)^* z)} = \overline{\ell(y^* x^* z)}$$
$$= \overline{\langle x^* z, Ay \rangle} = \langle Ay, x^* z \rangle$$
$$= \langle Ay, \pi(x^*)z \rangle.$$

Hence $A \in \pi(\mathcal{A})^c$ by (6). Further for $x \in \mathcal{A}$ ($1^* = 1$ will be used),

$$\ell(x) = \tilde{\ell}(x) = \tilde{\ell}(x1) = \langle x, A1 \rangle = \langle x1, A1 \rangle$$
$$= \langle 1, x^* A1 \rangle = \langle 1, Ax^* 1 \rangle = \langle 1, Ax^* \rangle$$
$$= \overline{\langle Ax^*, 1 \rangle} = \overline{(\varphi(Ax^*))},$$

giving the desired representation.

Conversely, if $\ell(x) = \overline{\varphi(Ax^*)}$ for $A \in \pi(\mathcal{A})^c$, then

$$|\ell(yx)| = |\varphi(Ax^* y^*)| = |\langle Ax^* y^*, 1 \rangle|$$
$$= |\langle x^* Ay^*, 1 \rangle| = |\langle Ay^*, x \rangle| \le \|Ay^*\| \, \|x\|,$$

by the CBS inequality. Hence $\ell << \varphi$.

For uniqueness, if $A_1 \in \pi(\mathcal{A})^c$ is another element satisfying the same equation, then for x, y in $\mathcal{A}$ one has

$$
\begin{aligned}
\langle A_1 y, x \rangle = \langle x^* A_1 y, 1 \rangle &= \langle A_1 x^* y, 1 \rangle \\
&= \varphi(A_1 x^* y) = \overline{\ell(y^* x)} = \varphi(A x^* y) \\
&= \langle A y, x \rangle.
\end{aligned}
$$

Since x, y are arbitrary, this implies $A = A_1$. Finally if ℓ is positive,

$$
\langle x, A x \rangle = \ell(x^* x) \geq 0 \quad \Rightarrow A \text{ is positive .}
$$

Thus all assertions are proved. $\square$

If $\mathcal{A}$ is as above and $\mathcal{B} \subset \mathcal{A}$ is a $*$-subalgebra with identity and φ is a faithful state, let P_B be the orthogonal projection of $\mathcal{H}(= \bar{\mathcal{A}})$ onto $\bar{\mathcal{B}}$. Let π be the $*$-representation of $\mathcal{A}$ as before and $\pi_B = \pi|\mathcal{B}$ which is a $*$-representation with domain $= D(\pi_B) = \mathcal{B} \subset \bar{\mathcal{B}}$. We define the conditional expectation on $\mathcal{A}$ relative to $\mathcal{B}$ as that element $E(x|\mathcal{B})$ in $D(\pi_B^*) \subset \bar{\mathcal{B}}$ satisfying for given $x \in \mathcal{A}$,

$$
\varphi(yx) = \varphi(\pi_B^* y E(x|\mathcal{B})) \qquad \text{for all } y \in \mathcal{B} . \tag{8}
$$

Theorem 5. *If $(\mathcal{A}, *, \varphi)$ is as given above and $\mathcal{B} \subset \mathcal{A}$ is a $*$-subalgebra with identity, then $E(x|\mathcal{B})$ exists uniquely for $x \in \mathcal{A}$, and $E(\cdot|\mathcal{B})$ is the orthogonal projection P_B defined above. Further,*

$$
\varphi(E(x|\mathcal{B})) = \varphi(x) \quad \text{and} \quad E(xy|\mathcal{B}) = \pi_B^*(y) E(x|\mathcal{B}), \quad x \in \mathcal{A}, \ y \in \mathcal{B} . \tag{9}
$$

Proof. Consider the orthogonal projection P_B defined above. Then $P_B = P_B^*$. For $x \in \mathcal{A}$ and y, z in $\mathcal{B}$ we have on using $P_B(\mathcal{B}) = \mathcal{B}$,

$$
\begin{aligned}
|\langle \pi(z)y, P_B x \rangle| = |\langle zy, P_B x \rangle| = |\langle P_B zy, x \rangle| &= |\langle zy, x \rangle| \\
&= |\langle y, z^* x \rangle| \leq \|y\| \ \|z^* x\| .
\end{aligned}
$$

Hence $P_B x \in \mathcal{D}_{\pi^*(z)}$ for all $z \in \mathcal{B}$ and so it is in their intersection. Thus $P_B x \in D_{\pi_B^*} \subset \bar{\mathcal{B}}$. To see that (8) holds, let $y \in \mathcal{B}$ be arbitrary. Then

$$
\begin{aligned}
\varphi(\pi_B^*(y) P_B x) = \langle P_B x, (\pi_B^* y)^* \rangle \\
= \langle P_B x, y^* \rangle \\
= \langle x, y^* \rangle = \varphi(yx) .
\end{aligned}
$$

Hence $P_\mathcal{B}$ is a desired map.

For unicity, suppose $E(\cdot|\mathcal{B})(= P_\mathcal{B})$ and $E_1(\cdot|\mathcal{B})$ be two such mappings. Then for any $x \in \mathcal{A}$, $y \in \mathcal{B}$ we have

$$\begin{aligned}
\langle E(x|\mathcal{B}), y \rangle &= \langle E(x|\mathcal{B}), \pi_\mathcal{B}(y).1 \rangle \\
&= \langle (\pi_\mathcal{B}(y))^* E(x|\mathcal{B}), 1 \rangle \\
&= \langle \pi_\mathcal{B}^*(y^*) E(x|\mathcal{B}), 1 \rangle \\
&= \varphi(\pi_\mathcal{B}^*(y^*) E(x|\mathcal{B})) = \varphi(y^* x), \qquad \text{by (8)}, \\
&= \varphi(\pi_\mathcal{B}^* y^* E_1(x|\mathcal{B})) = \langle E_1(x|\mathcal{B}), y \rangle.
\end{aligned}$$

Hence $E_1(x|\mathcal{B}) = E(x|\mathcal{B}) \in \bar{\mathcal{B}}$, for all $x \in \mathcal{A}$ so that they are the same.

Taking $y = 1$ in (8), we get the first part of (9). For the second part, let $y, z \in \mathcal{B}$ and $x \in \mathcal{A}$. Consider the identity

$$\begin{aligned}
\langle E(yx|\mathcal{B}), z \rangle &= \varphi(\pi_\mathcal{B}^*(z) E(yx|\mathcal{B})) = \varphi(zyx) \\
&= \varphi(\pi_\mathcal{B}^*(zy) E(x|\mathcal{B})) \\
&= \varphi(\pi_\mathcal{B}^*(z) \pi_\mathcal{B}^*(y) E(x|\mathcal{B})), \quad \text{since } \pi_\mathcal{B}^* \text{ is a representation}, \\
&= \langle \pi_\mathcal{B}^*(y) E(x|\mathcal{B}), z \rangle .
\end{aligned}$$

Hence $E(yx|\mathcal{B}) = \pi_\mathcal{B}^*(y) E(x|\mathcal{B})$ so that (9) holds. $\square$

We observe now if $\mathcal{A} \supset \mathcal{B}$ are von Neumann algebras each having the identity, and $\varphi(\cdot)$ is a faithful normal trace, the hypothesis of Theorem 5 is satisfied. Consequently the existence part of Proposition 3 is established. We can complete the rest of its proof now.

Proofs of (i) – (vi) of Proposition 3. Writing $E^\mathcal{B}(\cdot)$ as $E(\cdot|\mathcal{B})$, the conditional expectation on $\mathcal{A}$ onto $\mathcal{B}$, we have for $x \in \mathcal{A}$, since $E(x|\mathcal{B}) \in \mathcal{B}$,

$$\begin{aligned}
0 \leq E((x - E(x|\mathcal{B}))^*(x - E(x|\mathcal{B})) &= E(x^* x|\mathcal{B}) - E(x^* E(x|\mathcal{B})) \\
&\quad - E(E(x|\mathcal{B})^* x) + E(x|\mathcal{B})^* E(x|\mathcal{B}) \\
&= E(x^* x|\mathcal{B}) - E(x|\mathcal{B})^* E(x|\mathcal{B}),
\end{aligned}$$

since by (9) $E(xy|\mathcal{B}) = E(x|\mathcal{B})y$ for $y \in \mathcal{B}$ and $E(yx(\mathcal{B}) = yE(x|\mathcal{B})$. This is (i), and (ii) is a part of (9).

For (iii), by the monotonicity of $E(\cdot|\mathcal{B})$, if $\sup_\alpha x_\alpha = x$ then

$$\sup_\alpha E(x_\alpha|\mathcal{B}) \leq E(x|\mathcal{B}) .$$

Also by (9) and the normality of the trace functional, $\varphi(\cdot)$ here, one has

$$
\varphi(\sup_\alpha E(x_\alpha|\mathcal{B})) = \sup_\alpha \varphi(E(x_\alpha|\mathcal{B}))
$$
$$
= \sup_\alpha \varphi(x_\alpha) = \varphi(\sup_\alpha x_\alpha)
$$
$$
= \varphi(E(\sup_\alpha x_\alpha|\mathcal{B})) = \varphi(E(x|\mathcal{B})) .
$$

Hence $\varphi(E(x|\mathcal{B}) - \sup_\alpha E(x_\alpha|\mathcal{B})) = 0$. Since $\varphi(\cdot)$ is faithful (iii) follows from this equation. Also (iv) is immediate again from (9) and the properties of φ. Next (v) follows from (i) since $E(\cdot|\mathcal{B})$ is an orthogonal projection. In fact

$$
\varphi(E(x|\mathcal{B})^* E(x|\mathcal{B})) = \varphi(|E(x|\mathcal{B})|^2)
$$
$$
\leq \varphi(E(x^* x|\mathcal{B})) = \varphi(x^* x) = \varphi(|x|^2) .
$$

So $\|E(x|\mathcal{B})\| \leq \|x\|$ and $x_n \to x$ in $\mathcal{H}$ implies $\|E(x_n|\mathcal{B}) - E(x|\mathcal{B})\| \leq \|x_n - x\| \to 0$ and this is sufficient to conclude weak continuity also.

Finally for (vi), since $E(x|\mathcal{B}) \in \mathcal{B}$, it is clear that

$$
\|E(x|\mathcal{B}) - x\| \leq \inf\{\|x - y\| : y \in \mathcal{B}\} .
$$

For the opposite inequality, let $z \in \mathcal{B}$. Then, using $\varphi = \varphi \circ E(\cdot|\mathcal{B})$ of (9),

$$
\varphi(z^*(x - E(x|\mathcal{B}))) = \varphi(E(z^*(x - E(x|\mathcal{B}))))
$$
$$
= \varphi(z^*(E(x|\mathcal{B}) - E(x|\mathcal{B}))) = \varphi(0) = 0
$$

Hence for any $z \in \mathcal{B}$,

$$
\|x - z\|^2 = \|x - E(x|\mathcal{B}) + E(x|\mathcal{B}) - z\|^2
$$
$$
= \|x - E(x|\mathcal{B})\|^2 + \|E(x|\mathcal{B}) - z\|^2 +
$$
$$
2 \operatorname{Re} \varphi((x - E(x|\mathcal{B}))^*(E(x|\mathcal{B}) - z))
$$
$$
= \|x - E(x|\mathcal{B})\|^2 + \|E(x|\mathcal{B}) - z\|^2 + 0
$$
$$
\geq \|x - E(x|\mathcal{B})\|^2 .
$$

Hence

$$
\inf\{\|x - z\| : z \in \mathcal{B}\} \geq \|x - E(x|\mathcal{B})\| .
$$

This together with the earlier inequality establishes (*vi*) and hence all assertions of the proposition hold. □

Remark. The interrelations between $\pi(\mathcal{A})^c$, D_{π^*} and $\pi(\mathcal{A})'$ [and useful subsets of D_{π^*}] are discussed in Gudder and Hudson (1978) in detail, along with their use in quantum mechanical applications. We do not include them here as they are not needed for our present work. Further extension for semi-finite normal states φ appears possible to include the existence theorem for conditional expectations due to Takesaki (1972).

A characterization of conditional expectations, in the present context, among contractive projections on a von Neumann algebra can be considered. The following result, due to J. Tomiyama (1957), is a noncommutative analog of Theorem 7.3.7 with $\rho(\cdot) = \|\cdot\|_2$ there.

Theorem 6. *Let $\mathcal{B} \subset \mathcal{A}$ be W^*-algebras containing an identity. If $T : \mathcal{A} \to \mathcal{B}$ is a contractive projection (so $T1 = 1$), then T is a conditional expectation, i.e., $Tx = E(x|\mathcal{B})$, $x \in \mathcal{A}$.*

It should be observed that, in case $\mathcal{A} = L^\infty(\Omega, \Sigma, \mu)$ and $\mathcal{B} = L^2(\Omega, \Sigma_0, \mu)$ with μ_{Σ_0} localizable, the operator $T : \mathcal{A} \to \mathcal{B}, T^2 = T$ maps $L^2(\Sigma)$ onto $L^2(\Sigma_0)$, $\|T\| \leq 1$. So the above characterization is the analog of $T : L^2(\Sigma) \to L^2(\Sigma_0)$ but *not* of $L^\infty(\Sigma) \to L^\infty(\Sigma_0)$. In the latter case $Tf = E(fh|\Sigma_0)$ for some $h \in L^\infty(\Sigma)$ with $E(h|\Sigma_0) = 1$ a.e. [This point was noted somewhat inelegantly in Rao (1975), p. 375.] However, the result of Theorem 6 does not follow from the commutative theory and needs different techniques. We shall not include the details here. It will be interesting to extend the other characterizations of conditional expectations from the commutative theory. This may illuminate the subject with possible uses in similar areas.

11.5 Some applications of noncommutative conditioning

In this final section we present a few results on martingale convergence and "sufficiency" in this setting as applications of the work discussed in the preceding section.

Definition 1. Let $\{\mathcal{B}_\alpha, \alpha \in I\}$ be an increasing net of W^*-algebras each $\mathcal{B}_\alpha$ containing the identity and $E(\cdot|\mathcal{B}_\alpha)$ exists for each α. [This is automatic if there is a *finite* normal faithful state φ on $\mathcal{A}$, but otherwise $\varphi|\mathcal{B}_\alpha$ should be semifinite.] Then $\{E(x|\mathcal{B}_\alpha, \alpha \in I\}$ is an increasing *martingale (supermartingale)* if for each $\alpha < \beta$ in I

$$E(E(x|\mathcal{B}_\beta)|\mathcal{B}_\alpha) = (\geq)E(x|\mathcal{B}_\alpha) . \tag{1}$$

Decreasing (super) martingales are defined similarly.

We first give an example of such a martingale before proving a general convergence theorem for nets.

Example 2. Let $\{P_n, n \geq 1\}$ be a sequence of mutually orthogonal projections in a W^*-algebra $\mathcal{A}$ and suppose that $I = \sum\limits_{n=1}^{\infty} P_n$. Let

$$Tx = \sum_{n=1}^{\infty} P_n x P_n \quad , \quad x \in \mathcal{A}, \tag{2}$$

the series converging in the norm topology of $\mathcal{A}$. If $\{P_n, n \geq 1\}'$ is the comutant of $\{P_n, n \geq 1\}$, then $\mathcal{B} = \mathcal{A} \cap \{P_n, n \geq 1\}'$ is a W^*-algebra contained in $\mathcal{A}$, and we note that $T = E(\cdot|\mathcal{B})$ is the conditional expectation of $\mathcal{A}$ onto $\mathcal{B}$. To see this, since T is clearly linear, $TI = I$, and $(Tx, x) \geq 0$ [this is so because $(P_n x P_n, x) = (P_n^2 x P_n^2, x) = (P_n x P_n, P_n x P_n) \geq 0$ for each n], we verify the other parts of Definition 4.2. Now $T(\mathcal{A}) \subset \mathcal{B}$ since $P_n T x = P_n^2 x P_n = P_n x P_n^2 = T x P_n$ and $T x \in \mathcal{B}$. If $y \in \mathcal{B}$ then y commutes with each P_n, and hence

$$P_n y x P_n = y P_n x P_n, \ (n \geq 1) \Rightarrow T(yx) = y \, Tx .$$

Similarly $T(xy) = (Tx)y$, and $T^2 x = T(Tx) = Tx$ since $T|\mathcal{B} = I$. If $\tau(\cdot)$ is a state on $\mathcal{A}$, then

$$\tau(Tx) = \sum_{n=1}^{\infty} \tau(P_n x P_n) = \sum_{n=1}^{\infty} \tau(x P_n)$$

$$= \tau(x \sum_{n=1}^{\infty} P_n) = \tau(x) .$$

If $T(x^* x) = 0$, then $P_n T(x^* x)P_n = 0$ for all n, and hence

$$0 = P_n \left(\sum_{k=1}^{\infty} P_k x^* x P_k \right) P_n = P_n x^* x P_n = (x P_n)^*(x P_n), \ n \geq 1.$$

This implies $x P_n = 0$ and $0 = x \sum\limits_{n=1}^{\infty} P_n = xI = x$ so T is faithful. If $y \in \mathcal{H}(= \bar{\mathcal{A}})$, let $\xi_n = P_n y$ so that $\|y\|^2 = \sum\limits_{n=1}^{\infty} \|\xi_n\|^2$ and then

$$((Tx)y, y) = \sum_{n=1}^{\infty} (P_n x P_n y, y) = \sum_{n=1}^{\infty} (x \xi_n, \xi_n) = \tau(x)$$

where $\tau(\cdot)$ is the trace functional defined uniquely by the sum. This shows that $x_n \uparrow x \Rightarrow \tau(x_n) \nearrow \tau(x)$ and hence T is also normal. Thus $T = E(\cdot|\mathcal{B})$, as asserted.

We use this "concrete" conditional expectation to generate a martingale. Thus if $\mathcal{P}$ is any commutative family of orthogonal projections on $\mathcal{H}$ and $\mathcal{F} \subset \mathcal{P}$ is a finite subfamily containing O and I, consider

$$x_{\mathcal{F}} = \underset{y \in \tilde{\mathcal{F}}}{\Sigma}\ yxy\ , \tag{3}$$

where $\tilde{\mathcal{F}}$ is the Boolean algebra generated by $\mathcal{F}$. If $\mathcal{B}_{\mathcal{F}} = \mathcal{A} \cap \mathcal{F}'$, then $E(\cdot|\mathcal{B}_{\mathcal{F}}) : x \mapsto x_{\mathcal{F}}$ defines a conditional expectation by the preceding computation, since (3) is a special form of (2). If we partially order $\mathcal{P}$ by inclusion so that $\mathcal{F}_1 \subset \mathcal{F}_2 \Rightarrow \mathcal{B}_{\mathcal{F}_1} \supset \mathcal{B}_{\mathcal{F}_2}$, we get $\{E(x|\mathcal{B}_{\mathcal{F}}), \mathcal{F} \subset \mathcal{P}\}, x \in \mathcal{A}$, to be a decreasing martingale, and $\cap \mathcal{B}_{\mathcal{F}} = \mathcal{P}' \cap \mathcal{A}$. We shall now find conditions for convergence of such martingales. This is the content of the next result.

Theorem 3. *Let $\{E(x|\mathcal{B}_\alpha),\ \alpha \in J\}, x \in \mathcal{A}$, be a directed indexed martingale where $\{\mathcal{B}_\alpha, \alpha \in J\}$ is a monotone net of W^*-subalgebras of $\mathcal{A}$ for which $E(\cdot|\mathcal{B}_\alpha)$ exists for each $\alpha \in J$. (We always assume that each $\mathcal{B}_\alpha$ has the identity.) Then*

(i) for increasing $\mathcal{B}_\alpha$, we have $\lim_\alpha E(x|\mathcal{B}_\alpha) = E(x|\mathcal{B})$ in norm, where $\mathcal{B}$ is the smallest W^*-algebra containing $\cup \mathcal{B}_\alpha$, and

(ii) for decreasing $\mathcal{B}_\alpha$, if $\mathcal{B} = \underset{\alpha}{\cap} \mathcal{B}_\alpha$ and if $E(\cdot|\mathcal{B})$ exists, then $\lim_\alpha E(x|\mathcal{B}_\alpha) = E(x|\mathcal{B})$ is norm.

Proof. (i) If $y = E(x|\mathcal{B})$ for $x \in \mathcal{A}$, then $E(y|\mathcal{B}) = y$ so we may assume for simplicity that $\mathcal{B} = \mathcal{A}$, and show that $\|x - E(x|\mathcal{B}_\alpha)\| \to 0$ where if φ is the faithful normal state, relative to which the $E(\cdot|\mathcal{B}_\alpha)$ exist, we have $\|y\|^2 = \varphi(y^*y)$. If $d_\alpha = \inf\{\|x - y\| : y \in \mathcal{B}_\alpha\}$, then $\mathcal{B}_\alpha \subset \mathcal{B}_{\alpha'}$ for $\alpha < \alpha' \Rightarrow d_\alpha \geq d_{\alpha'} \geq 0$. Let $d = \lim_\alpha d_\alpha$. Since $\cup \mathcal{B}_\alpha$ is dense in $\mathcal{A}$, by property (vi) of Proposition 4.3, we must have $d = 0$ as desired.

(ii) By hypothesis $E(\cdot|\mathcal{B})$ exists where $\mathcal{B} = \underset{\alpha}{\cap} \mathcal{B}_\alpha$. To show that $\|E(x|\mathcal{B}) - E(x|\mathcal{B}_\alpha)\| \to 0$ we may assume that $P(x|\mathcal{B}) = 0$. In fact, if $x_1 \in \mathcal{A}$ and $x = x_1 - E(x_1|\mathcal{B})$, then $E(x|\mathcal{B}) = 0$ and since $E(E(\cdot|\mathcal{B})|\mathcal{B}_\alpha) = E(\cdot|\mathcal{B})$ we have

$$E(x|\mathcal{B}_\alpha) = E(x_1|\mathcal{B}_\alpha) - E(E(x_1|\mathcal{B})|\mathcal{B}_\alpha) = E(x_1|\mathcal{B}_\alpha) - E(x_1|\mathcal{B}) \to 0\ .$$

So the original statement is equivalent to showing $E(x|\mathcal{B}_\alpha) \to 0$. Thus let $E(x|\mathcal{B}) = 0$ and consider $\mathcal{S}_\alpha = \{E(x|\mathcal{B}_\beta) : \beta > \alpha\}$ for $x \in \mathcal{A}$. Then $\mathcal{S}_\alpha$ is nonempty and $\|E(x|\mathcal{B}_\alpha)\| \leq \|x\|$, by Theorem 4.5. So $\mathcal{S}_\alpha$ is in a ball of radius $\|x\|$, and for $\alpha_1, \alpha_2, \cdots \leq \beta$, $\bigcap_{i=1}^{n} \mathcal{S}_{\alpha_i} \supset \mathcal{S}_\beta$. But $\mathcal{A}$ is a W^*-algebra. Now its balls are weak *-compact (i.e. if $\mathcal{A}_*$ is the predual of $\mathcal{A}$, then it is $\sigma(\mathcal{A}_*, \mathcal{A})$-compact), and so is $\bar{\mathcal{S}}_\beta$, its closure. But $\mathcal{S}_\alpha \subset \{E(x|\mathcal{B}_\alpha), x \in \mathcal{A}\}$, and it has a cluster point y which belongs to the set $\{ \ \}$ for all α, i.e. $y \in \bigcap_\alpha \{E(x|\mathcal{B}_\alpha) : x \in \mathcal{A}\} \subset \{E(x|\mathcal{B}) : x \in \mathcal{A}\}$. Hence $E(y|\mathcal{B}) = y$. However, we also have by the initial simplification,

$$0 = E(E(x|\mathcal{B}_\alpha)|\mathcal{B}) = E(x|\mathcal{B}) \ .$$

so that $y = E(y|\mathcal{B}) = 0$, the weak cluster point of $\{E(x|\mathcal{B}_\alpha), \alpha \in I\}$.

Since $\varphi = \varphi \circ E(\cdot|\mathcal{B})$, consider for $\beta > \alpha$,

$$
\begin{aligned}
\varphi(E(x|\mathcal{B}_\beta)^* E(x|\mathcal{B}_\beta)) &= \varphi(E(E(x|\mathcal{B}_\alpha)|\mathcal{B}_\beta)^*(E(E(x|\mathcal{B}_\alpha)|\mathcal{B}_\beta))) \\
&\leq \varphi(E(x|\mathcal{B}_\alpha)^* E(x|\mathcal{B}_\alpha)), \quad \text{by Proposition 4.3. (i),} \\
&= \varphi(E(E(x|\mathcal{B}_\alpha)^* E(x|\mathcal{B}_\alpha)|\mathcal{B})) \\
&= \varphi(E(E(x|\mathcal{B}_\alpha)^* x|\mathcal{B}_\alpha)), \quad \text{by Proposition 4.3 (ii),} \\
&= \varphi(E(x|\mathcal{B}_\alpha)^* x). \quad\quad\quad\quad\quad\quad (4)
\end{aligned}
$$

By the weak *-continuity of τ on bounded sets and the fact that the set $\{E(x|\mathcal{B}_\alpha), \alpha \geq \alpha_0\}$ clusters at 0, we conclude that the right side of (4) tends to 0, so the left side and hence $E(x|\mathcal{B}_\alpha) \to 0$ in the same sense. But in W^*-algebras this and strong convergence on bounded sets agree (cf. Dixmier (1962), Prop. 58, p. 58). This is the assertion. $\square$

Remark. The above proof is adapted from Arveson (1967). It was also proved in a slightly diferent form by Tsukada (1983).

The convergence statement for the problem raised at the end of Example 2 is contained in the following:

Corollary 4. *Let $\mathcal{A}$ be a W^*-algebra, $\mathcal{P}(\subset \mathcal{A})$ be an abelian family of (orthogonal) projections and $\mathcal{B} = \mathcal{P}' \cap \mathcal{A}$ which admits a conditional expectation $E(\cdot|\mathcal{B})$ relative to a given state. Then for each $x \in \mathcal{A}, x_\mathcal{F} \to E(x|\mathcal{B})$ in the strong topology, where $x_\mathcal{F}$ is defined by (3).*

Analogous results for certain generalized conditional expectation operators have been proved by Hiai and Tsukada (1984, 1987).

There are also possibilities for the almost everywhere convergence statements for martingales. For this we introduce the relevant concepts in:

Definition 5. Let $\{T_n, n \geq 1\}$ be a sequence in a W^*-algebra $\mathcal{A}$ with a faithful normal state τ. Then the sequence

(a) converges in *measure (almost everywhere)* to T in $\mathcal{A}$ if for each $\varepsilon > 0$ there is a sequence $\{P_n, n \geq 1\} \subset \mathcal{A}$ of projections such that $\|(T_n - T)P_n\| < \varepsilon$ as $\tau(P_n) \to 1$ for $n \to \infty$ (and as $P_n \uparrow I$ for $n \to \infty$);

(b) converges *almost uniformly* if for each $\delta > 0$, one can find a projection P in $\mathcal{A}$ such that $\tau(P) \geq 1 - \delta$, and $\|(T_n - T)P\| \to 0$ as $n \to \infty$.

With these concepts we can establish the following analogs of the classical measure theory for finite W^*-algebras. These are shown in Padmanabhan (1967).

Theorem 6. *Let $\mathcal{A}$ be a finite W^*-algebra (i.e. there is a state τ with $\tau(I) = 1$), and $\{x_n, n \geq 1\} \subset \mathcal{A}$. Then $x_n \to x$ almost everywhere iff it converges almost uniformly. If the sequence (only) converges in measure to x, and if there is a $0 \leq y \in \mathcal{A}$ integrable (i.e. $\|y\|_1 < \infty$) such that*

$$-y \leq Re\,(x_n),\ Im(x_n) \leq y \tag{5}$$

where $Re\,x_n = \frac{1}{2}(x_n + x^*)$ and $Im(x_n) = \frac{1}{2i}(x_n - x_n^*)$, and $\mathcal{B} \subset \mathcal{A}$ is a W^*-algebra for which $E(\cdot|\mathcal{B})$ exists [which is automatically true in finite algebras as here], then $E(x_n|\mathcal{B}) \to E(x|\mathcal{B})$ in norm.

This result is just the Egorov and the dominated convergence statements of the classical theory. The roles of points (or sets) and measures of the latter subject are played by the projections and states in the present case. We shall not include a proof here. The corresponding statement for martingale convergence is as follows.

Proposition 7. *Let $\mathcal{B}_n \subset \mathcal{B}_{n+1} \subset \mathcal{A}$ be a sequence of W^*-subalgebras with $\varphi : \mathcal{A} \to \mathbb{R}$ as a state such that $E(\cdot|\mathcal{B}_n)$ exists for each n. Let $x_n \in \mathcal{A}$ be such that $E(x_{n+1}|\mathcal{B}_n) = x_n, n \geq 1$, so that $\{x_n, \mathcal{B}_n, n \geq 1\}$ is a martingale sequence on $(\mathcal{A}, \varphi)$. If $\sup_n \|x_n\|_2 < \infty$ where $\|x_n\|_2^2 = \varphi(x_n^* x_n)$, then the martingale converges almost everywhere and in norm to x in $\mathcal{A}$ where $E(x|\mathcal{B}_n) = x_n$, $n \geq 1$.*

This and certain other results on martingale convergence are given by Cuculescu (1971). Supermartingales in this context are defined with $E(x_{n+1}|\mathcal{B}_n) \leq x_n, n \geq 1$, and similarly one can obtain several results based on the classical case. However the techniques of proofs are different, and we refer the reader to the above paper for details.

It may be instructive here to consider a martingale of scalar functions on a vector measure space, where the measures have values in a special algebra. The new problems will be clarified by that example which is close to the classical theory and yet depending on a new Radon-Nikodým theorem. The following discussion complements the work of Section 6.5. Recall from that section that the space of all real continuous functions on a compact Stone space S, denoted $C(S)$, is a Stone algebra [with uniform norm], and suppose that it has the countable chain property, i.e., each bounded subset of $C(S)$ has a countable subset with the same supremum. This is always satisfied if $C(S)$ is isometrically isomorphic to $L^\infty(P)$. Suppose (Ω, Σ) is a measurable space [as in our earlier work on which P is defined], and $\nu : \Sigma \to C(S)$ be a vector measure which is positive and σ-additive in the order topology. We consider the space $L^\infty(\Omega, \Sigma, \nu)$ and suppose that there is an algebra homomorphism $\pi : C(S) \to L^\infty(\Omega, \Sigma, \nu)$ such that

$$\int_\Omega \pi(f)g \, d\nu = f \int_\Omega g \, d\nu, \quad f \in C(S), \quad g \in L^\infty(\Omega, \Sigma, \nu). \quad (6)$$

Such a ν is called a modular measure relative to π. If $\nu = P^\mathcal{B}$ for a σ-subalgebra $\mathcal{B} \subset \Sigma$, then (6) is satisfied and we have a modular measure. By Proposition 6.5.3, for each $g \in L^1(\Omega, \Sigma, \nu)$, the conditional expectation of g, namely $E(g|\mathcal{B})$, exists, is in $L^1(\Omega, \mathcal{B}, \nu)$, and satisfies $\int_A f d\nu = \int_A E(f|\mathcal{B})d\nu$, $A \in \mathcal{B}$. We now establish the following martingale convergence result.

Theorem 8. *Let (Ω, Σ) be a measurable space, $\nu : \Sigma \to C(S)$ be a modular measure relative to π where $C(S)$ is a Stone algebra satisfying the countable chain condition and $\pi(C(S)) \subset L^\infty(\Omega, \mathcal{B}_1, \nu)$, with $\mathcal{B}_1 \subset \mathcal{B}_2 \subset \cdots \subset \Sigma$ being σ-subalgebras. Then for each $f \in L^1(\Omega, \Sigma, \nu)$, letting $f_n = E(f|\mathcal{B}_n)$ $(\in L^1(\Omega, \mathcal{B}_n, \nu))$, we have $\lim_n f_n = E(f|\mathcal{B})$, a.e. relative to ν, where $\mathcal{B} = \sigma(\bigcup_n \mathcal{B}_n)$.*

Proof. The argument is similar to the classical one due to E.S. Anderson and B. Jessen, and the latter is given in detail in (Rao (1979),

p. 122). We present the essentials with simple modifications here for completeness.

If $f_n \to f$ is false on a set of "positive ν-measure", i.e., if $F = \{\omega : \liminf_n f_n(\omega) < \tilde{f}(\omega)\}, \tilde{f} = E(f|\mathcal{B})$, then $\nu(F) > 0$ in that $\|\nu(F)\| > 0$. But if $F_{ab} = \{\omega : \liminf_n f_n(\omega) \leq a < b \leq \tilde{f}(\omega)\}$ then $F = \cup\{F_{ab} : a < b, \text{ rational}\}$, and so $\nu(F_{ab}) \neq 0$ for some $a < b$.

Let $H_a = \{\omega : \liminf_n f_n(\omega) \leq a\}$ and $H_n = \{\omega : \inf_{r \geq 1} f_{n+r}(\omega) < a + \frac{1}{n}\}$, so that $H_a = \overset{\infty}{\underset{n=1}{\cap}} H_n$. Then $H_a, H_n \in \mathcal{B}$. We express H_n as a disjoint union in the following way. Let $H_{n1} = \{\omega : f_{n+1}(\omega) < a + \frac{1}{n}\}$ and for $r > 1$,

$$H_{nr} = \{\omega : f_{n+r}(\omega) < a + \frac{1}{n}, \text{ and } f_{n+j}(\omega) \geq a + \frac{1}{n} \text{ for } 1 \leq j \leq r-1\}.$$

Note that $H_{nr} \in \mathcal{B}_{n+r}$, $H_n = \overset{\infty}{\underset{r=1}{\cup}} H_{nr}$ (disjoint union), and $H_{nr} \subset \{\omega : F_{n+r}(\omega) < a + \frac{1}{n}\}$. If $A \in \overset{\infty}{\underset{n=1}{\cup}} \mathcal{B}_n$, then $A \in \mathcal{B}_{n_0}$ for some n_0 and for all $n \geq n_0$ and $r \geq 1$, we have $H_{nr} \cap A \in \mathcal{B}_{n+r}$. Since the dominated and monotone convergence theorems hold for ν-integrals (see Sec. 6.5), and since $\tilde{f} = E(f|\mathcal{B})$ is ν-integrable, we have

$$\int_A \tilde{f} \chi_{H_n} d\nu = \sup_m \int_A \overset{m}{\underset{r=1}{\Sigma}} \tilde{f} \chi_{H_{nr}} d\nu. \tag{7}$$

Also

$$\int_A \tilde{f} \chi_{H_{nr}} d\nu = \int_{A \cap H_{nr}} \tilde{f} d\nu = \int_{A \cap H_{nr}} f_{n+r} d\nu, \text{ by the martingale property,}$$

$$\leq (a + \frac{1}{n}) \int_A \chi_{H_{nr}} d\nu . \tag{8}$$

Substituting (8) in (7) we get, on using $H_n = \overset{\infty}{\underset{r=1}{\cup}} H_{nr}$,

$$\int_A \tilde{f} \chi_{H_n} d\nu \leq (a + \frac{1}{n}) \int_A \chi_{H_n} d\nu, \quad n \geq n_0 . \tag{9}$$

Taking limits as $n \to \infty$, and using the fact that $H_n \downarrow H_a$ we obtain

$$\int_A (a - \tilde{f}) \chi_{H_a} d\nu \geq 0, \quad A \in \overset{\infty}{\underset{n=1}{\cup}} \mathcal{B}_n . \tag{10}$$

Just as in the standard measure theory we can conclude that (10) holds for all $A \in \mathcal{B}$ itself. This may also be verified directly as follows.

Let g be the integrand in (10), and consider $\mathcal{C} = \{A \in \mathcal{B} : \int_A g \, d\nu \geq 0\}$. Then by (10), $\underset{n}{\cup} \mathcal{B}_n \subset \mathcal{C} \subset \mathcal{B}$. Consider all algebras contained in $\mathcal{C}$ that include the left algebra. Partially order them by inclusion. Then by Zorn's lemma, there is a maximal such algebra $\mathcal{D}$ satisfying $\underset{n}{\cup} \mathcal{B}_n \subset \mathcal{D} \subset \mathcal{C}$. We claim that $\mathcal{D} = \mathcal{B}$ so that $\mathcal{C} = \mathcal{B}$ proving our assertion. Indeed let $\mathcal{D}^* = \{A : A = \lim_n A_n, A_n \in \mathcal{D}\}$, where this means $\chi_A = \lim_n \chi_{A_n}$. By the dominated convergence for this vector integral we get (since $A \in \mathcal{B}$ obviously)

$$\int_A g \, d\nu = \int_S g \chi_A \, d\nu = \lim_n \int_S g \chi_{A_n} \, d\nu = \lim_n \int_{A_n} g \, d\nu \geq 0 \, ,$$

so that $A \in \mathcal{C}$. It is evident that $\mathcal{D}^*$ is closed under intersections and differences so that it is an algebra and by the above computation $\mathcal{D}^* \subset \mathcal{C}$ and is a σ-algebra. By the maximality of $\mathcal{D}$, we have $\mathcal{D}^* = \mathcal{C}$ and $\mathcal{C} = \sigma(\mathcal{C}) = \mathcal{B}$, so that $\mathcal{D}^* = \mathcal{B} = \mathcal{C}$ as desired.

Thus (10) holds for $A \in \mathcal{B}$. Since $F_{ab} \subset H_a$ this implies

$$a\nu(F_{ab}) \geq \int_{F_{ab}} \tilde{f} \, d\nu \geq b \, \nu(F_{ab}).$$

Since $a < b$, this can hold only if $\nu(F_{ab}) = 0$. Then $\liminf_n f_n = \tilde{f}$ a.e. Similarly $\limsup_n f_n = \tilde{f}$ can be established (or apply the above result to $-f$ and deduce from it). Hence $\lim_n f_n = \tilde{f} = E(f|\mathcal{B})$, a.e. $\square$

Remark. We thus see that the Radon-Nikodým and the Dominated Convergence theorems are needed for the new structures. The above result is due to Wright (1969b) who developed the necessary theory of these order integrals.

Recall that the vector integrals of Wright (i.e., with those of Stone algebra valued measures) were considered in Chapter 6 to study sufficiency in the undominated case. Thus conditioning in *-algebras of this and the preceding sections may also be of use in a similar study. To see that such a desire is realizable, we now define sufficiency in *-algebras, following Hiai-Ohya-Tsukada (1983), to round out our introduction to noncommutative conditioning, with this application to (abstract) statistical theory.

Definition 9. Let $\mathcal{A}$ be a *-algebra with identity and $\mathcal{S}$ be the set of all faithful normal states on $\mathcal{A}$. A *-subalgebra $\mathcal{B}$ of $\mathcal{A}$ with identity is said to be *sufficient* (in the weak sense) for a fixed set $\mathcal{S}_1 \subset \mathcal{S}$ of states, each normalized with value 1 at the identity, if the following condition holds: For each $x \in \mathcal{A}$, and a positive $\varphi \in \mathcal{S}_1$ there is a fixed sequence $\{y_n, n \geq 1\}$ of elements in $\mathcal{B}$ such that (the symbols are again recalled below)

$$E_\varphi(x|\mathcal{B}) = s - \lim_{n \to \infty} \pi_\varphi(y_n)x_\varphi^o . \tag{11}$$

$\mathcal{B}$ is *sufficient in the strong sense* if there is a single expectation $E(\cdot|\mathcal{B})$ such that $\varphi \circ E(\cdot|\mathcal{B}) = \varphi$ for all $\varphi \in \mathcal{S}_1$.

In (11) π_φ is a *-representation of $\mathcal{A}$ on a Hilbert space $\mathcal{H}_\varphi$ with a dense domain D_{π_φ} and a cyclic vector x_φ^o such that $D_{\pi_\varphi} = \pi_\varphi(\mathcal{A})x_\varphi^o$, $\mathcal{H}_\varphi = \overline{\pi_\varphi(\mathcal{A})x_\varphi^o}$, and $\varphi(x) = \langle x_\varphi^o, \pi_\varphi(x)x_\varphi^o \rangle$ for all $x \in \mathcal{A}$. This construction of the triple $(\mathcal{H}_\varphi, \pi_\varphi, x_\varphi^o)$ is usually called the *Gel'fand-Naĭmark-Segal* (or GNS) construction of the underlying $\mathcal{A}$. According to Theorem 4.5, the operator $E_\varphi(\cdot|\mathcal{B})$ exists and satisfies $\varphi \circ E_\varphi(\cdot|\mathcal{B}) = \varphi$ if φ restricted to $\mathcal{B}$ is semifinite. If there is a $\varphi_0 \in \mathcal{S}_1$ which dominates all other states of $\mathcal{S}_1$ then one can establish the following result as in Section 6.3.

Proposition 10. Let $\mathcal{S}_1 (\subset \mathcal{S})$ be a dominated family of states and $\mathcal{B}$ be a *-subalgebra with identity in $\mathcal{A}$. Then the following are true:

> (a) if $\mathcal{B}$ is sufficient for $\mathcal{S}_1$, then the same is true for the convex hull of $\mathcal{S}_1$, and if $\mathcal{C}$ is a sufficient *- subalgebra of $\mathcal{B}$ for $\mathcal{S}_1|\mathcal{B}(= \varphi_\mathcal{B} : \varphi \in \mathcal{S}_1)$ then $\mathcal{C}$ is sufficient for $\mathcal{S}_1$ itself, and
>
> (b) if $\mathcal{B}$ is a sufficient W^*-subalgebra of a W^*-algebra $\mathcal{A}$ and $\mathcal{B} \subset \mathcal{C} \subset \mathcal{A}$, $\mathcal{C}$ is a W^*-algebra, then $\mathcal{C}$ is also sufficient for $\mathcal{S}_1$.

The next result is analogous to Theorem 6.3.4 and is adapted from Hiai, Ohya and Tsukada (1983), (the above notation is again used):

Theorem 11. Let $\mathcal{B}$ be a *-subalgebra, with identity, of $\mathcal{A}$ and $\mathcal{S}$ be the set of all finite states of $\mathcal{A}$, and $\mathcal{S}_1 \subset \mathcal{S}$ be given. Suppose there is a $\varphi_0 \in \mathcal{S}$ (φ_0 can be in $\mathcal{S} - \mathcal{S}_1$) dominating each element of $\mathcal{S}_1$. Then $\mathcal{B}$ is sufficient (weakly) relative to $\mathcal{S}_1$ iff $\frac{d\varphi}{d\varphi_0}\pi_{\varphi_0}(\mathcal{B})x_{\varphi_0}^o \subset \overline{\pi_{\varphi_0}(\mathcal{B})x_{\varphi_0}^o}$, for all $\varphi \in \mathcal{S}_1$.

Proof. Let $\mathcal{B}$ be weakly sufficient for $\mathcal{S}_1$. Then for each $x \in \mathcal{A}$, there exists a $y_n \in \mathcal{B}$, $n \geq 1$, such that (11) holds. Consequently $\psi = s-\lim_n \overline{\pi_{\varphi_0}}(y_n)x_{\varphi_0}^o$ exists and belongs to $\overline{\pi_{\varphi_0}(\mathcal{B})x_{\varphi_0}^o} = \bar{\mathcal{B}}_{\varphi_0}$ (say). But for any $\varphi \in S_1, z = \frac{d\varphi}{d\varphi_0}$ exists by Theorem 4.4. If φ_B, φ_{0B} are restrictions of the states φ, φ_0 to B, let $\hat{z} = \frac{d\varphi_B}{d\varphi_{0B}}$ which again exists since φ_{0B} is finite on B. Consider for any $y \in B$ the following:

$$
\begin{aligned}
\langle z\pi_{\varphi_B}(y)x_{\varphi_0}^0, \pi_{\varphi_0}(x)x_{\varphi_0}^0 \rangle &= \varphi(y^*x), \text{ by definition,} \\
&= \langle \pi_\varphi(y)x_\varphi^0, E_\varphi(x|B) \rangle \\
&= \lim_n \varphi(y^*y_n) \\
&= \langle \hat{z}\pi_{\varphi_0}(y)x_{\varphi_0}^0, E_{\varphi_0}(x|B) \rangle, \quad \text{by (11),} \\
&= \langle \hat{z}\pi_{\varphi_0}(y)x_{\varphi_0}^0, \pi_{\varphi_0}(x)x_{\varphi_0}^0 \rangle.
\end{aligned} \tag{12}
$$

Since x in A is arbitrary (12) implies

$$
z\pi_{\varphi_0}(y)x_{\varphi_0}^0 = \hat{z}\pi_{\varphi_0}(y)x_{\varphi_0}^0 \in \bar{B}_{\varphi_0}, \quad y \in B. \tag{13}
$$

This shows that $z B_{\varphi_0} \subset \bar{B}_{\varphi_0}$.

In the opposite direction, suppose this inclusion holds so that $\frac{d\varphi}{d\varphi_0} B_{\varphi_0} \subset \bar{B}_{\varphi_0}$, $\varphi \in S_1$. Let $x \in A$, and consider $y_n \in B$ such that $E_{\varphi_0}(x|B) = s - \lim \pi_{\varphi_0}(y_n)x_{\varphi_0}^0$. Since φ is dominated by φ_0, the right side sequence determines a vector ψ in $\bar{B}_\varphi$. Consider for any $y \in B$,

$$
\begin{aligned}
\langle \pi_\varphi(y)x_\varphi^0, E_\varphi(x|B) \rangle &= \varphi(y^*x) \\
&= \langle \frac{d\varphi}{d\varphi_0}\pi_{\varphi_0}(y)x_{\varphi_0}^0, E_{\varphi_0}(x|B) \rangle \\
&= \lim_n \varphi(y^*y_n) = \langle \pi_\varphi(y)x_\varphi^0, \psi \rangle.
\end{aligned} \tag{14}
$$

Since x in A is arbitrary, we have from (8) $E_\varphi(x|B) = \psi = s - \lim \pi_\varphi(y_n)x_\varphi^0$. Hence B is weakly sufficient for S_1, as asserted. □

For strongly sufficient case the above result takes a simpler form:

Corollary 12. S_1 *is a dominated set of states, then a* * *-subalgebra* B *of* A, *having an identity, is strongly sufficient iff* B *is weakly sufficient and* $\varphi = \varphi(E_{\varphi_0}(\cdot|B))$ *for all* $\varphi \in S_1$, φ_0 *being the dominating state.*

Proof. Only the last identity remains to be verified. For this consider with the y_n of (11),

$$
\begin{aligned}
\varphi(E_{\varphi_0}(x|\mathcal{B})) &= \langle \frac{d\varphi}{d\varphi_0} x^0_{\varphi_0}, \ \pi_{\varphi_0}(E_{\varphi_0}(x|\mathcal{B}))x^0_{\varphi_0}\rangle \\
&= \lim_n \langle \pi_{\varphi_0}(y_n)x^0_{\varphi_0}, \ \pi_{\varphi_0}(E_{\varphi_0}(x|\mathcal{B}))x^0_{\varphi_0}\rangle \\
&= \lim_n \varphi_0(y_n^* E_{\varphi_0}(x|\mathcal{B})) \\
&= \lim_n \varphi_0(y_n^* x) = \langle \frac{d\varphi}{d\varphi_0} x^0_{\varphi_0}, \ \pi_{\varphi_0}(x)x^0_{\varphi_0}\rangle \\
&= \varphi(x), \quad x \in \mathcal{A}.
\end{aligned}
$$

The converse simply reverses the steps. $\square$

The following example connects the above corollary with Theorem 6.4.3.

Example. Let (Ω, Σ) be a measurable space and Π a set of probability measures on Σ. We then recall that a σ-subalgebra $\mathcal{G}$ of Σ is sufficient for Π if for each $A \in \Sigma$ there is a measurable function $g : \Omega \to \mathbb{R}^+$ such that $E_P^{\mathcal{G}}(\chi_A) = g$ a.e. $[P]$, for all $P \in \Pi$, where $E_P^{\mathcal{G}}$ is the conditional expectation on (Ω, Σ, P) relative to $\mathcal{G}$. Let $\mathcal{A}(\mathcal{B})$ be the set of all complex simple functions on Ω which are measurable relative to $\Sigma(\mathcal{G})$. Then $\mathcal{A}$ and $\mathcal{B}$ become *-algebras under pointwise multiplication and conjugation. Let $\varphi_P(f) = \int_\Omega f \, dP$, $f \in \mathcal{A}$, so that φ_P is a state on $\mathcal{A}$ and let $\mathcal{S}_1 = \{\varphi_P : P \in \Pi\}$. If $\mathcal{H}_P = L^2(\Omega, \Sigma, P)$, $\pi_{\varphi_P}(f)g = fg$, $g \in \mathcal{A}$, and $x^0_{\varphi_P} = 1$ then $\{\mathcal{H}_P, \pi_{\varphi_P}, 1\}$ becomes a GNS-representation of $\mathcal{A}$, $E_P(\cdot|\mathcal{B})$ becomes $E_P^{\mathcal{G}}(\cdot)$ and the sufficiency of $\mathcal{G}$ for Π becomes the sufficiency of $\mathcal{B}$ for $\mathcal{S}_1$ respectively in Theorem 6.4.3 and in Theorem 11 above.

Other applications to entropy and information theory can be made extending the commutative theory, and some of this was already done by the above authors. We shall conclude this study by noting the fact that noncommutative conditioning and its applications can be carried forward following the material in earlier chapters. Thus there is a great potential for further work which can have applications in quantum mechanics as well as in other areas.

11.6 Bibliographical notes

In this chapter we presented some applications of conditioning if the spaces have additional algebraic structure. The work of Section 2 is primarily related to Moy's (1954) paper. It is the averaging identity that is prominent in this analysis. The relation between positive contractive projections on function algebras and averaging operators is clarified by Theorem 3.1 which is due independently to Seever (1966) and Lloyd (1966), and an extension to a nonpositive case to Wulbert (1971). They were also discussed in Rao (1975).

The algebraic structure of the particular function spaces indicated that one may continue the corresponding study of (noncommutative) conditioning as in function algebras. Indeed this extension was essentially used in quantum physics as well as in structural analysis of operator algebras. The work is again based on the Kolmogorov model via the Radon-Nikodým theory, appropriate to the situation, and hence the noncommutative integration. The latter was pioneered by Segal (1953) and extended by Kunze (1959), Stinespring (1959) and (to finite W^*-algebras) by Padmanabhan (1967). We sketched a (minimum) necessary part in Section 4 following a short cut formulated and developed by Gudder and Hudson (1978). Theorems 4.4 and 4.5 are adapted from the latter paper.

The noncommutative conditional expectation was first introduced by Umegaki (1954) who thereafter applied the concept in several different directions and inspired some of his students and associates. The presentation in the text is adapted from Arveson (1967) whose study attests to the basic role played by this conditioning in an analysis of certain classes of operator algebras. As applications of the noncommutative conditioning we included some martingale convergence theorems. The strong martingale convergence is studied by Hiai and Tsukada (1984, 1987) in a more general case and the pointwise convergence by Cuculescu (1971). The corresponding work with vector measures, still in the commutative case (as a prelude), is due to Wright (1969b).

The brief account in the last part of Section 5 on sufficiency is to show how the ideas of Chapter 6 can be extended to the noncommutative case. The work here followed Hiai, Ohya, and Tsukada (1983) who also studied the subject on (relative) entropy and information theory, following the lead of H. Umegaki. All these areas are active and interesting. They demand new methods to extend the subject treated in the preceding chapters and present challenging problems for future research. The area of noncommutative probability is still in formative stages and it depends and uses the theory of C^*-algebras and W^*-algebras. For a readable account one can start with Arveson (1976) and proceed to Sakai (1971) and then to Dixmier's (1969) classic. Probabilists need to study this area from their vantage point and hopefully our introduction here induces and invites them to future research in this potentially interesting area.

REFERENCES

Akcoglu, M.A. and R.V. Chacón (1965). "A convexity theorem for positive operators," *Z. Wahrs.* **3**, 328–332.

Andô, T. (1966). "Contractive projections on L^p-spaces," *Pacific J. Math.* **17**, 391–405.

Andô, T. and I. Amemiya (1965). "Almost everywhere convergence of prediction sequence in L^p, $1 < p < \infty$," *Z. Wahrs.* **4**, 113–120.

Aniszezyk, B., J. Burzyk, and A. Kamiński (1987). "Borel and monotone hierarchies and existence of Rényi's probability spaces," *Colloq. Math.* **51**, 9–25.

Aronszajn, N. (1950). "Theory of reproducing kernels," *Trans. Amer. Math. Soc.* **68**, 337–404.

Arveson, W.B. (1967). "Analyticity in operator algebras," *Amer. J. Math.* **89**, 578–642.
(1976), *An Invitation to C^*-Algebra*, Springer-Verlag, New York.

Bahadur, R.R. (1954). "Sufficiency and statistical decision functions," *Ann. Math. Statist.* **25**, 423–462.
(1955). "Statistics and subfields," *Ann. Math. Statist.* **26**, 490–497.

Berberian, S.K. (1965). *Measure and Integration*, Macmillan Co., New York.

Blackwell, D. (1942). "Idempotent Markov chains," *Ann. Math.* **43**, 560–567.

Blackwell, D. and L.E. Dubins (1963). "A converse to the dominated convergence theorem," *Ill. J. Math.* **7**, 508–514.

Blackwell, D. and C. Ryll-Nardzewski (1963). "Nonexistence of everywhere proper conditional distributions," *Ann. Math. Statist.* **34**, 223–225.

Bochner, S. (1947). "Stochastic processes," *Ann. Math.* **48**, 1014–1061.

(1955). *Harmonic Analysis and the Theory of Probability*, Univ. of Calif. Press, Berkeley and Los Angeles, CA.

Bourbaki, N. (1966). *Elements of Mathematics: General Topology, Part I*, Herman (Paris) and Addison-Wesley Publishing Co., Reading, MA.

(1968). *Elements of Mathematics: Theory of Sets*, Herman (Paris), and Addison-Wesley Publishing Co., Reading, MA.

Brody, E.J. (1971). "An elementary proof of the Gaussian dichotomy theorem," *Z. Wahrs.* **20**, 217–226.

Bruckner, A.M. (1971). "Differentiation of integrals." *Amer. Math. Monthly*, **78**, (November, Part II), 51 pp.

Burkholder, D.L. (1961). "Sufficiency in the undominated case," *Ann. Math. Statist.* **32**, 1191–1200.

(1962). "On the order structure of the set of sufficient subfields," *Ann. Math. Statist.* **33**, 596–599.

Burkill, J.C. (1924). "Functions of intervals," *Proc. London Math. Soc.* **22**, 275–310.

(1924). "The expressions of area as an integral," *Proc. London Math. Soc.* **22**, 311–336.

Čelidze, V.G. and A.G. Džvaršeišvili (1989). *The Theory of the Denjoy Integral and Some Applications*, World Scientific Publishing Co., Singapore.

Chow, Y.S. (1960). "Martingales in a σ-finite measure space indexed by directed sets," *Trans. Amer. Math. Soc.* **97**, 254–285.

Chow, Y.S., S. Moriguti, H. Robbins, and S.M. Sammuels (1964). "Optimal selection based on relative rank (the 'secretary' problem)," *Israel J. Math.* **2**, 81–90.

Chow, Y.S., H. Robbins and D. Siegmund (1971). *Great Expectations: The Theory of Optimal Stopping*, Houghton-Mifflin Co., Boston.

Chow, Y.S., and H. Teicher (1978). *Probability Theory*, Springer-Verlag, New York.

Cramér, H. (1946). *Mathematical Methods of Statistics*, Princeton University Press, Princeton, NJ.

Cramér, H. and M.R. Leadbetter (1967). *Stationary and Related Stochastic Processes,* J. Wiley and Sons, New York.

Császár, Á. (1955). "Sur la structure des espaces de probabilité conditionellé," *Acta Math. Acad. Sci. Hung.* **6**, 337–361.

Cuculescu, I. (1971). "Martingales on von Neumann algebras," *J. Multivar. Anal.* **1**, 17–27.

Day, M.M. (1962). *Normed Linear Spaces*, Springer-Verlag, New York.

DeFinetti, B. (1932). "La prévision: ses lois logique, ses sources subjectives," *Ann. de l'H. Poincaré*, **7**, 1–68.

DeGroot, M.H. (1970). *Optimal Statistical Decisions*, McGraw-Hill, New York.
(1986). *Probability and Statistics* (2nd ed.), Addison-Wesley, Reading, MA.

Dellacherie, C. and P.-A. Meyer (1988). *Probability and Potential, C (Potential Theory for Discrete and Continuous Semigroups)*, North-Holland, Amsterdam, The Netherlands.

Dinculeanu, N. (1967). *Vector Measures*, Pergamon Press, London, UK.

Dinculeanu, N. and I. Kluvanek (1967). "On vector measures," *Proc. London Math. Soc.* **17**, 505–512.

Dixmier, J. (1969). *Les algèbres d'opérateurs dan l'espace Hilbertian (Algebres de von Neumann).* (2nd ed.), Gautier-Villar, Paris.

Doob, J.L. (1953) *Stochastic Processes*, Wiley and Sons, New York.
(1963). "A ratio operator limit theorem," *Z. Wahrs.* **1**, 288–294.

Douglas, R.G. (1965). "Contractive projections in an L^1-space," *Pacific J. Math.*, **15**, 443–462.

Dunford, N. and J.T. Schwartz (1958, 1963). *Linear Operators, Parts I and II*, Wiley-Interscience, New York.

Dynkin, E.B. (1951). "Necessary and sufficient statistics for a family of probability distributions," *Uspehi Math. Nauk (N.S.)*, **6**, (41), 68–90. (English Translation, AMS Translation (1961), 17–40).
(1961). *Theory of Markov Processes*, Prentice-Hall, Englewood Cliffs, NJ (English Translation).

Edgar, G.A. and L. Sucheston (1992). *Stopping Times and Directed Processes*, Cambridge University Press, New York.

Ennis, P. (1973). "On the equation $E(E(X|Y)) = E(X)$," *Biometrika*, **60**, 432–433.

Feller, W. (1960). *An Introduction to Probability Theory and its Applications*, Vols. I, II, Wiley and Sons, New York.

Fisher, R.A. (1922). "On the mathematical foundations of theoretical statistics," *Roy. Soc. Phil. Trans., Ser A.* **222**, 309–368.

Frolik, Z. and J. Pachl (1973). "Pure measures," *Comment. Math. Universität Carolinae*, **14**, 279–293.

Ghosh, J.K., H. Morimoto, and S. Yamada (1981). "Neyman factorization and minimality of pairwise sufficient subfields," *Ann. Statist.* **9**, 514–530.

Gilbert, J.P. and F. Mosteller (1966). "Recognizing the maximum of a sequence," *J. Amer. Statist. Assoc.* **61**, 35–73.

Gnedenko, B.V. and A.N. Kolmogorov (1954). *Limit Distributions for Sums of Independent Random Variables*, Addison-Wesley, Redding, MA (English Translation).

Gould, G.G. (1965). "Integration over vector valued measures," *Proc. London Math. Soc.* **15**, 193–225.

Gretsky, N.E. (1968). "Representation theorems on Banach function spaces," *Mem. Amer. Math. Soc.* **84**, 56 pp.

Gudder, S.P. and R. L. Hudson (1978). "A noncommutative probability theory," *Trans. Amer. Math. Soc.* **245**, 1–41.

Halmos, P.R. and L.J. Savage (1949). "Application of Radon-Nikodým theorem to the theory of sufficient statistics," *Ann. Math. Statist.* **20**, 225–241.

Hausdorff, F. (1962). *Set Theory* (2nd ed.), Chelsea Publishing Co., New York.

Hayden, R. (1977). "Injective Banach lattices," *Math. Z.* **156**, 19–47.

Hayes, C.A. and C.Y. Pauc (1970). *Derivation and Martingales*, Springer-Verlag, Berlin.

Hewitt, E. and L.J. Savage (1955). "Symmetric measures on cartesian products," *Trans. Amer. Math. Soc.* **80**, 470–501.

Heyer, H. (1982). *Theory of Statistical Experiments,* Springer-Verlag, New York.

Hiai, F., M. Ohya, and M. Tsukada (1981). "Sufficiency, KMS condition and relative entropy in von Neumann algebras," *Pacific J. Math.* **96**, 99–109.
(1983). "Sufficiency and relative entropy in *-algebras with applications in quantum systems," *Pacific J. Math.* **107**, 117–140.

Hiai, F. and M. Tsukada (1984). "Strong martingale convergence of generalized conditional expectations in von Neumann algebras," *Trans. Amer. Math. Soc.* **282**, 791–798.
(1987). "Generalized conditional expectations and martingales in noncommutative L^p-spaces," *J. Operator Theory* **18**, 265–288.

Hilbert, D. (1900). "Mathematical problems," *Bull. Amer. Math. Soc.* **8** (1902), 437–479, (Translation).

Hunt, G.A. (1966). *Martingales et Processus de Markov*, Dunod, Paris.

Ionescu Tulcea, C.T. (1949). "Mesures dans les espaces produits," *Atti Acad. Naz. Lincei Rend. el. Sci. Fis. Math. Nat* (8) **7**, 208–211.

Ionescu Tulcea, A. and C. Ionescu Tulcea (1964). "On the lifting property IV: Disintegration of measures," *Ann. Inst. Fourier Grenoble,* **14**, 445–472.

(1969) *Topics in the Theory of Lifting,* Springer-Verlag, Berlin.

Jessen, B., J. Marcinkiewicz, and A. Zygmund (1935). "Notes on the differentiability of multiple integrals," *Fund. Math.* **25**, 217–234.

Johnson, D.P. (1970). "Markov process representation of general stochastic processes," *Proc. Amer. Math. Soc.* **24**, 735–738.

(1979). "Representation of general stochastic processes," *J. Multivar. Anal.* **9**, 16–58.

Kac, M. and D. Slepian (1959). "Large excursions of Gaussian processes," *Ann. Math. Soc.* **30**, 1215–1228.

Kakutani, S. (1941). "Concrete representations of abstract M-spaces, (A characterization of the space of continuous functions)," *Ann. Math.* **42**, 994–1024.

(1944). "Two dimensional Brownian motion and harmonic functions," *Proc. Imp. Acad., Tokyo,* **20**, 706–714.

(1948). "On the equivalence of infinite product measures," *Ann. Math.* **49**, 214–224.

Kallianpur, G. (1959). "A note on perfect probability," *Ann. Math. Statist.* **30**, 169–172.

Kamiński, A. (1985). "On extensions of Rényi's conditional probability spaces," *Colloq. Math.* **49**, 267–294.

Knapp, A.W. (1965). "Connection between Brownian motion and potential theory," *J. Math. Anal. Appl.* **12**, 328–349.

Kolmogorov, A.N. (1930). "Untersuchungen uber den Integralbegriff," *Math. Ann.* **103**, 654–696.

(1933). *Foundations of Probability* (2nd ed. 1955 translation) Chelsea Publishing Co., New York.

Kransnoselskii, M.A. and Ya.B. Rutickii (1958). *Convex Functions and Orlicz Spaces* (English 1961), Noordhoff, Groningen, The Netherlands.

Krauss, P.H. (1968). "Representation of conditional probability measures on Boolian algebras," *Acta Math. Acad. Sci. Hung.* **9**, 224–241.

Kulakova, V.G. (1981). "Positive projections on symmetric KB-spaces," *Proc. Steklov Inst. of Math.* **155** (English translation 1983), 93–100.

Kunita, H. and T. Watanabe (1967). "Some theorems concerning resolvents over locally compact spaces," *Proc. 5th Berkeley Symp. on Math. Statist. and Prob.* **2**, *Part* II, 131–164.

Kunzi, R.A. (1959). "Lp-Fourier transforms on locally compact unimodular groups," *Trans. Amer. Math. Soc.* **89**, 519–540.

LeCam, L. (1986). *Asymptotic Methods in Statistical Decision Theory,* Springer-Verlag, New York.

Lloyd, S.P. (1966). "A mixing condition for extreme left invariant means," *Trans. Amer. Math. Soc.* **125**, 461–481.
(1974). "Two lifting theorems," *Proc. Amer. Math. Soc.* **42**, 128–134.

Loève, M. (1955). *Probability Theory,* (3rd ed. 1963) D. Van Nostrand Co., Princeton, NJ.

Losert, V. (1979). "A measure space without the strong lifting property," *Math. Ann.* **239**, 119–128.

Luschgy, H. and D. Mussmann (1985). "Equivalent properties and completion of statistical experiments," *Sankhyā, Ser A.* **47**, 176–195.

Manard, H.B. (1972). "A Radon-Nikodým theorem for operator valued measures," *Trans. Amer. Math. Soc.* **173**, 449–463.

Marczewski, E. (1953). "On perfect measures," *Fund. Math.* **40**, 113–124.

Markov, A.A. (1906). *Calculus of Probability* (in Russian) (4th ed. 1924), Moscow.

McShane, E.J. (1962). "Families of measures and representations of algebras of operators," *Trans. Amer. Math. Soc.* **102**, 328–345.

Meyer, P.-A. (1966). *Probability and Potentials*, Blaisdell Publishing Co., Waltham, MA.

Millington, H. and M. Sion (1973). "Inverse systems of group-valued measures," *Pacific J. Math.* **44**, 637–650.

Montgomery, D. and L. Zippin (1955). *Topological Transformation Groups*, Interscience Publishers, New York.

Moy, S.-C. (1954). "Characterizations of conditional expectation as a transformation on function spaces," *Pacific J. Math.* **4**, 47–64.

Neuts, M.P. (1973). *Probability*, Allyn and Bacon, Boston, MA.

Olson, M.P. (1965). "A characterization of conditional probability," *Pacific J. Math.* **15**, 971–983.

Ornstein, D.S. (1968). "On the pointwise behavior of iterates of a self-adjoint operator," *J. Math. Mech.* **18**, 473–478.

Pachl, J. (1975). "Every weakly compact probability in compact," *Bull. Acad. Pol. Sci. Mat.* **23**, 401–405.

Padmanabhan, A.R. (1967). "Convergence in measure and related results in finite rings of operators," *Trans. Amer. Math. Soc.* **128**, 359–378.

Parthasarathy, K.R. (1967). *Probability Measures on Metric Spaces*, Academic Press, New York.

Parzen, E. (1954). "On uniform convergence of families of sequences of random variables," *Univ. Calif. Publ. in Statist.* **2**, 23–54.

Pfanzagl, J. (1978). "Conditional distributions as derivatives," *Ann. Prob.* **7**, 1046–1050.

Pitcher, T.S. (1957). "Sets of measures not admitting necessary and sufficient statistics or subfields," *Ann. Math. Statist.* **28**, 267–268.

(1965). "A more general property then domination for sets of probability measures," *Pacific J. Math.* **15**, 597–611.

Ramachandran, D. (1981). "A note on regular conditional probabilities in Doob's sense," *Ann. Prob.* **9**, 907–908.

Ramamoorthi, R.V. and S. Yamada (1983). "On union of compact statistical structures," *Osaka J. Math.* **20**, 257–264.

Rao, M.M. (1965). "Conditional expectations and closed projections," *Indag. Math.* **27**, 100–112.
    (1967). "Abstract Lebesgue-Radon-Nikodým theorems," *Ann. Mat. Pura ed Appli.* (4), **76**, 107–132.
    (1969). "Stone-Weierstrass theorems for function spaces," *J. Math. Anal. Appl.* **25**, 362–371.
    (1970a). "Projective limits of probability spaces," *J. Multivar. Anal.* **1**, 28–57.
    (1970b). "Generalized martingales," in *Contributions to Ergodic theory and Probability*, Lect. Notes in Math. **160**, 241–261.
    (1970c). "Linear operations, tensor products, and contractive projections in function spaces," *Studia Math.* **38**, 133–186, Addendum, *ibid*, **48**, 307–308.
    (1973). "Remarks on a Radon-Nikodým theorem for vector measures," *Proc. of Vector and Operator Valued Measures and Applications*, Academic Press, New York, 303–317.
    (1974). "Inference in stochastic process -IV: Predictions and projections," *Sankhyā, Ser A.* **36**, 63–120.
    (1975). "Conditional measures and operators," *J. Multivar. Anal.* **5**, 330–413.
    (1976). "Two characterizations of conditional probability," *Proc. Amer. Math. Soc.* **59**, 75–80.
    (1979a). *Stochastic Processes and Integration*, Sijthoff and Noordhoff, Alphen aan den Rijn, The Netherlands.
    (1979b). "Bistochastic operators," *Comment. Math.* **21**, 301–313.
    (1981). *Foundations of Stochastic Analysis,* Academic Press, New York.

(1984). "Probability," *Encylopedia of Physical Science and Tech.,* **11**, Academic Press, 289–309.

(1984). *Probability Theory with Applications*, Academic Press, New York.

(1987). *Measure Theory and Integration*, Wiley-Interscience, New York.

(1988). "Paradoxes in conditional probability," *J. Multivar. Anal.* **27**, 434–446.

(1992). "Exact evaluation of conditional expectations in the Kolmogorov model," UCR Tech. Rept. **18**, (Revised version, *Indian J. Math.* (1993), to appear).

Rao, M.M. and Z.D. Ren (1991). *Theory of Orlicz Spaces*, Marcel Dekker, Inc., New York.

Rao, M.M. and V.V. Sazonov (1993). "A projective limit theorem for probability spaces and applications," *Theor. Prob. Appl.* (to appear).

Ray, D.B. (1959). "Resolvents, transition functions, and strongly Markovian processes," *Ann. Math.* **70**, 43–78.

Rényi, A. (1955). "On a new axiomatic theory of probability," *Acta Math. Acad. Sci. Hung.* **6**, 285–335.

(1956). "On conditional probability spaces generated by a dimensionally ordered set of measures," *Theor. Prob. Appl.* **1**, 55–64.

(1970) *Foundations of Probability*, Holden-Day, San Francisco, CA.

Revuz, D. and M. Yor (1991). *Continuous Martingales and Brownian Motion*, Springer-Verlag, New York.

Rice, S.O. (1958). "Distribution of the duration of fades in radio transmission: Gaussian noise model," *Bell System Tech. J.* **37**, 581–635.

Romanovski, P. (1941). "Integrale de Denjoy dans les espaces abstraits," *Recueil Mat. (Mat. Sbornik) NS.* **9** (51), 67–120.

Rosenberg, R.L. (1968). "Compactness in Orlicz spaces based on sets of probability measures," Unpublished Ph.D. thesis, Carnegie-Mellon University.

(1970). "Orlicz spaces based on families of measures," *Studia Math.* **35**, 15–49.

Rota, G.-C. (1960). "On the representation of averaging operators," *Rend. Seminario Mat. Univ. Padova,* **30**, 52–64.

(1962). "An 'alternierende verfahren' for positive operators," *Bull. Amer. Math. Soc.* **68**, 95–102.

(1964). "Reynolds operators," *Proc. Symp. Appl. Math. (Amer. Math. Soc.)* **16**, 70–83.

Ryll-Nardzewski, C. (1953). "On compact measures," *Fund. Math.* **40**, 125–130.

Sakai, S. (1971). C^*-*Algebras and* W^*-*Algebras*, Springer-Verlag, New York.

Saks, S. (1935). "On the strong derivations of functions of an interval," *Fund. Math.* **25**, 235–252.

Sard, A. (1963). *Linear Approximation*, Amer. Math. Soc. Surveys, **9**, Providence, RI.

Sazonov, V.V. (1962). "On perfect measures," *Izv. Akad. Nauk SSSR, Ser. Mat.* **26**, 391–414 (*Translation* AMS (2) **48**, 229–254.)

Schwartz, L. (1973). "Surmartingales régulières à values mesures et désintégrations régulieres d'une mesure," *J. d'Analyse Math.* **26**, 1–168.

Seever, G.L. (1966). "Nonnegative projections on $C_0(X)$," *Pacific J. Math.* **17**, 159–166.

Segal, I.E. (1953). "A noncommutative extension of abstract integration," *Ann. Math.* **57**, 401–457, Correction, *ibid* **58**, 595–596.

Shiryayev, A.N. (1978). *Optimal Stopping Rules*, Springer-Verlag, New York (English Translation).

Soloman, D.W. (1969). "Denjoy integration in abstract spaces," *Mem. Amer. Math. Soc.* **85**, 69 pp.

Starr, N. (1966). "Operator limit theorems," *Trans. Amer. Math. Soc.* **121**, 90–115.

Stinespring, W.F. (1959). "Integration theorems for gages and duality for unimodular groups," *Trans. Amer. Math. Soc.* **90**, 15–56.

Stroock, D.W. and S.R.S. Varadhan (1979). *Multidimensional Diffusion Processes*, Springer-Verlag, New York.

Takesaki, M. (1972). "Conditional expectations in von Neumann algebras," *J. Functional Anal.* **9**, 306–321.
(1983). *Structure of Factors and Automorphism Groups*, CBMS and AMS, No. **51**, Providence, RI.

Tjur, T. (1974). *Conditional Probability Distributions*, Lect. Notes **2**, Institute of Statistics, Univ. of Copenhagen, Denmark.
(1980). *Probability Based on Radon Measures*, J. Wiley and Sons, New York.

Tomiyama, J. (1957–59). "On projection of norm one in W^*-algebras," I, *Proc. Japan Acad.* **33**, 608–612; II *Tôhoku Math. J.* **10**, 204–209; III *ibid*, **11**, 125–129.

Tsukada, M. (1983). "Strong convergence of martingales in von Neumann algebras," *Proc. Amer. Math. Soc.* **88**, 537–540.
(1985). "The strong limit of von Neumann subalgebras with conditional expectations," *Proc. Amer. Math. Soc.* **94**, 259–264.

Umegaki, H. (1954–62). "Conditional expectation in an operator algebra," I, *Tôhoku math. J.* **2**, 177–181; II, *ibid* **8**, 86–100; III, *Kodai Math. Sem. Rep.* **11**, 51–74; IV, *ibid* **14**, 59–85.

Wright, J.D.M. (1968). "Applications to averaging operators of the theory of Stone algebra valued modular measures," *Quarterly J. Math. (Oxford)* (2) **19**, 321–332.
(1969a). "Stone algebra valued measures and integration," *Proc. London Math. Soc.* **29**, 107–122.
(1969b). "Martingale convergence theorems for sequences of Stone algebras," *Proc. Glasgow Math. Soc.* **10**, 77–83.
(1969c). "A Radon-Nikodým theorem for Stone algebra val-

ued measures," *Trans. Amer. Math. Soc.* **139**, 75–94.

Wulbert, D.E. (1969). "Averaging projections," *Ill. J. Math.* **13**, 689–693.

Zaanen, A.C. (1967). *Integration* (2nd ed.), North-Holland Publishing Co., Amsterdam, The Netherlands.

NOTATION

$\mathbb{R}(\mathbb{C})$ = real (complex) number field.

CBS (= Cauchy-Bunyakovskii-Schwarz) inequality

$A^c = \Omega - A$, 5

$(\mathcal{A}, {}^*, \varphi)$, 380

$B(I)$, 185

$B(\Omega, \Sigma)$, 181

$\mathcal{B} \subset \tilde{\mathcal{B}}$, a.e. $[\mathcal{M}]$, 193

$\mathcal{C}_c(\Omega)$, 145

$\mathcal{D}\eta\mathcal{A}$, 378

Δ, 79

$E_\mathcal{P}(\cdot)$, 5

$E_B(\cdot) = E(\cdot|B)$, 5

$E^{\mathcal{B}}(\cdot) = E(\cdot|\mathcal{B})$, 26

$E(\cdot|X = a) = E^{\mathcal{B}}(\cdot)(a)(= E^{\sigma(X)}(\cdot)(a))$, 85

$\mathcal{E}_p(\mathcal{M})$, 202

$\mathcal{E}_\Phi(\Theta, \mathcal{M})$, 204

$F_n^{\varepsilon_n(\alpha)}$, 125

$F(\cdot|B), \varphi(\cdot|B)$, 112

$f \overset{w}{=} g$, 144

$\mathcal{F}(\tau_j)$, 295

χ_A, 5

$L_{loc}^1(\mu)$, 224

$\varprojlim(\Omega_i, g_{ij})$, 271

$L^p(\mathcal{M})$, 210

$L^p(\mathcal{A}, \tau)$, 379

$\mathcal{L}_I(S, \mathcal{S}, \nu)$, 144

$\mathcal{M}(\cdot), \mathcal{N}(\cdot)$, 209

M^φ, 234

M^ρ, 247

$\mathcal{M}(\Sigma)$, 225

$M(S, \mathcal{S})$, 144

$N(0, \sigma^2)$, 165

$\nu << \mu$, 80

$N_{\Theta,\Phi}^{G}(\cdot)$, 205

$\tilde{\Omega}_d$, 139

$(\Omega, \Sigma, \frac{P}{Q})$, 305

$P(\cdot|B)$, 4

$P_{\mathcal{B}}(\cdot) = (P|\mathcal{B})(\cdot)$, 6

$P^{\mathcal{B}}(\cdot) = P(\cdot|\mathcal{B})$, 27

$P(A|B_\alpha^\beta)$, 114

$P(A|X) = P^{\sigma(X)}(A)$, 317

$p(\xi, s; A, t)$, 321

$\pi(\mathcal{A})^c$, 380

$q * q = q^2, q^n = q * q^{n-1}$, 335

$R_{\mu_1 * \mu_2}$, 344

$\mathcal{R}_m$, 160

$\Sigma(B)$, 5

supp, 143

$T\eta\mathcal{A}$, 378

$\|T\|_p$, 378

$U_p(\mathcal{M})$, 202

$\{X_t, \mathcal{B}_t, t \in T\}$, 294

$\vee = \max$, 186

$\wedge = \min$, 247

INDEX